Regional Science Matters

Peter Nijkamp • Adam Rose • Karima Kourtit
Editors

Regional Science Matters

Studies Dedicated to Walter Isard

Springer

Editors
Peter Nijkamp
Karima Kourtit
Department of Spatial Economics
VU University
Amsterdam
The Netherlands

Adam Rose
University of Southern California Price School of Public Policy
Los Angeles
CA, USA

ISBN 978-3-319-07304-0 ISBN 978-3-319-07305-7 (eBook)
DOI 10.1007/978-3-319-07305-7
Springer Cham Heidelberg New York Dordrecht London

Library of Congress Control Number: 2014955456

Printed on acid-free paper

Springer is part of Springer Science+Business Media (www.springer.com)

Contents

Regional Science: What Matters? Which Matters?

1

Peter Nijkamp, Adam Rose, and Karima Kourtit

1.1 Overview

It is often taken for granted that the year 1956 heralds the birth of regional science. In this year Walter Isard's classic *Location and Space-Economy* was published. Admittedly, Isard had already published some noteworthy articles in preceding years. In 1949, he published a first major contribution to the foundations of early regional science, in his article 'The General Theory of Location and Space Economy', in the *Quarterly Journal of Economics*. His article 'Interregional and Regional Input-Output Analysis: A Model of a Space-Economy', published in1951 in the *Review of Economics and Statistics* (vol. 33, pp. 318–328), was also a remarkable scholarly achievement. The same holds for his article 'Location Theory and Trade Theory: Short-run Analysis', published in 1954 in the *Quarterly Journal of Economics* (vol. 68, pp. 305–320).

But the real break-through occurred with the publication of his seminal opus on *Location and Space-Economy*. The subtitle of this volume illustrates typically the great ambition of Isard, as he is in search of *A General Theory Relating to Industrial Location, Market Areas, Land Use, Trade, and Urban Structure*. This book is a scientific highlight, as it seeks to offer a critical analysis of past contributions by various scholars to spatial issues and to lay the foundation for an integrated perspective on the complex space-economy. The main goals of Isard's first comprehensive analysis of locational questions in a multi-faceted space-economy are

P. Nijkamp • K. Kourtit (✉)
Department of Spatial Economics, Faculty of Economics and Business Administration, VU University Amsterdam, Room 4A-33, De Boelelaan 1105, 1081 Amsterdam, The Netherlands
e-mail: p.nijkamp@vu.nl; k.kourtit@vu.nl

A. Rose
Price School of Public Policy, University of Southern California (USC), RGL 230, 650 Childs Way, Los Angeles, CA 90089, USA
e-mail: adam.rose@usc.edu

P. Nijkamp et al. (eds.), *Regional Science Matters*, DOI 10.1007/978-3-319-07305-7_1,

convincingly expressed in the Preface to this book, where he unfolds the new horizons of regional science:

> Over the historical record, the process of social development and economic growth has been for the most part cumulative though at times seemingly erratic. On occasion it has been recessive. Some of its stages have been forced to start at their origins more than once. Moreover, these stages have started independently at diverse places on the earth at different times. Together with the manifold cultural, political, social, and economic forces which have evolved this fact has led, over the centuries, to marked regional differentiations and to tremendous disparities in the welfare of the peoples of the several regions of the world.
>
> Such a record of man's adaptation to, and interaction with, his physical environment suggests that a comprehensive theory of society or economy should embrace both time and space dimensions. It should be able to unravel the dynamic interplay of forces not only currently but also over the long past. Its propositions should be testable against the background of historic development in the several regions of the world and concomitantly should offer explanation of this development. Its hypotheses should encompass the influence of past events and intermingling of forces upon existing economic and social structure and conditions. Ideally, its conceptual framework should enable one to anticipate the course of future development, given certain premises and judgments.

Photo taken by Bill Staffeld, College of AAP, Cornell University (The editors wish to thank David Boyce for his great support in obtaining an original photograph and personal signature of Walter Isard.)

> Unhappily, the state of the social science disciplines leaves very much to be desired in the way of achieving such a comprehensive theory. The dynamic frameworks of these disciplines, frameworks which are designed to catch the effect of what some characterize as the "time variable", are rather crude and even at times naïve. Still more so are their spatial and regional frameworks (Isard 1956, pp. vii–viii).

The methodology of regional science was further developed by Isard in subsequent works, in particular, in his *Methods of Regional Analysis* (Isard 1960). The latter book—containing 784 (!) pages—has served for years as a toolbox for regional scientists all over the world, and brought harmonized approach in the multidisciplinary approach of regional science.

Isard's search for an integrated and general theory of spatial development led also to the publication of two less well-known, but extremely original books, viz. his ambitious volume on *General Theory, Social, Political, Economic, and Regional with Particular Reference to Decision-Making Analysis,* published in 1969, and a novel book on spatial-ecological issues, with the title *Ecologic-Economic Analysis for Regional Development,* published in 1972. All these books were without any doubt landmarks in the intellectual contributions of Isard to regional science. But it goes without saying that—in addition to the wealth of books—Isard has also published numerous seminal articles in many journals. These share all the same features: analytical rigour, a sound conceptual basis, and a high degree of original thinking. His publications have influenced generations of regional scientists for almost half a century. Walter Isard is not only the founder of regional science but acted as the discipline's guiding light for much of its development.

In a way, regional science is a relatively recently developed discipline. But at the same time it draws on established disciplines, such as economics, sociology, planning and political science. Its novelty lies mainly in its capability to combine different analytical frameworks from different disciplinary backgrounds, although a quantitative economic theme has often dominated the toolbox of regional science. Needless to say that the emergence of regional science has sometimes encountered fierce resistance, as was extensively documented by Isard (2003) in his book on *History of Regional Science and the Regional Science Association International.* The great achievement of regional science has been its potential to bring together scholars from different disciplines from all over the world, through major conferences, focused workshops, and advanced publications in major journals.

Scientific Contributions of Walter Isard to Regional Science

Walter Isard's research interests in regional and urban phenomena were formed during his graduate studies, and led to his first major book, *Location and Space Economy* (1956). Next, he initiated research on the economic and social consequences of atomic power and industrial complexes, and intensified his research on methods of regional analysis, including industry location analysis, interregional interindustry analysis, interregional linear programming models, and migration and gravity models, among others. These investigations resulted in his second major book, *Methods of Regional Analysis: an Introduction to Regional Science* (1960), in association with several others. The final chapter of this book depicted of how these methods might be used together to forecast the future state of a system of regions. (In 1998, Isard and five co-authors published a thoroughly updated version of this work as *Methods of Interregional and Regional Analysis.*) During the 1960s Isard turned to more theoretical pursuits, which he presented in his third major book, *General Theory, Social, Political, Economic, and Regional with Particular Reference to Decision-Making Analysis* (1969), co-authored

(continued)

with Tony Smith and several others. In his fourth major work, Isard and Panos Liossatos examined theoretical questions related to *Spatial Dynamics and Optimal Space-time Development* (1979). He referred to these four books as his 'quadrilogy.'

Concurrently with the writing of the third and fourth books, Isard and his students undertook a major interindustry study of the Philadelphia region and conducted empirical research related to environmental problems. Throughout his career, Isard also pursued policy interests related to conflict management and resolution, disarmament and peace science. Several of his books, which number more than 20, as well as many of his 300 published papers, concerned topics in peace science.

Isard's contributions were clearly more related to interregional constructs and relationships than intraurban ones. The general focus of his research concerned systems of cities and regions.

David Boyce
Northwestern University, Evanston, USA

Isard, W. 1956. *Location and Space-Economy*: *A General Theory Relating to Industrial Location, Market Areas, Land Use, Trade, and Urban Structure*, New York: The Technology Press of Massachusetts Institute of Technology and John Wiley and Sons, Inc.

Isard, W., in association with D. F. Bramhall, G. A. P. Carrothers, J. H. Cumberland, L. N. Moses, D. O. Price, E. W. Schooler. 1960. *Methods of Regional Analysis: an Introduction to Regional Science*, New York: The Technology Press of Massachusetts Institute of Technology and John Wiley and Sons, Inc.

Isard, W., in association with T. E. Smith, and P. Isard, T. H. Tung, M. Dacey. 1969. *General Theory, Social, Political, Economic, and Regional with Particular Reference to Decision-Making Analysis*, Cambridge: The M. I. T. Press.

Isard, W. and P. Liossatos, with assistance by Y. Kanemoto and P. C. Kaniss. 1979. *Spatial Dynamics and Optimal Space-time Development*, New York: North-Holland Publishing Co.

Isard, W., I. J. Azis, M. P. Drennan, R. E. Miller, S. Saltzman, E. Thorbecke. 1998. *Methods of Interregional and Regional Analysis*, Brookfield, Vermont: Ashgate Publishing Co.

1.2 Location Analysis

The genesis of regional science can be found in location analysis, not only of firms, but also of households, and production facilities. The '*where*' question is indeed central in most regional and urban analyses. This spatial issue has prompted an avalanche of analytical—both conceptual and applied—contributions over the past six decades. Indeed, the complex functioning of the space-economy cannot be properly understood without profound insights into the geographic location behaviour of economic agents. As argued in a recent review of location theory by Gorter and Nijkamp (2001), location analysis is the key to understanding the emergence of urban agglomerations, industrial concentrations, regional growth and spatial disparities.

Clearly, a location decision may be based on an individual choice, but in most cases we observe interactive spatial decisions caused by spatial externalities. The origins of location analysis in a multi-actor spatial system date back to the seminal contributions of Von Thünen, Weber, Hotelling, Palander, Christaller, Lösch and many others [see for an overview Ponsard (1983)]. Their analyses were combined by Isard in a more coherent and consistent framework, based on rational behavioural assumptions on spatial actors.

This comprehensive approach led also to a better understanding of the mechanism of regional and urban growth, as space is not neutral, but offers a productivity–enhancing potential to actors in an open and competitive spatial system. In this regard, also spatial interactions (e.g., interregional input-output linkages, interregional trade) play a critical role, as these shape the conditions for spatial growth and spatial disparity phenomena. Against this background, also transport costs—and, in a more general sense, distance friction costs—exert a space-organizing effect on the behaviour of actors. The current popular debate on 'the death of distance' (Cairncross 1997) or on 'the flat world' (Friedman 2005) cannot be well understood without a profound knowledge of the foundations of location theory. This is also clearly illustrated in the emergence of the 'new economic geography' [see e.g. Fujita et al. (1999)] which draws essentially—in a more stylized form—on the fundamentals offered by Isard.

It goes without saying that there is a bi-directional effect between prices of goods, production factors and land on the one hand and spatial locational patterns on the other hand: location determines prices and prices determine locations. This prompts the need for a spatial economic equilibrium approach whose roots go back to Lösch, but whose integrated conceptualization was developed by Isard. The coherent analysis of the complex and interdependent decisions of firms and households in geographic space calls indeed for a more comprehensive framework that has to be built on spatial general equilibrium modelling, as has been shown by the 'new economic geography'. Clearly, the synergetic interactions between actors—on both the demand and supply side of a space-economy—may exhibit both fast and slow dynamics and hence prompt the need for spatial complexity models [see also Reggiani and Nijkamp (2009)].

On a final note, regional (and urban) policy is concerned with re-shaping the geographic-economic landscape. This means developing incentives to change the behaviour of actors in space. It is widely accepted that regional policy is not based on a 'command and control' strategy, but on the provision of 'sticks and carrots' that offer a more socially satisfactory outcome of a dynamic space-economy. Due insight into locational motives is a prerequisite for any appropriate regional development policy.

Walter Isard: A Visionary Scientist

Walter Isard's seminal contribution was the introduction of geographical space into economic theory. He developed the general theory of Interregional General Equilibrium of which Walrasian theory is a special case. He also introduced the time variable into spatial economics borrowing ideas from physics. His work in input-output analysis is also well known and he used this concept in trade flows. The outline of the channels of synthesis in the last chapter of his "Methods of Regional Analysis" book hints at his long time interest in the General Equilibrium model, which he formulated later. His work on gravity and potential models opened new avenues in transportation research. Isard was not only interested in theoretical model building. He had also a great interest in empirically testing models.

In addition to his academic achievements, he had extraordinary organizational and administrative abilities, rare for an academic of his stature. He established a graduate program in Regional Science, built the Regional Science Association world-wide, and mentored a set of students who became renowned scholars in their own right. He established the leading journal in the field, the Journal of Regional Science. Although Walter did not work on econometrics and economic development directly, he was very enthusiastic about these topics. His encouragement to one of his students inspired the creation of a new area of study, Spatial Econometrics. His support to scholars in developing countries led also directly to new approaches in Regional Science in those countries. Isard realized that the development of theories and methods is not enough. He knew we would need data from all around the world to validate theories. This conviction was why he was very supportive of Geographical Information System (GIS). Isard's breadth of knowledge and interest was particularly notable. He had great ideas in diverse fields like topology, fuzzy set theory, metropolitan studies, nuclear energy, and many others. Isard spent the later portion of his career in Peace Science where, in my judgment, his contributions were equally significant as his seminal work in Regional Science.

Manas Chatterji
Binghamton University, Binghamton, USA

1.3 Spatial Interaction

Regional science is concerned with spatial structures and evolution of a complex space-economy. It is driven by top-down and bottom-up movements, while it exhibits at the same time slow and fast dynamics. In all cases, distance friction (physical, economic, social, cognitive, or cultural) plays an intrinsic role, as was argued by an early predecessor of regional science, Janôwski (see also the English book translation, 2013).

It is noteworthy that the spatial economy is not a result of a random process, but emerges out of rational and interactive decisions of a multiplicity of economic actors, on both the supply and demand side. For example, the emergence of large-scale industrial concentrations near accessible places and the growth in interregional or international trade are two interlinked factors, as is also argued by the 'new economic geography'. The structure of the space-economy may sometimes look rather complex, but has a deep underlying economic foundation that is shaped by economic motives and responses which evolve over time. One of the founding fathers of structured land use, for example, was Johann Heinrich von Thünen, who was able to combine locational market advantages with transport and production patterns so as to derive a consistent relationship between land rent and hierarchical ordering of industrial activities in space.

Regional Science and Its Surroundings

Professor Walter Isard was not only the founder of regional science and organiser of the world research movement developing in this field. What is still underestimated is the impact of his ideas on the development of two other areas: he was one of the forerunners of the research on the complexity of social systems and on space-time socio-economic development.

According to Professor Isard, regions, or their systems, are living organisms composed of a variety of units: political, economic, social, and cultural. Their interdependent behaviour patterns are determined by psychological, institutional, natural and technical factors. Regional science focuses on spatial aspects of the behaviour of those organisms and their component units. In search of answers to questions that arise when examining the object of study thus defined, regional science relies on the conceptions and methods of many other sciences, in particular economy, geography, political science, sociology, and ecology. It seeks to explain how new structures form and develop, and also how the new structures modify the behaviour patterns of individual units. In doing this, it employs methods for the study of nonlinearity, microstructural and macrostructural systems, as well as deterministic and probabilistic ones.

Initially, Professor Isard's conceptions were connected with general equilibrium theory. In his later works he advocated seeking ways of giving this theory a dynamic character. He admitted that it was not enough to introduce time variables to a system of equations describing static states. Each unit should be defined as a dynamic system already at the stage of assumptions. The behaviour of macrosystems should be derived from probability theory applied to a large number of interacting units.

Ryszard Domański
Poznan University of Economics, Poznan, Poland

The spatial economy does—as mentioned above—not only have a rational structure, but also a systematically organized spatial interaction system (transport, mobility, trade, telecommunications). The basis for a systemic analysis of spatial Interactions in relation to centres of human activity and distance friction costs originates from Newtonian gravity theory (the 'universal law of gravitation'). The gravity model has served as a conceptual and analytical framework for many applied flow analyses in the social and spatial sciences, e.g. in the domain of migration, trade, tourism, and commuting flows. This quantitative modelling approach has become a mainstream toolkit in regional and urban economics, in economic geography, and in trade and transport economics. A thorough analysis of the underlying principles of Newton's gravity concept has revealed that the application of this concept to spatial flows of any kind is supported by two principles, viz. scale (or agglomeration) advantages and distance friction costs (both physical and social). Consequently, the use of gravitational principles in quantitative interaction modelling can be justified on theoretical economic grounds (the so-called 'social physics'), a notion also supported by the 'new economic geography' [see also Nijkamp (2013)].

The popularity of the gravity principle—and related principles such as entropy—in the spatial sciences has also prompted the recognition and explanation of a hierarchical structure in the organization of geographical space. This is not only reflected in the Christaller-Lösch central place theory, but also in subsequent statistical regularities and principles, known as the Zipf law and its companion, the Gibrat law. Both laws have been the subject of intensive quantitative research on spatial pattern analysis in the past decade and have highlighted the existence of hierarchical systems in spatial economics and related domains, such as industrial organization.

A newly emerging question is the impact of cyber space. Clearly, a new trend which has promoted heated debates on spatial friction issues is the emergence of the digital world [see for details Cairncross (1997)]. The notion of cyber space has led to many speculative remarks on the future of the spatial economy, such as the 'death of distance' hypothesis. It turns out, however, that ICT development has generally not led to a vanishing importance of space; the digital world is not a substitute for the material world, but reinforces in many cases existing physical and geographical structures determined by cost frictions. Consequently, agglomeration structures and physical transport patterns have not shown a clear decline nor a fundamental shift over the past decades. In conclusion, the gravity approach will most likely continue to be an important analytical navigation instrument guiding our understanding of the complex space-economy. This issue has prompted a new strand of fascinating literature, sometimes called cyber geography, internet geography, or virtual geography [see also Batty (1997), Gorman and Malecki (2000), or Kitchin (1998)].

Location and Space-Economy After Half a Century

Nearly half a century has passed since the late Professor Isard published his seminal work, *Location and Space-Economy*, which provided the theoretical basis for a new field called regional science. This fascinating book aimed at nothing less than initiating the development of a *general theory of location and space-economy*, embracing the total spatial array of economic activities. His basic ideas on the general theory of location and space-economy are summarized in Chap. 2, emphasizing the following three points (pp. 53-54):

First, "the then prevailing paradigm of general equilibrium analyses...based on perfect competition is not comprehensive enough to incorporate the particular effects of transport and spatial costs on the distribution of economic activities in space." Second, "because of the monopoly elements which are almost invariably present in spatial relations, a broadly defined *general theory of monopolistic competition* can be conceived as identical with the general theory of location and space-economy." Third, "the *evolutionary approach* of Weber and others...should be helpful in suggesting lines along which this general theory may be nurtured to embody dynamic relations."

What remarkable insights these are nearly 50 years ago! Since then, as we know, location theory and spatial economics, including the so-called new economic geography initiated in the early 1990s, have progressed largely along the lines suggested by Isard.

Location and Space-Economy is a book that fascinates everybody with the author's noble mission to reclaim space as a core element of economic analysis, but it is never easy to comprehend the real meaning of what is written in each condensed page and in each long footnote. What one learns from reading the book depends on the types of questions one asks and on how deeply one poses them. In this respect, *Location and Space-Economy* reminds me of a famous Zen garden in Kyoto.

Masahisa Fujita, Konan University, and Research Institute of Economy, Trade and Industry, Kyoto, Japan

1.4 The Broader Spatial Context

Regional science has found its genesis in the awareness that a thorough understanding of the complex space-economy needs a multidimensional analytical approach in which various disciplines are merged, or at least related to each other. In the early writings of Isard, we come across the recognition that the physical geography of a region (e.g., lakes, mountains) exerts a critical influence on its efficiency or productivity, due to its impact on transport friction costs. The step from physical geography to the environment in a broader sense then seemed a plausible one [see also Rose et al. (2014)]. And indeed, Isard was able to include in regional science the notion of ecology in a very innovative way, by addressing the spatial-ecological interfaces from two angles: (1) he was able to incorporate environmental externalities (e.g., air pollutants, water pollutants) systematically in regional and interregional input-output analysis, based on Leontief's seminal contributions to input-output analysis. But Isard did not only include environmental flow data, but was also able to relate production and consumption flows to environmental capital data [see Isard (1972)]; (2) he was also aware of the conflictual nature of economic objectives and environmental (and in broader sense, resource) objectives. In this regard, the use of multi-objective decision theory was seen by him as an important analytical instrument to map out and to understand conflicts and discrepancies in regional or urban policy strategies. His orientation towards conflict management and peace research in later stages of his career can well be understood from the great many dilemma's in regional and urban policy-making.

The interfaces between regional science and ecological science have extensively been studied by Isard in various books and many articles. And also in this context, he was able to combine his rigorous analytical insights with topical issues at that time, such as air pollution or water pollution (Isard and van Zele 1975). It is interesting to note that in his integrated regional-ecological analysis, he was not only addressing environmental effects of regional production or consumption patterns, but also the implications of introducing policy handles (e.g., pollution abatement programmes). This approach led also to a solid analysis framework with an applied orientation in regional science.

The real-world space-economy comprises indeed a system of open regions, with different degrees of accessibility and connectivity. Spatial unevenness is consequently a necessary constituency in regional science. To understand the fundamental and often persistent causes of spatial disparities is a great challenge. Isard has offered a great intellectual contribution to comprehensive insights into the complex mechanisms shaping regional developments and spatial imbalances. At the same time, he also looked into policy and intervention tools so as to cope with undesirable spatial evolutionary processes. His work was both driven by scholarly curiosity (or serendipity) and real-world concern. He has been a role model for regional scientists all over the globe.

1.5 The Menu of This Volume

The menu of this volume has been composed and prepared by former university colleagues, general professional colleagues, and students of Walter Isard from all over the globe. Several of the papers were presented at sessions in memory of Walter at the North American Regional Science Council (NARSC) Meetings in 2012 and 2013, organized by Tschangho John Kim, Marcus Berliant, and Adam Rose. In addition, many of the co-authors include a new generation of regional scientists, indicating Walter's reach over time. We note the popular concept of "6 Degrees of Separation" that contends everyone is connected within six acquaintances. In Walter's case, the network is much tighter, probably requiring only three personal linkages, not only in this volume but in all of regional science.

The first course of this menu is a starter which zooms in on spatial systems analysis and the raison d'etre of regional science.

The next course offers a great variety of contributions on one of the core issues of regional science, namely regional growth and distribution.

In the third course we can enjoy refreshing contributions to regional policy and uncertainty, an issue that has always been at the heart of regional science.

The dessert is a surprise course that focuses the attention on one of the most prominent and amazing phenomena in regional science, viz. urban agglomerations.

All articles included in this volume are an original tribute to a real scholar who has changed our views on the complex space-economy. Walter Isard's research reflects an undisputed highlight in the history of regional science.

References

Batty M (1997) Virtual geography. Futures 29:337–352

Cairncross F (1997) The death of distance. Harvard Business School Press, Cambridge, MA

Friedman TL (2005) The world is flat. Farrar Straus & Giroux, New York

Fujita M, Krugman P, Venables A (1999) The spatial economy. MIT Press, Cambridge, MA

Gorman S, Malecki EJ (2000) The networks of the internet. Telecommun Pol 24:113–134

Gorter C, Nijkamp P (2001) Location theory. In: Hanson S (ed) International encyclopedia for the social and behavioural sciences. Elsevier, Amsterdam, pp 9013–9019

Isard W (1949) The general theory of location and space economy. Q J Econ 63(4):476–506

Isard W (1951) Interregional and regional input-output analysis: a model of a space-economy. Rev Econ Stat 53:318–328

Isard W (1954) Location theory and trade theory: short-run analysis. Q J Econ 68:305–320

Isard W (1956) Location and space-economy. MIT Press, Cambridge, MA

Isard W (1960) Methods of regional analysis. MIT Press, Cambridge, MA

Isard W (2003) History of regional science and the regional science association international. Springer, Berlin

Isard W, van Zele R (1975) Practical regional science analysis for environmental management. Int Reg Sci Rev 1:1–25

Isard W (in association with Choguill C, Kissin J, Seyforth R and Tatlock R) (1972) Ecologic-economic analysis for regional development. Free Press, New York

Isard W (in association with Smith TE, Isard P, Tung TH and Dacey M) (1969) General theory, social, political, economic, and regional, with particular reference to decision-making analysis. MIT Press, Cambridge, MA

Janowski B (2013) On distances as a factor of cultural development. Bogucki, Poznan

Kitchin R (1998) Cyberspace. Wiley, New York

Nijkamp P (2013) The universal law of gravitation and the death of distance. Rom J Reg Sci 7 (2):1–10

Ponsard C (1983) History of spatial economic theory. Springer, Berlin

Reggiani A, Nijkamp P (eds) (2009) Complexity and spatial networks. Springer, Berlin

Rose A, Folmer H, Nijkamp P (2014) Walter Isard's contributions to environmental economics and ecological economics. Int Reg Sci Rev 37:96–106

Part I

Spatial Systems Analysis

The Spatial Economy: A Holistic Perspective 2

Peter Nijkamp and Waldemar Ratajczak

Abstract
This paper offers a concise holistic-historical perspective on regional science. It positions regional science among various related disciplinary approaches and presents a comprehensive image of its multi-faceted architecture.

2.1 Structure and Connectivity in Space

Regional science has a history of more than half a century old. It has found its roots in the explicit recognition of the importance of space in social science research. Space is not only an attribute of any phenomenon on earth (in terms of its geographical coordinates), but it also functions as a facilitator for or barrier to human interactions (Isard 1956, 1960).

Right from the outset there has been a debate on the semantics of 'regional science'. Regional science is a scientific endeavour that addresses the indigenous role of space (including regions, cities, environment, infrastructure, communication) in various social-science oriented issues, such as socio-economic disparities, regional and urban growth, or transport logistics. In the literature the question has sometimes been raised whether regional science should not be replaced by 'spatial science', a term that also nowadays is still often used in several countries. As argued by Barnes (2004), the term 'spatial science' (or perhaps 'spatial analysis') would

P. Nijkamp (✉)
Department of Spatial Economics, VU University Amsterdam, 1081 Amsterdam, The Netherlands

Adam Mickiewicz University, Poznan, Poland
e-mail: p.nijkamp@vu.nl

W. Ratajczak
Faculty of Geographical and Geological Sciences, Institute of Socio-Economic Geography and Spatial Management, Adam Mickiewicz University, Poznan, Poland
e-mail: walrat@amu.edu.pl

P. Nijkamp et al. (eds.), *Regional Science Matters*, DOI 10.1007/978-3-319-07305-7_2,

have been a more appropriate description than 'regional science', but to avoid confusion with the rapidly emerging space technology in the 1950s and 1960s—addressing outer space studied by physical scientists and engineers—, the second best choice was 'regional science' (even though Isard's seminal work on '*Location and Space-Economy*' in 1956 still refers to space as the main scientific orientation). The predominant agenda item of regional science research was to complement the '*wonderland of no dimensions*' in conventional economics with a spatial plane. Admittedly, this goal was already set earlier by German location theorists in the first part of the twentieth century (for instance, Alfred Weber, August Lösch, Andreas Predöhl) and by predecessors in the nineteenth century (e.g. Johann-Heinrich von Thünen).

The central feature of regional science however, was to provide a rigorous analytical framework—often derived from mathematical theorizing—to connect several social science disciplines with the goal to develop testable theoretical structures and space-relevant concepts that were lending themselves to general applicability.

An important methodological question is clearly whether 'space' adds only an interesting geographical component to a given phenomenon (in terms of its coordinates, e.g.) or whether space is an intrinsic object of study. In the first case, cartographic mapping (e.g., geo-science, spatial information science) may bring along new representational and exploratory scientific results. In the second case, the notion of region, city, locality, connectivity or interaction is an essential ingredient of spatial research. Such a spatial unit is the playing field for economic, social, technological, institutional or regulatory forces. The history of regional science has shown that both research orientations have been extremely fruitful and have led to path-breaking findings, in which theory development and applied analysis were running in tandem. In this way, also a bridge could be built between economics, geography, demography, planning, political science and public administration. In conclusion, a rich variety of real-world social science phenomena can be projected into geographical space (thus avoiding that such phenomena collapse in one point), but space is not only a passive attribute. There is an increasing awareness that geographical space plays also an *active* role: it determines industrial location, commuting patterns of residents, connectivity of cities, accessibility of inner cities, transport and migration flows, and so forth (see also Coffey 2003; Mulligan 2003).

This endeavour has turned out to be very successful (see Isard 2003). Regional science conferences have become mass meetings of often more than 1,000 participants, the number and volume of regional science publications has shown a rapid rise over the past decades, and regional science has become a respected scientific approach, not in contrast to—but in tandem with—related disciplines, such as regional economics, urban economics, geography, transportation science, demography, planning, political science etc.

In the rich history of regional science various intriguing methodological research questions have emerged that directly touched on the '*raison d'etre*' of regional science. Examples are: What is a relevant spatial scale of analysis in regional science? Should the focus of regional analysis be on geographic entities (e.g., cities, settlements, industrial complexes, regions, interregional trade or transport) or on the behaviour of the economic objects in a concrete and given geographical space? If spatial movements and interactions take time, what is the relationship between space and time? If all spatial phenomena are linked together—but nearby things more than distant things [the so-called Law of Tobler (1970)],—what are then the essential spatial connectivity (interaction or communication) principles? If a given geographic space acts as both a barrier and an opportunity, what does this mean for our explanatory analysis? And how are modern concepts from networks and complexity related to regional dynamics? Such questions do not have easily available answers. And therefore, the present paper has a modest scope: its aim is to seek for an operational framework for understanding the multiple and mutually interwoven dimensions of the structure and evolution of regional and urban phenomena as well as their interactions, in particular behavioural spatial patterns and governance modes.

This paper is organized as follows. Section 2.2 will offer a concise and selective description of some key issues in regional science. This will be followed by the design of a systemic conceptual framework for analyzing regional phenomena, based on a so-called Triple-Layer representation. Then, Sect. 2.4 will be devoted to a holistic perspective on the spatial economy, including space-time complexity issues. The paper will conclude with retrospective and prospective remarks.

2.2 Positioning Regional Science

Regional science originates from both *economics* and *geography*. On the economic side, its early roots can already be found in labour specialization and spatial labour division, in industrial organization and spatial industrial concentration, and in international trade and interregional transport [see for a review, Paelinck and Nijkamp (1980) and Ponsard (1983)]. Early predecessors were—next to classical economists—amongst others Johann-Heinrich von Thünen, Alfred Weber, Tord Palander, August Lösch and Andreas Predöhl.

Regional science finds its origin also in geography, in particular economic geography. The latter discipline derives its name from Emile Levassseur (1828–1911), a French economist and geographer (see Boureille 1998; Leroux and Hart 2012). Economic geography studies the spatial distribution of individual and economic activities—and their interdependencies—at various geographic scales, and their evolution over time (see Knox and Marston 2001; Clark et al. 2003; Warf 2006). Seminal contributions from economic geography to the foundations of regional science were inter alia provided by Walter Christaller, with his path-breaking conceptualization of the hierarchical distribution of places. Seminal contribution can be found in Isard (1956, 1979).

Regional science offers essentially a multidisciplinary perspective on the interface of both economics and economic geography. Regional science—sometimes also coined spatial science or spatial economics—studies the *where, what, how* and *for whom* questions in a complex space-economy. In so doing, it also employs notions from related disciplines, like transportation science, decision theory, political science, demography, sociology and the like. Regional science has very often exhibited a strong methodological—often statistical-econometric—orientation. Examples can be found inter alia in O'Sullivan (1981), McCann (2005), Brakman et al. (2009) and Capello and Nijkamp (2009).

Location theory has always formed the heart of regional science, first in the sense of industrial localization, but later on also in terms of residential location and facilities location. This has spurred an overwhelming volume of advanced studies on the foundations of the space-economy, including its complex space-time interactions. A concise review of the principles of location theory can be found in Gorter and Nijkamp (2001). In the course of time, location theory has been integrated in a much wider context of dynamic evolution of regions and cities. To a large extent, it has been mainstreamed with analytical spatial economics and economic geography, on the basis of a strong quantitative orientation (see also Fischer and Nijkamp 2013).

In recent years, we have witnessed an increasing popularity of the so-called New Economic Geography (NEG). NEG serves to find an interface between analytical economic geography and mainstream economics from a more conceptual theoretical perspective, with a clear emphasis on formal modelling exercises which take into account the heterogeneity in geography causing spatial disparities and imperfect competition. It places, therefore, particular emphasis on spatial agglomeration forces and regional growth convergence. Its first orientation was towards centre-periphery models, integrated regional and urban systems, and models of geographic agglomeration and trade, while later on the attention shifted to monopolistic competition issues, heterogeneity of production inputs, diversity in geographic space, spatial spread of industries, and heterogeneous preferences of economic actors.

According to Ascani et al. (2012) there are four characteristic features of NEG that distinguishes this discipline from regional science and related methodological frameworks:

- Increasing returns to scale in relation to the spatial unevenness of economic activity (explaining spatial concentration of industrial and human activity).
- Monopolistic competition in relation to the absence of perfectly competitive markets (inducing large industrial conglomerates).
- Transport costs in relation to locational choices, e.g. the 'iceberg' transport costs phenomenon (stimulating, for instance, multi-plant companies).
- External economies in relation to labour market pooling or technological spillover effects (inducing a spatial concentration of labour force and industries).

All in all, NEG has built up a solid formal apparatus, although its empirical and evidence-based underpinning is still feeble compared to modern applied modelling experiments in regional science. Its orientation is predominantly towards 'stylized facts'. It leans more on economics than on analytical economic geography, while it has provided only new insights in a limited domain of economic geography (in particular, locational analysis).

2.3 A Triple-Layer Representation of the Spatial Economy

The notion of space is fraught with many conceptual problems, both ontological and operational (see, for example, Russell 1976; Harvey 1990; Lefebvre 1991; Hillier 2007; Braun 2011). Space can be understood in a variety of contexts:

- Mathematical (abstract or absolute), in the sense of the metric (distance) that defines it;
- Physical, as a characteristic property of matter;
- Geographical, as a result of the evolution of nature (real space) outside a human influence;
- Social (relational), defined by relative distances (or inversed proximity) produced by human communities in the course of history;
- The entire surface of the earth, or parts of it (real space).

Space is the centrepiece of regional science. According to Ponsard (1988): "…its introduction does not mean corrections in detail; its introduction changes everything. Because space is not economically neutral".

There is a growing awareness among regional scientists that geographic space is not only a passive space (a projection of activities onto a two-dimensional plane), but increasingly also an active space. Several scholars have even agreed that in the past decades space has been de-humanized and objectified or that space was often seen as a dead or immobile entity based on positive measurements rather than on actions (see Graham 1997; Soja 1989). A more pronounced role of space—in terms of dynamic space or space-time evolution—has therefore been advocated in more recent years (see Thrift 1996). *The 'where'* question is more and more replaced by a '*how*' and '*why*' question on the geography of human activity. Furthermore, the role of point-based physical geography in the location of socio-economic activities has diminished, in favour of an increasingly important role of spatial interactions and communication. Consequently, even though the 'death of distance' hypothesis has largely been proven to overstate the importance of the emerging 'virtual reality', connectivity and accessibility—both physical and virtual—have become key concepts in modern regional science, next to traditional concepts like agglomeration and urbanization.

The currently popular NEG (see Fujita et al. 1999) reiterates these considerations by drawing attention to the close interwovenness between agglomeration and trade. The authors offer an attempt to re-track traditional economic geography (and regional science), while they also aim to build a new economics of space, based on a few rigorous economic principles. The authors start by observing a regrettable division line between mainstream economics and the economics of location and aim to build a bridge on the basis of a few simple (certainly not universally valid) concepts, in particular the imperfect competition model marked by increasing returns to scale [originating from Dixit and Stiglitz (1977)]. In an open (multi-region or multi-country) system various types of spatial agglomeration patterns may emerge, depending *inter alia* on transport costs, forward and backward linkages, and immobility of resources. The authors illustrate their arguments by referring to core-periphery phenomena, agriculture, urban systems, city size, transportation, international trade, and industrial clustering. Their study forms a good manifestation of the rigour of solid economic analysis for the explanation of the spatial patterns and evolution of economic activity. The shadow side of this study is that the authors have missed out significant parts of the rich history of regional economics. Some examples may clarify this weakness in their study. First, any exposition on agglomeration advantages ought to start off from the basic economic principles laid down in growth pole, growth centre, or attraction pole theories developed by scholars such as Boudeville, Perroux, Richardson, or Klaassen. Now the book begins with an anecdotal story on the geographical concentration of second hand bookshops in St. Martins's Court in London, but wasn't it Chamberlin who already before World War II pointed out similar phenomena in Harvard Square in Cambridge, Massachusetts? Furthermore, are there in our modern times no agglomeration disadvantages? The history of regional science offers many insightful and solid cornerstones for location principles.

A second example: the two pillars of regional economics are certainly formed by agglomeration economies and generalized transportation costs. Much emphasis is laid by the authors on the economic analysis of urbanization and scale advantages, but less on transportation costs. In our era of ICT development where many economists advocate the 'death of distance', it would have been necessary to pay more attention to both psychological and virtual distance costs, and their implications for the spatial organization of our world. The reference to Samuelson's iceberg metaphor may be insufficient to explain the emergence of global urban networks as a new geographical phenomenon.

And thirdly, there is no doubt a need for unifying principles, but the above study misses out some relevant cornerstones of (applied) spatial price and equilibrium theory, so convincingly and rigorously developed by regional economists such as Takayama, Labys, Berliant, Nagurney, Hewings, or Oosterhaven. In addition, there are also important contributions from spatial endogenous growth theory, recently developed in the spirit of Romer and Lucas by spatial economists such as Bröcker, Button, Poot, Stough, and many others. Similarly, recent fundamental contributions to spatial dynamics and complexity theory by Puu, Sonis, Reggiani, Camagni, and many others would certainly have deserved more attention.

Finally, to call central place theory in the Christaller-Lösch tradition at best a classification scheme is a misconception of Lösch' equilibrium theory and of the wealth of literature which has been published ever since (Lösch 1940). A reference to the fundamental regional and trade models of Tinbergen (1964) on the relevance of hierarchical economic principles for spatial economic equilibrium would have been appropriate in this context. Nevertheless, this study is an important milestone in the history of regional science, as it aims to build a bridge to mainstream economics. Indeed, regional science has over the past decades developed into a full-fledged and rigorous scientific approach to the spatial economy. It has immensely enriched our understanding of spatial complexities.

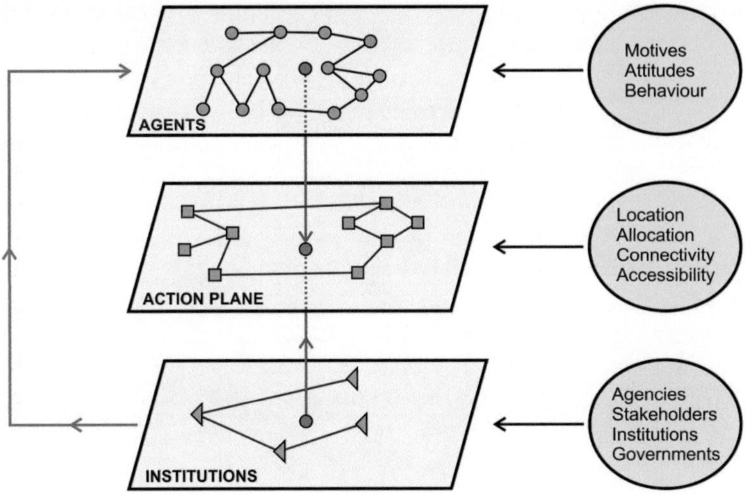

Fig. 2.1 A triple-layer activity space of the spatial economy

It should be added that the interplay of space and time is critical for a proper understanding of regional and urban dynamics. This interplay may exhibit both slow and fast dynamics, a space-time evolution based on interactive forces among all agents in space. This calls for the design and application of both conceptual and operational space-time models. Furthermore, several new contributions can be found in the interpretation of both space and time as continuous dimensions, rather than as discrete entities. As a result of advances in mathematical analysis of and statistical data bases on regional systems, we witness also a further generalization in spatial analysis by the inclusion of continuous space next to discrete space (see also Oud et al. 2012). The level of sophistication of spatial analysis has no doubt drastically increased over the past decades [see e.g. the Handbook of Regional Science by Fischer and Nijkamp (2013)].

The playing ground of actors in regional science is occupied by economic agents and institutions. Thus, the real-world spatial (regional or urban) action platform in regional science exhibits patterns of location, allocation, accessibility and connectivity. This leads altogether to a Multi-level Activity Space that can be mapped out in three layers (see Fig. 2.1).

The central middle layer in Fig. 2.1 represents the spatial action platform where all forces (horizontal and vertical) come together. This is a complex force field influenced by individual agents (space users) and collective governance bodies. The role of the latter bodies may be partial and sectoral ('management in space') or comprehensive and integrated ('management of space'). Clearly, in a dynamically evolving system also space-time elements may have to be added. Agents and institutions altogether then create a dynamic interactive spatial-economic force field.

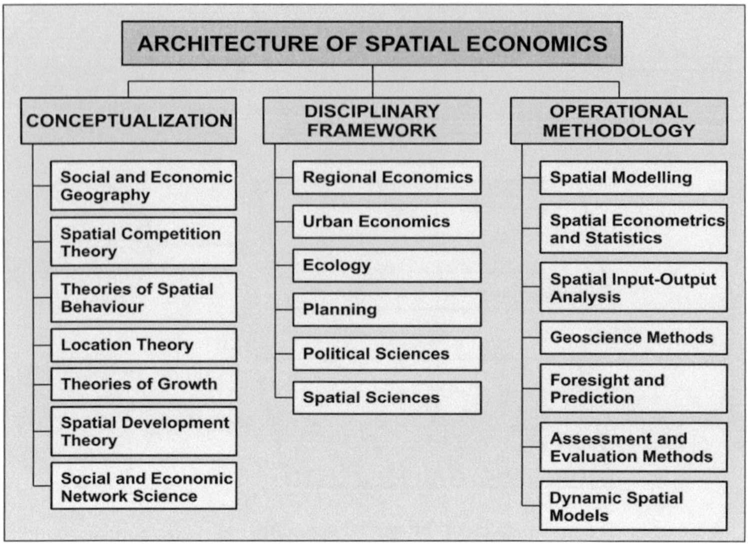

Fig. 2.2 Spatial economics and its architecture

2.4 A Holistic Map of Regional Science

In the course of history, many regional science scholars have corroborated the role of space in explaining economic processes (Garrison 1959; Barnes and Gregory 1997). This merger of space and economics has resulted into a new branch of economics, complementary to economic geography and often coined spatial economics. It focuses the attention in particular on spatial analysis, based on a modern toolbox of quantitative research. This holds for many domains (such as urban economics or transportation economics) and issues (such as location theory, growth theory, theories on spatial behaviour, spatial competition, development theory).

Clearly, spatial economics is partly overlapping with regional science and partly with socio-economic geography. From this perspective, NEG has a rather narrow scope, as it only addresses the economic aspects of geography (see Dymski 1996; Martin 1999; Thomas 2002; Bosker et al. 2007). It should be noted that, whether the term spatial economics is used or the term socio-economic geography—or in a broader sense, regional science—, all these scientific orientations are concerned with the spatial pattern and interaction of systems of production, distribution or consumption (or more generally, human activities) in a spatial context, including the management, planning and forecasting of spatial development. The methodological architecture of spatial economics is depicted in Fig. 2.2.

The focus on human activity in space does not mean an exclusive orientation towards the economic geography of our planet. Human activities takes place in a broad force field and has to positioned in a broader context, in which many other

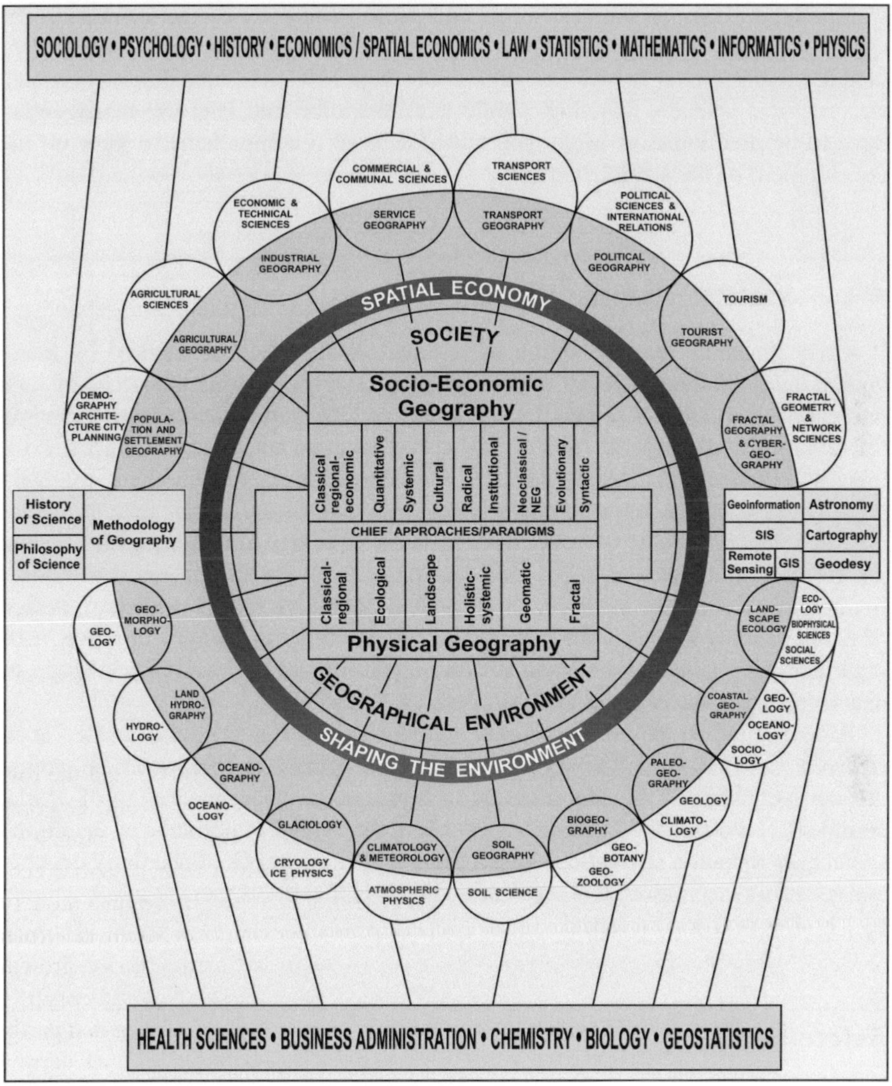

Fig. 2.3 The spatial economy and its relations with complementary sciences. *NEG* new economic geography, *SIS* spatial information systems, *GIS* geographic information systems

disciplines and scientific domains play a role. This viewpoint calls essentially for a comprehensive holistic approach in which regional science, geography and other spatial disciplines are embedded (see also Goodchild et al. 2000; Goodchild and Janelle 2004).

This complex interdependent force field is mapped out in Fig. 2.3. Clearly, such a conceptual framework is rich in scope and depicts the interfaces between real-world spatial phenomena and a variety of scientific orientations.

Finally, a focus on the real-world economic geography of our world, i.e. the spatial economy, prompts a variety of issues on management and governance of land use and scarce space (O'Loughlin 2000; Soja 1997). Against this background, also political science, planning, public administration and land use management have to be mentioned as important constituents of a comprehensive view on the development of the spatial economy.

2.5 Retrospect and Prospect

It seems plausible that the future of regional science will be marked by many uncertainties on the dynamics of the spatial economy. Prominent sources of such uncertainties are: global population dynamics and its spatial distribution (including the urban-rural divide), the future of urbanization in an era where the megatrend is towards urbanized settlement patterns, the complementary (sometimes mutually supportive) interface between physical-material and virtual-digital spatial interaction, and the complexity of governance systems in an age of unprecedented spatial transformation in our world. All such phenomena call for advanced research tools in regional science, for instance, on individual-collective spatial behaviour, design of early warning systems, for critical transitions ('tipping points') in space, self-organizing or resilient systems models on adaptability and vulnerability in space, or data mining in case of large-scale or massive databases.

After the above exposition, it goes without saying that regional science is an '*Unvollendete*': there will always be more secrets behind the horizon which prompt our curiosity. Issues like the analysis of continuous space, the nature of spatial complexity, the future of data-driven models, the spatial importance of the digital society, or dynamic space-time interactions will be a source of scholarly concern and scientific inspiration. It is predictable that regional science in the future will not be a boring or dismal science!

References

Ascani A, Crescenzi R, Iammarino S (2012) New economic geography and economic integration: a review, WP1/02, search working paper. London School of Economics, London

Barnes TJ (2004) The rise (and decline) of American regional science. J Econ Geogr 4:107–129

Barnes T, Gregory D (1997) Space, spatiality and spatial structure. In: Barnes T, Gregory D (eds) Reading human geography: the poetics and politics of inquiry. Arnold, London, pp 232–243

Bosker M, Brakman S, Garretsen H, Scharm M (2007) Adding geography to the new economic geography. Working paper no 2038, Center for Economic Studies/ifo Institute for Economic Research, Munich

Boureille B (1998) The notion of space in the economic work of Emile Levasseur. In: Bellet M, L'Harmet C (eds) Industry, space and competition: the contribution of economists of the past. Edward Elgar, Cheltenham, pp 187–198

Brakman S, Garretsen H, van Marrewijk C (2009) The new introduction to geographical economics. Cambridge University Press, Cambridge, MA

Braun GO (2011) Towards understanding urban processes and structures. In: Mierzejewska L, Wdowicka M (eds) Contemporary problems of urban and regional development. Bogucki Wydawnictwo Naukowe, Poznan, pp 91–121
Capello R, Nijkamp P (eds) (2009) Handbook of regional growth and development theories. Edward Elgar, Cheltenham
Clark GL, Feldman MP, Gertler MS (eds) (2003) The oxford handbook of economic geography. Oxford University Press, Oxford
Coffey WJ (2003) God tricks or barness tricks? Comments on 'what's wrong with American regional science?'. Can J Reg Sci 26(1):27–29
Dixit AK, Stiglitz JE (1977) Monopolistic competition and optimum product diversity. Am Econ Rev 67(3):297–308
Dymski GA (1996) On Krugman's model of economic geography. Geoforum 27(4):439–452
Fischer MM, Nijkamp P (eds) (2013) Handbook of regional science. Springer, Berlin
Fujita M, Krugman P, Venables AJ (1999) The spatial economy. MIT Press, Cambridge, MA
Garrison WL (1959) Spatial structure of the economy: I. Ann Assoc Am Geogr 49(2):232–239
Goodchild MF, Janelle DG (2004) Spatially integrated social science. Oxford University Press, Oxford
Goodchild MF, Anselin L, Appelbaum RP, Harthorn BH (2000) Toward spatially integrated social science. Int Reg Sci Rev 23(2):139–159
Gorter C, Nijkamp P (2001) Location theory. In: Hanson S (ed) International encyclopaedia for the social and behavioural sciences. Elsevier, Amsterdam, pp 9013–9019
Graham S (1997) Cities in the real-time age. Environ Plan A 29:105–127
Harvey D (1990) Between space and time: reflections on the geographical imagination. Ann Assoc Am Geogr 80(3):418–434
Hillier B (2007) Space is the machine: a configurational theory of architecture. Space Syntax, London
Isard W (1956) Location and space-economy. MIT Press, Cambridge, MA
Isard W (1960) Methods of regional analysis. MIT Press, Cambridge, MA
Isard W (1979) Spatial dynamics and optimal space-time development. Elsevier, Amsterdam
Isard W (2003) History of regional science and the regional science association international. Springer, Berlin
Knox PL, Marston SA (2001) Places and regions in global context: human geography. Prentice Hall, Upper Saddle River, NJ
Lefebvre H (1991) The production of space. Blackwell Publishing, Oxford
Leroux R, Hart DM (2012) French liberalism in the 19th century: an anthology. Routledge, New York
Lösch A (1940) Die Räumliche Ordnung der Wirtschaft. Gustav Fischer, Jena
Martin R (1999) The new 'geographical turn' in economics: sore critical reflections. Camb J Econ 23(1):65–91
McCann P (2005) Urban and regional economics. Oxford University Press, Oxford
Mulligan GF (2003) What's wrong with economic geography. Can J Region Sci 26(1):33–37
O'Loughlin J (2000) Geography as space and geography as place: the divide between political science and political geography continues. Geopolitics 5(3):126–137
O'Sullivan P (1981) Geographical economics. Macmillan, London
Oud H, Folmer H, Patuelli R, Nijkamp P (2012) Continuous time modelling with spatial dependence. Geogr Anal 44:29–48
Paelinck J, Nijkamp P (1980) Operational theory and method in regional analysis. Farnborough, Lexington, MA
Ponsard C (1983) History of spatial economic theory. Springer, Berlin
Ponsard C (ed) (1988) Analyse economique spatial. Presses Universitaires de France, Paris
Russell B (1976) Human knowledge: its scope and limits. Simon and Schuster, New York
Soja E (1989) Postmodern geographies. Verso, London
Soja E (1997) The socio-spatial dialectic. In: Barnes T, Gregory D (eds) Reading human geography: the poetics and politics of inquiry. Arnold, London, pp 244–256

Thomas I (2002) Transportation networks and the optimal location of human activities: a numerical geography approach. Edward Elgar, Cheltenham

Thrift N (1996) Inhuman geographies. In: Thrift N (ed) Spatial formation. Sage, London, pp 256–311

Tinbergen J (1964) Sur un Modéle de la Dispersion Geógraphique de l'Activité Economique. Revue d'Economie Politique (Special Issue):30–44

Tobler WR (1970) A computer movie simulating urban growth in the Detroit region. Econ Geogr 46:234–240

Warf B (ed) (2006) Encyclopaedia of human geography. Sage, Newbury Park, CA

Interregional and International Trade: Different Causes, Different Trends?

3

Paul Krugman

Abstract

It is argued in this paper that there are significant differences between interregional and international trade, mainly because of the existence of increasing returns at a regional level, next to comparative advantages. Consequently, location matters, although interregional trade seems to be lagging GDP growth, most likely as a result of a structurally declining regional specialisation.

Our charge, in writing for this memorial volume, is to produce new analytical work. And in what follows I'm going to try to do a bit of that. As an introduction, however, I want to tell the story of two old books.

First, and not surprisingly, there's Walter Isard's pioneering opus *Location and Space-Economy* (1956), which founded the field of regional science. As Masa Fujita pointed out some years ago (Fujita 1999), it was and is a remarkable work, anticipating much of what would be laboriously rediscovered decades later by mainstream economists. In particular, Isard—then still in his 1920s—pointed out, correctly, that a theory of the spatial economy would necessarily have to be at least in part a theory of monopolistic competition. Sure enough, when some of us realized that the tools of the New Trade Theory could be harnessed to the discussion of economic geography, monopolistic competition was crucial to the project.

In my own case, however, I'm sorry to say, Isard's classic was not the proximate source of inspiration—I came to his work late. Instead, I drew inspiration from an even older work, Bertil Ohlin's (1933) *Interregional and International Trade*. As you might guess from the title, Ohlin argued throughout for an integrated view, one that saw trade between, say, different regions of the United States as essentially similar to trade between different countries within Europe. He concluded by declaring that "The theory of international trade is nothing but *internationale Standortslehre*" (the international study of location.) I took that as my motto,

P. Krugman (✉)
Department of Economics, Princeton University, Princeton, NJ 08544, USA
e-mail: pkrugman@Princeton.EDU

P. Nijkamp et al. (eds.), *Regional Science Matters*, DOI 10.1007/978-3-319-07305-7_3,

arguing (and attempting to model) the point that the same forces the New Trade Theory had diagnosed as drivers of international trade could also be seen as drivers of interregional trade, specialization, and agglomeration.

Yet if you think about it, Ohlin and Isard were in a way offering opposed points of view. Ohlin was saying that we could think of interregional trade employing the same approaches we used to think about international trade—and while Ohlin made mention in passing of the role of industry clusters in causing trade, his main models of international trade, like those of almost everyone at the time, were models of perfect competition. Isard was saying that such models would not and could not make sense of the location of activity in space within countries, that a new framework, one that involved monopolistic competition, was needed.

So who was right? In what follows, I'll try to make the case that the answer, mostly, is that Isard was right. Certainly the perfectly competitive models of the Old Trade Theory won't do for regional science, while they do a decent job of making sense of much though not all trade. As a practical matter, interregional and international trade are on average quite different, although there is an overlap.

I'll also try to make two empirical points. One is that international trade has become *less* like interregional trade over the past couple of decades. The other is that because the two kinds of trade are different, they have had different destinies: international trade has soared in an era of "hyperglobalization" [to use the term of Subramanian and Kessel (2013)], while interregional trade has not, and may even be lagging overall economic growth.

3.1 America Is Flat

Traditional trade theory starts with the notion of comparative advantage, which we may define loosely as the view that countries trade to take advantage of their differences. Ricardo emphasized differences in productivity, which he may have considered the result of differences in climate, national character etc. but which we would nowadays usually attribute to technology. Ohlin, following on Heckscher, emphasized the role of differences in resources, and under the defter modeling hand of Paul Samuelson this became the famous factor proportions model, in which countries export goods whose production intensively uses their abundant factors. (Still a mouthful after all these years.)

It's a beautiful model. But—ironically given the title of Ohlin's book—it breaks down if you try to apply it to interregional trade. The factor proportions theory requires that there be differences in relative resource abundance among regions in order for them to trade, and if we add in transport costs those differences in abundance (a) have to be sufficiently large (b) will lead to differences in factor prices. But if factors are mobile—one of the defining differences between international and interregional economics—they will move to where their return is highest, narrowing the factor-price differences and eliminating the reason to trade. As Mundell (1957) pointed out long ago, in Ohlin-type models factor mobility and trade are substitutes, and if there are barriers to trade such as transport costs—which there must be for space to matter—factor mobility will crowd out trade.

What about technology differences? Since we don't have any good way to model these formally, there's nothing inside the model to say that technologies can't be different between, say, Chicago and New York. It's pretty obvious, however, that in practice high mobility of people and easy communication are likely to make such technology differences transient. With apologies to my New York Times colleague Tom Friedman, the world isn't really flat—but America is, with people, capital, and ideas flowing very easily from one region to another.

This in turn means that you can't explain interregional trade flows, let alone agglomeration and divergent regional growth, in terms of comparative advantage. Instead, you need to appeal to some form of increasing returns, in which regions specialize because there are inherent advantages to large-scale production. Mathematically, increasing returns can be consistent with perfect competition if they take the form of pure positive external economies; in practice, plausible stories about local specialization, even if they are about explaining industry clusters rather than the location of individual plants, just about always involve indivisibilities and some kind of monopolistic competition. So Isard was right in saying that spatial economics—the economics of location and regional activity within a national economy—required a different kind of model than what trade theorists were doing in the 1950s.

Now, you might say that this is all well and good, but we had a major rethinking of international trade in the 1980s, with a new emphasis on increasing returns and imperfect competition. Did this restore the unity between regional science and international trade?

The answer is, partially but not completely—and the subjects have been diverging again with the rise of modern globalization, as I will now try to explain.

3.2 Trade: Everything Old Is New Again

The intellectual path to the "New Trade Theory" of the 1980s began with empirical observations in the 1960s, inspired mainly by developments in Europe. Both Balassa (1966) and Grubel and Lloyd (1971) observed that the formation of the European Common Market in 1958 was followed by rapid growth in manufactures trade among Western European countries. And this rapid growth posed a puzzle.

First, why were these countries trading so much? They were similar in resources and technology; you might say that Europe in the 1960s was (and still is) flat in the same way I've just described as being true for America, even though labor mobility wasn't as high. (Actually, the mobility of guest workers from southern Europe and beyond, who were effectively the marginal source of labor supply in a number of countries, may have made labor mobility effectively quite high even though Western European workers weren't themselves mobile.) So where were the sources of comparative advantage?

One might try to infer European patterns of comparative advantage by looking at the industrial composition of trade—but when Balassa and Grubel/Lloyd tried to do this, they simply deepened the puzzle. For the main form of trade growth within Europe after 1958 was "intra-industry" trade: countries selling each other seemingly similar products, at least as measured by standard industrial classifications. What was going on here?

The Balassa-Grubel-Lloyd answer was, increasing returns and monopolistic competition: countries were specializing in different products, often different products within industries, so as to take advantage of economies of scale, and this specialization was giving rise to international trade. At the time their insight made little impact on international trade theory, because nobody knew how to embed increasing returns and monopolistic competition in formal models. But a decade later we had the tools to do that, and the New Trade Theory was born.

One way to think about that theory—not the way we thought about it at the time, but one that makes sense in retrospect—was that it argued that a significant part of world trade was actually similar in both form and motivation to interregional trade. Or to put it a bit differently, when it came to trade within Europe, or between the United States and Canada, Ohlin's dictum about the identity of international trade theory and location theory was exactly right. Trade between the U.S. and Canada is a lot like trade between Michigan and New York, trade between Norway and France a lot like trade between southern England and Scotland.

But this was only true for trade among similar countries. Meanwhile, a funny thing happened to the world economy after 1980, and even more so after 1990 or so: the center of gravity of world trade shifted, with trade among advanced countries more or less stagnating relative to GDP while trade between advanced and developing countries soared. And there was every reason to believe that the new growth in trade involved much more old-fashioned comparative advantage than the trade that inspired the New Trade Theory.

Why? Because more and more of world trade takes place between countries at very different levels of development, with correspondingly different resources, factor prices, and technology. Take one measure suggested by Subramanian and Kessler (2013), the average income level of exporters to advanced economies—that is, per capita GDP of exporters weighted by their shares in imports. As they show, between 1990 and 2010 the average income level of exporters to Europe fell from roughly 100 to 75 %; the corresponding decline for the United State was from 70 to 50%. This is telling us that a rapidly growing share of both European and U.S. imports was coming from countries with much lower incomes, wages, human and physical capital, etc. than the importing nations.

We can be reasonably sure that such trade was driven not by increasing returns but by comparative advantage. To take an example: while there are doubtless important increasing returns in the apparel industry, as demonstrated by the one-time concentration of that industry in a single district of Manhattan, the fact that Bangladesh exports apparel and the U.S. imports apparel clearly reflects the interaction between Bangladeshi labor abundance and apparel-industry labor intensity.

Now, within this overall pattern of comparative-advantage-driven trade, there is a fascinating economic geography story. Take the case of China: as Walter Isard would surely have reminded us, China is not a dimensionless point in space, it is a collection of locations—and Chinese export industries, like U.S. industries a century ago, are famously clustered, with one town dominating world production of buttons, another world production of cigarette lighters, yet another accounting for a large share of underwear production, and so on. I've argued (Krugman 2009) that the advantages of such clustering play an important role in generating overall gains from trade, even though the international pattern of trade is determined by old-fashioned comparative advantage.

So location isn't irrelevant, even to comparative-advantage trade. Yet the key point is that the forces driving trade between America and China are different from the forces driving trade between Atlanta and Chicago. And this difference may, in turn, explain the very divergent recent trends in interregional and international trade.

3.3 Soaring Globe, Stagnating Regions

I have already made use of Subramanian and Kessler's term "hyperglobalization," which mainly refers to the dramatic growth in international trade since about 1990.

Figure 3.1 gives you an idea of what we're talking about here. The figure shows the ratio of an index of world manufactures trade to world manufacturing production, with 1950 = 1 (data from United Nations 1962 and World Trade Organization 2013). If you look at the data up through around 1990, you see a point many international economists, myself included, used to make frequently—that much of the growth in trade since World War II, impressive as it was, only represented a return to pre-World War I levels. Actually, I used to say that the crucial technologies underlying globalization were the steamship, the railroad, and the telegraph—everything since had been a minor improvement.

But since 1990 trade has risen far beyond anything seen in the past. As I have already stressed, much of this trade growth involved exports from developing nations. Much of it also involved vertical specialization, with different pieces of a given final good produced in different countries.

What drove this new era of globalization? Despite containerization, it's surprisingly hard to find a clear downward trend in the costs of ocean transportation (Hummels 2007). Containerization may, however, have reduced effective shipping times by making it much easier to get goods onto and off liners. Meanwhile, air transportation became much cheaper. At the same time, there were dramatic policy changes—not in advanced countries, which did the great bulk of their liberalization before 1980, but in developing countries, which in the late 1980s and early 1990s shifted from inward-looking, import-substituting industrialization strategies to outward-looking policies.

The result of these forces was a broad relocation of international production, with labor-intensive activities—including the labor-intensive aspects of vertically disintegrated production, even for skill- or capital-intensive goods—moving to emerging nations, and hence a dramatic rise in overall trade.

Fig. 3.1 Index of world manufactures trade to world manufacturing production (1950 = 1). *Source*: Data from United Nations 1962 and World Trade Organization 2013

Table 3.1 Domestic shipments by transport mode, 2011

Truck	10,225,111.30
Rail	373,930.16
Water	158,396.12
Air	150,535.50
Other and unknown	2,549,256.90

Source: Federal Highway Administration

The important point I want to make for this paper is that these forces were more or less irrelevant to interregional trade. Changes in trade policy obviously weren't an issue, because interstate trade is constitutionally free from barriers. Progress in transportation technology was arguably far less important too. Table 3.1 shows the 2011 breakdown of U.S. interstate shipments of goods by transport mode (Federal Highway Administration); by value, shipments overwhelmingly move by truck. (Many ton-miles move by train, but that's largely low-value products, especially coal.) And given the absence of large wage differentials within the United States, the ability to slice up the value chain and outsource labor-intensive segments wasn't going to lead to a significant rise in interregional trade.

Thus one might suspect that the explosion of international trade was not matched by comparable growth in interregional trade. Is this suspicion true?

We can do a quick and dirty test by looking at data from the Freight Analysis Framework, which tracks shipments of goods between U.S. states as well as state exports to and imports from abroad. Specifically, we can look at the inflation-adjusted changes in domestic shipments, exports, and imports from 1997 to 2011; this doesn't cover the whole period of hyperglobalization, but is nonetheless illuminating. Table 3.2 shows what happened. Imports grew much faster than

Table 3.2 Growth in shipments, 1997–2011

Exports	46.5
Imports	108.7
Total trade	81.2
Domestic shipments	25.6
GDP	36.6

Source: Federal Highway Administration and author's calculations

exports, as the United States went into trade deficit, but both imports and exports grew much more than domestic shipments, with the overall volume of international trade—exports plus imports—growing three times as much.

So international and interregional trade not only look different, they seem to be growing increasingly different, and are on different trajectories. International trade is exploding upwards; interregional trade seems, if anything, to be growing more slowly than GDP. What might account for that last observation?

3.4 Homogenizing Regions?

At this point in history, we are accustomed to the notion not just that trade is always growing, but that it always grows faster than income, as the world gets smaller. In reality, international trade fell relative to world GDP for much of the twentieth century, not just because of protectionism but also because of a rise in real transport costs (Estaveodoral et al. 2003) Still, international trade has risen much faster than GDP for several generations.

Yet interregional trade appears, at least on a quick examination, to be lagging GDP growth. What might account for this lag?

One answer is that the Freight Analysis Framework only tracks trade in physical goods, even as we increasingly become a service economy. It's certainly possible that interregional trade in services—driven by improvements in communications technology (hello, Internet) has indeed grown rapidly.

But I'd like to suggest another possibility: maybe U.S. cities and regions have less reason to trade with each other than they once did. As Kim (1995) has shown, U.S. regional specialization, as measured by the difference in industrial mix, seems to have peaked circa 1920 and been declining ever since. He suggests that this decline reflects convergence of factor endowments, which in turn is due to factor mobility. I wouldn't entirely disagree, but I'd also suggest that technological change has weakened some of the forces that formerly supported industry clusters. For example, while thick markets in specialized skills are still a factor in, say, the concentration of high-tech firms in Silicon Valley, the modern economy may well have made the labor requirements of different sectors much more similar.

Put it this way: drive across America, and you see many, many seemingly identical office parks. Within these parks are offices filled with workers in cubicles—we all work in little boxes, and we all look just the same. (I don't know about the ticky-tacky aspect.) And maybe that similarity goes deeper than mere appearance. To the extent that Americans are doing pretty much the same thing everywhere, the rationale for specialization and interregional trade is reduced.

OK, that's just a hypothesis, which would take much more testing to fill out. Yet surely trying to test this idea would deepen our understanding of the economy—which brings me back to this short paper's starting point. Walter Isard pioneered a field by observing, and preaching, the importance of location in space to understanding economies. His observation remains as true as ever.

References

Balassa B (1966) Tariff reductions and trade in manufactures among the industrial countries. Am Econ Rev 56(3):466–473

Estevadeordal A, Frantz B, Taylor M (2003) The rise and fall of world trade, 1870–1939. Q J Econ 118(2):359–407

Federal Highway Administration, Freight analysis framework (2011), at http://www.ops.fhwa.dot.gov/freight/freight_analysis/faf/

Fujita M (1999) Location and space-economy at half a century: revisiting professor Isard's dream on the general theory. Ann Reg Sci 33:371–381

Grubel HG, Lloyd PJ (1971) The empirical measurement of intra-industry trade. Econ Rec 47(4): 494–517

Hummels D (2007) Transportation costs and international trade in the second era of globalization. J Econ Perspect 21(3):131–154

Isard W (1956) Location and space-economy. MIT Press, Cambridge, MA

Kim S (1995) Expansion of markets and the geographic distribution of economic activities: the trends in U. S. regional manufacturing structure, 1860–1987. Q J Econ 110(4):881–908

Krugman P (2009) Increasing return in a competitive advantage world. In: Stern RM (ed) Comparative advantage, growth, and the gains from trade and globalisation. World Scientific, Singapore, pp 43–51

Mundell R (1957) International trade and factor mobility. Am Econ Rev 47(3):321–335

Ohlin B (1933) Interregional and international trade. Harvard University Press, Cambridge, MA

Subramanian A, Kessler M (2013) The hyperglobalization of trade and its future. Working paper # 3, global citizen foundation

Refining the Isard Multiregional Input–Output Model: Theory, Operationality and Extensions

4

Jiyoung Park and Harry W. Richardson

Abstract

Isard demonstrated as early as 1951 that traditional (national) I–O models are inadequate because they cannot capture the effects of linkages and interactions among regions. To examine the full (short-term) impacts of unexpected events, such as terrorist attacks or natural disasters on the U.S. economy, the economic links among states (e.g. interregional trade, based on detailed shipments data) had also be analyzed. IMPLAN provided State input–output data. Few MRIO models have been implemented (since Polenske's attempt in 1980), but the potential increased after 1993 when the Commodity Flow Surveys (CFS) were introduced. This chapter introduces our model, the National Interstate Economic Model (NIEMO) which is an attempt to operationalize Isard's ideas fully, and covers 52 regions (the 50 States, Washington DC, and the "rest of the world") and 47 (29 commodity and 18 service) sectors. This was made possible by the expansion of the data base and the dramatic improvement in computer capacity. Much of the chapter explains how the model was built. A key step was the creation of conversion tables to make IMPLAN and CFS data compatible, and two sets of coefficients (industrial trade and regional interindustry). Model extensions were also introduced, TransNIEMO (to integrate the national highway

J. Park (✉)
Graduate School of Governance, Sungkyunkwan University, Suseon Hall #61503, Chongno-gu, Seoul 110-745, South Korea

Department of Urban and Regional Planning, University of Buffalo, The State University of New York, Buffalo, NY 14214-3087, USA
e-mail: jp292@buffalo.edu; jp292@skku.edu

H.W. Richardson
College of Architecture and Design, Autonomous University of the State of Mexico (AUEM), Toluca, Mexico
e-mail: professorhwr@gmail.com

P. Nijkamp et al. (eds.), *Regional Science Matters*, DOI 10.1007/978-3-319-07305-7_4,

network), FlexNIEMO (to relax the standard I–O assumption of fixed production coefficients), and adding induced impacts to indirect impacts in interregional trade analysis.

4.1 Background to Multiregional I–O Construction

Many economists and planners are interested in evaluating the socioeconomic impacts of various disruptions. Occasionally, they use geographically detailed input–output (I–O) models. Isard demonstrated as early as 1951 that traditional (national) I–O models are inadequate because they cannot capture the effects of linkages and interactions among regions. To examine the full (short-term) impacts of unexpected events, such as terrorist attacks or natural disasters, on the U.S. economy, the economic links among states should be considered. Multiregional input output models (MRIOs) include interregional trade tables and avoid some of the problems associated with excessive aggregation. Building an operational MRIO for all the states of the U.S.,[1] however, requires highly detailed interstate shipments data.

Although Chenery (1953) and Moses (1955) had formulated a relatively simplified MRIO framework in response to the earlier discussions by Isard (1951), data problems persisted and have stymied most applications. The non-existence and/or rarity of useful interregional trade data are the most problematic issues. Intraregional and interregional data must be comparable and compatible, yet the currently available shipments data between states are only sporadically available and difficult to use.

It is not surprising, then, that few MRIO models have been constructed or widely used in the U.S. The best known are the 1963 U.S. data sets for 51 regions and 79 sectors published in Polenske (1980) and the 1977 U.S. data sets for 51 regions and 120 sectors released by Jack Faucett Associates (1983), then updated by various Boston College researchers and reported in 1988 (Miller and Shao 1990).

More recently, there have been some attempts to estimate interregional trade flows using Commodity Flow Survey (CFS) data (Liu and Vilain 2004; USDOT, FHWA 2002; Jackson et al. 2006; Park et al. 2009). The U.S. Commodity Transportation Survey Data on inter-regional trade flows have been available since 1977

[1] This paper focuses only on the US economy. However, interregional I-O models have been developed in many parts of the world. Most of them have been built up from Isard's (and Leontief's) ideas. None of them have the degree of spatial and sectoral disaggregation found in NIEMO, the model discussed in this paper. Samples of the research are listed here: Akita and Kateoka (2002), Bonfiglio (2005), Hulu and Hewings (1993), Ichimura and Wong (2003), IDE-JETRO (2000, 2005), Kratena et al. (2013), Okamoto et al. (2005), Perez et al. (2008), Sargento (2009), Thijs (1983), Trinh et al. (2013), Turner et al. (2011), Zhang and Shi (2009). Most of these have been written in the last decade. The examples are from Asia and Europe: Japan, China, Korea, Vietnam, Indonesia, the European Union, Netherlands, the United Kingdom, Spain, Portugal and Romania. A large majority of the studies are empirically based and we were unable to discover a theoretical innovation that needed to be discussed in the main text of this paper.

but reporting was discontinued for some years. For the years since 1993, this data deficit can be met to some extent with the recent (CFS) data from the Bureau of Transportation Statistics (BTS), but these data are incomplete with respect to interstate flows. Based on the currently available CFS data, Jackson et al. (2006) used IMPLAN data (from the Minnesota Implan Group) to adjust the incomplete CFS reports by adopting gravity models constrained via distance and by making other adjustments. Along similar lines and using the same basic data sources, we elaborate Park et al. (2009), who suggested a different estimation approach that relied on an AFM (adjusted flow model) and a DFM (doubly-constrained Fratar model). To proceed in this way, it was first necessary to create conversion tables to reconcile the CFS and IMPLAN sectors. The detailed procedure bridging the CFS and IMPLAN sectors are explained in the study of Park et al. (2009).

This chapter focuses on four versions of a National Interstate Economic Model (NIEMO): NIEMO itself, which is a standard but operational multiregional input–output model (MRIO); TransNIEMO which adds on a national highway network to handle interregional freight flows; FlexNIEMO which constructs a supply-side MRIO that permits relaxation of the fixed production coefficients assumption; and finally a version that analyzes both intrastate and interstate induced impacts.

The NIEMO model revives an approach adopted in the late 1970s and the early 1980s (Polenske 1980; Jack Faucett Associates 1983), the development of a MRIO (multiregional input–output) model. We combine State level data from the IMPLAN input–output models with the estimated interregional trade flows based on the Commodity Flow Survey (CFS) using an approach developed in Park et al. (2009). NIEMO is aggregated to 47 economic sectors over 52 regions (50 States, Washington, D.C., and the Rest of the World). This results in an MRIO matrix with almost six million cells. Construction of the model involves substantial data assembly and considerable data manipulation.

NIEMO is a multiregional input–output model that is fully operational. The idea for such a model has a long history stretching back to Isard's suggestion of the "ideal interregional model" (Isard 1951, 1960) and Leontief's valiant but failed attempt to operationalize a variant of the model in the 1960s. The importance of sub-national models has long been recognized. Aggregation accounts for the loss of information, especially when positive effects in one area cancel negative impacts in another. It is also clear that most politicians have a keen interest in local constituencies. To say that NIEMO has succeeded where Leontief failed is not an immodest statement, but rather a reflection of the improvements in databases and computing capacity over the past 30 years. However, building bridges among the various data sources has been a substantial task.

NIEMO is not an exact replica of the original design as conceived by Isard and Leontief. Rather, NIEMO rests on the successful integration of state-level input–output information with data from the Commodity Flow Survey (CFS). The NIEMO approach is valuable because it uses only secondary data sources yet represents an innovative regional science procedure. Since 1993, the CFS has provided the most comprehensive single data source for U.S. freight movement flows. The data were collected every 5-year by the Bureau of Transportation

Statistics and the Census Bureau using a sample survey of industries through the Economic Census (USDOT, RITA and BTS 2005). Although the CFS is useful due to the wide range of commodity shipments and multimodal movement data, some user groups are not satisfied with the CFS details which have the incomplete coverage by commodity sectors and regions and the inability to fully capture foreign imports and exports (Southworth 2005; Park et al. 2009; Giuliano et al. 2010). Furthermore, the CFS estimates only cover less than 75 % of all the annual freight tons delivered in the U.S. This is because the CFS survey drops many establishments classified as farms, forestry, fisheries, construction, transportation, governments, foreign establishments, services, and most retail activities (ORNL 2000).

Nevertheless, Park et al. (2009) managed to estimate interstate trade flows. They developed two computation modules: an Adjusted Flow Model (AFM) and a Doubly-constrained Fratar Model (DFM). The systematic approach relies on the secondary data of CFS and IMPLAN. To reconcile different definitions and classifications of the commodities among multiple data sources, a new commodity sector scheme of 47 sectors (29 commodity, 18 service) was developed, referred to as the "USC Sectors." With various applications applying this initial version of NIEMO that excluded interstate services trade, the NIEMO's compatibility of the state-to-state trade flows and the flows between the states and the rest of world for the 29 USC commodity sectors were all readily testable (Park and Gordon 2005).

We use NIEMO to estimate total economic impacts including indirect impacts. As many of our case studies show, we have developed and applied NIEMO to estimate the impacts of various natural and man-made disasters. While many types of economic approaches including benefit-cost analysis, single-region I–O modeling, social accounting models, partial equilibrium models, and computable general equilibrium (CGE) models, are available for estimating economic impacts (Rich et al. 2005a, b), none of these tools provide national economic impacts at the level of individual states. For example, our studies, estimating the indirect and induced effects of impacts associated with capacity losses at the twin ports of Los Angeles-Long Beach showed that two-thirds of the impacts leaked outside the Southern California region. Without an interstate model such as NIEMO, we would have had no idea where these leakage effects might occur.

While a mathematical method to estimate interregional trade flows by Canning and Wang (2005) permits an empirical performance test, our foundation of estimating interregional trade flows is an approach based on secondary data sources. It uses CFS and IMPLAN, but adopts a different approach to estimating missing CFS data, a doubly-constrained Fratar model (DFM) to update the MRIO matrix to a base year.

Constructing NIEMO requires two basic kinds of tables: industrial trade coefficients tables and regional interindustry coefficients tables. While trade tables by industry are difficult to construct because of incomplete information in the CFS data, the interindustry tables present fewer problems because reliable data are available from IMPLAN at the state and industry levels. For details of the procedure used to estimate values for the empty cells in the trade flow matrix, see Park et al. (2009). Once initial trade flow matrices are estimated for sectors based on a reconciliation of CFS and IMPLAN data, these can then be iteratively refined via a Fratar model. However, the conventional Fratar model (Sheffi 1985) cannot estimate the diagonal (intrastate flow) values, so NIEMO incorporates a DFM (doubly-constrained Fratar Model) to supplement the off-diagonal flow estimates from the standard Fratar model, providing consistent estimates for both on- and off-diagonal values.

Usually, input–output models measure the direct (final demand), indirect (intermediate input flows) and induced (secondary consumption associated with direct and indirect employment) effects of changes in economic activity. In applying NIEMO we chose, in most cases, to measure the direct and indirect impacts only, for two reasons. First, it is a convention in MRIO to ignore induced effects (Miller and Blair 1985), presumably because induced consumption is less likely to cross interregional, e.g. interstate boundaries. Second, although there are local induced impacts associated with local indirect effects, the local and the imported indirect impacts are typically allocated in a MRIO model via some assumption that falls short of the accurate allocations of the "ideal" interregional input–output model (Isard 1951). For example, imported inputs of an origin sector are allocated to a particular destination sector in the same proportion as local inputs.

In pursuing our research goals, the choice of approaches involved difficult trade-offs. The use of linear economic models is justified by various factors, including the richness of the detailed results made possible at relatively low cost; NIEMO, for example, has approximately six million multipliers. The principal insight that drives our research is that, with some effort, it is possible to integrate data from IMPLAN's state-level input–output (I–O) models with the CFS data and other sources for all individual States, making it possible to build an operational multi-regional input–output model.

The drive behind the development of NIEMO was two-fold: to assess the interstate impacts of events analyzed with our regional model [SCPM (the Southern California Planning Model)]; and to allow us to extend the range of problems that may be studied at the interstate level.

4.2 Building NIEMO

The U.S. Commodity Transportation Survey (CTS) on inter-regional trade flows have been available since 1977 but reporting was discontinued for some years. For the years since 1993, this data deficit can be met to some extent with the recent (CFS) data from the Bureau of Transportation Statistics (BTS), but these data are incomplete with respect to interstate flows. A key step was to create conversion tables to reconcile the CFS and IMPLAN sectors.

The primary requirements for building an interstate model for the U.S. of the Chenery-Moses type are two sets of data, regional coefficients tables and trade coefficients tables (Miller and Blair 1985). Models of this type can be used to estimate inter-state industrial effects as well as inter-industry impacts on each state,

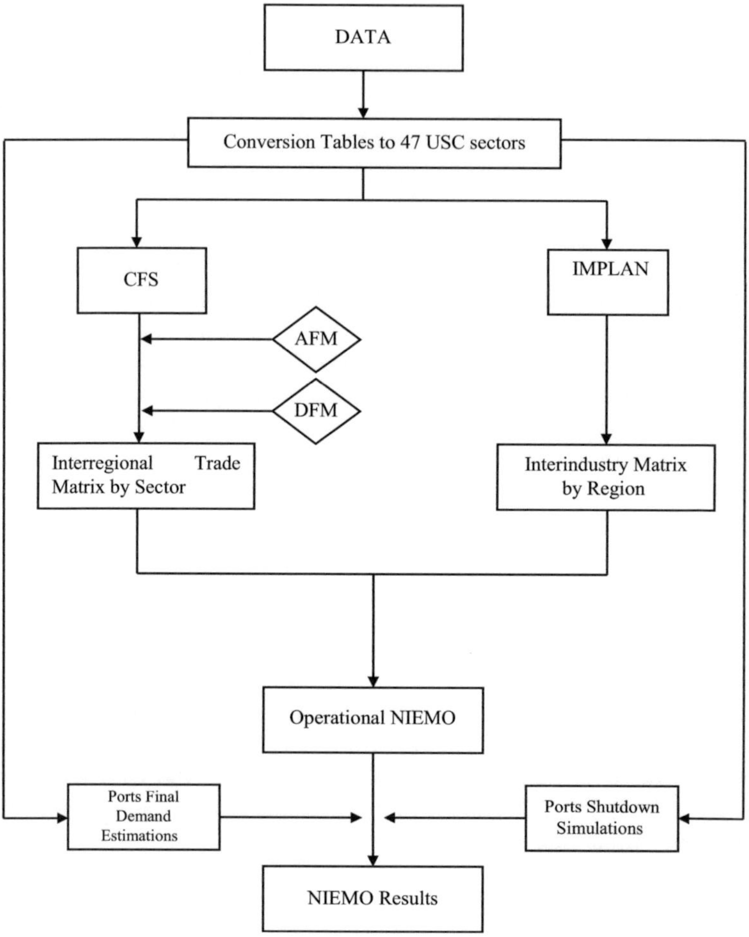

Fig. 4.1 NIEMO modeling steps as applied to a ports case. *Source*: Richardson et al. (2014)

based mainly on the two data sources: regional I–O tables that provide intra-regional industry coefficients for each state and interregional trade tables to provide comparable trade coefficients. This implies the creation of three types of matrices: an intraregional interindustry matrix, the interregional trade matrix, and the combined interregional, interindustry matrix (in other words, the NIEMO matrix). Before creating the matrices, however, the data reconciliation problem has to be addressed.

The main steps involved in building and testing NIEMO are shown in Fig. 4.1. We developed the 47 USC sectors into which many of the other classification systems can be converted. Figure 4.2 shows the industrial code conversion matrix among the many data sources used in this study.

CODE	USC	SCTG	BEA	NAICS	IMPLAN	SIC	HS	SITC	WCUS
USC		*C, E*	*C, E*	*C, E*	*C, E*	*C, W*	*C, E*	*C, W*	*C, W*
SCTG	*C, E*		*C, E*	*C, E*	*C, E*	*P*	*C, E*	*C, W*	*C, W*
BEA	*C, E*	*C, E*		*A*	*A*	*P*	*A*	*P*	*P*
NAICS	*C, E*	*C, E*	*A*		*A*	*C, W*	*C, E*	*P*	*P*
IMPLAN	*C, E*	*C, E*	*A*	*A*		*P*	*C, E*	*P*	*P*
SIC	*C, W*	*P*	*P*	*C, W*	*P*		*P*	*P*	*P*
HS	*C, E*	*C, E*	*A*	*C, E*	*C, E*	*P*		*C,W*	*C,W*
SITC	*C, W*	*C, W*	*P*	*P*	*P*	*P*	*C, W*		*C, E*
WCUS	*C, W*	*C, W*	*P*	*P*	*P*	*P*	*C, W*	*C, E*	

Notes
C: Created
A: Available from other sources
P: Possible to create
E: Bridge allocations evenly made, where necessary, without any weights
W: Bridge allocations made with plausible weights
Code Descriptions
USC: USC sectors newly created
SCTG : Standard Classification of Transported Goods
(http://www.bts.gov/cfs/sctg/welcome.htm)
BEA: Bureau of Economic Analysis
(http://www.bea.doc.gov)
NAICS : North American Industry Classification System
(http://www.census.gov/epcd/www/naics.html)
IMPLAN: IMPLAN 509-sector codes
SIC : Standard Industrial Classification
(http://www.osha.gov/oshstats/sicser.html)
HS : Harmonized System
(http://www.statcan.ca/trade/htdocs/hsinfo.html)
SITC: Standard International Trade Classification available from WISERTrade
(http://www.wisertrade.org/home/index.jsp)
WCUS: Waterborne Commerce of the United States
(http://www.iwr.usace.army.mil/ndc/data/datacomm.htm)

Fig. 4.2 Industrial code conversions. *Source*: Park et al. (2007)

The detailed conversion processes occasionally involved case-by-case reconciliations. Inevitably, some conversions involved mapping one sector into more than one. The light-gray cells of Fig. 4.2 represent one-to-one allocations. The dark-gray cells denote bridge allocations with plausible weights specified on a case-by-case basis.

A major problem in developing an inter-state interindustrial model stems from the fact that it is difficult to obtain data describing trade flows among states in the U.S. Since 1993, however, CFS data have been used, in spite of the fact that there are still problems such as high sampling variability and disclosure rules limiting the use of individual company data. The existence of many unreported values requires relying on other data sources to approximate completeness. It is not surprising, therefore, that since the work by Polenske (1980) and Faucett Associates (1983), there has been no comprehensive inventory of MRIO flows in the U.S.

The CFS reports trade flows between U.S. states for 43 SCTG sectors while the industry and commodity sectors (the data file of the IMPLAN Version 2) include 509 sector estimates, available for all states. The CFS trade flows data include both foreign and domestic trade. This means that all commodities coming into a U.S. port are listed as *outbound from that port* and inbound to the next destination; likewise, all commodities going to a port from anywhere in the U.S. are outbound from the origin and inbound to the port. For these reasons IMPLAN are added to the IMPLAN Total Commodity Output tally. In the current application, the 1997 CFS data were used as a baseline and updated to 2001 year using 2001 IMPLAN data. The years are being updated to a more recent period.

Differences among alternative industry classification systems from different data sources make data reconciliation especially difficult in the absence of standardized and tested conversion bridges. The estimation of trade flows from CFS, therefore, required intermediate conversion steps between the SCTG code system used in the CFS and the IMPLAN system of sectors, not always one-to-one matched pairs. Figure 4.3 shows the data reconciliation steps when creating a SCTG-IMPLAN conversion bridge enabling the aggregation of 509 IMPLAN sectors to 43 SCTG sectors.

The following paragraphs and equations summarize the NIEMO model. The traditional Leontief demand-side model is expressed as

$$\mathbf{X} = \mathbf{LINV} \times \mathbf{F} \tag{4.1}$$

where $\mathbf{X}$ is the $m \times 1$ total output vector for m sectors

$\mathbf{F}$ is the $m \times 1$ vector of final demand from private consumers, government, investment, and net exports of outputs from m sectors, and

$\mathbf{LINV}$ is a Leontief inverse matrix, $(\mathbf{I} - \mathbf{A})^{-1}$, where $\mathbf{A}$ is the $m \times m$ matrix of technical coefficients that captures interindustry relationships in terms of backwards linkages between m sectors and $\mathbf{I}$ is the identity matrix.

The inverse matrix in Eq. (4.1) is referred to as the demand-driven I–O model. The demand driven version of NIEMO can be expressed similarly as

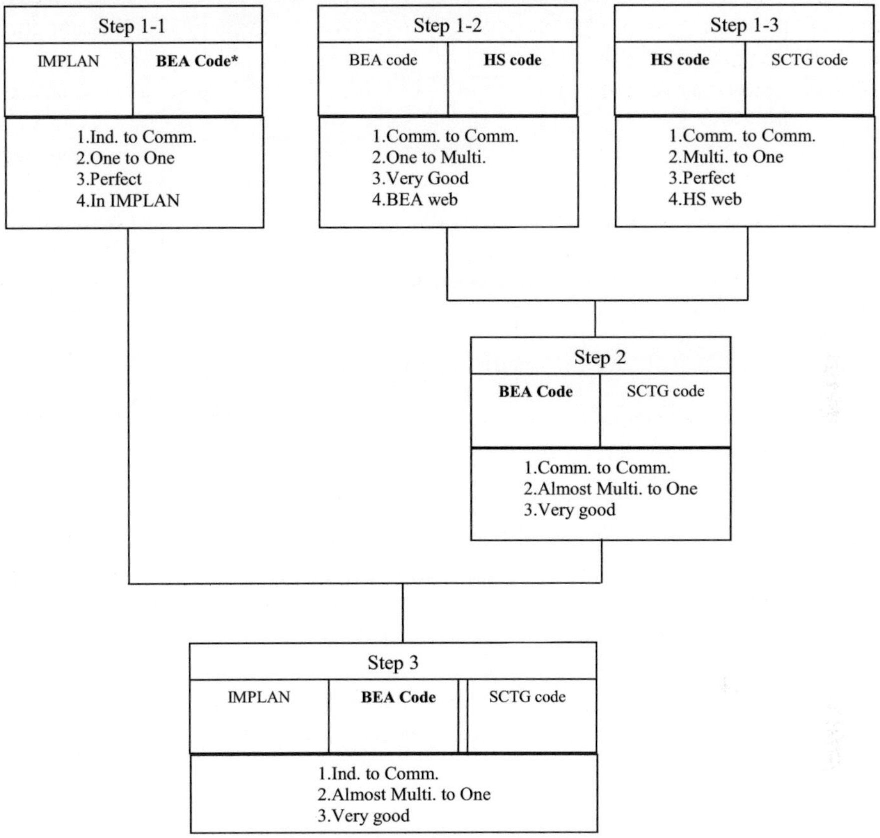

< NOTE >

*Bold: Used as Bridge Code
1: Ind. = Industry, Comm.= (Commodity)
2: One =One sector, Multi. =Sectors more than one
3: (Merged) Data Status
4: Source and Abbreviation
BEA: Bureau of Economic Analysis (http://www.bea.doc.gov)
SCTG: Standard Classification of Transported Goods (http://www.bts.gov/cfs/sctg/welcome.htm)
HS : Harmonized System (http://www.statcan.ca/trade/htdocs/hsinfo.html)

Fig. 4.3 Data reconciliation steps between SCTG and IMPLAN. *Source*: Park et al. (2007)

$$\mathbf{X}_{NIEMO} = \mathbf{LINV}_{NiEMO} \times \mathbf{F}_{NIEMO} \tag{4.2}$$

where $\mathbf{X}_{NIEMO}$ is the $nm \times 1$ total output vector for USC sectors $m(=1,\ldots,47)$ in each region $n(=1,\ldots,52)$

$\mathbf{F}_{NIEMO}$ is the $nm \times 1$ vector of region-specific final demand and foreign exports, and

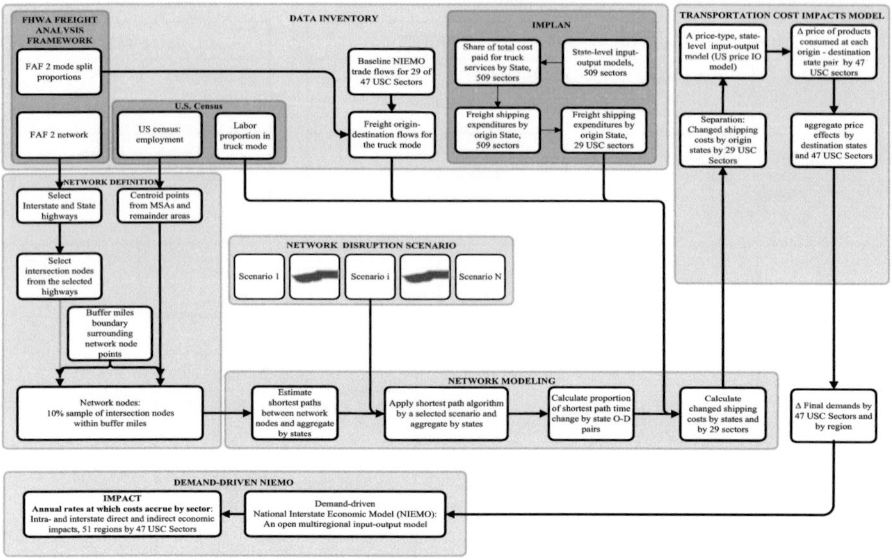

Fig. 4.4 Framework of national-level TransNIEMO. *Source*: Cho et al. (2014) TransNIEMO: economic impact analysis using a model of consistent interregional economic and highway network equilibria, CESIfo, WP 4601

$\mathbf{LINV}_{NiEMO} = (\mathbf{I} - \mathbf{W})^{-1}$, where $\mathbf{W}$ is an $nm \times nm$ matrix that combines technical coefficients with coefficients describing interregional trade flows. Note that the matrix $\mathbf{W}$ is defined as

$$\mathbf{W} = \mathbf{T}_{NIEMO} \times \mathbf{A}_{NIEMO} \tag{4.3}$$

where $\mathbf{A}_{NIEMO}$ is a block diagonal matrix of technical coefficients linking input commodities to output industries within each region $n(=1,\ldots,52)$, and $\mathbf{T}_{NIEMO}$ is an $nm \times nm$ collection of diagonal matrices describing interregional trade flows. The matrix $\mathbf{T}_{NIEMO}$ is defined as

$$\mathbf{T}_{NIEMO} = \mathbf{T} \times \left(\hat{\mathbf{T}}_{\mathbf{R}}\right)^{-1} \tag{4.4}$$

where $\mathbf{T}$ is an $nm \times nm$ matrix of trade flows, and

$\hat{\mathbf{T}}_{\mathbf{R}}$ is an $nm \times nm$ matrix formed by diagonalizing the $1 \times nm$ row vector $\mathbf{T}_{\mathbf{R}}$, consisting of the column sums of $\mathbf{T}$.

Note that we identify losses of foreign exports and final demand losses or gains as different types of direct impacts, and that these vary across scenarios. Foreign exports losses and government expenditures define a region-specific vector of direct impacts. Final demand losses or gains define vectors of regionally distributed direct impacts. Our scenarios involve two mechanisms for distributing impacts across regions. First, we use the standard MRIO procedure to distribute calculated region-specific final demand impacts across states.

$$\mathbf{F}'_{NIEMO} = \mathbf{T}_{NIEMO} \times \mathbf{F}_{NIEMO} \tag{4.5}$$

Distributing final demand losses resulting from the complete elimination of activity in a given sector is more complicated. This requires modifying the commodity trade coefficients matrix $\mathbf{T}_{NIEMO}$ to delete domestic exports from the quarantine area, for example, California in one of our case studies [see Lee et al. (2012)]. We set the entries for the California row vector, for example, describing USC Sector 1 *Live animals and fish, meat, seafood* in the matrix $\mathbf{T}_{NIEMO}$ to zero. In addition, outbound state flows from this sector are redistributed to origins in other states based on existing flow proportions. This defines a modified matrix $\mathbf{F}_{MDFNIEMO}$ that is used to allocate final demand losses.

$$\mathbf{F}'_{MDFNIEMO} = \mathbf{T}_{MDFNIEMO} \times \mathbf{F}_{MDFNIEMO} \tag{4.6}$$

This provides three types of direct impacts: region-specific direct impacts $\Delta\mathbf{F}_{NIEMO}$; and, via Eqs. (4.5) and (4.6), regionally distributed impacts $\Delta\mathbf{Y}'_{NIEMO}$ and $\Delta\mathbf{F}'_{MDFNIEMO}$. Total economic impacts may be estimated as

$$\Delta\mathbf{X}_{NIEMO} = \mathbf{LINV}_{NIEMO}\left[\Delta\mathbf{F}_{NIEMO} + \Delta\mathbf{F}'_{NIEMO} + \Delta\mathbf{F}'_{MDFNIEMO}\right] \tag{4.7}$$

4.3 TransNIEMO

TransNIEMO involves three sub-models (Fig. 4.4), a national highway network model, a transportation cost impact model, and NIEMO (a demand-driven multiregional input–output model as described above). Applications of TransNIEMO also require appropriate substantial data preparation. The model is applied to generate 1-year results, but because NIEMO is linear it is a simple matter to down-scale the results to shorter periods. The three major research steps associated with the three sub-models are discussed below.

The highway network model can be applied to problems such as combining the highway networks with bridge or tunnel disruption scenarios. A user equilibrium (UE) model is applied twice for each test: first to develop a baseline and second by applying the scenario. The user equilibrium approach is appropriate when there is significant congestion on the network. When dealing with freight flows on highway networks among metropolitan regions, applying the UE algorithm is reasonable. The results from applying the UE algorithm include the times and the distances from origin regions to destination regions. We assume that trip durations are related to truckers' labor costs and distance is associated with the other variable costs besides labor. The results from the network model simulations are used as inputs into the transportation cost impact model. The first empirical application of TransNIEMO, in which geography boundaries are limited to California and Arizona, is found in Park et al. (2011a).

While NIEMO and its related models are spatially disaggregated only to the state level, the transportation nodes for freight modes are the major metropolitan areas, which are the dominant centers of economic activity. Furthermore, in most states there is one or more major metropolitan area. Hence, the interstate freight flows had to be expanded to the flows among 114 nodes (specifically "economic centroids") to make them comprehensible. The non-metropolitan regions in selected cases also account for a significant proportion of state gross domestic product and freight O-D movements. Although local governments are mostly responsible for transportation infrastructure planning within their jurisdictions, most highways serve areas beyond their boundaries.

FAF provides a comprehensive data set but not all of the data are directly applicable to our research problems because services are also included in the annual flows among NIEMO sectors. Service values need to be excluded from the model.

Therefore NIEMO freight flows are used as freight flow input values. However, NIEMO does not account for transportation mode. Consequently, truck proportions from FAF are used to apportion NIEMO-estimated trade flows to obtain truck shipments. These are then loaded onto the highway network.

We also require data on the shipping costs associated with all flows. Total shipping costs between states are estimated using data from NIEMO, IMPLAN, and FAF as follows

$$ShipCost_{ij}^{k} = TTradeValue_{ij} \times TruckP_{ij} \times TruckCostValue_{i}^{k} \tag{4.8}$$

where

$ShipCost_{ij}^{k}$ are aggregated shipping costs from state i to j by commodity sector k.

$TTradeValue_{ij}$ are total trade values obtained from NIEMO for 49×49 states (Hawaii and Alaska are omitted for obvious reasons).

$TruckP_{ij}$ are truck proportions of total trade calculated by applying truck output values divided by total output values obtained from FAF data. Data for 114 MSAs by 114 MSAs are aggregated to 49 states by 49 states.

$TruckCostVaule_{i}^{k} = \frac{PurTruckSer_{i}^{k}}{TIndValue_{i}^{k}}$ are truck costs per value. These are the truck cost proportions in origin states obtained from IMPLAN. IMPLAN's sectors are aggregated to 29 USC *commodity* sectors. $PurTruckSer_{i}^{k}$ are the total value of purchased services by the trucking sector and $TIndValue_{i}^{k}$ are the total output of industry sectors.

Increased time and distance proportions are estimated by applying the user equilibrium (UE) model. Time changes and distance changes are separately modeled in Eqs. (4.6) and (4.7).

$$\Delta T_{ij}^{k} = ShipCost_{ij}^{k} \times \left(PTimeCost_{ij} \times PLaborCost\right) \tag{4.9}$$

where

ΔT_{ij}^{k} are increased costs caused by the increased time of travel.

$PTimeCost_{ij}$ are the proportions of time changes calculated as total increased time divided by total baseline time. Data for 114 MSAs by 114 MSAs flows are aggregated to 49 states by 49 states.

PLaborCost is the proportion of labor costs in the operations of the transportation industry (0.65).

$$\Delta D_{ij}^{k} = ShipCost_{ij}^{k} \times \left(PDistCost_{ij} \times PVarCost\right) \tag{4.10}$$

where

ΔD_{ij}^{k} are increased costs associated with increased shipping distance.

$PDistCost_{ij}$ are proportions of distance change calculated as total increased distance divided by total baseline distance. Data for 114 MSAs by 114 MSAs are aggregated to 49 states by 49 states.

PVarCost is the assumed proportion of variable costs in operation of the transportation industry (0.35).

Total increased shipping costs are estimated by adding the two increased costs, time and distance. Equation (4.11) shows the procedure for estimating increased shipping costs.

$$\Delta ShipCost_{ij}^{k} = \Delta D_{ij}^{k} + \Delta T_{ij}^{k} \tag{4.11}$$

where

$\Delta ShipCost_{ij}^{k}$ are increased shipping costs from origin state i to destination state j for industry sector k resulting from an event. In the short run, shipping costs are assumed to be non-decreasing. In the event of an emergency, sellers can pass on higher costs. They may cut prices because of competitive pressures, but probably only in the longer run.

The increased shipping costs, $\Delta ShipCost_{ij}^{k}$, are passed forward and lead to increased prices at destinations resulting in lower consumer expenditures. Dietzenbacher (1997) has shown that the supply-driven I–O model is more meaningful for estimating price increases than the Leontief price I–O model when *absolute* costs in value-added sectors are available. An application of a supply-driven I–O model is summarized in equation (4.12).

$$\Delta CS_j^k = \sum_{i=1}^{51} \left(\Delta ShipCost_{ij}^k \times InvMtrxO_i\right) \tag{4.12}$$

where

ΔCS_j^k are decreased consumer expenditures at destination j and industry sector k

$\sum_{k=1}^{47} \Delta CS_j^k$ are direct impacts by states; and

$InvMtrxO_j = (I - B_j)^{-1}$ is a 47×47 inverse matrix, where B is the direct output-based technical coefficients matrix in destination state j.

The reduced consumer expenditures associated with increased shipping costs drive reductions in household final demand. We assume that there are no substitution effects and final demand is directly affected by the reduced consumer expenditures. Equation (4.13) applies the demand-driven NIEMO to estimate the state-by-state economic impacts resulting from these reductions in household final demand.

$$\Delta X_j^k = \mathbf{LINV}_{NIEMO} \times \left(-\Delta CS_j^k\right) \tag{4.13}$$

where

ΔX_j^k are decreased final outputs in destination states j and industry sector k.

$\sum_{k=1}^{47} \Delta X_j^k$ are total impacts by states and $\sum_{j=1}^{51} \Delta X_j^k$ are total impacts by sector.

A simple equation in Berwick and Farooq (2003, Appendix A) was applied to calculate truckers' labor cost per mile (*TRC*) as

$$TRC = LR + \frac{(D/Speed + Waittime) * LRPH}{D} \tag{4.14}$$

where

LR = Labor (Wage) Rate Per Mile, given as 0.493 (\$/mile).

D = Trip Distance, given as 100 (miles),

Speed = 65 (mile/h),

$Waittime$ = Wait Time, given as 1 (h), and

$LRPH$ = Labor (Wage) Rate per Hour, given as 17 (\$/h).

We modified the current *LRPH* is close to \$17 while the literature assumed \$10 per hour. Accordingly, we could obtain \$0.9 per mile truckers' labor cost. Also, from the literature, other variable costs were given as \$0.48 per mile. Therefore, we estimated labor cost to be 65 % of total variable cost as,

$$0.65 = \frac{0.9}{(0.9 + 0.48)} \tag{4.15}$$

A full nationwide TransNIEMO was applied to study the economic impacts of three hypothetical road closure scenarios: two bridge service disruptions on the Mississippi River and one tunnel service disruption in the Rocky Mountains.

4.4 FlexNIEMO

FlexNIEMO is used to construct period-to-period versions of NIEMO. The approach developed recently by Park et al. (2011b) allows the fixed coefficients in the input–output world to be continuously modified, reflecting previous economic events and changes in interindustry relations. For example, a problem using the supply-side model is how to reflect demand-side adjustments during the recovery. With the supply-side FlexNIEMO, some of the major shortcomings inherent in the I–O model can be overcome. Based on supply and demand-driven economic input–output models suggested by Park et al. (2011b) and Gordon et al. (2009), where the entire detailed modeling approaches are found, we summarize the procedure here.

The FlexNIEMO approach begins with estimating total impacts. Let $X^s(t)$ and $X^d(t)$ be the total input row vector and the total output column vector respectively for various commodity and service sectors. Dropping off the subscript *NIEMO* from the previous equations, the total input vector is the sum of inter-industry sales ($A(t)$ $X^s(t)$) and total final demands ($F(t)$), and the total output vector is the sum of inter-industry purchases ($B(t)X^d(t)$) and total value added factors ($W(t)$).

$$X^s(t) = A(t)X^s(t) + F(t) \tag{4.16}$$

$$X^d(t) = B(t)X^d(t) + V(t) \tag{4.17}$$

where $A(t)$ is a $X^S(t)$-based requirement matrix composed of a technical flows matrix for industries within a region, $Z^{(t)}$, and a block diagonal matrix of inter-regional trade flows, $C^s(t)$. It is, defined as

$$A(t) = \Gamma_t[Z(t), C^s(t)] \tag{4.18}$$

A column vector $F(t)$ represents region specific final demand changes. Similarly, $B(t)$ is a $X^d(t)$-based requirement matrix composed of a technical flows matrix for industries, $Z^{(t)}$, and a block diagonal matrix of inter-regional trade flows, $C^d(t)$, defined as

$$B(t) = \Theta_t\left[Z(t), C^d(t)\right] \tag{4.19}$$

In addition, $V(t)$ is a row vector of region specific changes in value added. Note that Γ_t and Θ_t are matrix functions that are used to update the requirement matrices based on the technical flows matrix and the trade flows matrix. Under various scenarios, two vectors $F(t+k)$ and $V(t+k)$ can be acquired or computed for period $(t+k)$. Following the detailed procedures suggested by Park et al. (2011b), new vectors $F(t+k)$ and $V(t+k)$ and a new technical flows matrix $Z(t+k)$ will be derived for the next period.

4.5 Local and Domestic Trade Induced Impacts

As pointed out earlier, most applications of NIEMO focus solely on direct and indirect impacts, primarily because the secondary consumption associated with induced impacts rarely crosses State borders. However, we can expand the model to measure two types of induced impacts: local (i.e. intrastate) and those associated with domestic interstate trade (Park et al. 2013).

This requires the following manipulations of the data:

$$\mathrm{D}^{\mathrm{IMPLAN}} = \left(\mathrm{INDU\_R}^{\mathrm{IMPLAN}}\right) \times (\mathrm{Dom\_Trade})^{\mathrm{IMPLAN}} \tag{4.18}$$

where

$\mathrm{D}^{\mathrm{IMPLAN}}$ = domestic trade effects ascribed to the induced effects that occurred in a state that experienced direct impacts; this is defined as the product of $\mathrm{INDU\_R}^{\mathrm{IMPLAN}}$ and $(\mathrm{Dom\_Trade})^{\mathrm{IMPLAN}}$;

$\mathrm{INDU\_R}^{\mathrm{IMPLAN}}$ = ratio of induced effects to direct and indirect effects via IMPLAN in a state that experienced the impact; and

$(\mathrm{Dom\_Trade})^{\mathrm{IMPLAN}}$ = total domestic trade effects for that state via IMPLAN.

Note that if $\mathrm{D}^{\mathrm{IMPLAN}} > (\mathrm{Dom\_Trade})^{\mathrm{IMPLAN}}$, then $\mathrm{D}^{\mathrm{IMPLAN}} = (\mathrm{Dom\_Trade})^{\mathrm{IMPLAN}}$.

The whole 509 IMPLAN sector system is translated to the 47 USC Sector system and used for an input value set required for the second NIEMO run. Based on the USC Sector system, we redefine

$\mathrm{EXP}^{\mathrm{NIEMO}} = \mathrm{D}^{\mathrm{IMPLAN}}$ if the domestic trade is exports; and

$\mathrm{IMP}^{\mathrm{NIEMO}} = \mathrm{D}^{\mathrm{IMPLAN}}$ if the domestic trade is imports.

Therefore, based on the total impact prompted by local induced effects within the directly impacted state which is defined as Tr_IDCIMP, $\mathrm{Tr\_ IDCIMP}^{\mathrm{exp}}$ represents induced domestic exports effects and $\mathrm{Tr\_ IDCIMP}^{\mathrm{imp}}$ represents induced domestic imports effects. These were calculated as

$$\mathrm{Tr\_IDCIMP}^{\mathrm{exp}} = \mathbf{LINV}_{NIEMO} \times \mathbf{T}_{NIEMO} \times \mathrm{EXP}^{\mathrm{NIEMO}} \tag{4.19}$$

$$\mathrm{Tr\_IDCIMP}^{\mathrm{imp}} = \mathrm{IMP}^{\mathrm{NIEMO}} \times \mathbf{ST}_{NIEMO} \times \mathbf{SLINV}_{NIEMO} \tag{4.20}$$

where
$\mathbf{SLINV}_{NIEMO} = (\text{I-SNIEMO})^{-1}$
SNIEMO =the technical and trade coefficients in the supply-driven NIEMO; and
$\mathbf{ST}_{NIEMO}$ = the trade coefficients in the supply-driven NIEMO.

In the final step, we estimate *net* induced impacts which equal total induced impacts from the application of IMPLAN but via subtracting the induced trade impacts. This is also a part of Type II's induced effects to be added to the original NIEMO results. The net induced impacts resulting from the IMPLAN sector system are again aggregated to the USC sector system.

Conclusions

This chapter presents a methodological review of NIEMO (the National Interstate Economic Model) that is used for nationwide economic studies, as opposed to SCPM (the Southern California Planning Model) that is used for applications in Southern California. NIEMO is the most detailed and most comprehensive attempt to implement Isard and Leontief's dream of an operational MRIO (multiregional input–output model), although there have been other contributions along the way by Chenery, Moses, Polenske, Jackson (Jackson et al. 2006) and others. As widely noted, the most serious limitation of IO modeling is to obtain the fixed coefficients; the standard IO models ignore substitution opportunities prompted by market signals and uncertainties about the impact of economic sector losses. The FlexNIEMO model described above shows initial attempts to overcome this obstacle. Moreover, there are quantitative approaches in information theory that may improve NIEMOs' coefficients. Furthermore, flexible approaches can be applied to estimating economic input–output coefficients for areas smaller than the state level. This is useful in capturing the non-linearity of input–output coefficients because it can use local employment compensation information for the value added and final demand consumption sectors, This type of analysis can be used to account for local economies of scale, impossible with standard IO models. The required economic data sources in the U.S. are available from the Bureau of Economic Analysis, Consumer Expenditure Survey, Economic Census, U.S. Census of Population,; and more. This more disaggregated approach (both spatial and sectoral) is currently under development and could reinvigorate IO analysis.

However, we have also attempted to go beyond the standard concept of the MRIO by extending it in three important directions. First, by developing TransNIEMO we added a highway network that enabled us to explore the transportation as well as the spatially disaggregated economic impacts of any exogenous shock, such as a terrorist attack or a natural disaster. Second, as already suggested, we have attempted to deal with one of the major criticisms of MRIO, the fixed production coefficients assumption, by developing FlexNIEMO with a capacity for changing interindustry coefficients by introducing a supply-side component. Third, we devised a method of tackling a familiar critique of MRIOs, their inability to measure induced as well as indirect impacts with respect to interregional trade. Although there is still a need for further progress with these approaches, a more refined conceptualization of MRIO is moving in the right direction.

References

Akita T, Kateoka M (2002) Interregional interdependence and regional economic growth: an interregional input-output analysis of the Kyushi Region (1965-90). Rev Urban Reg Develop Stud 14(1):18–40

Berwick M, Farooq M (2003) Truck costing model for transportation managers. Upper Great Plains Transportation Institute, North Dakota State University. http://www.mountain-plains.org/pubs/html/mpc-03-152/index.php

Bonfiglio A (2005) The impact of EU accession in Romania. An analysis of regional development policy effects by a multiregional I-O model. European Association of Agricultural Economists, Copenhagen

Canning P, Wang Z (2005) A flexible mathematical programming model to estimate interregional input-output accounts. J Reg Sci 45:539–563

Chenery HB (1953) Regional analysis. In: Chenery HB, Clark PG, Pinna VC (eds) The structure and growth of the Italian economy. U.S. Mutual Security Agency, Rome, pp 98–139

Cho J, Gordon P, Moore II JE, Pan Q, Park JY, Richardson HW (2014) TransNIEMO: economic impact analysis using a model of consistent interregional economic and highway network Equilibria. CESifo Working Paper No. 4601

Dietzenbacher E (1997) In vindication of the ghosh model: a reinterpretation as a price model. J Reg Sci 37:629–651

Giuliano GG, Peter QP, Park JY, Wang LL (2010) Estimating freight flows for metropolitan area highway networks using secondary data sources. Netw Spatial Econ 10(1):73–91

Gordon P, Park JY, Richardson HW (2009) Modeling input-output impacts with substitutions in the household sector: a numerical example. Econ Model 26(3):696–701

Hulu E, Hewings GJD (1993) The development and use of interregional input-output models for Indonesia under conditions of limited information. Rev Urban Reg Develop Stud 5:135–152

Ichimura S, Wong H-J (2003) Interregional input-output analysis of the Chinese Economy. Scientific World, Singapore

IDE-JETRO (2000) The transnational inter-regional input-output table between China and Japan. Tokyo

IDE-JETRO (2005) The transnational inter-regional input-output table between China, Japan and Korea. Tokyo

Isard W (1951) Interregional and regional input-output analysis: a model of a space economy. Rev Econ Stat 33:318–328

Isard W (1960) Methods of regional science. Wiley, New York

Jack Faucett Associates, Inc (1983) The multiregional input-output accounts, 1977: introduction and summary, vol I (final report). Department of Health and Human Services, Washington, DC

Jackson RW, Schwarm WR, Okuyama Y, Islam S (2006) A method for constructing commodity by industry flow matrices. Ann Reg Sci 40:909–920

Kratena K et al (2013) FIDELIO 1: Fully Interregional Dynamic Econometric Long-term Input-Output Model for the EU2. European Commission, Luxembourg

Lee B, Park JY, Gordon P, Moore JE II, Richardson HW (2012) Estimating the state-by-state economic impacts of a foot-and-mouth disease attack. Int Reg Sci Rev 35(1):26–47

Liu LN, Vilain P (2004) Estimating commodity inflows to a substate region: using input-output data: commodity flow survey accuracy tests. J Transport Stat 7(1):23–37

Miller RE, Blair PD (1985) Input-output analysis: foundations and extensions. Prentice-Hall, Upper Saddle River, NJ

Miller RE, Shao G (1990) Spatial and sectoral aggregation in the commodity-industry multiregional input-output model. Environ Plan A 22:1637–1656
Moses LN (1955) The stability of interregional trading patterns and input-output analysis. Am Econ Rev 45:803–832
Oak Ridge National Laboratory (ORNL) (2000) Freight USA. Highlights from the 1997 commodity flow survey and other sources. Report prepared for the Bureau of Transportation Statistics US Department of Transportation, Washington, DC 20590
Okamoto N, Sano T, Inomata S (2005) Estimation of International Input-Output Analysis by Non-Survey Method. IDE-JETRO
Park JY, Gordon P (2005) An evaluation of input–output aggregation error using a new MRIO model. Regional Science Association International 52nd Annual Conference, Las Vegas, NV
Park JY, Gordon P, Moore II JE, Richardson HW, Wang L (2007), Simulating the state-by-state effects of terrorist attacks on three major US ports: applying NIEMO (National Interstate Economic Model). In: Richardson HW, Gordon P, Moore II JE (eds) The economic costs and consequences of terrorism. Edward Elgar, Cheltenham, pp 208–234
Park JY, Gordon P, Moore JE II, Richardson HW (2009) A two-step approach to estimating state-by-state commodity trade flows. Ann Reg Sci 43(4):1033–1072
Park JY, Gordon P, Moore II JE, Richardson HW (2011b) Constructing a flexible national interstate economic model. 19th International Input-Output Conference
Park JY, Cho J, Gordon P, Moore JE II, Richardson HW, Yoon S (2011b) Adding a freight network to a national interstate input-output model: a TransNIEMO application for California. J Transport Geogr 19(6):1410–1422
Park JY, Moore JE II, Richardson HW (2013) The gulf oil spill and economic impacts: extending the national interstate economic model (NIEMO) to account for induced impacts. J Homel Secur Emerg Manage 10(1):1–14
Perez J, Dones M, Llano C (2008) An interregional impact analysis of the EU Structural Funds in Spain (1995–1999). Universidad Autonoma de Madrid, Madrid
Polenske KR (1980) The U.S. multiregional input-output accounts and model. DC Health, Lexington, MA
Rich KM, Miller GY, Winter-Nelson A (2005a) A review of economic tools for the assessment of animal disease outbreaks. Revue Scientifique Et Technique-Office International Des Epizooties 24:833–845
Rich KM, Winter-Nelson A, Miller GY (2005b) Enhancing economic models for the analysis of animal disease. Revue Scientifique Et Technique-Office International Des Epizooties 24: 847–856
Richardson HW, Park JY, Pan Q, Moore JE II (eds) (2014) National economic impact analysis of terrorist attacks and natural disasters. Edward Elgar, Cheltenham, UK
Sargento ALM (2009) Regional input-output tables and models. Ph D Dissertation, Faculty of Economics, University of Coimbra, Portugal
Sheffi Y (1985) Urban transportation networks: equilibrium analysis with mathematical programming methods. Prentice-Hall Inc, Upper Saddle River, NJ
Southworth F (2005) Filling gaps in the US commodity flow pictures: using the CFS with other data sources. CFS Conference 2005 Workshop by Oak Ridge National Laboratory, Boston, MA. http://www.ornl.gov/~webworks/cppr/y2001/pres/123616.pdf
Thijs TR (1983) Some interregional input-output analysis. Revue d'Economie Regionale et Urbaine, #2
Trinh B, Hung DM, Huan NV (2013) Vietnam inter-regional input-output analysis: the bi-regional and 8-regional cases of Vietnam. J Contemp Manage 2(4):11–20
Turner K, Gilmartin M, McGregor PG Swales JK (2011) An integrated IO and CGE approach to analysing changes in environmental trade balances. University of Stirling Economics Discussion Paper 2011–4, Stirling, Scotland

US Department of Transportation (USDOT) Federal Highway Administration (2002) Freight analysis framework (FAF) Overview. Washington DC, http://www.ops.fhwa.dot.gov/freight/freight_analysis/faf/index.htm

US Department of Transportation (USDOT), Research and innovative technology administration (RITA) and Bureau of transportation statistics (BTS) (2005) Transportation Statistics Annual Report. Washington, DC, http://www.bts.gov/publications/transportation_statistics_annual_report/2005/pdf/entire.pdf

Zhang Z, Shi M (2009) Spatial linkage of manufacturing industries in China: based on inter-regional input-output analysis. Proceedings of the 17th international conference on input-output techniques. http://www.iioa.org/conferences/17th/papers.html

Time in Economic Theory

5

Åke E. Andersson

Abstract

Much of spatial economic theory is based on general economic equilibrium theory, although there are problems in a direct addition of a spatial dimension. The most striking is the lack of an analysis of the role of non-material and material public capital (or infrastructure) in the deduction of a static equilibrium structure or patterns of growth and development of economies. In this paper I demonstrate how different aspects of time can improve our understanding of dynamics of economies.

In this paper I furthermore show that a proper refocusing on the time dimension can also shed light on the structure of economies in space. Four approaches are necessary for such a synthesis.

1. Subdivision of products and systems of production according to their different and always positive durability, implying that everything produced is capital.
2. Subdivision of products according to the time used in their production
3. Subdivision into private and public goods, allowing for non-linearity.
4. Allowing for differences in time scales of economic processes.

With these distinctions it can be shown that the economic development in time **and** space is determined by the impact of economies of scale, duration of the production process, durability of products and the—relative to most other kinds of capital—much slower growth of public capital (i.e. material and non-material infrastructure).

This technical paper is closely related to my popular and thus non-technical paper published as Andersson Å.E. (2013) Time, space and capital. In: Andersson DE (ed) Advances in Austrian economics—the spatial market process. Emerald, Bingley

Å.E. Andersson (✉)
Department of Economics, Finance and Statistics, Jönköping International Business School, Sandgatan 31, 311 34, Falkenberg, Sweden
e-mail: Ake.Andersson@jibs.hj.se

P. Nijkamp et al. (eds.), *Regional Science Matters*, DOI 10.1007/978-3-319-07305-7_5,

5.1 Time in Economics

Walter Isard was concerned with the importance of correctly analyzing the role of the dimensions of time and space in economics as witnessed by his research together with Liossatos (Isard and Liossatos 1979). My intention with this paper is to shed some more light on how the many aspect of time ought to be represented in spatial and non-spatial economic theory.

First, the theoretically most convenient way is to represent time as a continuous variable, as is common in the modeling of many dynamic economic processes and especially in growth analysis. This implies that the processes are modeled as ordinary differential equations or in two dimensional space as partial differential equations, as shown by Puu (2003).

Second, a procedure that is popular in applied economic models, is to represent the dynamic economic processes as a discrete set of periods (e.g. weeks, months, quarters or years).

Third, a quite novel approach in dynamic economic theory is to subdivide the dynamic analysis into substantially different interactive **time scales** of the economic processes.

A quite different and economically realistic aspects of time is the representation of *each good in terms of durability (or alternatively the rate of depreciation) and each production process in terms of its' duration.*

5.2 Time as the Essential Element of Capital

More than a century ago in a critique of the Marxian definition of capital as accumulated labour Knut Wicksell (1966, 1967) proved that the time use or *duration of a production process* determines the value of capital.

This had earlier been demonstrated in Böhm Bawerk's numerical tables describing roundabout processes [Böhm Bawerk (1959–1921); Burmeister (1974); Morgenstern (1935); Marschak (1934); Dorfman (1959); Hicks (1970); Hicks (1973)].

The mathematician Wicksell realized that the numerical tables used by Böhm Bawerk could be densely represented as a mathematical maximization problem. This became the famous wine storing problem. See also Jevons (1871–1970). He assumed that the value of the wine would be growing with the time of storage. During the storage time a natural biological process using solar energy and the activity of yeast would contribute to the growing value of the wine, finally to be determined by the willingness to pay for the matured wine. The limiting factor on the time of storage is the opportunity cost of storage, including the loanable funds rate of interest.

In his model V(T) is the net value of the wine, if it is brought to the market at time T.

The present value (PV) of stopping the maturing by storage at time T is determined as:

$$\text{Maximize PV} = \mathrm{V(T)}e^{-rT}; \tag{5.1}$$

The necessary condition of optimality of storing time is:

$$\mathrm{dV/V = r}; \tag{5.2}$$

Optimal economic duration of the process thus implies that the storing should be stopped when:

The rate of growth of value equals the rate of interest.

This rule of thumb is not a special dynamic case, only relevant for point input, point output decision problems. The following dynamic optimization model shows that the value maximization condition also holds for harvesting sequences over time of some renewable biological resource (e.g. wine in some plantation, trees in a natural forest, or fish in the sea) as shown by the following model.

Maximize harvesting income = Maximize harvesting income = $\int_0^{\infty} p\mathrm{u(t)x(t)}\, e^{-rt}\mathrm{dt}$;

Subject to the growth condition:

$$\mathrm{dx/dt = ax(t) - bx(t)^2 - u(t)x(t)}$$

u(t) = the rate of harvesting at time t

x(t) = the stock of the biological resource at time t

We assume that the price is kept at a constant level.

Maximizing the Hamiltonian

$$\mathrm{H = pu(t)x(t)e^{-rt} - \lambda\big(ax(t) - bx(t)^2 - u(t)x(t)\big)}; \tag{5.3}$$

leads to an optimal rate of harvesting at each instance of time.

One of the conditions of optimality requires that the rate of harvesting is determined when

$$\mathrm{d}\lambda/\lambda = \mathrm{r} \tag{5.4}$$

Again we get the rule of thumb of optimal harvesting when:

The rate of growth of value should equal the rate of interest at the value maximizing rate of harvesting.

The Marxian labor theory of value was finally proved to be wrong. Not only is land (i.e. natural resources needed) an indispensable fundamental factor of production beside labor, as had already been shown by Johann Heinrich von Thünen (1826, 1960). Also the flow of time itself had now been shown to be a crucial variable in the determination of the dynamics of capital value.

5.3 Durability, Depreciation of Capital and the Rate of Economic Growth

The average durability of capital goods (i.e. all goods) of the economy is determined as the ratio between the value of capital as a stock and the value of production as a flow and has the *dimension of time* (Hawkins 1948; Hawkins and Simon 1949; Bródy 1970).

This implies that the average durability of the all goods (i.e. total capital) is determined as the ratio of the aggregate value of capital to the aggregate value of production. Under certainty, the optimal depreciation δ per year is 1/T. With an uncertain durability with known mean value the optimal rate of depreciation (and the implicit durability, T) is a constant fraction of the net asset value (Lev and Theil 1978).

Assume that the outputs of production processes in the economy are subdivided into currently used production and production for future use (i.e. investment). Current production requires inputs in fixed proportions, represented by input-output coefficients. Investment (I) is determined by an accelerator relation. I = Bgx, where I and x are vectors, g is the rate of growth and the matrix B gives the fixed capital requirements per unit of increase of production. b_{ij} divided by a_{ij} equals the durability of the good, $T_{i,}$ where T_i is the durability of good i. B = T′A where T′ is a diagonal matrix of durability of goods (i = 1,. . .,n):

$$x = Ax + gBx = (I + gT)Ax; \tag{5.5}$$

The primal condition of a general equilibrium production structure and rate of growth.

$$p = pA + rpB = pA(I + rT); \tag{5.6}$$

The dual price structure and rate of interest condition of a general growth equilibrium.

where x = production vector

p = price vector

A = n × n semi-positive input/output matrix

B = n × n semi-positive capital/output matrix = T'A.

g = the maximal rate of growth at the general growth equilibrium

r = the minimal rate of interest at the general growth equilibrium

A unique equilibrium growth and interest rate with an associated pair of equilibrium quantity and price vectors can be proved to exist [with the use of optimization procedures or the use of Perron's or Nikaido's theorems (Andersson and Beckmann p. 26, pp. 204–206 and p.235 and Debreu and Herstein 1953)].

Thus: The rate of interest equals the rate of growth in an economically sustainable equilibrium of this deterministic growing economy.

The larger the durability of any one product, ceteris paribus, the lower would be the rate of interest and growth. A compensating reduction of the use of current inputs is the only way of maintaining equilibrium rates of interest and growth, if any goods durability is increased.

5.4 von Neumann and the Birth of Mathematical General Equilibrium Economics

The models of growth shown in Eqs. (5.5) and (5.6) are special cases of the general equilibrium theory formulated by the mathematicians John von Neumann (1937) and Abraham Wald (1936, 1951). They developed the theory, when collaborating in a Vienna colloquium in the 1930s on the mathematics of general equilibrium theory as formulated by Walras (1874) and Cassel (1918, 1932). Wald proved the existence of a static general equilibrium and von Neumann proved the existence of a dynamic general equilibrium of a growth model based on a simpler model, formulated in Cassel's textbook. In Cassel's model the equilibrium rate of growth is determined by the ratio of the savings ratio and the capital-output ratio. Von Neumann proceeded to generalize this model into a theory of an economically sustainable dynamic general equilibrium, based on his saddle point theorem, proved in the 1920s.

He introduced time into his equilibrium growth theory in two ways:

First, he formulated the basic model in terms of discrete period dynamics.

Second, the durability of all products were introduced in an inverse form as constant rates of depreciation between periods, which is consistent with the assumption of a deterministic economic system, as shown by Lev and Theil (ibid).

Von Neumann assumed joint production in order to treat depreciation and durability efficiently in his model, as for example in the process of making paper in which wood, energy and machines are used as inputs at the start of the process. At the end of the paper making process a **joint product** vector of outputs consisting of paper, store of energy and of machines, which have depreciated and thus have become smaller in capacity. Formally the model is given by Eq. (5.7).

$$\begin{gathered}\mathbf{q}^{\mathrm{T}}\mathbf{B} \geq \alpha\mathbf{q}^{\mathrm{T}}\mathbf{A} \\ \mathbf{B}\mathbf{p} \leq \beta\mathbf{A}\mathbf{p} \\ \mathbf{q}^{\mathrm{T}}(\mathbf{B} - \alpha\mathbf{A})\mathbf{p} = 0 \\ \mathbf{q}^{\mathrm{T}}(\mathbf{B} - \beta\mathbf{A})\mathbf{p} = 0 \\ \mathbf{q} \geq 0; \mathbf{p} \geq 0\end{gathered} \tag{5.7}$$

Where $\mathbf{q}$ = vector of outputs
$\mathbf{p}$ = vector of prices
$\alpha = 1 +$ rate of growth
$\beta = 1 +$ rate of interest
$\mathbf{A}$ = mn matrix of inputs
$\mathbf{B}$ = mn matrix of outputs

The model allows for joint production and substitution of inputs.

Von Neumann proved that for the economic system (5.7) a sustainable equilibrium exists and is a saddle-point solution determining the equilibrium price and quantity vector. At the equilibrium point the minimum rate of interest β and the maximum sustainable rate of growth α are equalized.

von Neumann's theory does not presume utility maximization by the decision makers. This assumption probably stems from Gustav Cassel, who dismissed individual utility functions as necessary for the existence of demand functions.

Von Neumann and Wald had by then initiated modern mathematical economics with the use of saddle point and fixed point theorems. These ideas were later to be used in game theory as created by John von Neumann and Oscar Morgenstern (1944) and much later in general equilibrium theory as reformulated by Debreu (1959). von Neumann's introduction of inequalities in the formulation of saddle point theory also became one of the main preconditions for the development of linear and non-linear programming theory. The other important set of mathematical theorems to be used as a basis of programming theory was the book *Inequalities* by Hardy, Littlewood and Polya (HLP) (1933). Hardy, Littlewood and Polya proved all theorems behind the Constant Elasticity of Substitution (CES) function, which was much later to be used in the formulations of neoclassical economic growth theory (mostly without any references to HLP). Spatial analysis based on these theories later became important in theoretical and applied Regional Science.

5.5 Growth, Institutions, Uncertainty and Risk in the Determination of the Rate of Interest and the Value of Capital

It has been shown in the former section that a dynamic equilibrium in a deterministic economy requires equality of the growth rate and the rate of interest.

However, Wicksell (1914) and later Keynes showed that institutional factors may make the rate of interest deviate from this rule. The interest rate, as charged for loanable funds is a macroeconomic variable, determined by central banks, often using the interest rate as an instrument of monetary policies, sometimes in an international game between different governments. This monetary policy determined rate of interest can thus easily deviate from the "natural rate of interest" as determined by general equilibrium requirements. Such a deviation would then lead to inflation, deflation or unemployment, depending on the sign and size of the deviation and the institutional conditions ruling in the region.

Beside disregarding monetary institutions, most of the early analysis of the relation between the rates of growth and interest was based on deterministic equilibrium modelling. However, it is quite obvious that there could be uncertainty about the future among decision makers, for example as a consequence of variations in the conditions influencing production or demand. There would then be a required risk compensation in the form of a higher rate of interest in order to bridge the gap between lenders and borrowers. In the real world there will always be some uncertainty about the future (Knight 1934). This implies that a general dynamic economic equilibrium with uncertainty among decision makers requires that *the rate of interest is kept above the rate of economic growth*. How much above depends on the level and character of uncertainty.

5.6 Uncertainty and the Value of the Aggregate Stock of Capital

Early capital theorists had in vain tried to develop a consistent method of aggregation of the smallest units of durable goods into a consistent aggregate capital stock (Hayek 1941). Very often the starting point was the assumption that each little unit of durable good would have to be valued at some given unit price. The unsolved (and unsolvable) problem was how to determine the right micro level durable goods prices to be used in the aggregation procedure.

Paradoxically, the problem of capital aggregation can be resolved, as soon as we accept the necessary risk of all capital investments, organized into production units. These risks are revealed in the pricing of firms in financial markets and especially in the stock market. A firm, traded in the stock market, is essentially an already aggregated value of all the different capital goods of the firm, including information and knowledge capital in disembodied and embodied forms.

The theory of the stock market as a capital value determining machinery was initially formulated by Markowitz (1952) and further developed by Modigliani and Miller (M-M) (1958), Sharpe (1964), Lintner (1965), and Mossin (1966).

This modern financial market theory claims that the total equilibrium values of capital of all traded firms is determined (as an average over some period of observation) in the markets for securities and bonds, taking expected returns, perceived risk (as a measure of uncertainty) and the real rate of interest into consideration.

The generic claim is that the capital market is M-M-efficient, implying that the total value of all capital allocation opportunities can be captured by the expected return r(m) and the risk or standard deviation of returns (β(m)) for the market portfolio of all traded instruments. The value of a firm *as an aggregate of material and non-material capital* is determined in a similar way as a combination of expected returns and risk.

The risk-free or deterministic capital value would give $\beta(0)=0$ with $r(0)=g$. Any other portfolio would imply a rate of returns (natural interest rate) higher than the natural rate of growth.

From this follows the conclusion that the heterogeneous capital value, aggregated by the firm and valued in the stock market, when divided by the scale of production of the firm would generate the average durability of the capital, invested in the firm.

5.7 The Importance of Organization of Capital

The theory of the firm as formulated by Oliver Williamson (1981) and others, being based on Ronald Coase's transaction cost assumptions (Coase 1937), is a useful starting point for an analysis of the formation of firms. However, it says nothing about the best spatial and other allocation of material and human capital in the organization of the firm.

A haphazard arrangement of the carriers of human capital and the machinery and other material capital will not give the same high level of output as a profit maximizing organization. However, it could be hard or sometimes impossible to find such an optimal organization. It can be shown that maximizing the global profitability between a large number of such discrete interdependent human and material capital objects can often not be found, even with the aid of powerful computers.[1] For a firm with only 50 groups of employees with their machinery to be allocated to 50 different tasks there are in fact more than one trillion possible patterns of assignment employees to tasks. With quadratically represented interaction advantages, there are usually a large number of local profit maxima in this class of problems and the search for the global maximum is thus very hard.

The quadratic optimal assignment problem of Koopmans and Beckmann can be approached as in the following integer programming model, proposed by Andersson and Kallio (1982).

$$\begin{aligned} &\text{Maximize } x' \text{ S x} + \text{Rx} \\ &\text{Subject to} \sum (j) x_{ij} \leq 1 \text{ Specialist groups available} \\ &\qquad\qquad \sum (i) x_{ij} \geq 1; \text{Tasks to be fulfilled} \\ &\qquad\qquad \text{x} = (0 \text{ or } 1) \end{aligned} \tag{5.8}$$

S is typically a non-definite matrix giving the positive or negative advantages of collaborating (possibly at a distance) between each pair of employees and R gives the revenue effects of each individual if operating a task on her own.

Andersson and Kallio (ibid.) developed a computer algorithm that would efficiently search for a local optimum, when started from randomly selected starting points. The numerical procedures found a number of local optima, with quite different organization patterns. For problems with many tasks and groups of specialists the number of such local optima could be extremely large. In such a situation there is no guarantee that a global optimum would be found in finite computer time.

[1] If we assume indivisible units of machines and humans and that the productivity of a machine or a human (x(i)) depends on interaction with (x(j)) and if these interaction net benefits can be captured by the quadratic form x'Cx, then there is no simple incentive mechanism or computerized search algorithm that would provide the route to a global maximum for most interaction matrices C (Koopmans and Beckmann 1959).

However, the probability of finding solutions close to the global maximum is vastly increased if different decision agents are simulated to be experimenting in different ways with their organization of production. Competitive search will then (if started at some ridge to be defined below) in the long run reveal the agent with superior organization in terms of profitability.

An evolutionary procedure mimicking a competitive, evolutionary search in problems like (5.8) has been developed by Stuart Kaufmann and his associates (1996). They called it the *Patch Procedure,* where a patch can be a predetermined team of employees with some given equipment. The computation experiences were summarized as follows:

The results hint at something deep and simple about why flatter, decentralized organizations may function well: contrary to intuition, breaking an organization into "patches" where each patch attempts to optimize for its own selfish benefit, even if that is harmful to the whole, can lead, as if by an invisible hand, to the welfare of the whole organization. The trick, as we shall see, lies in how the patches are chosen. We will find an ordered regime where poor compromises for the entire organization are found, a chaotic regime where no solution is ever agreed on, and a phase transition between order and chaos where excellent solutions are found rapidly (Kauffmann, p. 147).... He concludes: *Therefore, as a general summary, it appears that the invisible hand finds the best solution if the coevolving system of patches is in the ordered regime rather near to the transition to chaos* (Kauffman, p.264).

It is clear that the evolutionary search must start in a rather special position on a ridge between order and chaos that might be hard to find. However, if it would be found, at the end of such an evolutionary process the superior firms with their structure of teams will have a capital value far above what would be indicated by their book-value of purchased machines and human capital.

The part of the capital value that cannot be easily accounted for as book value is often in accounting practice called Good Will Value. That value is always included in the valuation of the competing firms in the stock market.

5.8 Durability of Products and Patterns of Location of Production

The problem of the spatial structure of production is determined by the sustainable scale of firms and the total and spatial distribution of demand for their products. The sustainable scale of a firm is determined by the minimum of long term total average cost, including capital, transaction and transport (or logistics) costs. A long term equilibrium of the firm requires the price that can be charged to correspond to this minimal long run average cost.

Durability of the goods and the duration of production processes will have an important impact on the spatial structure of production. In order to determine the impact of durability of duration on the spatial structure of production we need to specify the long run average production cost function (C) and dependence on the scale of production, C(x). A common assumption is to disaggregate total production cost into fixed cost (F), and variable cost (V(x)). Fixed cost is the cost of all material and non-material capital of the production unit and is thus independent of the scale of operation as soon as the production unit has been established. For simplicity we assume that the firm is the production unit.

The duration (τ) of the production process increases the amount of capital needed and thus of the fixed cost of production, i.e. $F(\tau)$ with $F'(\tau) > 0$. This is especially pronounced in knowledge intensive production, needing a long period of research and development before actual production can occur. Typical examples are firms of the airplane producing industry, the movie industry, and the large pharmaceutical firms. Such firms regularly invest more than a fifth of the sales value in creation and innovation of new products and associated production equipment.

The variable cost is normally monotonously increasing with the scale of operation up to the capacity limit of the capital of the firm. In the sequel we assume that the optimal scale of operation is smaller than or equal to that upper limit. The simplest variable cost function is the linear case $V(x) = vx$. The total production cost function would be $P = F + vx$; and the average production cost function would thus be $C = F(\tau)/x + v$, where τ = duration of the production process.

Transport and transactions or logistics costs depend on the deliveries and other contacts between firm and customers. With a scattered distribution of customers in space around the firm, total transport and transaction cost would increase progressively with the increase in the scale of production and sales. Thus average cost of logistics (L), i.e. transactions and transport, would be increasing with the scale of operations. Assume $L = kx$. The term k can be decomposed into cost per unit of shipments, a, and the frequency of contacts, which is inversely depending on the durability, T, of the product. The average logistics cost is thus $L = (a/T)\,x$. Thus, the larger the durability, the lower is the average logistics cost.

The total average cost A equals the sum of average production cost C and average logistics cost L:

$$A = F(\tau)/x + c + (a/T)\,x; \tag{5.9}$$

Minimization of A implies that the optimal scale of production of the representative firm is:

$$\boldsymbol{x}(\boldsymbol{opt}) = \sqrt{\frac{\boldsymbol{F}(\tau)\boldsymbol{T}}{\boldsymbol{a}}}; \tag{5.10}$$

The optimal scale of production of the firm is thus increasing with increasing **duration** of the production process and also with increasing **durability** of the product.

The optimal number of firms is determined by the total scale of the market. The maximal total market scale is today the world market, to the extent that it is integrated by information and transport networks. The existence of an integrated world market is a precondition for perfect—or at least free—competition for many tradable goods.

The total number of firms in the world market, *N*, for a good is then determined as:

$$N = \text{Total demand}/x(\text{opt}); \tag{5.11}$$

Total demand is determined by the distribution of consumers in space and the minimal A, facing each consumer. However, with any spatial consumer distribution, the total number of firms would be decreasing with the **duration of production process and with the durability of the good** being analyzed.

An increased fixed cost, as influenced by an increased duration of the production process and durability of the good produced reinforce each other in decreasing the number of firms if the demand of the world market is given. In some cases the number of firms could be so severely constrained, that the assumption of perfect or free competition cannot be upheld even if the product is globally traded. Examples are trains, ships, airplanes and nuclear reactors, which are produced only in a few locations, serving a global market. These firms have an extremely long duration of production from the initial research stage through many stages of laboratory experiments.

The number of firms is thus determined by the procedure given above, but not the geographical locations of firms For that a connection with the theory of location and trade is needed. A theory of location and trade summarizing the contributions by Ricardo, von Thünen, Heckscher and Ohlin, Isard and Beckmann is the variational inequality model as formulated by Anna Nagurney (1999). In her model demand at each location and supply in each location is represented dually by the prices announced in the locations. An increased flow of a good from a location to another requires the price difference to be larger than the sum of logistics (including interest) cost, associated with a unit trade flow between the two locations, possibly at different instances of time.

The pattern of location and trade flows comes to an equilibrium when each good price difference is equal to (or smaller than) the sum of transaction and transport costs. As we have seen above the durability of each good determines their logistics cost. The larger the durability of the good the smaller is this cost. Trade will increase until there are no price differences between different locations for the limiting case of extremely large durability of a good. For goods of extreme durability and low logistical cost the law of one price must rule. An example is the pricing and trading in currencies.

5.9 Infrastructure: Capital That Is Durable and Public

Equilibrium theory and associated models have provided the fundaments for modern theoretical and applied economics. But they are inadequate in at least one important respect. These theories and models are not compatible with the dynamics of durable public goods (i.e. tangible and intangible infrastructure). Examples of infrastructure are fundamental values, constitutions, scientific knowledge, transportation and communication networks. In general equilibrium theory infrastructure is a mainly implicit, but always exogenously determined and stable stage on which the economic games are played. A dynamic theory of the interdependent evolution of the infrastructure and general economic equilibrium theory has been lacking.

The reason for the omission of a link between GET and infrastructure theory is quite clear. The necessary mathematical foundation for such an interdependency analysis did not exist before the 1970s. The first attempt to analyze catalytic and other collective/public phenomena was by Reneé Thom in his *Structural stability and morphogenesis: an outline of a general theory of models* (1989). In this book, originally published in French in 1972, he showed how collective phenomena could be modelled with singularity theory and applied to biological phenomena, such as the simultaneous blooming of a certain species by the influence of the slowly rising temperature, acting as a collective/public good.

Related to Thom's bifurcation theory is Synergetics, formulated by Hermann Haken as a way of solving some hard dynamic, non-linear problems in physics (Haken 1977). Haken showed that system predictability can often be achieved by subdividing dynamic processes according to their widely separated time scales. A general equilibrium of the combined dynamic system then becomes a possibility, if a few slowly changing variables are causally impacting a large collective of rapidly changing variables.

The institutional and material infrastructure can be defined to be such a collectively impacting (or public) variable moving on a qualitatively slower time scale than the private goods allocated in the markets. Thus any economic entity is defined to be an infrastructure if it is:

- simultaneously used by many firms or households and
- very durable, compared with other goods.

The following dynamic model of a market economy illustrates the power of subdividing the variables of the economic system into widely different time scales.

The dynamics of the markets for ordinary goods is determined as in general economic equilibrium theory by excess demand differential equations determining the price trajectories:

$$dp/dt = f(p, A); \quad (5.12)$$

where

p = a vector of prices of ordinary market goods including factor services (possibly in different regions),

A = a vector of infrastructure accessible in different regions

The development of infrastructure (as e.g. represented by accessibility values) can be represented by the equation:

$$s(A)dA/dt = m(p, A); \tag{5.13}$$

where s(A) represents the very large durability of infrastructure, indicating that s (A) is a very small, positive number, possibly in the order of 0.01 or lower.

This implies that in the time frame of the other variables of this system dA/dt can be set approximately equal to zero, **most of the time** (but not always). The fast and slow processes will rarely be synchronized and the whole system will then go into a period of creative destruction, eventually to come into rest at a new economic structure.

We thus have a dynamic system:

$$dp/dt = f(p, A^*), \tag{5.14a}$$

to be solved for an equilibrium, i.e. with f = 0, subject to the **temporary constraint**:

$$m(p, A^*) \leq 0, \tag{5.14b}$$

where A* indicates a given level of infrastructure in all parts of the economy.

For systems of this kind we can apply Tikhonov's theorem (Sugakov 1998):

Assume a dynamic system of N ordinary differential equations, which can be divided into two groups of equations. The first group consists of m fast equations, the second group consists of m + 1, . . ., *N* slow equations.

Tikhonov's theorem states that such a system has an equilibrium solution under certain economically reasonable conditions:

For each position of the slow subsystem, representing the dynamics of infrastructure, the fast general equilibrium market price subsystem has plenty of time to stabilize. Such an approximation is called adiabatic. (For a proof see Sugakov 1998)

In the very long run dA/dt cannot be assumed to be approximately equal to zero and thus the infrastructure would have substantially changed. The structure of prices and quantities of goods, as determined by f(p,A) could then cease to be as well behaved as in the short term dynamics, given by (Eq. 5.12).

The system would in the very long term have all the bifurcation properties, typical of non-linear, interactive dynamic systems. However, between periods of change of the economic structure, there could be periods of stable General Economic Equilibrium.

Most neoclassical economists have become skeptical about the possibility to mathematically model the dynamics of economic systems. Modern mathematical theory of dynamic systems supports this view. Chaos is the generic outcome of a non-linear economic system if all interactive economic variables are moving on the same time scale. General Equilibrium Theory, as formulated by e.g. Arrow and Hurwisz (1957), Debreu (1959) and others, is thus in fact not general enough to be expandable into a well behaved dynamic economic systems theory (and even less into combined spatial and dynamic systems).

However, I have shown above that this impossibility can be resolved if the dynamic models of the economy contain proper distinctions between the time scales of markets for goods, and the slow changes of the infrastructural stage on which the markets operate.

Conclusions

The theories and observations of the role of different time dimensions are fundamental to our understanding of economic processes in time and space. The most important time dimensions in dynamic and spatial economic theory are:

Duration of production.

Durability of goods and the inverse—the rate of depreciation.

Differences in time scales between infrastructure and market goods.

The choice of duration of a production process is important both for the temporal and spatial structure of production. With the increasing importance of large costs of scientific and industrial research and technological development this issue has become increasingly important.

But duration of the production process must be complemented by the durability of the goods produced. All goods are durable although to different degree—and all goods are consequently capital—and durability of capital is thus an irreducable determinant of many aspects of the economy as a growing spatial system.

The durability of goods and duration of production processes contribute in determining capital-output ratios, optimum rates of interest and growth and the spatial extent and pattern of the competitive markets.

The extreme durability and public nature of institutions, knowledge, networks and other infrastructure provides the basis of a new theory of complex dynamic spatial economic system. Within such an evolving system a general equilibrium of prices can exist and be stable.

References

Andersson ÅE (2013) Time, space and capital. In: Andersson DE (ed) Advances in Austrian economics—the spatial market process. Emerald, Bingley

Andersson ÅE, Kallio M (1982) A mathematical programming approach to land allocation in regional planning. IIASA, Laxenburg

Arrow KJ, Hurwicz L (1957) On the stability of the competitive equilibrium. Department of economics technical report 46. Stanford University

Böhm-Bawerk E (1959–1921) History and critique of interest theories. Libertarian Press, South Holland, IL (First published as Kapital und Kapitalzins.) Volume 1: History and critique of interest theories, 1884. Volume 2: Positive theory of capital, 1889. Volume 3: Further essays on capital and interest was first published as appendixes to volume 2 of the 1909–1912 edition, and was printed in a separate volume in 1921

Bródy A (1970) Proportions, prices and planning. North Holland, Amsterdam

Burmeister E (1974) Synthesizing the Neo-Austrian and alternative approaches to capital theory: a survey. J Econ Lit 12:413–456

Cassel G (1918) The theory of social economy, 1932nd edn. Harcourt, Brace and Company, New York
Coase R (1937) The nature of the firm. Economica 4(16):386–405
Debreu G, Herstein IN (1953) Nonnegative square matrices. Econometrica 21:597
Debreu G (1959) Theory of value an axiomatic analysis of economic equilibrium. Yale University Press, New Haven
Dorfman R (1959) Waiting and the period of production. Q J Econ 73(3):351–372
Haken H (1977) Synergetics, an introduction. Nonequilibrium phase-transitions and self-organization in physics, chemistry and biology. Springer, Berlin
Hayek FA (1941) The pure theory of capital. Routledge and Kegan Paul, London
Hawkins D (1948) Some conditions of macroeconomic stability. Econometrica 309–322
Hawkins D, Simon HA (1949) Note: Some conditions of macroeconomic stability. Econometrica 17:245–224
Hicks J (1970) A Neo-Austrian growth theory. Econ J 80:257–281
Hicks J (1973) Capital and time. Clarendon Press, Oxford
Isard W, Liossatos P (1979) Spatial dynamics and optimal space-time development. North-Holland, Amsterdam
Jevons WS (1871/1970) The theory of political economy. Penguin Classics, London
Kauffman S (1996) At home in the universe. The search for laws of self-organization and complexity. Penguin, London
Koopmans TC, Beckmann MJ (1959) Assignment problems and the location of economic activities. Econometrica 15:53–76
Knight F (1934) Capital, time, and the interest rate. Economica 257–286
Lev B, Theil H (1978) A maximum entropy approach to the choice of asset depreciation. J Accounting Res 16(2):286–293
Lintner J (1965) The valuation of risk assets and the selection of risky investments in stock portfolios and capital budgets. Rev Econ Stat 73:13–37
Markowitz H (1952) Portfolio selection. J Fin 12:71–91
Marschak J (1934) A note on the period of production. Econ J, 146–151
Modigliani F, Miller M (1958) The cost of capital, corporation finance, and the theory of investment. Am Econ Rev 48(3):261–297
Mossin J (1966) Equilibrium in a capital asset market. Econometrica 34(4):768–783
von Neumann J, Morgenstern O (1944) Theory of games and economic behavior. Princeton University Press, Princeton, NJ
Morgenstern O (1935) Zur Theorie der Produktionsperiode. Zeitschrift für Nationalökonomie, (Band VI):196–208
Nagurney A (1999) Network economics: a variational inequality approach. Kluwer, Amsterdam
von Neumann J (1937) Über ein ökonomisches Gleichungssystem und eine Verallgemeinerung des Brouwerschen Fixpunktsatzes. In: Menger K (ed) Ergebnisse eines Mathematischen Kolloquiums, 8: 73, 83. Deuticke, Leipzig, Translated as: A Model of General Economic Equilibrium. Review of Economic Studies (1945–1946), 13: 19
Puu T (2003) Attractors, bifurcations and chaos. Springer, New York
Sharpe WF (1964) Capital asset prices: a theory of market equilibrium under conditions of risk. J Fin 19(3):425–442
Sugakov VI (1998) Lectures in synergetics, vol 33, Series on Nonlinear Science, Series a. World Scientific, Singapore
Thom R (1989) Structural stability and morphogenesis: an outline of a general theory of models. Addison-Wesley, Reading, MA
von Thünen JH (1826) Der isolierte Staat in Beziehung auf Landwirthschaft und National-Ökonomie, oder Untersuchungen über den Einfluß, den die Getreidepreise, der Reichthum des Bodens und die Abgaben auf den Ackerbau ausüben. Hamburg

von Thünen JH (1960) The isolated state in relation to agriculture and political economy. Second part. In: Dempsey BW (ed) The frontier wage. The economic organization of free agents. Chicago

Wald A (1936, 1951) On some systems of equations of mathematical economics. Zeitschrift für Nationalökonomie (translated in Econometrica 19(4), 368–403

Walras L (1874, 1954, Engl. transl.)(Éléments d'économie politique pure, ou théorie de la richesse sociale) Elements of Pure Economics, or the theory of social wealth, transl. W. Jaffé, London: Allen and Unwin

Wicksell K (1966) Föreläsningar i nationalekonomi. Gleerups, Femte upplagan, Lund

Wicksell K (1967) Lectures on political economy. Reprints of economic classics. Augustus M. Kelley Publishers, New York

Wicksell K (1914) Lexis och Böhm-Bawerk. Ekonomisk Tidskrift Årg 16, häfte 11, pp. 322–334

Williamson OE (1981) The economics of organization: the transaction cost approach. Am J Sociol 87:548–577

The Regional Density Function and the Definition of Regional Boundaries

6

John B. Parr and Darryl Holden

Abstract

The population density function is usually applied at the level of the urban or metropolitan area. In this chapter, however, it is examined at the scale of the region, namely, an economic region of a nodal (rather than homogeneous) type. The regional density function to be employed has an inverse power form. This is shown to differ significantly from the negative exponential structure typically associated with urban areas. The primary purpose of the investigation is to explore the possibility of using the density function as a means of determining the boundary between adjacent economic regions. Initially, the boundary is examined in terms of a point and then, more conventionally, as a line. After discussing several possible extensions, the results are compared with findings of other theoretical models concerned with the specification of boundaries, but based on wholly different approaches.

6.1 Introduction

Segmentation of the economic space of a nation into discrete units represents an important cornerstone of regional economic analysis. While it is generally acknowledged that the results of a given investigation are likely to be sensitive to the regionalization scheme employed, there tends to be an uncritical acceptance of the validity of the regions used in official statistics, whether published or

J.B. Parr (✉)
Urban Studies, School of Social and Political Sciences, University of Glasgow, Glasgow G12 8QQ, UK
e-mail: John.Parr@glasgow.ac.uk

D. Holden
Department of Economics, University of Strathclyde, Sir William Duncan Building, Glasgow G4 0GE, UK
e-mail: d.holden@strath.ac.uk

P. Nijkamp et al. (eds.), *Regional Science Matters*, DOI 10.1007/978-3-319-07305-7_6,

unpublished. This is partly a recognition of the fact that the assembly of statistical data is a costly and time-consuming process, so that the independent generation of an alternative set of regional data for a particular analysis is usually out of the question. It is also due to the investigator being sufficiently anxious to proceed with the analysis in question that the validity of the regionalization scheme becomes a secondary concern. However, because of a myriad of factors, ranging from historical decisions to politically-motivated pressure, there can be no assurance that official regions represent a meaningful set of functional economic units.

It is against this general background that we wish to propose a further basis for determining the extent of a region. Our approach relies on the density function, which describes the decrease in population density with the distance from a central point. The concept of the density function, introduced by Clark (1951), has proven valuable in our understanding of urban spatial structure. The density function has also been applied at the regional level, though less commonly (Barkley et al. 1996; Nairn and O'Neill 1988; Parr 1985; Wang 2001). The discussion to follow has two broad purposes: first, to consider the nature and form of the regional density function; and second to demonstrate the manner in which it may be employed in the determination of regional boundaries. Unlike most applications of the regional density function, no a priori definition of the boundary is assumed. Rather, the regional boundary is shown to result from a particular application of the density function. The next section outlines certain key elements of the approach.

6.2 The Economic Region and the Regional Density Function

Although not all writers are comfortable with the notion of an economic region (Wren 2009), it continues to represent a frequently-employed unit of analysis. The significance of the economic region has been discussed in some detail by various authors, notably Isard (1956a), Lösch (1944/1954, pp. 103–137) and Teitz (1962).

6.2.1 The Nodal Region

There are, of course, a number of valid definitions of the economic region (Meyer 1963), but attention will be confined here to perhaps the most common one, the nodal region, a region that also lends itself to analysis in terms of the density function. The nodal region has been widely used in regional analysis (Brown and Holmes 1982; Nystuen and Dacey 1961), as well as in numerous public-policy interventions. The nodal region is considered here at the scale of a territory dominated economically by a major metropolitan area having a population above some minimum level. Such a nodal region thus comprises a metropolitan part or node, and also a more extensive non-metropolitan part, representing the hinterland. This latter part contains a rural population as well as an (often much larger) urban population, located in a hierarchical network of centers. A nodal region of this scale has been referred to as a "city-region" (Dickinson 1947) and also as a "polarized region" (Boudeville 1966).

A defining feature of such a nodal region is the substantial economic interaction between the two parts with respect to trade, commuting, capital flows, migration, etc. It must be emphasized, however, that an economic region of this type cannot be regarded as a closed system or self-sufficient entity, and the economic interrelations with other regions and nations are typically very well developed. To capture this interaction between the two parts of the nodal region, we employ the regional density function. This extends far beyond the boundary of the metropolitan area on which the region is centered. It will be shown that the form of the regional density function differs from that of the more familiar urban density function.

6.2.2 The Bogue Contribution

In order to provide an historical context for the analysis, we consider an influential work by Bogue (1950). This examined 67 regions (termed "metropolitan communities") for the US, using data from the 1940 Census. With certain exceptions each region was based on a metropolitan area (a "metropolis") having a central-city population in excess of 100,000 (Bogue 1950, pp. 16–17), and conformed to the definition of a nodal region at the scale outlined above. The boundary between any two adjacent regions was identified as the perpendicular bisector of a line connecting their respective metropolitan areas, so that any region not situated on a coast nor adjacent to an international border would have the form of a Thiessen polygon. In view of the breadth of the study this approximation was reasonable. However, one of the principal objectives of this chapter is to provide an alternative approach to the regional boundary.

The Bogue study constructed distance-density plots for all 67 regions. These were collapsed into generalized composite plots based on: (a) different components of population (urban, rural non-farm, rural farm); (b) different sectors emanating from the metropolitan area (intermetropolitan, subdominant, local); (c) different geographical sections of the US (North East, North Center, South, West); (d) different size classes of central city of the metropolitan area. The findings, which were presented to good effect in graphical and tabular form, revealed remarkable regularities. These pointed to the existence of a regional density function, as the following comments confirm: "On the average, as the distance from the metropolis increases, the number of persons per square mile decreases" and "the distance [-density] pattern for the suburban and for the most distant zones [of the region] are only different aspects of the same phenomenon" (Bogue 1950, p. 31).

Isard (1956b, p. 68) has stated that of the various writers on human ecology "Bogue has most explicitly considered the distance variable within the framework of the metropolitan regional analysis." To date the Bogue study is probably the only nation-wide attempt at applying the concept of the density function at the regional scale.

6.2.3 Bases for the Regional Density Function

In general terms the density function may be viewed as closely related to the overall rent gradient within a competitive land market. The association between population density and land value has been noted by Alonso (1964) and Evans (1973) among others. The rent gradient is usually considered in an urban or metropolitan context, but there is no reason why it cannot be employed at the regional scale.[1] Of importance in this latter connection is the interaction between the non-metropolitan hinterland and the metropolitan area. The metropolitan area represents a center of demand for the output of the non-metropolitan hinterland such as agricultural commodities, raw materials, semi-manufactured goods, etc. It is also a center for the direct and indirect supply of goods and services to households and firms throughout this hinterland.

As a center of demand the metropolitan area plays the role assigned to it by von Thünen (1842/1966), while as a center of supply the metropolitan area acts in the manner outlined by Lösch (1944/1954). In both cases, however, transportation costs cause interaction with the metropolitan area to decrease with distance. The regional density function may thus be seen as a reflection of the varying spatial interaction within the region (Kanafani 1978). The phenomenon of decreasing interaction with distance from a metropolitan area is commonplace, and manifests itself over the hinterland in terms of diminishing land-use intensity, declining levels of commuting to the metropolitan area, as well as falling consumer and non-household purchases made there (Bogue, Chart 3–1). The question of interaction can also be approached in terms of accessibility (Song 1996).

6.3 The Form of the Regional Density Function

In discussing certain of his findings, Bogue (1950, p. 31) noted that "There is a fairly constant rate of change between the relative decrease in land occupancy [density] and relative increases in distance." This observation points to a regional density function having the form of an inverse power function (Vining 1955, pp. 154–157):

$$M(x) = Nx^{-n} \quad (N > 0; n > 0; 0 \leq x \leq R) \tag{6.1}$$

or

$$\ln M(x) = \ln N - n(\ln x) \tag{6.2}$$

where *M(x)* is the population density within a thin annular ring at distance *x* from the center of the region, i.e., the center of the metropolitan area. Under actual

[1] In textbooks on urban economics the rent gradient is sometimes shown as ending at the boundary of the city or becoming horizontal thereafter, suggesting some prevailing level of rent beyond the city.

conditions there is usually a variation of density within each ring, a feature that will be considered in a later section. The value of $N = M(1)$ is the central density or the density at one unit of distance from the center of the region, while n is the value of the slope of the density function.

The inverse power function may be contrasted with the negative exponential function, the form of the density function approximating the pattern of distance-density decay within the metropolitan area (Clark 1951). This may be expressed as

$$M(x) = C\exp(-bx) \quad (C > 0; b > 0; 0 \leq x \leq x') \tag{6.3}$$

or

$$\ln M(x) = \ln C - bx \tag{6.4}$$

where C is the density at the center of the metropolitan area and b is the slope, with x' representing the limit of the metropolitan area, as defined by some minimum density.

6.3.1 Parameters of the Regional Density Function

Returning to the inverse power function of Eq. (6.2), the graph of this function is shown as curve M in Fig. 6.1, where R refers to an unspecified distance from the center, and $M(R)$ is the density at this distance. Within the metropolitan part of the region the inverse power function tends to exaggerate density levels, so that the central-density parameter, N in Eq. (6.2), should be seen as an extrapolated value. However, N may be taken as an indication of the population of the metropolitan area which, in turn, may be regarded as an indication of the population of the region, though not of its territorial extent. In other words, across regions the value of N increases with regional population. This regularity is similar to the situation across metropolitan areas, where the value of the central density, C in Eq. (6.4), increases with metropolitan-area population (Clark 1951).

In the case of the slope parameter, n in Eq. (6.2), Bogue indicated at several points that across regions this would increase with the population of the metropolitan area, and thus with the population of the region. The following general explanation for this variation in slope across regions was offered: "the more rapid decline of settlement [density] with distance in the case of the larger units [regions with greater populations and thus with higher values of N] results from a much higher level of land occupancy in the inner zones and only a moderately higher level of land occupancy in the outer zones" (Bogue 1950, p. 33). We consider this apparent regularity in Sects. 6.4 and 6.5. It will be noted that across regions the increase in the slope of the density function with population stands in marked contrast to the situation across metropolitan areas, where the slope, b in Eq. (6.4), declines with population (Clark 1951; Weiss 1961).

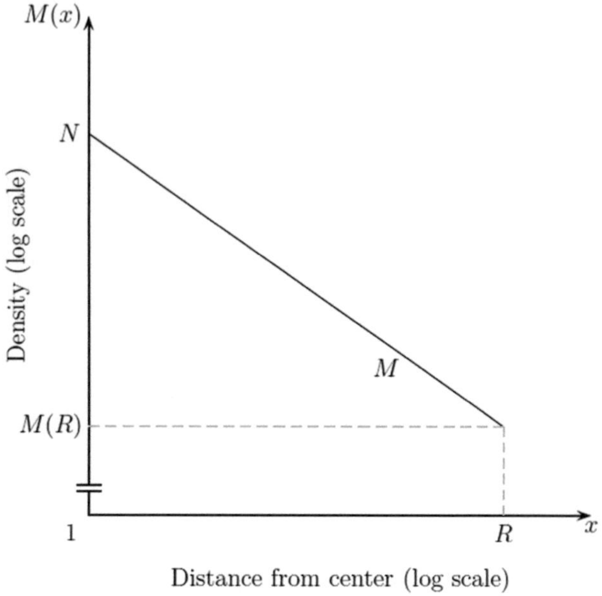

Fig. 6.1 Inverse power form of a regional density function

6.3.2 Additional Aspects

For the inverse power function the logarithm of density declines with logarithm of distance at a constant rate, as indicated in Eq. (6.2) and Fig. 6.1. The logarithm of density must therefore decline with "distance" at a decreasing rate, in contrast to the constant rate in the negative exponent function of Eq. (6.4). Evidence from the work of Bogue (1950), Clark (1968), Casetti (1969), Berry and Horton (1970), Mogridge and Parr (1997) among others indicates that beyond the metropolitan area the logarithm of density does, in fact, decline with distance at a decreasing rate rather than at a constant rate as in the negative exponential function.

A possible explanation for this departure from the negative exponential function is that the non-metropolitan part of the region contains not only a dispersed rural population that might be regarded as a "non-urban continuation" of the metropolitan area (the spatial structure of which could be expected to display a negative exponential form), but also a population located in urban centers having various economic functions such as service provision (based on both household and intermediate demand), transportation, mining, processing and fabricative manufacturing, etc. An important property of the inverse power function is that it is able to describe the pattern of density decline beyond the metropolitan area. Such a feature is of particular importance for our purposes, since a subsequent concern will be with conditions well beyond the metropolitan area.

It might be supposed that in an actual plot of density on distance the existence of these urban centres in the non-metropolitan part of the region would result in one or more local maxima or secondary peaks. In general, however, this is not the case.

Since the focus is on the mean density throughout each ring, there is a suppression or smoothing out of such peaks.[2] Obviously, in a regression involving a two-parameter function such as Eq. (6.2) any secondary peak is necessarily absent.

6.4 The Regional Boundary as a Point

In this section and the next, attention is focused on the possibility of using regional density functions to establish the boundary between two adjoining regions. We consider two regions A and B, with A' and B' representing their respective metropolitan areas, A' having the larger population. It is our contention that a point on the boundary between the two regions will be at the location where their respective regional density functions intersect. In other words, at the boundary of two regions the density is the same for both regions, reflecting comparable levels of interaction with A' and B'.

Attention in this section is focused on the relatively simple case, where the boundary is represented by a single point. In the following example T, the distance between A' and B', is 300 miles, and the single-point boundary occurs at location x^* along the line $A'B'$. We retain the above-mentioned Bogue generalization that the slope n across regions increases with increasing regional populations and therefore with higher values of N. This is reflected in the parameter values for the density functions of regions A and B, which are as follows: central densities ($N_A = 13{,}200$; $N_B = 1{,}900$) and slopes ($a = 1.15$; $b = 0.9$). The distance variables for regions A and B are expressed as x_A and x_B.

The boundary as a point is derived with the aid of Fig. 6.2. Here density is measured on the two logarithmically-scaled vertical axes, the left axis referring to region A and the right axis to region B. Each of the two horizontal axes measures the distance from metropolitan areas A' or B'. In order to obtain a solution, neither distance axis is scaled logarithmically, as was the case in Fig. 6.1 for the single region. The curves thus both decrease with distance at a decreasing rate, a feature of the inverse power function discussed in the preceding section. The two curves intersect at location x^*, which is 200 miles from metropolitan area A' and 100 miles from metropolitan area B'. The "reach" of metropolitan area A' is thus greater than that of metropolitan area B', so that $x_A^* > x_h > x_B^*$ where x_h is the halfway point between A' and B'. Further discussion of the point boundary, as defined by the equality of densities, is found in Appendix 6.1.

[2] Such a tendency was also reflected in a study of major metropolitan areas by Clark (1951, pp. 491–493). This used concentric rings "generally drawn at each mile radius." Out of the 35 cases examined, only five exhibited secondary peaks (due to the presence of sub-centers), and in each case this was relatively slight.

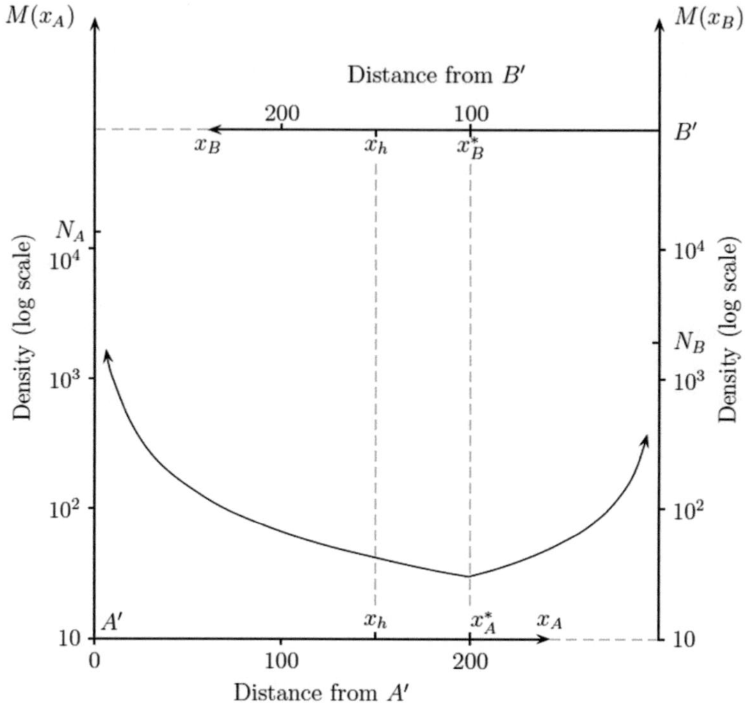

Fig. 6.2 Point boundary between two regions

6.5 The Regional Boundary as a Line

While this single-point perspective on the boundary indicates the approach adopted, it is too restrictive, and needs to be broadened. The boundary between two metropolitan area-based regions is now regarded as a line, the more conventional view of a boundary. At any point on this line the two regional density functions are equal in value, with the common value of density varying along this line.

6.5.1 The General Case

We again consider two regions A and B, which have A' and B' as their respective metropolitan areas. The density functions for the two regions again have unequal central densities ($N_A > N_B$) and unequal slopes ($a > b$). The boundary can be determined by first setting $M(x)$ in Eq. (6.1) equal for region A and region B, so that

$$\frac{N_A}{(x_A)^a} = \frac{N_B}{(x_B)^b} \quad (N_A > N_B; a > b),$$

or

$$\ln N_A - a(\ln x_A) = \ln N_B - b(\ln x_B)$$

Simple rearrangement leads to

$$\ln x_B - \frac{a}{b}\ln x_A = \frac{\ln N_B - \ln N_A}{b} \tag{6.5}$$

By taking a sufficient number of values for x_A in Eq. (6.5), and obtaining the implied values of x_B, the boundary is made explicit. Before this is done, however, we specify more completely the relationship between n and N.[3]

Previously, we have simply assumed a positive association between these two parameters. On the basis of the displays of Bogue (1950, Chart 6–1), together with the limited evidence for US regions from Berry and Horton (1970) and Clark (1968), we now assume that

$$n = s(\ln N) \quad (n > 0;\ s > 0) \tag{6.6}$$

The value of s is an empirically-determined constant, which may be assumed to hold for regions within a nation or within a particular section of a nation. For the US over the period 1940–1960 it was estimated that $s = 0.127$.

In our two-region case, Eq. (6.6) implies that for region A

$$a = s(\ln N_A) \tag{6.7}$$

and that for region B

$$b = s(\ln N_B) \tag{6.8}$$

so that

$$\frac{\ln N_A}{\ln N_B} = \frac{a}{b} \quad (= \rho) \tag{6.9}$$

where $\rho \geq 1$ That is, the ratio of the logarithms of central density is equal to the ratio of the slopes. Using Eqs. (6.7), (6.8), and (6.9) in Eq. (6.5) yields

[3] A relationship between parameters is not uncommon. Taking a longitudinal rather than a cross-sectional perspective for Paris over the period 1911–1968, Bussière (1972) demonstrated that the two parameters of the negative exponential function were closely related in a positive manner.

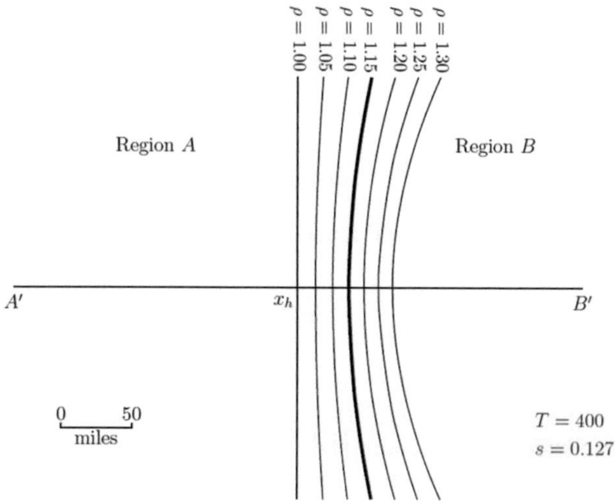

Fig. 6.3 Line boundary between two regions for different values of ρ

$$\ln x_B - \rho(\ln x_A) = \frac{1-\rho}{s} \tag{6.10}$$

This is a constrained version of Eq. (6.5), which is used to specify the boundary between the two regions.

An example is given in Fig. 6.3, where $T = 400$ (the mileage between A' and B') and $s = 0.127$ (the estimated constant for the US). The diagram indicates the boundaries between regions A and B for various values of ρ. When $\rho = 1$ we have the special case where the density functions of the two regions are identical, and the boundary between the two is shown as a straight line, perpendicular to the line $A'B'$ at x_h, the halfway distance between the two metropolitan areas. The bold line indicates the boundary for the case $\rho = 1.15$. This boundary intersects the line $A'B'$ where the distances from the two metropolitan areas are $x_A^* = 236$ and $x_B^* = 164$.

Two general features of the boundaries in Fig. 6.3 are of interest. First (when $\rho > 1$), the boundary is displaced from the halfway point toward the smaller metropolitan area B', and the greater the value of ρ, the more pronounced is this displacement. Second (again when $\rho > 1$), the boundary is convex to the larger metropolitan area A', and the greater the value of ρ, the greater is the extent of this convexity. A discussion of additional aspects of the boundary between regions A and B is contained in Appendix 6.2.

6.5.2 Two Modifications

In this approach to specifying the boundary, each density function assumes radial symmetry throughout its region, i.e., for every distance x there is no directional variation in $M(x)$. In the absence of such symmetry, $M(x)$ is simply a mean value, so that the position of the boundary necessarily becomes an approximation.[4] However, the general approach can be modified by examining conditions within a particular sector of each region instead of the whole region. This is illustrated by considering conditions in the vicinity of the inter-metropolitan axis $A'B'$.

The density function for region A now refers to conditions within a sector subtending a given angle, and extending from metropolitan area A' toward metropolitan area B'. For region B the corresponding "sectoral density function" is based on metropolitan area B', and extends in the direction of metropolitan area A'. The relevant part of the boundary between regions A and B now occurs at the intersection of the two sectoral density functions. This yields a more accurate estimate of the boundary, which will inevitably become less regular than a boundary shown in Fig. 6.3.

Having estimated the location boundary (with or without the assumption of radial symmetry), it is a relatively straightforward matter to move to the multi-regional case. In this second modification, the density functions for the various regions could be expected to have differing central densities and slopes. Here the overall boundary of a given region would be derived by determining the boundary between that region and each adjoining region, and disregarding the redundant parts of these boundaries.

6.6 Closing Comments

The notion of a regional density function has support from various strands of location theory. Here attention has been drawn to the possibility of using regional density functions as a method of determining the boundary between adjacent economic regions. The rationale for this approach is that at the boundary of two adjacent regions the interaction with the respective metropolitan nodes is of a comparable level, and also that population density is a proxy for this interaction. It will be recognized that there is an interesting parallel between the boundary in the urban density function and its regional counterpart considered here. When the density function is applied at the metropolitan area level, the boundary occurs where some minimum urban density is encountered. When, however, the density function is employed at the regional level, the boundary of the region occurs where the density function intersects a corresponding density function of an adjacent region. In both settings, therefore, it is an aspect of density that defines the relevant boundary.

The results of this density-function approach to regional boundaries are broadly in line with observed conditions, and are not in major conflict with the predictions of other theoretical models based on wholly different foundations. For example, the boundary derived above is similar to the boundary for most cases of the Economic

[4] The asymmetry at distance x may be due to the presence of one or more urban centers or to the existence of physical features such as mountains, deserts or large expanses of water.

Law of Market Areas, which is based on the prices of a good (or composite good) at two centers and the transportation costs of shipping that good from each center (Hyson and Hyson 1950; Isard 1956a, b; Norman 1993).[5] The boundary is also similar to the one associated with the Law of Retail Gravitation, which depends on the populations of two centers and the distance between these (Reilly 1953; Hoover 1971; Liang 2008). In both Laws the boundary is displaced from the halfway point towards the smaller center, and is also convex to the larger center. In the second Law, however, the boundary is always in the form of a circle or the arc of a circle, a restriction not present in the density-function approach considered above.

Under usual circumstances it would be customary to test the results of an approach (such as the one proposed here) against some objective standard. Unfortunately, because of the subjective nature of the concept of the region no benchmark of this type can exist. All that is available are the results of previous attempts at delimitation and regionalization. Nevertheless, we have at our disposal an additional method for determining the overall extent of an economic region. This should be seen as complementing rather than rivaling existing theoretical approaches to regionalization, whether for the purpose of analysis or the design of policy.

Acknowlegement The authors wish to express their thanks to N. Dean, A. Grimes and J. K. Swales for their comments on earlier versions of this chapter.

Appendix 6.1

Let T denote the distance from A' to B' in Fig. 6.2. Note that equality of densities requires Eq. (6.5) with $x_A + x_B = T$, giving

$$\ln(T - x_A) + \lambda - \rho(\ln x_A) = 0,$$

where $\lambda = (\ln N_A - \ln N_B)/b$ and $\rho = a/b$. We assume $N_A > N_B$ and $a > b$, implying $\lambda > 0$ and $\rho > 1$. We have:

1. There will be a unique x_A between 0 and T, say x_A^*, as in Fig. 6.2, such that $x_A = x_A^*$ and $x_B^* = T - x_A^*$, implying the equality of densities.
2. The value of x_A^*/T is unchanged by alteration of the units of area in $M(x)$ and of distance in x in Eq. (6.1).
3. The value of x_A^* is unchanged by alteration of the units of population embedded in $M(x)$ in Eq. (6.1).
4. The value of x_A^* exceeds $x_h = T/2$ if and only if

[5] For the Economic Law of Market Areas these spatial features are present in the following cases: (1) where center A' has a price advantage over center B' but no transportation-cost advantage; (2) where center A' enjoys a transportation-cost advantage but no price advantage; (3) where center A' has both a price advantage and a transportation-cost advantage. In the case where center A' has a transportation-cost advantage and center B' a price advantage these two features of the boundary are only present under certain conditions (Parr 1995).

$$\lambda - (\rho - 1)\ln(T/2) > 0. \tag{6.11}$$

5. The value of the term on the left in (6.11) is unchanged by the measurement changes identified in paragraph (2) and in paragraph (3).

Appendix 6.2

When $a = s\,(\ln N_A)$ and $b = s(\ln N_B)$, as in Eqs. (6.7) and (6.8), the λ in Appendix 6.1 will be given by $(\rho - 1)/s$. In view of this, we have the following modification of the results in Appendix 6.1, noting that $\rho > 1$ and $s > 0$:

6. The condition for x_A^* to exceed $x_h = T/2$ in (6.11) is

$$1 - s[\ln(T/2)] > 0 \tag{6.12}$$

independently of the value of ρ. However the comments of paragraph (5) do not apply here since, as a formal relationship, (6.6) is not maintained after the measurement changes identified in paragraph (2) or in paragraph (3). The usefulness of (6.6) as an approximation is a separate issue.

7. The value of x_A^* now depends on T, ρ, and s and is such that

$$\frac{\partial x_A^*}{\partial T} > 0 \quad \frac{\partial x_A^*}{\partial s} < 0.$$

Also we have

$$\frac{\partial x_A^*}{\partial \rho} = \frac{x_A^*\left(T - x_A^*\right)\left[1 - s\left(\ln x_A^*\right)\right]}{s\left[x_A^* + \rho\left(T - x_A^*\right)\right]}, \tag{6.13}$$

which has the same sign as $1 - s(\ln x_A^*)$.

8. The condition

$$1 - s(\ln T) > 0, \tag{6.14}$$

which holds for the $s = 0.127$ and $T = 400$ of Fig. 6.3, guarantees both (6.12) and $1 - s(\ln x_A^*) > 0$, so that the derivative in (6.13) is positive. This explains the rightward movement of the boundaries in Fig. 6.3 as ρ increases in value.

9. A boundary in Fig. 6.3 is convex to A', and eventually encircles B', if and only if

$$(\rho - 1)\ln(\rho - 1) - \rho(\ln\rho) + (\rho - 1)\left(\frac{1}{s} - \ln T\right) > 0. \tag{6.15}$$

As $(\rho - 1)\ln(\rho - 1) - \rho(\ln \rho)$ is negative for $\rho > 1$ the condition in (6.14) is necessary for (6.15) to hold. While (6.14) holds for $s = 0.127$ and $T = 400$, the condition in (6.15) does not hold for all of the ρ featured explicitly in Fig. 6.3. However, when

(6.14) holds, (6.15) will apply for sufficiently large ρ. For $s = 0.127$ and $T = 400$, the minimum ρ for (6.15) to hold is around 1.52.

10. A boundary in Fig. 6.3 is convex to A' as long as x_A, the distance from A' to the boundary, is less than or equal to the $\overline{x}_A$, given by

$$\overline{x}_A = \left(\frac{1}{\rho}\right)^{1/(2\rho-2)} \exp(1/s). \tag{6.16}$$

For x_A greater than $\overline{x}_A$ the boundary is concave to A'. Notice that (6.14) implies that the exp(1/s) of $\overline{x}_A$ exceeds T. For the boundaries displayed in Fig. 6.3, the values of $\overline{x}_A$ increase from $\overline{x}_A = 1,613$ for $\rho = 1.05$ to $\overline{x}_A = 1,697$ for $\rho = 1.3$. The boundaries are the parametric curves implied by increasing x_A from a minimum equal to the x_A^* of Appendix 6.1. In Fig. 6.3 the values of x_A^* increase from $x_A^* = 213$ for $\rho = 1.05$ to $x_A^* = 266$ for $\rho = 1.3$.

11. In (6.16) the proportion $(1/\rho)^{1/(2\rho-2)}$ goes to $\exp(-0.5) = 0.607$ as $\rho \to 1+$ and increases as ρ increases from 1.
12. It is possible for a boundary to be always concave to A' in Fig. 6.3. What is required is a value of x_A^* from Appendix 6.1 which is greater than or equal to the $\overline{x}_A$ of (6.16). An example is $T = 400$, $\rho = 1.1$ and $s = 0.18$, where $x_A^* = 202$ and $\overline{x}_A = 160$ are implied.

In Fig. 6.3, where $s = 0.127$ we have $\exp(1/s) = 2,628$ and, given the information in paragraph (11), the $\overline{x}_A$ of (6.16) must be at least 1,595. It follows that for a ρ where condition (6.15) does not hold, it is necessarily the case that $\overline{x}_A > x_A^*$ and the convexity of paragraph (10) applies.

13. When $n = s(\ln N)$, as in Eq. (6.6), we have

$$\ln M(x) = \ln N - n(\ln x) = \frac{n}{s}[1 - s(\ln x)]$$

for Eq. (6.2). It follows that $x_A = x_B = \exp(1/s)$ always satisfies Eq. (6.10) and so defines a point common to each of the family of boundaries defined by varying ρ for a given s. Note that the equality $x_A = x_B = \exp(1/s)$ implies that $M(x) = 1$, and defines a point on the perpendicular bisector of $A'B'$, i.e., a point on the $\rho = 1$ boundary. The vertical distance to the point is given by $0.5\sqrt{4\exp(2/s) - T^2}$. With $s = 0.127$ and $T = 400$ this is distance is 2,620, a calculation which is of theoretical rather than practical consequence.

References

Alonso W (1964) Location and land use: toward a general theory of land rent. Harvard University Press, Cambridge, MA

Barkley DL, Henry MS, Bao S (1996) Identifying ‘spread’ versus ‘backwash’ effects in regional economic areas: a density function approach. Land Econ 72:336–357

Berry BJL, Horton F (1970) Geographic perspectives on urban systems. Prentice-Hall, Englewood Cliffs, NJ
Bogue DJ (1950) The structure of the metropolitan community: a study of dominance and subdominance. Ann Arbor, MI
Boudeville JR (1966) Problems of regional economic planning. Edinburgh University Press, Edinburgh
Brown AL, Holmes J (1982) The delimitation of functional regions, nodal regions and hierarchies by a functional distance approach. J Reg Sci 11:57–72
Bussière R (1972) Static and dynamic characteristics of the negative exponential distribution. In: Wilson A (ed) Patterns and processes in urban and regional systems (London Papers in Regional Science, vol 3). Pion, London, pp 33–113
Casetti E (1969) Alternate urban population density models: an analytical comparison of their validity. In: Scott A (ed) Studies in regional science. London, Pion, pp 105–116
Clark C (1951) Urban population densities. J Roy Stat Soc A 114:490–496
Clark C (1968) Population growth and land use. Macmillan, London
Dickinson RE (1947) City, region and regionalism. Routledge and Kegan Paul, London
Evans AW (1973) The economics of residential location. Macmillan, London
Hoover EM (1971) An introduction to regional economics. Alfred A Knopf, New York
Hyson CD, Hyson WP (1950) The economic law of market areas. Q J Econ 64:319–327
Isard W (1956a) Regional science, the concept of the region, and regional structure. Papers and Proceedings of the Regional Science Association 2:13–26
Isard W (1956b) Location and space-economy. The Technology Press of MIT and Wiley, New York
Kanafani A (1978) Transportation and regional structure in underdeveloped regions. In: Funck R, Parr JB (eds) the analysis of regional structure: essays in honour of August Lösch (Karlruhe Papers in Regional Science, vol 2), London, Pion, pp 28–34
Liang S (2008) Research on urban influence domains in China. Int J Geogr Inform Sci 4: 1527–1539
Lösch A (1944/1954) Die räumliche Ordnung der Wirtschaft, 2nd edn. Jena, Gustav Fischer (trans: Woglom WH, Stolper as WF The Economics of Location. Yale University Press, New Haven, CT)
Meyer JR (1963) Regional economics: a survey. Am Econ Rev 53:19–54
Mogridge M, Parr JB (1997) Metropolis or region: on the development and structure of London. Reg Stud 31:97–115
Nairn AGM, O'Neill GJ (1988) Population density functions: a differential equation approach. J Reg Sci 28:81–102
Norman G (1993) Of shoes and ships and shredded wheat, of cabbages and cars: the contemporary relevance of location theory. In: Ohta H, Thisse J-F (eds) Does economic space matter? St. Martin's Press, New York, pp 38–68
Nystuen JP, Dacey MF (1961) A graph theory interpretation of nodal regions. Papers and Proceedings of the Regional Science Association 7:29–42
Parr JB (1985) A population density approach to regional spatial structure. Urban Stud 22:289–303
Parr JB (1995) The economic law of market areas: a further discussion. J Reg Sci 35:599–615
Reilly WJ (1953) The law of retail gravitation. Pilsbury, New York
Song S (1996) Some tests of alternatives accessibility measures: a population density approach. Land Econ 72:474–482
Teitz MT (1962) Regional theory and regional models. Papers and Proceedings of the Regional Science Association 9: 35–50
Vining R (1955) A description of certain aspects of an economic system. Econ Develop Cult Change 3:147–195
von Thünen JH (1842/1966) Der Isolirte Staat in Beziehung auf Landwirthschaft und Nationalökonomie (original spelling of the title). Rostock, Leopold (second edition) (trans: Wartenberg CM, with an editorial introduction by Hall P, as von Thünen's Isolated State). Pergamon Press, New York

Wang F (2001) Regional density functions and growth patterns in major plains of China, 1982–1990. Pap Reg Sci 80:231–240
Weiss HK (1961) The distribution of urban population and an application to a servicing problem. Oper Res 9:860–874
Wren C (2009) Onward the spatial: an essay on the nature and relevance of regional economics. Spatial Econ Anal 4:447–465

German Roots of Regional Science

7

Johannes Bröcker

Abstract

Walter Isard is the founder of the post-war tradition of Regional Science. He dates the birth of this tradition back to 1944 when he became familiar with the German literature on location. Based on these studies, he published his seminal *Quarterly Journal of Economics* paper *The General Theory of Location and Space-Economy* and later his famous book *Location and Space-Economy*. The present paper reassesses the German contributions. Starting from von Thünen it rushes through the history of thought until the 1930s of the last century. The paper agrees with Isard's emphasis on the fundamental insights offered by the German authors, but tries to revise some conventional wisdom about the German tradition. In particular, it argues that the so-called historical school offers more to learn from today than it is usually thought, while Weber, probably the most celebrated German spatial economist after von Thünen, chose a methodological approach that turned out not to be fruitful for later research. Finally, the paper explicitly dissents from Isard's praise of Weigmann who, for good reasons, is now completely forgotten.

7.1 Introduction

Walter Isard is the founder of the post-war tradition of Regional Science. He dates the birth of this tradition back to 1944 when, as a conscientious objector to military service, he was assigned to Civilian Public Service in a hospital and spent his time at night, when the patients happened to be asleep "... to translate into English the

J. Bröcker (✉)
Institute for Regional Research, Christian-Albrechts-Universität zu Kiel, Olshausenstr. 40, 24098 Kiel, Germany

Department of Economics, Christian-Albrechts-Universität zu Kiel, Olshausenstr. 40, 24098 Kiel, Germany
e-mail: broecker@economics.uni-kiel.de

P. Nijkamp et al. (eds.), *Regional Science Matters*, DOI 10.1007/978-3-319-07305-7_7,

German literature on location—the works of Lösch, Weigmann, Engländer, Predöhl and others" and was able to grasp fully the scope and depth of this literature. For the English speaking community, a summary of the main contributions was produced in Isard (1949)." (Isard 2003, 7). The mentioned review is Isard's *The General Theory of Location and Space-Economy* in the *Quarterly Journal of Economics* (Isard 1949) where he extensively reviews the works of three out of the four mentioned authors: Predöhl, Weigmann and Lösch. The other Germans referred to in the paper are Roscher and Schäffle, two prominent representatives of the so-called *Older Historical School*, Ritschl, a less prominent scholar close to the so-called *Younger Historical School* and, of course, Weber, Thünen and Launhardt, besides Lösch the three most famous German scholars in spatial economics.

One thus can compile an "Isard-list" of German scholars who have been seminal to the emergence of Regional Science. I put them in historical order, with years of birth and death, and with year of main publication in bold:

- Johann Heinrich von Thünen, 1783–1850, **1826**
- Albert Schäffle, 1872–1903, **1873**
- Wilhelm Roscher, 1817–1894, **1878**
- Wilhelm Launhardt, 1832–1918, **1882** and **1885**
- Alfred Weber, 1868–1958, **1909**
- Andreas Predöhl, 1893–1974, **1925**
- Oskar Engländer, 1876–1937, **1927**
- Hans Ritschl, 1897–1993, **1927**
- Hans Weigmann, 1897–1944, **1931** and **1935**
- August Lösch, 1906–1945, **1940**

Two important names are missing on this list: Walter Christaller (1893–1969) who published his famous book about central places in southern Germany in 1933 (Christaller 1933), and Erich Schneider (1900–1970) who made important contributions to the theory of spatial pricing (Schneider 1935). Both do appear in Isard's book *Location and space-economy* (Isard 1956), but only in the margin. Lösch is the author most extensively (on six pages) dealt with in the 1949 article. This is understandable, because Lösch's *Die räumliche Ordnung der Wirtschaft* (Lösch 1940) was the most ambitious attempt in the German literature to develop a comprehensive theory of location, and little was known about it in the Anglo-Saxon world at the time. The English edition only appeared 5 years later (Lösch 1954). Regarding the other authors, however, the focus of the paper is surprising, given that Isard aimed at summarising the essentials of German "Raumwirtschaft" (a term used by Isard (1956, p. 27) in German): five pages are devoted respectively to both, Weigman and Predöhl, the former being completely forgotten (for good reasons, as I will show) and the latter being almost forgotten today. Weber receives two pages, Thünen and Launhardt receive seven and four critical lines, respectively, and Roscher and Schäffle jointly just four footnote lines.

This article reconsiders the main contributions of German spatial economics until Isard's kick-off paper of Regional Science. There is not enough space for reviewing all authors in the above list. Thünen, Launhardt, Weber and Lösch are obviously a must, even though Isard left the first two more or less aside. Perhaps he thought them to be sufficiently known anyway, but I will show that their respective contributions to spatial economics are not sufficiently appreciated even until these days, and even certain discoveries regarding economics in general might not yet be fully perceived in the scientific community. Furthermore, regarding the Historical School, I will mention Schäffle only in passing, but devote a section to Roscher, because I want to convince empirical economic geographers to rediscover him as one of their early masters. Though influential on the development of spatial economics in Germany, he has been largely ignored since a century because of modern economics' verdict on the Historical School being a-theoretical and therefore useless. Finally, I will also deal with Predöhl and Weigmann, just because Isard devoted so much space to them. In the same year when Isard's paper appeared, Predöhl published a book in the spirit of the Historical School reporting interesting observations on the spatial dynamics of the global economy (Predöhl 1949), but his theoretical attempt of 1928 that Isard refers to did not turn out to be fruitful for progress in Regional Science. The writings of Weigmann before 1933 are confuse and those after 1933 cannot be called anything but awful. Regarding the other mentioned authors, I briefly mention Christaller in the context of Lösch and Central Place Theory. I skip Schneider due to space limitation; his work on spatial pricing, internationally well perceived at the time, would need an own treatment. Finally, I also skip Engländer (1924, 1926, 1927) and Ritschl (1927) because of minor originality. Readers interested in an extensive review and comprehensive list of references are referred to Ponsard (1983). Blaug (1997) gives a short historical overview which, in my opinion, overvalues Weber and undervalues Lösch.

7.2 Thünen

Marc Blaug, in his authoritative "Economic Theory in Retrospect", wonders about the fact that spatial economics "has always been the peculiar province of German economists." (Blaug 1997, p. 614). His conjecture is that this "historical curiosity" is just a coincidence to be explained by the fact that the 'father' of spatial economics, Johann Heinrich von Thünen, was a German. As Isard (1949, p. 479) pointed out, there may be other reasons related to the fact that geography was more important on the continent than in England, the hotspot of classical economics, in the nineteenth century. But no doubt, as Samuelson said in his masterful reconstruction of von Thünen's theory at Thünen's 200th anniversary, "Among geographers and location theorists, Thünen is the founding god." (Samuelson 1983).

Johann Heinrich von Thünen was born 1783 in Lower Saxony. After an education in agriculture he bought an estate (*Gut Tellow*) in north Germany and ran it profitably during his lifetime. In economics Thünen was an autodidact, naming Adam Smith as his main teacher. Thünen was not only an agricultural entrepreneur and private scholar, but also an active social reformer and innovator in agricultural technology. He died on *Tellow* 1850.

Von Thünen's famous contribution to spatial economics, elaborated in the first volume of *Der isolirte Staat in Beziehung auf Landwirtschaft und Nationalökonomie* (Thünen 1826) is a fully developed theory of land prices and land use in a monocentric economy, unfortunately not available in English until the first translation 1966 (Thünen 1966). Imagine a homogeneous infinite two-dimensional space with a city where all goods have to be sold or bought. Then, under perfect competition, an equilibrium allocation emerges where production of different commodities self-organises in rings around the city. Commodities with the highest per mile transport costs for the output of one acre occupy the most inner ring; those with lower costs lie further outside. Beyond the most outer ring the landscape ends in an uncultivated wilderness. The land rent per acre decreases from centre to periphery and approaches zero at the boundary to the wilderness. At any point its slope equals the transport cost per mile and acre of the good produced at the respective point.

This is probably the first consistent analysis within a theoretical framework that nowadays we would call neoclassical. There are well specified technologies and market structures. The behaviour of agents is derived from optimisation, and the market outcome is obtained as a market clearing equilibrium. From this masterpiece originates von Thünen's fame as a spatial as well as an agricultural economist.

An obvious disadvantage of his construction is that the existence of a centre is taken as given. The ultimate cause of spatial differentiation is exogenous. Land use and land prices or rents vary across space only because varying accessibility of the centre, the *raison de être* for this centre itself remains unexplained.

Thünen was, however, well aware of this limitation. In one of his notes published posthumous as volume 2, part 2, of *Der isolirte Staat* (Thünen 1863a, § 4, Sect. 9) he discusses how to remove this restriction. The chapter has the promising title "*On the configuration and distribution of cities in the isolated sate*", and it holds its promise. It gives a comprehensive account of centripetal and centrifugal forces, anticipating almost everything what the literature has offered in one and a half century since. The centrifugal forces are twofold, the costs of delivering goods produced in the city to the countryside and the costs of procuring agricultural goods. These costs lead to higher nominal wages in larger cities. The centripetal forces are the following. I describe them in modern terms to save space, such that the informed reader knows what Thünen is talking about. Longer citations would prove that what he explains in very clear plain words is exactly what the modern terms mean:

1. Scale economies on the firm level due to indivisibilities of machinery, equipment and factories
2. Scale economies due to the division of labour. The division is the deeper the larger the market (this point is of course taken from his master Adam Smith)
3. Economies of shopping
4. Labour market matching (causing an urban wage premium)
5. Risk sharing
6. Reduction of monopoly power and avoidance of hands-up situations and
7. Product diversity

This is a remarkable list. Interestingly, in the entire extensive literature on agglomeration and dispersion one does not find any reference to this chapter of von Thünen until Massa Fujita's citation in a working paper (Fujita 2011).

Comparing the list with the modern literature, its coverage is striking, and many modern authors neglect some important items in the list. In particular items 5 and 6 receive less attention in modern treatments than they should (Duranton and Puga 2004, being a notable exception). Thünen also drew the conclusion that horizontal interaction between firms reinforces the agglomeration tendency and leads to industrial clusters. And last but not least, he also predicted that the building of railway lines will favour the agglomeration in large cities. His arguments could serve as a concise summary of the predictions made by a continuous space version of a modern NEG model.

Though Thünen well understood the mechanisms behind concentration and dispersion, he did not try to extend his model to an endogenous agglomeration model. Today we know that within his framework he would not have succeeded. Starret's (1978) impossibility theorem tells us that an endogenous division of labour across space and positive transport costs are incompatible with homogenous space and price-taking. Agglomeration in homogeneous space implies scale economies which in turn imply imperfect competition. In Germany the theory of imperfect markets had to wait until Launhardt gave the first satisfying account of it (Launhardt 1885), being, like Thünen, far ahead of his time.

Admittedly, the list of agglomeration economies has one obvious and important omission: learning externalities. Though the second part of volume 2 of the Isolated State offers deep insights into human capital theory (Thünen 1863a, § 4, Sects. 9 and 10), from Thünen's rural perspective the role of cities in innovation does not seem to have been obvious enough. The twin pair agglomeration and innovation is however at the core of the writings of Wilhelm Roscher to whom I turn later.

Thünen's digressive style and the fact that his writings were for a long time not available in English have restrained economists in general and spatial economists in particular from fully exploiting his ideas. The next fundamental step in theory based on von Thünen was William Alonso's invention of the monocentric city model (Alonso 1964) that was seminal to an entire field of research, modern urban economics. It remains a mystery why this took a whole century though the idea is clearly and precisely stated in von Thünen: "Exploring more deeply the reasons why housing rents increase towards the city centre, we find them in labour savings as well as higher convenience and lower time cost of business; we thus find that housing rent and agricultural rent are governed by the same principles."[1]

Though Thünen's fame largely stems from his contribution to spatial and agricultural economics it is completely misleading to classify him merely as a spatial

[1] "Forschen wir den Ursachen, warum die Grundrente der Häuser nach der Mitte der Stadt hin immer mehr steigt, genauer nach; so finden wir diese in der Arbeitsersparung, der größeren Bequemlichkeit und der Verminderung des Zeitverlustes bei der Betreibung der Geschäfte; wir finden also, dass die Grundrente und die Landrente durch ein und dasselbe Prinzip reguliert werden" (Thünen 1826, p. 210).

or agricultural economist. His main concern was income distribution. In particular he was plagued by the Malthusian proposition that the wage of workers would always have a tendency to fall back to the minimum of existence, the level that does not allow the workers' families to live, but only to prevent them from dying. He gave the first satisfying answer to the positive issue of income distribution by developing the marginal productivity theory of income distribution, and he also suggested a normative answer engraved in his tombstone. It says that the natural wage is the geometric mean between the subsistence wage and the average product. This formula has been subject to much debate and misunderstanding, but space prevents me from touching this interesting debate [see Dorfman (1986), and Samuelson (1986)].

Nor am I able to pay homage to the many other achievements not mentioned so far. Thünen invented the modern style of abstract economic reasoning. He was the first to apply differential calculus in economics. He also was the first scholar whom, to use a modern term, we can call an econometrician. He never introduces a functional relation like pulling a rabbit out of the hat. He always calibrates it with real data, experimenting with different functional forms, among them a generalisation of the function reinvented 80 years later by Paul Howard Douglas and Charles Wiggins Cobb, the famous Cobb-Douglas function. Thünen had a clear perception of the theory of externalities and public goods, he developed first ideas on dynamic optimisation (when to cut a tree) (Thünen 1863b; English edition 2009), he wrote on the economics of politics (conscription and the cost of war), on population policies and more.

7.3 Roscher

Born 1817, Wilhelm Roscher grew up in Hannover and studied humanities in Göttingen. He is one of the most prominent representatives of the *Older Historical School*, a group of economists who were sceptical about the possibilities to derive laws of general validity from deductive reasoning. They set up a research program aiming at deriving regularities in the economic process from a careful study of historical developments and actual data. In this context they gathered, classified and systemised a huge body of empirical material. Despite their scepticism with regard to deductive methods, their interpretation of the data is by no means merely descriptive or "theory-free". As a matter of fact, in his interpretations Roscher developed a marginal productivity theory of factor pricing essentially not different from Thünen's or from the modern macro version of neoclassical factor price theory. Albert Schäffle, another representative of the *Older Historical School*, also contributed to spatial economics (Schäffle 1873).

Roscher published a short but extremely dense piece on regional economic development. It dates from 1865, but is accessible only as a chapter in his *Ansichten der Volkswirtschaft aus dem geschichtlichen Standpunkte*[2] (Roscher 1878). This

[2] Economic views from a historical perspective.

piece is rarely cited in modern literature; I have never seen any mention of it in the wide and flourishing empirical literature on clustering, knowledge, creativity, innovation and all that in space. I dare claiming that there is no fundamental idea in this literature that is not already in Roscher. Even Marshall's (1890) famous Chapter X of Book IV has less to offer in this regard than Roscher.

Gathering an enormous amount of information on industrial location, in particular for England and France, Roscher gives a detailed account of the factors determining the location choice within each and every industry. Though the style is narrative, a clear theoretical picture emerges during his tour through the economic geography. He has a clear perception about what Krugman (1991) calls first and second nature, illustrated with tons of concrete examples. Collecting the many remarks and observations in the text would offer a fairly comprehensive view on the theory of agglomeration. All the mechanisms summarised under the headings sharing, matching and learning by Duranton and Puga (2004) somewhere show up on the tour. Roscher is clearly much less convincing than Thünen with regard to deductive theory, but richer in terms of empirical knowledge about industrial organisation. Readers who, unlike me, are sceptical about neoclassical model building and trust more in a style of research trying to extract regularities from case studies as well as mass statistics are highly recommended to study carefully Roscher as one of their great masters. What is particularly impressive is Roscher's deep insight into the role of knowledge and creativity in the development of industry, which was lacking in Thünen's work. I just mention his wonderful exposition about the concentration of luxury industry in the largest cities, or his observations on knowledge spillovers, taking designers in Paris as a typical example.

7.4 Launhardt

The author carrying on some of Thünen's achievements was Wilhelm Launhardt. Reading Launhardt is a pleasure for an analytically trained economist of our days. He wrote in a compact, rigorous mathematical style, which was strange for academic economists in Germany at his time. Like some of the best economists of the nineteenth century—I just mention Jules Dupuit—Launhardt was a civil engineer. He was born in 1832 in Hannover and became professor there 1871. When the polytechnic was upgraded to a university in 1886, Launhardt became its first rector. His contribution to our field is threefold.

First, he reconstructed Thünen's spatial model in a rigorous mathematical way. One may use it one-to-one as a modern textbook version, and a better one than most of what I have seen. Launhardt also treats the case of endogenous prices with linear demand at the central market and solves an example numerically, simultaneously determining prices of goods, land use and land rent for the two-commodities case.

Launhardt's *second* contribution was about market areas with demand that is linear in the consumer's price and uniform across one- or two-dimensional Euclidean space (Launhardt 1885, § 27). Independently of Cournot, whose book he received only after his own went to the publisher, he solved the monopoly case. More remarkably, he also found the Bertrand-Nash equilibrium for a duopoly where the firms selling a homogeneous good are located at the opposite ends of a finite linear market. Demand is uniformly distributed along the line and perfectly inelastic. Each customer buys from the firm quoting the least delivered price. The delivered price is the mill price plus transport cost which is linear in distance (Launhardt 1885, § 28). His solution concept of "Friedensgrundlage" (base of peace) is precisely the Nash-Equilibrium; he characterises the price mark-ups over marginal cost chosen by the two firms as follows: "These mark-ups are the base of peace for the competition of the two entrepreneurs. Each would, if he reduced his price below or raised it above this level, while the competitor left his price unchanged, lose profit."[3] This is precisely a definition of the Nash equilibrium. He also solves an oligopoly case with an arbitrary number of firms that, however, all serve the spatial market from the same location.

Launhardt's innovativeness in price theory in general and spatial pricing and market area analysis in particular is incredible, but its impact on the literature has been limited. The Anglo-Saxon world did not notice him because Marshall did not mention him. An English translation of his 1885 book (Launhardt 1993) had to wait more than a century. Hotelling was obviously not aware of Launhardt when writing his famous *Stability in Competition* (Hotelling 1929). Even Lösch, though referring to his work in two of his many footnotes (Lösch 1940, pp. 55 and 68), did not make use of Launhardt's deep insights but based his market areas on Chamberlin (1933).

Launhardt's *third* contribution is transport cost minimization in the famous location triangle. The task is to find—in continuous Euclidean space—a firm's location minimizing the transport cost, if inputs have to be procured from two input locations and products have to be delivered to an output market. Locations of input sources and the output market, input and output quantities in tonnes, as well as transport costs per tonne-kilometre are given. The problem is equivalent to finding the point in a triangle minimizing the weighted sum of distances to the three vertices. A special case of the problem with uniform weights was posed already 1643 by Pierre de Fermat and solved by Evangelista Torricelli (the inventor of the barometer) in a ruler-and-compass construction (Martini 2001). In a wonderful brief and clear contribution to the journal of the German engineers' association (Launhardt 1882), Launhardt clearly states the generalized problem, derives the algebraic optimum conditions, presents a graphical solution method, extends the problem in several ways (more than three points, use of existing roads), and applies it to a practical case. The triangle reappears in the influential work of Alfred Weber. Weber should have referred to Launhardt, which he did not.

[3] "Die Gewinnsätze bilden die Friedensgrundlage für den Wettkampf der beiden Unternehmer. Jeder der beiden Unternehmer würde, sobald er über oder unter diesen Einheitsgewinn ginge, während der Gegner seinen Preis unverändert festhielte, am Gesamtgewinn verlieren" (Launhardt 1885, p. 162).

Launhardt does not give any reference to earlier literature in the cited paper. The solution may have been known in mathematics' academia at his times. The mathematician Jakob Steiner is reported to have presented the solution to the general problem, but the published references (Steiner 1837, 1881) only deal with the uniform weights version.

Launhardt also mentions a mechanical analogue to the problem: the optimal point is the equilibrium point of forces pulling it in the directions of the vertices with respective strengths proportional to the costs per tonne-kilometre. Remarkably, this observation had already been applied to optimal plant location as early as 1829 by two French engineers, G. Lamé and B.P.E. Clapeyron (Franksen and Grattan-Guiness 1989). It is almost sure that Launhardt never heard of their memorandum explicating the problem and its mechanical solution. This is just another example of the multiply rediscovered "law of multiple discoveries" (Kolata 2006).

7.5 Weber

Alfred Weber, brother of the more famous sociologist Max Weber, was born 1868 in Erfurt. He grew up in Berlin, where his father was a national-liberal city-deputy. His academic background is Humanities and Laws from which he switched to economics and got his doctoral degree and habilitation in Berlin under supervision of Gustav Schmoller, the main representative of the so-called Younger Historical School. His first chair was at the Karls-University in Prague from where he was called to Heidelberg 1907. He defended scientific freedom and was a professed anti-Nazi getting into open conflict with Nazi-power in March 1933 shortly before he retired (Demm 1986; Nutzinger 1997).

In 1909 Weber published a monograph outlining a comprehensive theory of the location of industries (Weber 1909; English edition 1929), later followed by a shorter addendum on the specific conditions of a capitalistic economy (Weber 1923). Though Weber's book became one of the most famous German contributions to location theory, modern spatial economics hardly takes anything from it to build on. Weber's background is the historical school, in particular Roscher's location studies, though he himself would have contradicted to be placed in such a context. The Historical School became under attack at the beginning of the century, and Weber disagreed with its empiricist approach. Opposed to the Historical School, he aimed at a theoretical foundation of general validity. His only references to authors before himself is to disqualify them all as foolish (Weber 1909, Exkurs I: Die bisherige Literatur), in particular Roscher and Schäffle, the other representative of the Historical School offering a special chapter on location (Schäffle 1873, §§ 265–267).

I subdivide the book in two stages. The first is rediscovering the previously mentioned location triangle. This is a clear, rigorously defined problem, and it is solved, though not by Alfred Weber himself but in an appendix by Georg Pick. Pick was professor of mathematics in Prague and probably known to Alfred Weber from his Prague days. He died 1942 in a concentration camp in Theresienstadt. Whether Weber knew Launhardt without citing him is unknown. At least I was able to check that Launhardt's book referring to his solution of the problem right in the introduction has been in the library in Heidelberg since it appeared 1885.

The second stage of the book extends the firm location problem in three ways:

1. First, for each input it allows firms to choose among different sources that sell the respective inputs at different, exogenously given prices.
2. Second, labour is introduced as a further input, with a wage rate exogenously varying across space.
3. Third, many firms instead of one are introduced, interacting due to the fact that joint locations generate cost savings by agglomeration economies. These economies are assumed, not derived from technologies and market structures.

On the one hand, as a model of a decentralised market the approach lacks the essentials: endogenous prices and quantities. On the other hand, as an optimisation problem the whole exercise is ill-posed, because it lacks an objective for the entire ensemble of interacting firms. If the deficiencies were fixed and numbers were given to all the exogenous functions and variables, a multi-firm location problem of this kind would still be difficult to solve even on a modern computer, not to speak about Weber's tiresome verbal casuistic.

Weber's work does in my opinion not deserve the attention it got in the literature. Though it is often cited, I am not aware of any surviving strand of literature that really built on him. Given what I said, this does not come as a surprise. The theory of firm location in the business literature refers to Weber in the context of the triangle (Hurter and Martinich 1989), but in this case it is Launhardt, not Weber who deserves the attention.

7.6 Lösch

The German literature on spatial economics culminates with the work of August Lösch, putting an idea that was first brought up the geographer Walter Christaller (1933) on a firm theoretical footing.

August Lösch was born 1906 in Württemberg, graduated in economics in Freiburg and got his doctoral and habilitation degrees in Bonn under Schumpeter's supervision. In 1945, immediately after the war ended, he died from starvation and a scarlet fever infection. His unwillingness to cooperate with the Nazi administration prevented him from having an academic career before 1945 (Zottmann 1949).

The first edition of Lösch's *Die räumliche Ordnung der Wirtschaft* appeared in 1940 (Lösch 1940). This monumental work stands below Launhardt in terms of theoretical level and rigour of its argument, but its theoretical ambitions are much wider, and to some extent the author was able to fulfil them. Lösch's choice of the subject may have been influenced by the prominent position that the subject achieved from Weber's book and from the writings of the Historical School, but Lösch's own approach is neither related to Weber's nor does it build on Thünen or Launhardt.

Lösch's colleague and close friend Wolfgang Stolper, in his introduction to the English edition (Stolper 1954), classified Lösch's contribution as an extension of the Walrasian system to a spatial economy. Isard expresses the same idea, when he interprets German "Raumwirtschaft" as an "almost inevitable" attempt to fuse space, a primary concern of the Historical School, with the general equilibrium analysis of the Lausanne school (Isard 1949, p. 479). Though claimed by two great minds of our discipline, both experts on Lösch's work, I dare to disagree with this interpretation. Actually Lösch made an attempt to outline a general equilibrium theory in his brief Chap. 8, which covers no more than eight pages of his book of more than 300 pages. The theory outlined is neither Walrasian nor general equilibrium, and it simply does not succeed; it does not reach the standards of general equilibrium theory of his time, set by Abraham Wald (1936), John von Neumann (1937) and others in the 1930s. The theory is not Walrasian because firms are not price-taking, it is not general equilibrium because there are no budget constraints, and demand is exogenous. The loose ends are too many. My personal interpretation is that Lösch just gave up this line of research because he rightly foresaw that no interesting and consistent conclusion on the spatial structure of the economy could be attained this way.

Lösch thus pursued a different line of reasoning, taking up the idea of a hierarchy of central places from the German geographer Walter Christaller (1933) and turning it from a descriptive to a deductive approach. Central places are the centres of differently sized overlapping market areas. Each network of market areas is derived as a partial equilibrium based on Chamberlin's monopolistic competition approach.

There are obvious objections against this method: superimposing partial equilibria does not make a general equilibrium. Cost payments of firms go to nowhere; demand comes from nowhere, no interdependence between markets of different commodities, no budget constraints and so on. Moreover, Lösch missed the difficulties implied by a direct application of Chamberlin's symmetry assumption to a spatial market (already noticed before by Kaldor 1935). But this deficiency could be corrected by later authors such as Beckmann and Thisse (1986), Eaton (1978) and Lipsey (Eaton and Lipsey 1976), Capozza and van Order (1978) and others, without affecting the substance of Central Place Theory. It reappears in the multi-industry continuous space version of New Economic Geography (Fujita et al. 1999).

These are the big stars of spatial economics in Germany until the end of World War II. As mentioned in the introduction, two other names on the "Isard-list", Weigmann and Predöhl, would not deserve an equally extended treatment, had not Isard devoted so much space to their respective works. This critical statement does not at all mean, however, that I put Predöhl and Weigmann, both publishing at roughly the same time and under the same political conditions, into one box. Despite certain concessions to the Nazi regime, Prodöhl kept is independence as a scientist (Bröcker 2014), while Weigmann became an ardent adherent and propagandist of Nazi ideology.

7.7 Weigmann

Hans Weigmann was born 1897 in Kiel and became professor in Rostock and Berlin. After 1939 he joined the armed forces as Nazi chief officer ("NS-Führungsoffizier"). He was posted missing in action and is assumed to have died 1944 in Belarus (Buddrus and Fritzlar 2007, 432–433).

Isard writes: "Weigmann's writings on general location theory are very difficult to comprehend, both because of a vague style and the complexity of the basic concepts." Isard nevertheless recommends these concepts to the reader as disclosing "... some of the most promising channels of exploration for further theoretical development." (Isard 1949, 489).

I absolutely disagree. Weigmann's concepts are not complex, they are confuse, without any serious theoretical background and with no relation to empirical observations. They are a labyrinth of empty phrases, published 1931 in the *Weltwirtschaftliches Archiv* (Weigmann 1931).[4] Isard also cites another brochure, *Politische Raumordnung*, dated 1935. The subtitle is *Gedanken zur Neugestaltung des deutschen Lebensraumes* (Weigmann 1935). The wording of the subtitle is typical for Nazi writings of the time. This is a Nazi pamphlet against the urban way of life, against the idea of individuality of human beings, and against the critical intellect. Weigmann propagates the notion of *Volksgemeinschaft*—the Nazi term used to emphasize the antagonism between the Weimar and third Reich society—as the basic concept of spatial planning. To Weigmann, the *Volksgemeinschaft* is the "... denial of the primacy of intellect and thus of urban life in general." Note that this is meant as praise, not condemnation! To substantiate this praise, he claims it to be "... one of the most important discoveries of modern sociology that thinking destroys the basis of social life." (Weigmann 1935, 24).

I would like to mention at this point in passing that the role of German "Raumforschung" in Nazi-Germany is another and sad chapter in the history of science. University researchers in spatial planning designed Heinrich Himmler's *Generalplan Ost*, a planning project based on displacement and murder of the non-German and Jewish population in the East. The leading person was the agricultural economist Konrad Mayer (Heinemann 2006; Rössler 1989, 2001; Rieter 2014), who after the war returned to a position as a respected professor in spatial planning in Hannover, and a leading person in the *Akademie für*

[4] Though a translation is difficult, I give an example: "At this occasion we discover the proper heuristic value of the Gestalt-view as it, by the order of Gestalt-importance, easily integrates into the entireness the *quale* of an individual circumstance whose special characteristic threatens to destroy the closeness of the physiognomic impression. It uses the notion of 'basic-Gestalt' by which a fixed point of reference for the individual assessment of significance is gained, which in other words substitutes a connected entireness in the way that not logical or temporal precedence of the *causa* but the heaviness of the Gestalt-weight ... is decisive for qualifying a circumstance as Gestalt-kernel or basic-Gestalt." (Weigmann 1931, 10) *Gestalt*, meaning something like shape or form, is a philosophical notion nowadays mainly appearing in psychology. *quale* is something that has its own quality, *causa* is Latin for "cause".

Raumforschung und Landesplanung, an important German spatial planning institution. Walter Christaller also belonged to the group and explicitly supported the plans. Central Places turned out to be a helpful concept to stabilise the Nazi power in the conquered areas. Still, Christaller also defended the scientific planning approach against leading Nazi ideologists in a planning debate around 1940 (see Rieter (2014), and further references there).

7.8 Predöhl

The other less known scholar most extensively reviewed by Isard is Andreas Predöhl, president of the Kiel Institute for the World Economy from 1934 until his dismissal by the British administration after the end of the war. 1953 he became professor in Münster. In a 1925 response to Weber, 1928 prominently published in the *Journal of Political Economy* (Predöhl 1928), he made an attempt to bring spatial economics back into the framework of neoclassical equilibrium theory. Criticizing Weber as having moved away from the basic concepts in economics was right. But in my opinion Predöhl's alternative was a cul-de-sac. The style was vague, not only according to today's standards but also according to the standards in equilibrium theory of his time. The main message is a common place about substitution. The impossibility of reconciling an endogenous location theory with the constant returns to scale version of neoclassical equilibrium was not touched upon. In later writings, Predöhl gathered interesting observations on the spatial structure of the global economy (Predöhl 1949). But this work is unrelated to the theoretical attempt just mentioned and reviewed by Isard. It is worth mentioning, however, that Predöhl was the missing link between pre-war and post-war spatial economics in Germany. His students from the University of Münster, Rolf Funck, Karin Peschel and others, revived the spatial economics tradition in the 1960s and founded the German section of the Regional Science Aassociation in 1965.

Conclusion

Isard deserves admiration for having forcefully brought back the spatial dimension into economics after World War II. It was largely ignored in the Anglo-Saxon literature, and under the dominating influence of this literature also in mainstream economics in general. Isard discovered the strong German tradition dealing with the issue and having brought up many interesting and fruitful ideas. Spatial economics today is a respected and integral part of mainstream economics, in particular since Krugman's groundbreaking paper showing, in a fully developed general equilibrium model, how endogenously a spatially differentiated structure emerges in an initially perfectly homogeneous world. With hindsight one can see more clearly than in the late 1940s, which elements of the German tradition turned out to be helpful and lasting, which ones just expressed wishes and ambitions for a general theory that never could be fulfilled, and which ones even proved to be misleading and hindering scientific progress.

Though spatial economics made much progress, there remains still something to learn from German "Raumwirtschaft". Some pieces are useless, indeed, not to speak about Weigmann's writings that became part of an ideology leading to a historical catastrophe. I do not see much lasting value in Weber's contribution, despite the fact that he was the most influential. Even as the grandfather of the operations research literature on location, not he but Launhardt is the one to whom the honour is due.

What remains are important, though largely non-coherent pieces.

1. The first is Thünen's land use and land rent theory, updated by Launhardt, a strong piece of what we nowadays would call neoclassical perfect competition equilibrium theory. It is admirable and undervalued in the literature. Almost unknown is Thünen's clear exposition on the endogenous forces determining the configuration and distribution of cities in homogeneous space. Krugman would probably have written less pejorative about the German tradition if he had studied Thünen's respective chapter.
2. Second, there is the Historical School, with interesting and lasting ideas and observations on agglomeration economies, innovation, learning in large cities and other issues that modern empirical economic geography focuses on. Empirical economic geography studying agglomeration and knowledge should refer to Roscher rather than Marshall as their master.
3. Third, there is Launhardt's early contribution to the theory of imperfect competition in space, almost unknown until these days,[5] carried on by Schneider.
4. And finally, fourth, there is Central Place Theory, in particular Lösch. I disagree with Krugman to put Lösch with Weber into one box called "Germanic Geometry" (Krugman 1995, 38). I only partly agree with his verdict that Lösch presents a planning rather than a market solution. On occasion, Lösch in fact confuses market equilibrium and planning optimum. But his approach is definitely a theory of markets, even if his application of Chambelin's model to a spatial economy has its deficiencies. I do fully agree with Krigman's "moral of the story", that "...Central place theory is a powerful organizing principle for looking at and thinking about urban systems—and in only slightly modified form it turns out to make sense in terms of a rigorous economic model" (Krugman 1995, 64).

The latter statement refers to the endogenous emergence of central places in a new economic geography framework in continuous space, with several industries and with land used for agricultural production (Fujita et al. 1999). If in addition one allows for several agricultural goods differing with regard to transport cost per mile and unit of land one would also obtain Thünen-like zones of agricultural land use. In this sense one could say that Thünen and Lösch are in modern theory *aufgehoben*. The word "aufgehoben" is a philosophical term used

[5] Dos Santos Ferreira and Thisse (1996), who discovered in Launhardt's book an elegant way to combine horizontal and vertical product differentiation, are an exception.

by Friedrich Hegel in a triple sense, as (1) "aufgehoben = annihilated", (2) "aufgehoben = preserved", and (3) "aufgehoben = raised to a higher level".

There is a rich German heritage of literature in spatial economics worth praising, and Germans themselves bear the blame for the fact that the rest of world stopped reading German and profiting from this heritage after 1945.

References

Alonso W (1964) Location and land use. Harvard University Press, Cambridge, MA

Beckmann MJ, Thisse J-F (1986) The location of production activities. In: Nijkamp P (ed) Handbook of regional and urban economics, vol 1. Elsevier, Amsterdam, pp 21–95

Blaug M (1997) Economic theory in retrospect, 5th edn. Cambridge University Press, Cambridge, MA

Bröcker J (2014) Deutsche Raumwirtschaftstheoretiker in der Zeit von 1933 bis 1945. In: Trautwein H-M (Hrsg.) Die Entwicklung der Raumwirtschaftslehre von ihren Anfängen bis in die Gegenwart. Studien zur Entwicklung der ökonomischen Theorie, vol 30. Duncker & Humblot, Berlin

Buddrus M, Fritzlar S (2007) Die Professoren der Universität Rostock im Dritten Reich: Ein biographisches Lexikon. de Gruyter, Berlin

Capozza DR, Van Order R (1978) A generalized model of spatial competition. Am Econ Rev 68: 896–908

Chamberlin E (1933) The theory of monopolistic competition. Harvard University Press, Cambridge, MA

Christaller W (1933) Die zentralen Orte in Süddeutschland: eine ökonomisch-geographische Untersuchung über die Gesetzmäßigkeit der Verbreitung und Entwicklung der Siedlungen mit städtischen Funktionen. Fischer, Jena

Demm E (ed) (1986) Alfred Weber als Politiker und Gelehrter. Steiner-Verl.-Wiesbaden-GmbH, Stuttgart

Dorfman R (1986) Comment: P.A. Samuelson, Thünen at two hundred. J Econ Lit 24:1773–1776

Dos Santos Ferreira R, Thisse J-F (1996) Horizontal and vertical differentiation: the Launhardt model. Int J Ind Organ 14:485–506

Duranton G, Puga D (2004) Micro-foundations of urban agglomeration economies. In: Henderson J, Thisse J-F (eds) Cities and geography, handbook of regional and urban economics, Chapter 48, vol 4. Elsevier, Amsterdam, pp 2063–2117

Eaton BC (1978) Freedom of entry and the existence of pure profit. Econ J 88:455–469

Eaton BC, Lipsey RG (1976) The non-uniqueness of equilibrium in the Löschian location model. Am Econ Rev 66:77–93

Engländer O (1924) Theorie des Güterverkehrs und der Frachtsätze. Fischer, Jena

Engländer O (1926) Standort. In: Elster L (ed) Handwörterbuch der Staatswissenschaften, Bd. 7. Fischer, Jena, pp 857–869

Engländer O (1927) Kritisches und Positives zu einer allgemeinen reinen Lehre vom Standort. Zeitschrift für Volkswirtschaft und Sozialpolitik 5:435–505

Franksen O, Grattan-Guiness I (1989) The earliest contribution to location theory? Spatio-economic equilibrium with Lamé and Clapeyron 1829. Math Comput Simulat 31:195–220

Fujita M (2011) Thünen and the new economic geography. RIETI discussion paper series 11-E-074

Fujita M, Krugman PR, Mori T (1999) On the evolution of hierarchical urban systems. Euro Econ Rev 43:209–251

Heinemann I (2006) Wissenschaft und Homogenisierungsplanungen für Osteuropa. Konrad Meyer, der "Generalplan Ost" und die Deutsche Forschungsgemeinschaft. In: Heinemann I, Wagner P

(eds) Wissenschaft - Planung - Vertreibung: Neuordnungskonzepte und Umsiedlungspolitik im 20. Jahrhundert, Beiträge zur Geschichte der Deutschen Forschngsgemeinschaft
Hotelling H (1929) Stability in competition. Econ J 39:41–57
Hurter AP, Martinich JS (1989) Facility location and the theory of production. Kluwer, Boston, MA
Isard W (1949) The general theory of location and space-economy. Q J Econ 63:476–506
Isard W (1956) Location and space-economy. MIT Press, Cambridge, MA
Isard W (2003) History of regional science and the regional science association international. Springer, Berlin
Kaldor N (1935) Market imperfection and excess capacity. Economica 2:33–50, New Series
Kolata G (2006) Pity the scientist who discovers the discovered. New York Times, 5 February
Krugman P (1991) Increasing returns and economic geography. J Polit Econ 99:483–499
Krugman P (1995) Development, geography, and economic theory. The Ohlin lectures. MIT Press, Cambridge, MA
Launhardt W (1882) Die Bestimmung des zweckmäßigsten Standortes einer gewerblichen Anlage. Zeitschrift des Vereins Deutscher Ingenieure 26:109–116
Launhardt W (1885) Mathematische Begründung der Volkswirthschaftslehre. Engelmann, Leipzig. English edition: Launhardt W (1993) Mathematical principles of economics (trans: Launhardt W) E. Elgar, Aldershot
Luanhardt W (1993) Mathematical principles of economics. Edward Elgar, Aldershot
Lösch A (1940) Die räumliche Ordnung der Wirtschaft. Fischer, Jena
Lösch A (1954) The economics of location. Yale University Press, New Haven, CT
Marshall A (1890) Principles of economics. Macmillan, London
Martini H (2001) Fermat-Torricelli problem. In: Encyclopaedia of mathematics (online edition). Springer, Heidelberg. http://eom.springer.de
von Neumann J (1937) Über ein ökonomisches Gleichungssystem und eine Verallgemeinerung des Brouwerschen Fixpunktsatzes. Ergebnisse eines mathematischen Kolloquiums 8:73–83
Nutzinger H-G (1997) Alfred Weber als Vertreter der "inneren Emigration". In: Hagemann H (ed) Zur deutschsprachigen wirtschaftswissenschaftlichen Emigration nach 1933. Metropolis, Marburg, pp 535–548
Ponsard C (1983) History of spatial economic theory. Springer, Berlin
Predöhl A (1928) The theory of location in its relation to general economics. J Polit Econ 36:371–390
Predöhl A (1949) Außenwirtschaft. Grundriss der Sozialwissenschaft. Vandenhoeck & Ruprecht, Göttingen
Rieter H (2014) Ökonomen im Dienste der nationalsozialistischen Raumplanung und „Raumforschung". In: Trautwein H-M (Hrsg.) Die Entwicklung der Raumwirtschaftslehre von ihren Anfängen bis in die Gegenwart. Studien zur Entwicklung der ökonomischen Theorie, vol 30. Duncker & Humblot, Berlin
Ritschl H (1927) Reine und historische Dynamik des Standortes der Erzeugungszweige. Schmollers Jahrbuch für Gesetzgebung, Verwaltung und Volkswirtschaft 51:813–870
Roscher W (1878) Ansichten der Volkswirthschaft aus dem geschichtlichen Standpunkte, 3rd edn. Winter, Leipzig
Rössler M (1989) Applied geography and area research in Nazi society: central place theory and planning, 1933 to 1945. Environ Plan D Soc Space 7:419–431
Rössler M (2001) Geography and area planning under national socialism. In: Szöllösi-Janze M (ed) Science in the third Reich. Berg, Oxford, pp 59–78
Samuelson PA (1983) Thünen at two hundred. J Econ Lit 21:1468–1488
Samuelson PA (1986) Yes to Robert Dorfman's vindication of Thünen's Natural-Wage Derivation. J Econ Lit 24:1777–1785
Schäffle AEF (1873) Das gesellschaftliche System der menschlichen Wirthschaft, 3rd edn. Laupp, Tübingen
Schneider E (1935) Bemerkungen zu einer Theorie der Raumwirtschaft. Econometrica 3:79–105
Starrett D (1978) Market allocations of location choice in a model with free mobility. J Econ Theor 17:21–37

Steiner J (1837) Über den Punkt kleinster Entfernung. Bericht über die zur Bekanntmachung geeigneten Verhandlungen der Königlich Preußischen Akademie der Wissenschaften zu Berlin, pp 1144–1145

Steiner J (1881) Gesammelte Werke, 2 Bde. Ed. K. Weierstraß. Berlin, Reimer

Stolper WF (1954) August Lösch in memoriam. In: Stolper W (ed) Lösch, August: the economics of location. Yale University Press, New Haven, CN, pp 7–11

von Thünen JH (1826) Der isolirte Staat in Beziehung auf Landwirthschaft und Nationalökonomie. Perthes, Hamburg

von Thünen JH (1863a) Der isolirte Staat in Beziehung auf Landwirthschaft und Nationalökonomie, Zweiter Theil: Der naturgemäße Arbeitslohn und dessen Verhältniß zum Zinsfuß und zur Landrente - 2. Abth. Leopold, Rostock

von Thünen JH (1863b) Der isolirte Staat in Beziehung auf Landwirthschaft und Nationalökonomie, Dritter Theil: Grundsätze zur Bestimmung der Bodenrente, der vortheilhaftesten Umtriebszeit und des Werths der Holzbestände von verschiedenem Alter für Kieferwaldungen. Leopold, Rostock

von Thünen JH (1966) Isolated state; an English edition of Der Isolierte Staat, 1st edn. Pergamon Press, Oxford

von Thünen JH (2009) The isolated state in relation to agriculture and political economy: part iii: principles for the determination of rent, the most advantageous rotation period and the value of stands of varying age in pinewoods. Palgrave Macmillan, Basingstoke

Wald A (1936) Über einige Gleichungssysteme der mathematischen Ökonomie. Zeitschrift für Nationalökonomie 7:637–670

Weber A (1909) Über den Standort der Industrien, Teil1: Reine Theorie des Standorts; mit einem mathematischen Anhang von Georg Pick. Mohr, Tübingen

Weber A (1923) Industrielle Standortslehre (Allgemeine und kapitalistische Theorie des Standorts). Grundriss der Sozialökonomik, 2. Aufl., Kapitel III: S. 58–86. Mohr, Tübingen

Weber A (1929) Alfred Weber's theory of the location of industries. University of Chicago Press, Chicago, IL

Weigmann H (1931) Ideen zu einer Theorie der Raumwirtschaft. Weltwirtschaftliches Archiv 34: 1–40

Weigmann H (1935) Politische Raumordnung: Gedanken zur Neugestaltung des deutschen Lebensraumes. Hanseatische Verlags-Anstalt, Hamburg

Zottmann A (1949) Dr. habil. August Lösch gestorben am 30.Mai 1945. Weltwirtschaftliches Archiv 62:28–31

Part II

Regional Growth and Distribution

The Process of Regional Economic Growth: A Case Study of Washington State

8

Richard S. Conway Jr

Abstract

Nearly 60 years ago, North and Tiebout engaged in a vigorous debate on the nature of regional economic growth that is still relevant today. The discourse led to the formulation of the economic base model, a simple but powerful means of explaining, analyzing, and forecasting the process of regional economic growth.

The objective of this study is to explain how the Washington state economy has grown and evolved over the past 50 years in light of the theory advanced by North and Tiebout. Drawing upon Washington's long history of regional economic modeling and analysis, the study addresses several questions, among them: Are exports the key to regional economic growth? To what extent do people follow jobs? Are the regional unemployment rate, wage rates, and consumer prices determined in the regional or national market? What are the causes of regional economic cycles?

This case study shows that the historical behavior of the Washington economy is largely consistent with the economic base theory of regional growth. The study presents a classical picture of an open regional economy operating in a much larger arena: a trade-dependent economy with little, if any, control over its unemployment rate, wages, and prices because of mobile factors of production.

8.1 Introduction

Nearly 60 years ago, North (1955, 1956) and Tiebout (1956a, b) engaged in a vigorous debate on the nature of regional economic growth, a discussion that is still relevant today. Their remarkable interchange covered a wide range of topics: definition of a region, open economy, stages of growth, economic base, comparative advantage, specialization, footloose industries, residentiary sector, labor and

R.S. Conway Jr (✉)
Dick Conway & Associates, 2323 Eastlake Avenue East, Seattle, WA 98102, USA
e-mail: dickconway@cs.com

P. Nijkamp et al. (eds.), *Regional Science Matters*, DOI 10.1007/978-3-319-07305-7_8,

capital mobility, employment multipliers, economies of scale, import substitution, technological development, transportation costs, and infrastructure. More importantly, the discourse led to the formulation of the economic base model, a simple but powerful means of explaining, analyzing, and forecasting the process of regional economic growth.[1]

The objective of this study is to explain how the Washington state economy has grown and evolved over the past 50 years in light of the theory advanced by North and Tiebout. In simple terms, do the facts fit the theory? The analysis draws upon various types of regional economic models—for example, an economic base model (Conway 2006), an input-output model (Bourque and Conway 1977), an interindustry econometric model (Conway 1990), and a structural time-series forecasting model (Conway 2001)—and the historical data bases that underpin them.

Among the questions addressed are the following: Are exports (domestic and foreign) the key to regional economic growth? To what extent has the economic base become service oriented? With regard to interregional migration, do people follow jobs? Are the regional unemployment rate, wage rates, and consumer prices determined in the regional or national market? What are the causes of regional economic cycles? What are the implications of the study's findings to our understanding of the regional growth process (the validity of the economic base theory), regional modeling (the applicability of demand-oriented models), and regional development policy (the efficacy of industrial recruitment programs in reducing local unemployment)?

At the outset, it should be emphasized that there are two caveats to this study. First, it describes the economic behavior of a region within the United States, where there are no national boundaries restricting the movement of labor or capital. Thus, the findings of the study may not pertain to other parts of the world. Second, even with respect to the United States, this is a case study. There has been no attempt to compare the Washington experience with that of other states. Indeed, it is doubtful that a study like this can be wholly replicated in other states. Few states, if any, have Washington's 60-year history of intense regional analysis—theory, modeling, and applications—that began with North and Tiebout. At the same time, there is no reason to believe that, with regard to the role of exports and the mobility of labor and capital, other regional economies behave much differently than Washington.

8.2 Economic Base Theory of Regional Growth

North, who dismissed the prevailing stages of growth theory as an inadequate explanation of the Pacific Northwest experience, initiated the exchange. He contended that exports were the key to economic growth. What a region exported was determined by its comparative advantage (i.e., its ability to produce

[1] The economic base concept originated in earlier studies of urban economies. See Andrews (1953) and Hoyt (1954).

commodities that were competitively priced for markets outside the region). Accordingly, he divided the regional economy into two sectors: the export (basic) sector and the residentiary (nonbasic) sector. While the export sector produced goods and services for the rest of the world, the residentiary sector served the regional market. But the fate of the residentiary sector depended on the success of the export sector. He presumed that capital and labor were mobile.

Tiebout did not dispute North's fundamental line of reasoning, but he thought that the theory was incomplete. Tiebout argued that there was more to regional economic growth than exports. He said that understanding regional economies, like understanding the national economy, required a broad theory of income determination. He further believed that there were autonomous aspects to the regional growth process, for example the construction of a federal military installation. Tiebout also contended that the size of the region was a factor in its growth. As the regional economy expanded, economies of scale permitted import substitution, which he viewed as a second path to growth. He maintained that the residentiary sector did not play a purely passive role in the growth process. For example, the efficiency of the nonbasic sector, including local government, was critical to the region's economic competitiveness in national and world markets. Finally, Tiebout believed that there was no "ideal region," a geographical area in which business, households, and government operated as a "unified economy."

While North and Tiebout did not resolve all of their differences, they did agree on the fundamental principles of the economic base theory:

1. A regional economy can be divided into two sectors, one producing exports (basic sector) and the other producing goods and services for the local market (nonbasic sector).
2. Export growth, bringing money into the region, causes an expansion of the nonbasic sector through a multiplier (respending) process.
3. A stable relationship exists between the basic and nonbasic sectors.
4. Labor and capital are mobile.

Expressed mathematically, the output variant of the economic base model is given by

$$X_{total} = X_{nonbasic} + X_{basic}$$

$$X_{nonbasic} = rX_{total}$$

$$X_{total} = (1 - r)^{-1}X_{basic} \text{ or}$$

$$\Delta X_{total} = (1 - r)^{-1}\Delta X_{basic}$$

where $(1-r)^{-1}$ is the output multiplier. The solution equation (the third equation) of the economic base model is similar to the solution equation of a multi-sector input-output model.

Even a simple economic base model can be used for forecasting and impact analysis.[2] In practice, however, regional economic models are typically complex systems of simultaneous equations. Following Tiebout's call for a "broad theory of income determination," they predict an array of interrelated economic variables, such as output, employment, and income. Some models add a demographic dimension, using economic variables to forecast net migration and population. Dynamic regional econometric models are designed to not only estimate the total impact on the economy of a given change (e.g., an increase in aircraft production) but also to portray how the impact plays itself out over time.

Whether working with simple or complex models, the economic base theory of regional growth presents several challenges to analysts, as evident in the North-Tiebout debate: defining the region; deciding what in addition to exports to include in the economic base; obtaining reasonable measurements of exports and basic employment and income; predicting changes in the economic base; and figuring out the specific relationship between basic and nonbasic activities. It should be stressed that these issues do not necessarily invalidate the economic base theory nor do they preclude the construction of useful models based on it.

8.3 Exports and Regional Growth

As posited above, the output variant of the economic base model implies that a change in exports will elicit a proportional change in output. For example, a 10 % increase in exports would eventually result in a 10 % increase in total regional output, all else being equal. If this kind of stability were observed in reality, it would lend credence to North's hypothesis that exports are the key to regional economic growth.

The 1963, 1972, 1987, 1997, and 2007 Washington input-output studies (Bourque and Weeks 1969, Bourque and Conway 1977, Chase et al. 1993, Conway et al. 2004, and Beyers et al. 2010) constitute a unique set of detailed regional economic accounts spanning more than four decades. Among other things, these survey-based tables provide estimates of the state's economic base over time. In this case, the economic base is defined to include exports to the rest of the United States and foreign countries (including expenditures by visitors to Washington) and

[2] Simple economic base models are most appropriate for uncomplicated economies, such as rural counties, where the major impact of a change in the economy stems from the induced income and consumption effects. But the models themselves can be complex. The economic base model used to predict the impact of shutting down the Goldendale aluminum smelter in Klickitat County, Washington (Conway 2006) is structured to yield one income multiplier but nine private industry and public sector employment multipliers. The total impact is expressed in terms of employment and labor income by industry, personal income, and population. The model consists of 156 lines of calculations on a spreadsheet. Twenty-one lines are required to estimate basic income from eight exporting industries, government, transfer payments, and the so-called residence adjustment (the income of residents working outside the county less the income of non-residents working inside the county).

Table 8.1 Washington output, exports, and imports (billions of dollars)

	1963	1972	1987	1997	2007
Total output	12.5	26.0	133.3	274.4	603.7
Total exports	4.9	10.8	53.0	111.7	246.7
Percent of output	39.2	41.5	39.8	40.7	40.9
Total imports	4.6	9.8	48.1	114.5	277.6
Percent of output	36.8	37.7	36.1	41.7	46.0

Note: Total exports include industry sales to the federal government. Total imports include imports by households, the investment sector, and state and local government
Source: Washington input-output tables

federal government expenditures. Like exports, federal expenditures result in money flowing into the state economy. This in turn leads to additional rounds of spending—the multiplier process—and the creation of jobs in the nonbasic sector. Examples of federal basic activities in Washington include the operation of a navy port, the administration of public forest lands, and interstate highway construction.

The input-output tables provide strong empirical support for the economic base theory of regional growth (Table 8.1). Indeed, the tables reveal a remarkably stable relationship between Washington exports (U.S. exports, foreign exports, and federal expenditures) and total output (the value of production in all industries). Between 1963 and 2007, exports and output in current dollars grew at almost identical annual rates, 9.3 and 9.2 %, respectively. Consequently, the export-output ratio never strayed far from 0.4. This implied that the non-basic output-total output ratio (r in the above economic base model) stayed around 0.6 and that the Washington aggregate output multiplier remained close to 2.5 (=1/[1.0–0.6]) for nearly 50 years.

The constancy of the export-output ratio is a little surprising in light of the fact that the input-output measurements were taken at quite different points in the business cycle. In 1972, the state economy was limping along, trying to recover from the Boeing Bust, the worst recession since the Great Depression. In 2007, due to the housing bubble, the economy was enjoying its best year of the decade.

The stability of the Washington export-output ratio is not only consistent with the fundamental principles underlying the economic base model but it also gives the model predictive power. In 1963, for example, if one had known the export-output ratio and had an accurate forecast of exports in 2007, it would have been possible to predict total output in 2007 with very little error (specifically, 4.2 %). Given a projection of labor productivity (output per worker), one could have also made a quite reasonable forecast of employment.

While the input-output data support the economic base theory of growth, they provide no evidence of import substitution. As noted previously, Tiebout saw import substitution as a second possible way for a regional economy to develop. In effect, import substitution would enhance the nonbasic sector of the economy, thereby increasing the regional output and employment multipliers. This would enable the regional economy to grow without expanding the economic base.

Table 8.2 Composition of Washington exports (percent)

	1963	1972	1987	1997	2007
U.S. exports	62.1	67.7	61.8	60.1	60.4
Foreign exports	9.7	18.1	28.6	32.8	33.0
Federal government sales	28.2	14.2	9.6	7.1	6.6
Total exports	100.0	100.0	100.0	100.0	100.0
Resources	6.0	5.0	2.2	1.6	3.3
Manufacturing	84.1	70.6	64.7	53.9	48.8
Nonmanufacturing	9.9	24.4	33.2	44.4	47.9
Total exports	100.0	100.0	100.0	100.0	100.0

Source: Washington input-output tables

The import-output ratio has fluctuated more than the export-output ratio, but it has shown no overall tendency to fall over time. If anything, the import-output ratio appears to be rising, indicating a declining propensity to purchase locally produced goods and services. This observation does not totally rule out the possibility of import substitution. It may be that, while import substitution is taking place in parts of the economy, it is being offset by increased importing elsewhere.

It should be noted, however, that import substitution as an economic development strategy in developing countries as well as regions of the United States has met with limited success (Conway 1989). Regional markets are often too small to make import substitution a viable option. In general, the trend in regional production seems to be moving toward greater specialization and away from increased self-sufficiency.

The constancy of the export-output ratio does not imply that there has been no significant change in Washington's exports. Two distinct trends are apparent (Table 8.2). The first is the increasing reliance on foreign markets. Foreign exports as a share of total exports (including sales to the federal government) rose from 9.7 % in 1963 to 33.0 % in 2007. Figures show that, on a per capita basis, Washington foreign exports are among the highest in the nation. Moreover, the rapid expansion of international trade is the principal reason why Washington has grown faster than the United States since 1960, according to an analysis with the Washington Projection and Simulation Model (Conway 1997).[3]

The second trend is a shift of exports to services. In 2007, nonmanufacturing industries accounted for 47.9 % of Washington exports, nearly five times the share in 1963. While these numbers tend to overstate the importance of services, since a significant portion of service exports are trade and transportation margins associated with commodity exports, they do indicate that the state's economic

[3] Originally developed in 1977, the Washington Projection and Simulation Model (WPSM) is an interindustry econometric model designed for regional forecasting and analysis (Conway 1983, 1990). A similar model commissioned by the state of Hawaii in 1977 is still in use today (Conway 1977). In 1989, a prototype of the Chicago Regional Econometric Input-Output Model (CREIM) was constructed following WPSM's specification (Conway et al. 1989).

base is evolving in the direction of software development, medical research, producer services, and tourism.

Richardson (1985) in a survey of input-output and economic base multipliers wrote that "the literature would need to be much more convincing than it has been hitherto for a disinterested observer to resist the conclusion that economic base models should be buried, and without prospects of resurrection."

The Washington data strongly suggest that the economic base theory is not wrong, as Richardson seems to imply. But the theory as expressed by North and Tiebout is incomplete. All it ever purported to do was to explain how a regional economy, in particular its nonbasic industries (principally, trade, services, and local government), would react to a change in its exporting industries. In this regard, the economic base theory is a conditional theory of growth. Except for the notion of comparative advantage, it does not attempt to explain why a change in exports occurs. As North recognized, that is left to location and related theories.

Richardson's harsh criticism notwithstanding, the economic base theory is still a good way of looking at regional growth:

> If you plunk an aircraft industry down on an empty plain, will someone build a grocery store? Probably. If you plunk a grocery store down on an empty plain, will someone build an aircraft industry or any other industry? Not likely.

It should also be said that, as a practical matter, the economic base concept is alive and well. All regional input-output models and most regional econometric models (e.g., Glickman 1977) are based on it. The concept and models continue to be useful because they provide good, though not total, predictive power.

8.4 Employment Change

One reason why the economic base theory has been dismissed as an inadequate explanation of regional growth is the empirical observation that most, if not all, employment growth occurs in nonbasic industries (e.g., trade, services, and local government). Less than 3 % of the more than 2.0 million jobs created in Washington since 1970 have been in manufacturing. As a share of total employment in the private sector, manufacturing fell from 19.8 % in 1970 to 8.7 % in 2010 (Table 8.3).

Does this mean that manufacturing, which is responsible for one-half of the state's exports, is losing its importance as a source of growth? Should economic development agencies abandon programs to retain and attract manufacturing enterprises? While one might be inclined to write off manufacturing as a fading sector, there are good reasons to resist the temptation. The first is the myth of de-industrialization. Noting the loss of manufacturing jobs nationally, some observers have lamented the decline of the United States as an industrial power. But today, using just one-fifth more workers, Washington manufacturers produce three times as much output as they did 40 years ago.

Table 8.3 Washington private output and employment

	1970	1980	1990	2000	2010
Output (bils. $00)					
Goods-producing	46.2	71.4	100.1	118.8	129.0
Resources	4.7	8.5	10.9	10.5	12.2
Construction	9.1	14.3	16.1	21.7	15.9
Manufacturing	32.4	48.6	73.1	86.6	100.9
Service-producing	46.4	82.1	114.9	215.1	262.8
Total private output	92.6	153.5	215.0	333.9	391.8
Percent					
Goods-producing	49.9	46.5	46.6	35.6	32.9
Resources	5.1	5.5	5.1	3.2	3.1
Construction	9.8	9.3	7.5	6.5	4.1
Manufacturing	35.0	31.7	34.0	25.9	25.8
Service-producing	50.1	53.5	53.4	64.4	67.1
Total private output	100.0	100.0	100.0	100.0	100.0
Employment (thous.)					
Goods-producing	381.2	517.5	624.6	667.2	576.8
Resources	86.7	105.6	110.5	105.1	104.2
Construction	68.9	121.0	158.9	214.5	198.1
Manufacturing	225.6	290.9	355.2	347.6	274.5
Service-producing	757.4	1,193.0	1,738.5	2,292.6	2,575.0
Total private employment	1,138.6	1,710.5	2,363.1	2,959.8	3,151.8
Percent					
Goods-producing	33.5	30.3	26.4	22.5	18.3
Resources	7.6	6.2	4.7	3.6	3.3
Construction	6.1	7.1	6.7	7.2	6.3
Manufacturing	19.8	17.0	15.0	11.7	8.7
Service-producing	66.5	69.7	73.6	77.5	81.7
Total private employment	100.0	100.0	100.0	100.0	100.0

Note: Employment includes wage and salary employees and proprietors
Source: Washington projection and simulation model data base

Even in relative terms manufacturing output has not declined substantially. In Washington, manufacturing's share of the total value of production in the private sector fell from 35.0 % in 1970 to 25.8 % in 2010. But the latter figure was depressed by the Great Recession. Over the subsequent 2 years, Boeing increased aircraft deliveries from 462 to 601, while other manufacturers boosted output by an estimated 15 %. As a consequence, manufacturing accounted for approximately 29 % of total private output in 2012.

As in agriculture, the decline in jobs in manufacturing has been due largely to the success of efforts to raise labor productivity. Between 1970 and 2010, manufacturing productivity in Washington rose at a 2.4 % annual rate, increasing from $143,600 (in 2,000 dollars) per person-year to $367,600. In contrast, productivity in services (business, health, education, and consumer services), which in terms of employment has been the fastest growing segment of the economy, rose from $61,300 per person-year to $102,100, an annual growth rate of just 1.3 %.

In general, a strict focus on employment may give a misleading impression of the nature of regional economic change:

> Consider an economy with two workers. The first manufactures one airplane per year, which is sold for a dollar. The income is used to purchase a hamburger, which is produced by the second worker. Ten years later, the aircraft worker, because of productivity gains, can build two airplanes, which earn him two dollars. He buys two hamburgers, which now require two workers to make, since the hamburger industry has not become more efficient. From the standpoint of employment, it appears that the hamburger industry is responsible for all of the growth, when in fact it is the aircraft industry that has caused the economy's output and income to double and employment to increase by one-half.

Some economists have overstated the importance of small businesses for this same reason. David Birch (1988) concluded that most of the net jobs created in Washington between 1983 and 1987 were attributable to firms with fewer than 100 employees. This observation failed to recognize that much of the growth in small businesses, especially in trade and services, was due to shorter work weeks and lagging labor productivity. As a consequence, the growth of these small businesses in terms of labor income (wages, salaries, and proprietors' income) was considerably less impressive. For example, between 1970 and 1990, eating and drinking places accounted for 7.4 % of the total employment growth in Washington but only 2.5 % of the total gain in labor income. In contrast, the aircraft industry (principally Boeing, the state's largest private employer) accounted for only 3.5 % of the employment growth but 5.5 % of the income growth. Thus, if one were to conduct an analysis of economic change using income rather than employment statistics, one would draw substantially different conclusions regarding the contributors to regional growth.

Birch's interpretation of employment growth also neglects to take into account the indirect job impact—the multiplier effect—taking place when large businesses expand. Between 1983 and 1987, Boeing directly accounted for 8.2 % of the job growth in Washington. According to simulations with the Washington Projection and Simulation Model, the aircraft industry employment multiplier (specifically, the short-run multiplier over this period of time) was 3.0. In other words, for each aircraft job, the industry indirectly created two other jobs in the economy. This finding implies that the aircraft industry accounted for nearly one-fourth of the jobs created during this period, including a significant fraction of the new employment found in small businesses, such as eating and drinking places.

Export industries can conceivably contribute to regional growth without increasing their own employment, as the airplane-hamburger story illustrates. Since the typical export industry (especially ones in manufacturing) experiences relatively large productivity and real wage gains over time, it will support an increasingly greater number of jobs in the economy per job in the industry, even if its output multiplier remains constant.

Simulations of the agriculture, lumber and wood products, and aircraft industries illustrate how multipliers can change over time. Consider lumber and wood

Table 8.4 Selected output, employment, and income multipliers (total change in economy per unit change in industry)

	1972	1977	1982
Agriculture			
Output ($82's/$82)	2.52	2.53	2.55
Employment (jobs/job)	2.70	2.74	2.80
Income ($82's/$82)	3.08	3.10	3.14
Lumber and wood products			
Output ($82's/$82)	2.98	3.20	3.31
Employment (jobs/job)	4.51	4.93	5.20
Income ($82's/$82)	3.67	4.12	4.44
Aircraft			
Output ($82's/$82)	1.78	1.80	1.82
Employment (jobs/job)	3.18	3.50	3.88
Income ($82's/$82)	2.20	2.31	2.41

Source: Conway (1991)

products, a highly integrated industry—forestry, logging, sawmilling, and finished wood manufacturing—with large output and employment multipliers. In the first simulation with WPSM, lumber and wood products exports are increased permanently by $1.0 billion (1982 dollars) in 1967 and the total impact is measured 5 years later. This provides estimates of the long-run output, employment, and labor income multipliers for 1972 (Table 8.4). Simulations are then repeated to obtain long-run multipliers for 1977 and 1982.

Between 1972 and 1982, each type of lumber and wood products multiplier increases in size. But the change in the employment multiplier, from 4.51 to 5.20 (15.3 %), is relatively greater than the change in the output multiplier, from 2.98 to 3.31 (11.1 %). This is also true for the output and employment multipliers for agriculture and aircraft. In the case of aircraft, however, a small change in the output multiplier (from 1.78 to 1.82) is accompanied by a large change in the employment multiplier (from 3.18 to 3.88) primarily because of the sizeable gain in the industry's labor productivity over the 10-year period.

Can an employment multiplier rise indefinitely? Though seemingly unlikely, the answer is conceivably yes. Prior to conducting an economic impact analysis of Microsoft (Conway et al. 2006), WPSM was updated and tested. A simulation of the aircraft industry's long-term multipliers showed that between 1982 and 2004 the estimated output multiplier fell from 1.82 to 1.79—the negative effect of outsourcing being slightly greater than the positive impact of rising aircraft wages—while the employment multiplier jumped from 3.88 to 4.96. Here is how the model explains the changes. In 1982, it took 5,900 workers to produce $1.000 billion worth of aircraft ($169,500 of output per worker) in 2000 dollars. At the same time, it took 17,000 workers to produce the $0.820 billion of goods and services ($48,200 per worker) needed to directly and indirectly support aircraft production and employees. The implied output multiplier was 1.82 (=[1.000 + 0.820]/1.000), while the implied employment multiplier was 3.88 (=[5,900 + 17,000]/5,900). Over the next 22 years, aircraft labor productivity grew at a 3.3 % annual rate, reaching $345,000 per worker, while trade and services labor productivity grew at a 1.6 % rate, reaching $68,700. Thus, in 2004, it took just 2,900 workers to produce $1.000 billion worth of aircraft and 11,500 workers to produce $0.790 billion of supporting goods and services. In this case, the implied output multiplier was 1.79, while the implied employment multiplier was 4.96.

It should be pointed out that alternative specifications of regional models give rise to the possibility that the results of impact analysis as well as our observations about how a regional economy behaves depend, perhaps not insignificantly, on the choice of models. For example, at the time that the simulations reported in Table 8.4 were run with WPSM, multiplier estimates were also made with an economic base model and an input-output model of the state economy. In 1982, the economic base model indicated that the employment multiplier for aircraft—as well as for all other basic industries—was 3.11. The input-output model estimate was 2.42. In both cases, the multipliers were smaller than the 3.88 estimate from WPSM.

The interindustry econometric model yields higher multipliers than the input-output model principally because of its greater degree of closure. Whereas the so-called Type II input-output multipliers do not capture the effects of induced investment or state and local government spending, WPSM recognizes that regional economic activity, as measured by changes in population, per capita income, and unemployment, does have an impact on such things as housing construction and public education. A simulation with WPSM shows that, when the links to the investment sector (primarily construction) and state and local government are severed (opened), the aircraft employment multiplier in 1982 drops to 2.14, much closer to the input-output estimate of 2.42. Neglecting the response of construction and state and local government in economic impacts, as does the Type II input-output model, is tantamount to writing off one-sixth of the economy.

8.5 Mobility and Equilibrium

The Washington Projection and Simulation Model is a demand-oriented model in the tradition of economic base models. Implicit in the theory is the presumption that the supply of resources, labor, capital, and other inputs to production can adjust to any level of demand without an adverse effect on production costs. This does not imply that WPSM is devoid of supply-side considerations. These are most notable in the labor force submodel, which effectively depicts the drawn out adjustment of labor supply to a rise or decline in regional economic activity.

Nevertheless, WPSM assumes that in the long run there are no supply constraints on growth and that growth causes no perceptible loss of regional competitive advantage. There are two reasons why the model does not attempt to portray regional production costs and comparative advantage. As a practical matter, regional data are insufficient—in terms of type, quantity, and quality—to adequately support testable models of regional production functions and comparative advantage.

More importantly, however, with the exception of natural resources (e.g., agricultural land, timber, and fish), the assumption of perfectly mobile factors of production in the long run seems quite reasonable for a small and open economy,

Table 8.5 Washington unemployment rate and net migration

	1968	1969	1970	1971	1972	1973	1974
Unemployment rate (%)	4.8	5.3	9.2	10.1	9.5	7.9	7.2
Net migration (thous.)	66.0	43.0	44.4	−0.4	−23.1	13.4	52.7

Source: Washington projection and simulation model data base

at least based on the economic history of Washington. While the state has grown two-fifths faster than the United States over the past 40 years, there is no evidence that it has lost competitive advantage because of higher production costs due to the inelasticity of the supply of labor or capital.[4]

Indeed, there is substantial evidence in the state's boom and bust history that labor supply is highly responsive to labor demand. When the economy expands, the improved conditions, as signaled by new jobs, lower unemployment (at least temporarily), and higher wages, trigger a migration of people into the region. On the other hand, when the economy performs poorly, people move out.

Between 1969 and 1971, for example, Washington lost 82,100 jobs during the Boeing Bust, boosting the unemployment rate to 10.1 %, almost twice the national rate (Table 8.5). In response to the downturn, 23,100 people on net left the state in 1972. The bust was followed by a 10-year expansion that created 662,200 new jobs and attracted 534,200 migrants on net into the state.

Figure 8.1 illustrates not only the mobility of labor but also the distinct time lag between employment change and net migration. The peaks and troughs of employment change consistently lead the peaks and troughs of net migration by 1 or

[4] At the risk of oversimplificaton, economic modeling of regions in the United States is divided into two camps: demand-oriented models based on the assumption that labor and capital are mobile; and supply-side models that assume the supply of labor and capital is inelastic. What regional economists presume about labor and capital mobility has a significant bearing on the formulation of their models and the results of their analyses. Early versions of the Massachusetts Economic Policy Analysis Model (Treyz et al. 1977 and Treyz and Duguay 1983), described as an endogenous wage determination model, yielded long-run employment multipliers close to one principally because of the inelasticity of labor supply. According to the model, if an industry expanded, it would generate jobs in other sectors through the traditional multiplier mechanism. However, because the labor force—net migration in particular—did not fully respond to the increased demand for labor, the new jobs permanently lowered the unemployment rate, drove up wages, increased business costs, and reduced the demand for the state's exports. This in turn led to a loss of jobs across the economy that offset most of the initial indirect employment gains. In a simulation with the Massachusetts model, the total impact of 10,000 new jobs in the electrical machinery industry amounted to only 13,786 jobs (including the initial 10,000 workers) 10 years later. The implied long-run employment multiplier was 1.38. Also contributing to the small size of the multiplier was a seemingly high estimate of the elasticity of demand for exports with respect to production costs. Termed the "location response," it was estimated to be −4.28. It would seem that a model with such low employment multipliers because of the negative feedback effects would have a difficult time explaining the Massachusetts Miracle in the late 1980s, which created 500,000 wage and salary jobs in a 6-year period, one sixth of the state's total payroll employment today. It certainly would not have predicted the economic boom in Washington between 1983 and 1990, which added 600,000 jobs or nearly one-fifth of today's total employment.

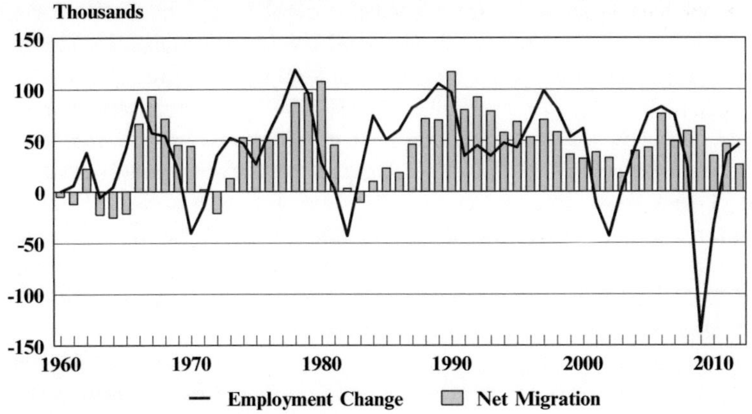

Fig. 8.1 Washington employment change and net migration

Table 8.6 Washington and U.S. employment and population growth rates (average annual percent change)

	Employment growth*			Population growth		
	Washington	U.S.	Difference	Washington	U.S.	Difference
1960–1970	3.3	2.8	0.5	1.9	1.3	0.6
1970–1980	3.5	2.5	1.0	2.0	1.1	0.9
1980–1990	2.7	1.9	0.8	1.7	1.0	0.7
1990–2000	2.6	1.8	0.8	1.9	1.3	0.6
2000–2010	0.6	0.2	0.4	1.3	0.9	0.4
1960–2010	2.5	1.8	0.7	1.7	1.1	0.6

Note: The wage and salary employment growth rates are calculated 1 year earlier (e.g., 1959–1969 and 1969–1979)
Source: Washington projection and simulation model data base

2 years. This indicates that, while people may offer all sorts of non-economic reasons (social, political, and environmental) for moving, they tend to follow jobs.[5]

The relationship between employment growth and net migration is strong enough to be predictable by a simple rule of thumb, as demonstrated in Table 8.6.

[5] The population equation in the Puget Sound Forecasting Model (Conway 2001, 2012), a structural time-series model, predicts the growth rate of regional population relative to the national rate based on the relative growth rate of regional employment. All variables in the model are expressed as the change in the natural log (a measure of each variable's growth rate). The relationship between population change and employment change is described by an eight-quarter polynomial distributed lag. The population forecasting equation, which also contains an ARMA term, has a good track record—a 0.2 % average absolute error for the 19 one-year-ahead predictions made in the fourth quarter of each year since 1993 (Conway 2013a)—because the principal explanatory variables (Puget Sound and U.S. employment) are lagged variables and thus known when making one-year-ahead forecasts.

Employment and population growth rates are calculated over 10-year intervals between 1960 and 2010 for Washington and the United States. Recognizing that employment change leads net migration, the employment growth rates are calculated 1 year earlier (e.g., 1959–1969 and 1969–1979). The rule of thumb is that the difference between the state and national employment growth rates will equal the difference between the state and national population growth rates, allowing for the 1-year lag. This is essentially true for the entire 50-year period. Washington's annual employment growth rate averaged 2.5 %, 0.7 % above the national rate, while its population growth rate averaged 1.7 %, 0.6 % above the national rate. This rule of thumb also holds over each 10-year period and each 5-year period, in spite of changing growth rates.

The strong tendency for people to follow jobs explains why Washington has not been able to maintain a permanently low unemployment rate despite relatively fast employment growth. In general, due to the mobility of labor, changes in the state's rate of net migration keep the state unemployment rate from wandering too far from the national rate (Fig. 8.2a).

The inclination of the Washington unemployment rate to maintain equilibrium with respect to the U.S. rate does not mean that the two rates will necessarily equalize. Washington has had on average a high unemployment rate—about one percentage point above the national rate—over the past 40 years. A relatively large number of workers in seasonal industries, such as agriculture, fishing, logging, and construction, and the attractiveness of Washington as a place to live (the "Mt. Rainer factor") have been conjectured as reasons for the above-average unemployment rate.

Moreover, this difference between state and national unemployment rates has persisted through good times and bad. Between 1975 and 1980, for example, the state expanded two times faster than the nation, creating 445,800 jobs. At the end of the period, however, the state unemployment still stood 0.7 percentage points above the national rate because of the influx of migrants. Two years later, in the throes of the 1981–1982 recession, the unemployment gap widened to 1.9 %. The Washington unemployment rate soared to 11.8 %, but it would have ventured even higher had it not been for a sharp drop in the state's rate of in-migration.[6]

A mobile population also tends to eliminate wage differentials between the region and the nation. If wages are bid up during a growth spurt, the gain will be temporary, as people move into the region and expand the supply of labor. As with the unemployment rate, wage rates are effectively set in the national market. Because of the existence of high-wage employers, such as Boeing and Microsoft,

[6] WPSM depicts the relationship between employment change and population change in a "roundabout" manner. The model presumes that the Washington labor market, through changes in the rate of migration, tends to maintain equilibrium with respect to the U.S. labor market. Specifically, Washington population is forecast by predicting employment, the labor force participation rate, and the unemployment rate. It is expected—and indeed borne out historically—that the Washington labor force participation rate and the unemployment rate follow their national counterparts in the long run. See Conway (1990) and Joun and Conway (1983) for a further discussion of this approach to modeling regional population change.

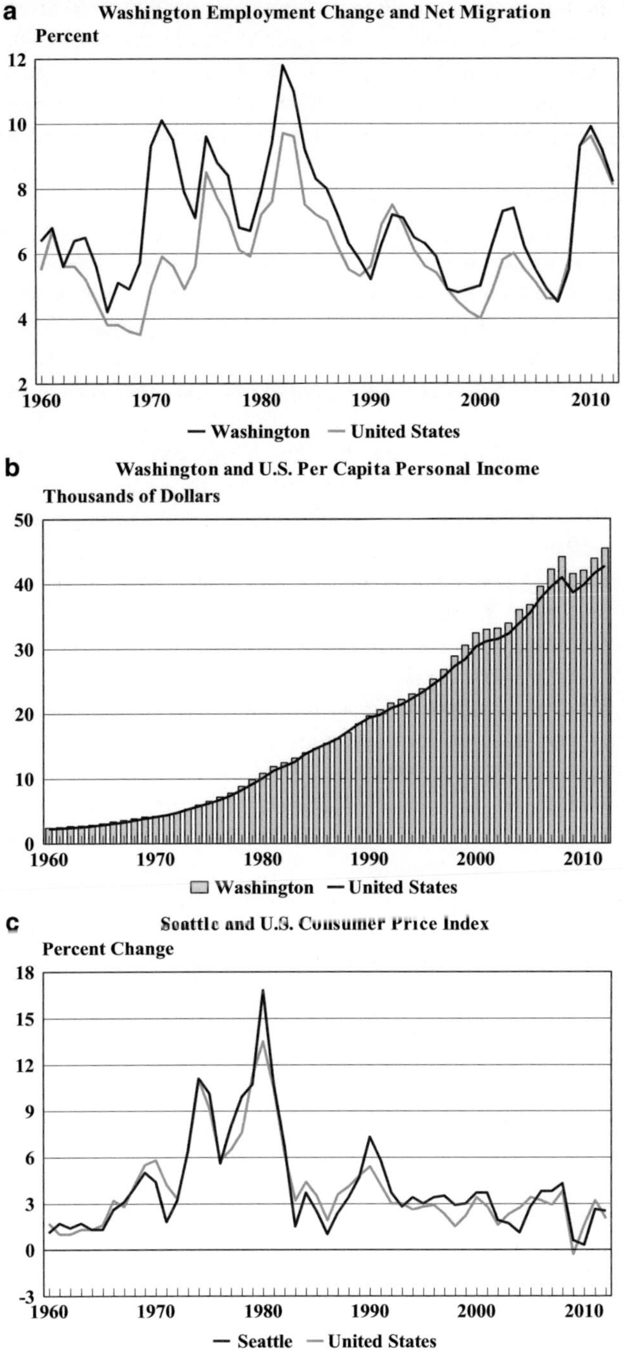

Fig. 8.2 Washington and U.S. economic equilibrium

Washington annual wage and salary disbursements per worker are about 5 % higher than the U.S. average. Since 1970, however, wages per worker in Washington and the United States have increased at the same average rate, 4.9 % per year. This is the principal reason why Washington and U.S. per capita incomes have risen more or less together, 5.9 and 5.8 %, respectively, over the same time period (Fig. 8.2b).

Although strong economic growth will often lift regional per capita income above its equilibrium level, there is no guarantee of it. Between 1983 and 1990, Boeing and its aerospace contractors added approximately 42,800 high-wage jobs at their Puget Sound facilities, contributing to an economic boom that created 702,400 jobs statewide (nearly one-fifth of today's employment). While the state grew one and one-half times faster than the nation during this period, its relative per capita income actually fell from 4.5 % above the national average in 1983 to only 1.5 % in 1990.

There appear to be three reasons for this unexpected outcome. First, aircraft's high wages had no measurable effect on pay rates in other industries, as a flood of migrants kept labor abundant. Second, through the multiplier process, the aircraft industry indirectly created an estimated 140,000 jobs, many of which were in low-wage industries like retail trade and personal services. Third, incomes in some of Washington's other high-wage industries, principally aluminum processing and lumber and wood products, declined during this period, at least in real terms, for reasons unrelated to the Boeing expansion.

As a final point on mobility and equilibrium, goods and services can move from place to place even more easily than labor. As a consequence, the prices of most products, like cars, computers, processed food, and haircuts, are essentially set in the national market. As it is with wage rates, Washington is a price-taker. One exception is the price of houses, which are not mobile. Prices vary across regions because of differences in transportation costs, but local and national prices tend to move together over time. Since 1970 the Seattle inflation rate has averaged 4.5 % per year, just 0.2 percentage points higher than the U.S. inflation rate (Fig. 8.2c).

There have been periods of time, however, when the Seattle inflation rate has diverged from the national inflation rate. In the 1990s, Seattle prices rose substantially faster (3.5 % per year) than U.S. prices (2.8 %). An analysis of the consumer price index by component revealed that local prices outpaced national prices in every major category except medical care. However, housing accounted for most of the difference between the Seattle and U.S. inflation rates. Because of a housing bubble fueled by speculative demand (e.g., home-flipping), Seattle area home prices rose 20 % in 1989 and 23 % in 1990. This had the effect of significantly boosting the imputed housing rent component of the consumer price index for several years in the 1990s.[7]

[7] The Puget Sound Forecasting Model provides strong evidence that prices are primarily determined in the national market. The equation forecasting the change in the Seattle consumer price index has two explanatory variables: the change in the U.S. consumer price index and the relative

8.6 Regional Economic Cycles

The term "economic cycle" conjures up an image of a wave that rises and falls with regularity. But that analogy is misleading, since economies mainly go up. The Washington economy experienced only six annual declines in employment between 1970 and 2010. During the six down years, the economy lost a total of 312,000 wage and salary jobs (186,100 in the Great Recession), while during the 34 up years, the economy created a total of 2,028,100 jobs.

Nevertheless, economies do cycle, though this is only apparent when measured in terms of growth rates (Fig. 8.3). Accordingly, an economic cycle might best be described as a period of slow growth followed by a period of rapid growth. Since 1970 Washington has had three full economic cycles (four going back to 1960). There would have been another complete cycle last decade if its recovery stage had not been aborted by the Great Recession.

To understand the nature of regional economic cycles, it is helpful to consider three causal factors: (1) external forces (namely, fluctuations in the U.S. and world economies); (2) volatile regional industries; and (3) the internal workings of the regional economy that tend to amplify the cycles.

U.S. economic cycles appear to follow the calendar. With the exception of the last decade, each cycle since 1960 has lasted about 10 years. Slow growth in the first 5 years of the decade has been followed by rapid growth in the last 5 years. The timing of U.S. economic cycles with respect to the calendar is of course coincidental, but it is no fluke that Washington economic cycles have followed the same temporal pattern. Since the United States is the largest market for Washington exports—and exports are a key to regional economic growth—there is a strong correlation between national and state output and employment growth. In other words, economic conditions in the United States, whether good or bad, tend to play themselves out in Washington.

This does not imply that national and regional economic cycles always look alike. A large and volatile industry, like aerospace, can significantly shape a regional economic cycle. In 1969, Boeing airplane orders dried up during a shallow national recession, resulting in the loss of 64,000 aerospace jobs over a 2-year period. Helped by an otherwise strong economy, Washington shed only 82,100 jobs during the Boeing Bust. Nevertheless, it suffered a much deeper downturn than the nation. In contrast, the state added 469,300 jobs between 1985 and 1990. With 37,900 new hires, Boeing and its aerospace subcontractors directly and indirectly accounted for one-third of the total employment expansion.

---

growth rate of population—the regional rate relative to the national rate—over the prior eight quarters. Currently, the estimated elasticity of the Seattle consumer price index with respect to the U.S. consumer price is 1.025 (t-value = 29.3). The population term captures the fact that the Puget Sound inflation rate may deviate from the U.S. rate if the regional economy is growing substantially faster or slower than the national economy. Such times are likely to have differential impacts on national and regional home prices and apartment rents. The long-run elasticity of the relative growth rate of population is 0.453 (t-value = 2.6).

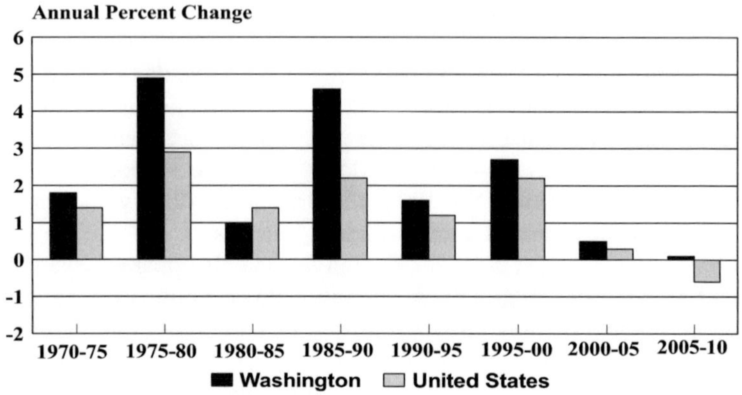

Fig. 8.3 Washington and U.S. employment

Regional economic cycles triggered by external forces are usually amplified by the internal workings of the economy. This occurs most commonly in industries that produce capital goods for local use, such as the homebuilding industry. In accordance with the accelerator principle, when a regional economy turns up, the demand for new homes increases, which in turn intensifies the expansion.[8] Since the demand for housing depends on the region's economic growth rate, it can be very unstable. For example, the number of Washington residential building permits fell 65.9 % between 1979 and 1982 only to increase 102.0 % between 1982 and 1985.

Another sector of the economy that is sensitive to economic cycles is state and local government (Conway 2013b). Lacking an income tax, Washington's biggest tax base is taxable retail sales. Since consumer durables and new residential and nonresidential construction comprise one-third of the tax base, sales taxes were hard hit by the Great Recession, greatly disrupting government operations. Between 2007 and 2010, while nominal Washington personal income rose 3.6 %, current-dollar taxable retail sales—and thus sales tax revenue—plummeted 15.3 %. Adjusted for inflation and population growth, Washington state government tax revenue declined 17.5 % between FY 2007 and FY 2012. This meant that state

[8] To depict the accelerator effect, estimated models of housing construction include both demand and supply factors. The explanatory variables in the housing construction equation in the original version of WPSM (Conway and Howard 1980) included population, income, an interest rate spread, construction costs, and housing stock. The forecasting equation also incorporated a speculative demand variable (the percentage change in the Seattle housing price less the mortgage rate), since it was observed that a rapid increase in house prices had triggered a housing demand for investment purposes (e.g., home-flipping). The speculative demand variable was not only statistically significant with the correct sign, but its inclusion aided the model in tracking the 1975–1977 Washington housing boom.

government had lost one-sixth of its ability to provide public goods and services—education, safety, healthcare, and infrastructure—for Washington residents.[9]

8.7 Open Economy

Tiebout was known for his quips. One dealt with social choice: "People vote with their feet." Another pertained to forecasting: "Forecasts are bound to be wrong, at least to a degree." The implication of the latter quip is that economists cannot promise accurate predictions, only reasonable ones.

Nevertheless, a strong test of a regional economic model is the accuracy of its ex ante forecasts. Since its inception, the prediction errors of the Puget Sound Forecasting Model (Conway 2001) have been tabulated and analyzed on a regular basis. The most recent evaluation entails measuring one and 2-year-ahead errors for the nineteen end-of-year forecasts made since 1993 (Conway 2013a). The purpose of this exercise is two-fold: to quantify the uncertainty of the predictions and to improve the specification of the forecasting model.[10]

The 100-variable econometric model generates forecasts for the Puget Sound region, a four-county area comprising three-fifths of the Washington economy. Variables are expressed in terms of their growth rates (the change in their natural log) to enhance the statistical robustness of the model. Following the economic base theory, employment in basic industries, such as durable goods manufacturing, is predicted using exogenous variables (e.g., the U.S. industrial production index),

[9] One concern about Washington's economic future is its dysfunctional state and local tax structure. Counting retail sales taxes (which include taxes on new construction), business and occupation taxes (much of which are passed on to consumers in the form of higher prices), and so-called sin taxes, the state is twice as dependent on sales taxes as the rest of the nation. As a consequence, the tax system is not only volatile but it is also regressive and inadequate. The Institute on Taxation and Economic Policy (2013) rates Washington as having the most unfair tax structure in the nation. Inadequacy refers to the inability of tax revenue to grow with the economy (as measured by personal income) without expanding the tax bases or raising the tax rates. The problem of inadequacy is evident in the forecasting equation for Washington taxable retail sales (Conway 2010). With each variable expressed in change-in-log form, the explanatory variables are personal income, the unemployment rate, and housing permits. Trend growth in taxable retail sales is projected by personal income, while cyclical changes are predicted by the unemployment rate and housing permits. While an adequate tax base would have an income elasticity of 1.00, the estimated elasticity is only 0.80 (t-value = 14.7) in the case of Washington taxable retail sales. This is the chief reason why Washington's state and local effective tax rate (state and local tax revenue as a percent of personal income) fell from 11.6 % in FY 1995 to 9.7 % in FY 2011, putting it well below the 10.6 % 40-year average for all U.S. state and local governments. This has created a tax policy bind: should Washington maintain the current tax bases and tax rates, causing a further decline in the adequacy of state and local tax revenue; or should it expand the tax bases and raise the tax rates, exacerbating the unfairness of its tax system?

[10] One weakness of the Computable General Equilibrium (CGE) model, the most popular supply-side model, is the difficulty of assessing its prediction errors. As Dixon and Rimmer (2009) put it, "how do I know the results (of the CGE models) are accurate?" After assessing the forecasts of USAGE, Dixon and Rimmer concluded that "USAGE output forecasts at the 500-commodity level comfortably outperform trends. On the other hand, average errors for these forecasts seem alarmingly high."

while employment in nonbasic industries, such as retail trade and state and local government, is predicted using endogenous variables (e.g., Puget Sound personal income and the unemployment rate). In addition to employment, the regional model forecasts the unemployment rate, personal income, wages and salaries, retail sales, population, the consumer price index, housing permits, home prices, and the apartment vacancy rate. The predictions of the exogenous variables come from the Blue Chip consensus forecast for the U.S. economy (Moore, monthly).

Over the past 20 years, the one-year-ahead absolute prediction error for employment has averaged 0.9 %. This compares favorably with the 1.2 % average error for real U.S. Gross Domestic Product (GDP) associated with the Blue Chip forecasts over the same period. These two variables are equally difficult to predict given that the ranges in their yearly growth rates since 1990 are similar. While real GDP growth rates have varied from −2.8 % in 2009 to +4.8 % in 1999, regional employment growth rates have varied from −4.0 % in 2009 to +5.1 % in 1997.

This does not mean that the regional employment prediction errors are all tightly clustered around 0.9 %. There are two errors—a 2.7 % error in 2001 and a 3.2 % error in 2009—that are significantly larger. Both errors are associated with the onset of a recession, reflecting another fact of life in forecasting: national and regional economists have difficulty predicting downturns.

Ironically, the large forecasting errors caused by the Great Recession provide an opportunity to illustrate the openness of the Puget Sound economy. Consider the following question: under what conditions would we have predicted the regional recession? A conditional prediction test provides an answer.

In an ex ante forecasting test conducted at the end of 2009 (Conway 2009), it was assumed that back in December 2005 there was perfect knowledge about the course of the U.S. economy, Boeing, and Microsoft—the exogenous variables in the regional forecasting model—through 2009. With the benefit of known exogenous variables, the model calibrated in December 2005 would have predicted the regional recession with relatively small errors (Table 8.7). At the end of the 4-year forecast period, for example, the conditional accumulated prediction error for employment was −0.6 %, much less in absolute terms than the 6.2 % actual accumulated error. The conditional accumulated prediction errors for personal income and population were −2.8 % and 0.7 %, respectively, both small miscues for a 4-year projection.[11]

A perhaps counterintuitive outcome of the Great Recession was its negligible impact on Puget Sound population growth. In September 2007, prior to the downturn, it was predicted that population would increase by 254,000 over the next

[11] During the Great Recession, Washington state government produced economic and tax revenue forecasts that not only exhibited extremely large errors but were also unreasonable. A state version of the Puget Sound Forecasting Model (Conway 2010) was built to provide alternative projections through FY 2013. The alternative predictions turned out to be significantly more accurate. A contributing reason for the relative precision of the alternative forecasts was the fact that the Blue Chip panel of national economists, which called for a slow national recovery between 2010 and 2013, essentially hit the mark.

Table 8.7 Puget Sound employment ex ante prediction error analysis

	Actual (% change)	12/05 Prediction (% change)	Prediction error (%)	Conditional prediction (% change)	Conditional prediction error (%)
2006	3.2	2.7	−0.5 (−0.5)	3.9	0.7 (0.7)
2007	2.9	2.4	−0.5 (−1.0)	2.4	−0.5 (0.2)
2008	0.9	2.3	1.4 (0.4)	−0.4	−1.3 (−1.1)
2009	−4.0	1.8	5.8 (6.2)	−3.5	0.5 (−0.6)

Note: The numbers in parentheses show the accumulated prediction errors to date
Source: The Puget Sound Economic Forecaster (Conway 2009)

6 years, reaching 3,823,900 in 2013. Based on the December 2013 forecast, the current population estimate in 2013 is 3,837,000, a difference of only 0.3 %.

Why did population growth stay the course during the Great Recession? As noted in Sect. 8.5, the Puget Sound population growth rate depends primarily on the difference between the regional and national employment growth rates, a difference that did not change as a consequence of the downturn. Six years ago it was forecast that between 2007 and 2013 the Puget Sound average annual employment growth rate would exceed the U.S. rate by 0.3 percentage points (1.5 % versus 1.2 %). The latest data indicate that due to the recession the actual job growth rates were much slower (0.1 % and −0.2 %, respectively) but that the difference was still 0.3 percentage points.

Although the Puget Sound Forecasting Model is premised on the economic base theory of regional growth, it has a much more complex logical structure than the three-equation output model discussed earlier. The fact that the Puget Sound Forecasting Model performed well in the ex ante prediction experiment attests not only to the efficacy of the model but also to its underlying theory of regional growth. In particular, the conditional predictions shows how the Puget Sound economy is intricately tied to the U.S. economy, a finding that is hardly surprising given that the nation is the region's largest export market. Reasonably accurate predictions of nonbasic employment support the concept of the multiplier process. Finally, the ex ante test sheds further light on the mobility of labor, revealing that population growth follows a very predictable path even in a severely contracting economic environment.

Conclusion

Washington is a good case study of the process of regional economic growth. It has an abundance of economic data, most notably the 44-year series of input-output tables, as well as a long history of regional economic modeling. The state economy has also exhibited strong but volatile growth over the past 50 years. In particular, Washington's ups and downs enable analysts to observe how the economy responds to changes in its environment (e.g., the Great Recession).

The case study presents a classical picture of an open regional economy operating in a much larger arena: a trade-dependent economy with little, if any, control over its unemployment rate, wages, and prices because of mobile factors of production.

The experience of the Washington economy is largely consistent with the economic base theory of regional economic growth as posited by North and Tiebout. Notwithstanding the theory's limitations, it is difficult to imagine understanding economic and demographic growth in Washington—or any other region in the United States for that matter—without first considering the performance of its economic base, whether it is embodied in agricultural commodities, manufactured goods, tourism, information technologies, producer services, military bases, or the income of retirees.

The case study also suggests why demand-oriented models remain more popular than supply-side models for practical applications, such as forecasting and impact analysis. This does not mean that supply considerations, whether they pertain to timber harvest restrictions, natural and cultural amenities, the educational attainment of the workforce, or infrastructure, are unimportant. What the case study does imply is that models that attempt to portray production costs, comparative advantage, and investment decision-making may be based on the unfounded assumption that regions have an inelastic supply of labor and capital and are therefore not price-takers.

Finally, this study points out some limitations of regional economic policies. Washington's expansions and contractions vividly demonstrate that, because of the mobility of labor, job growth per se has little or no impact on the regional unemployment rate or per capita income in the long run.

References

Andrews R (1953) The mechanics of the urban economic base: historical development of the base concept. Land Econ 29:161–67

Beyers WB, Baldwin M, Lin T, others (2010) 2007 Washington input-output model. Washington State Office of Financial Management

Birch DL (1988) Employment growth in Washington State. Washington State Department of Trade and Economic Development

Bourque PJ, Conway RS Jr (1977) The 1972 Washington input-output study. Graduate School of Business Administration, University of Washington

Bourque PJ, Weeks EE (1969) Detailed input-output tables for Washington State, 1963. Washington Agricultural Experiment Station, College of Agriculture, Washington State University

Chase RA, Bourque PJ, Conway RS Jr (1993) Washington state input-output study, 1987. Washington State Office of Financial Management

Conway RS Jr (1983) Applications of the Washington projection and simulation model. In: Dutta M, Hartline JC, Loeb PD (eds) Essays in regional economic studies. The Acorn Press, Durham, NC, pp 105–115

Conway RS Jr (1977) The Hawaii population and economic forecasting model. Hawaii Department of Planning and Economic Development

Conway RS Jr (1989) Import substitution program evaluation. Northwest Area Foundation

Conway RS Jr (1990) The Washington projection and simulation model: a regional interindustry econometric model. Int Reg Sci Rev 13:141–165

Conway RS Jr (1991) An empirical comparison of regional multipliers. In: Dewhurst JH, Jensen RC, Hewings GJD (eds) Regional input-output modelling. Avebury Academic Publishing Group, Aldershot, pp 178–195

Conway RS Jr (1997) Foreign exports and the Washington State economy, Washington State Department of Community, Trade, and Economic Development, Port of Seattle, Port of Tacoma, and Washington Public Ports Association

Conway RS Jr (2001) The Puget Sound forecasting model: a model of Ron Miller's hometown. In: Lahr ML, Dietzenbacher E (eds) Input-output analysis: frontiers and extensions. Palgrave Publishers, Hampshire, England, pp 431–450

Conway RS Jr (2006) The economic impact of the Washington state aluminum industry. Alcoa, Inc.

Conway RS Jr (2009) Special topic: regional forecasting (6). The Puget Sound Economic Forecaster 17 (December 2009), 6

Conway RS Jr (2010) Washington forecasting model notes. Washington State Economic and Revenue Forecast Council

Conway RS Jr (2012) Special topic: estimating population. The Puget Sound Economic Forecaster 20 (December 2012), 6

Conway RS Jr (2013a) Special topic: regional forecasting (7). The Puget Sound Economic Forecaster 21 (March 2013), 6

Conway RS Jr (2013b) The tax revenue problem. Washington State Governor's Council of Economic Advisors

Conway RS Jr, Hewings G, Israilevich P, Mahidhara R (1989) The Chicago Regional Econometric Input-Output Model (CREIM). Regional Economic Applications Laboratory (Federal Reserve Bank of Chicago and the University of Illinois)

Conway RS Jr, Henn C, Eicher T (2006) The Microsoft economic impact study. University of Washington Economic Policy Research Center

Conway RS Jr and others (2004) Washington input-output table, 1997. Washington State Office of Financial Management

Conway RS Jr, Howard CT (1980) A forecasting model for regional housing construction. J Reg Sci 20:1–10

Dixon PB, Rimmer MT (2009) Forecasting with a CGE model: does it work? General Paper No G-197, The Centre of Policy Studies, Monash University

Glickman NJ (1977) Econometric analysis of regional systems. Academic Press, New York

Hoyt H (1954) Homer Hoyt on the concept of the economic base. Land Econ 30(1954):182–186

Institute on Taxation and Economic Policy (2013) Who pays? A distributional analysis of all tax systems in the United States

Joun RYP, Conway RS Jr (1983) Regional economic-demographic forecasting models: a case study of the Washington and Hawaii models. Socio-Econ Plan Sci 17:345–353

Moore R (monthly) Blue Chip economic indicators. Aspen Publishers

North DC (1955) Location theory and regional economic growth. J Polit Econ 63:243–258

North DC (1956) Exports and regional growth: a reply. J Polit Econ 64:165–168

Richardson HW (1985) Input-output and economic base multipliers: looking backward and forward. J Reg Sci 25:607–661

Tiebout CM (1956a) Exports and regional economic growth. J Polit Econ 64:160–164

Tiebout CM (1956b) Exports and regional economic growth: rejoinder. J Polit Econ 64:169

Treyz G, Duguay GE (1983) Endogenous wage determination: its significance for state policy analysis models. In: Dutta M, Hartline JC, Loeb PD (eds) Essays in regional economic studies. The Acorn Press, Durham, NC, pp 116–130

Treyz G, Friedlander AF, McNertney EN, Stevens B, Williams RE (1977) The Massachusetts economic policy analysis model. The Econometric Model Project, University of Massachusetts

Integration, Contagion, and Income Distribution

9

Iwan J. Azis

Abstract

The 2008/2009 global financial crisis and the unprecedented policy response in advanced economies have a worldwide impact. The episode also led many to question the standard framework of economic thinking on regional integration, financial liberalization and their repercussions on income disparity. The paper argues that one needs to take a balanced view on integration—not just the benefits but also the risks. If regional integration leads to greater inequality, the expected growth and prospect of improved welfare can be diminished. Utilizing a general equilibrium framework, it is also shown how financial liberalization and the surge of capital inflows can produce not only financial instability but also worsening income disparity. By combining model-based results and theory-based ranking applied to the Asian case, and considering the benefits, opportunities, costs and risks of alternative policies, it is revealed that imposing levy on bank-led flows can be used to reduce instability and inequality. This type of macro-prudential policy reflects a departure from the 'First Best' to the 'Second Best' approach of liberalization, where the frictionless outcome of the former is seriously questioned.

9.1 Introduction

Whether we understand it or not, what began the latter part of last decade is changing global economic thinking. The subprime crisis, the resultant collapse of Lehman Brothers in the United States (US), and the subsequent Eurozone crisis humbled economists and regional scientists. Given that these two are the world's biggest, most advanced economies, it is not difficult to imagine the extent of

I.J. Azis (✉)
Asian Development Bank, and former professor and director of graduate study, Regional Science program, Cornell University, Ithaca, USA
e-mail: ija1@cornell.edu; iazis@adb.org

P. Nijkamp et al. (eds.), *Regional Science Matters*, DOI 10.1007/978-3-319-07305-7_9,

repercussions on the rest of the world not only in terms of macroeconomic impact but also on income inequality. The crisis also put the standard framework of economic thinking on the spot, where the validity of some fundamental concepts is now questioned. Three stand out: (1) the virtue of integration; (2) the risks of financial liberalization; and (3) how finance affects income inequality.

To deal with the first, I revisit the concept of regional integration by taking a more balanced view—not just the benefits but also the risks of integration. On the second and third, I offer a model to show the mechanism how seemingly unrelated financial phenomena are in fact closely interlinked with income inequality. I use the case of capital flows intermediated by banks (bank-led flows) in emerging Asia, given their importance to a region where integration has been accelerating since the late 1990s. Capital inflows helped the financial sector grow, but it can also increase the risk of instability and worsen the income inequality.

9.2 Regional Integration Revisited

9.2.1 Benefits and Opportunities of Integration

Proposition 1 Regional integration expands markets and input sources, better allocating resources across the region and accelerating economic growth.

Regional economic integration is used by countries to achieve national interests—only this time in concert with others. It expands national markets to the region. Like globalization, it can be thought of as an alternative to international embeddedness—or how one relates to the rest of the world. But unlike globalization, regional integration is geographical, and in some cases political.[1] It is institutionally stronger than globalization, as rules tend to be tighter and peer pressure can be more intense.

Expanding markets and input sources beyond national boundaries is one of the most compelling arguments for integration. With an expanded market for goods and services, for both output and inputs, higher economic growth and improved welfare can be expected (Fig. 9.1). Integration helps more efficient resource allocation across the region (or globally) in line with the principle of comparative advantage. If, as a result, productivity growth is enhanced, regional integration can accelerate economic growth and increase employment.

Asia's strong economic performance and resilience despite the recent global financial crisis was largely supported by the region's openness. The drop in external demand from Europe and the US, the 'traditional' market for Asia's final goods

[1] When integration is promoted for political reasons and to build trust, the political windfall that follows can also lead to significant economic benefits.

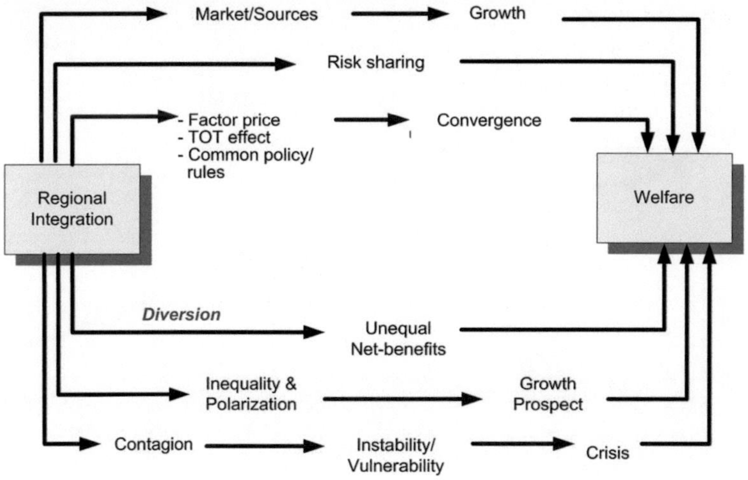

Fig. 9.1 Regional integration—a balanced view. *Source:* Author's construction

exports, has been substituted by increased exports to other Asian countries, as well as to emerging markets outside Asia (Fig. 9.2).[2]

Growing and deepening value chains play a major role in intraregional trade. While the region continues to rely on the global market, the shift reflects a trend of growing regional integration, a process that began in earnest following the 1997/1998 Asian financial crisis. Other factors may have played a role, but this trend certainly helped Asia maintain market expansion—even when world trade fell sharply in 2009.

Proposition 2 Regional integration can reduce income inequality between countries.

It is important to realize that integration may not generate the same benefits for all. Whether in trade, finance, or infrastructure, integration benefits some more than others. And when one measures its effects in the broader sense—beyond the original purpose of integrating—some countries can even lose.[3] So how the benefits of regional integration are distributed matters a great deal.

Most studies based on Europe's integration indicate the process coincides with a substantial decrease in income inequality between countries (Leonardi 1995;

[2] Asia's export share to other emerging markets outside Asia, particularly exports to the Middle East, grew fast, followed by Africa and Latin America. As a result, the export share to these regions increased from 8.4 % in January 2007 to 11.3 % in August 2012.

[3] Venables (2009) argued that the gains from integration are unevenly distributed. Some countries will lose from integration. In particular, the effects of preferential liberalization in regional integration will only benefit resource-poor countries, whereas non-preferential liberalization tends to benefit only resource-rich countries.

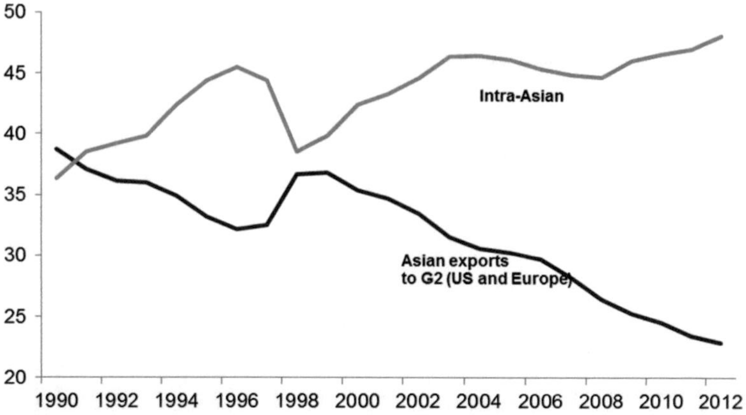

[1]Includes Brunei Darussalam; Cambodia; People's Republic of China; Hong Kong, China; Indonesia; Republic of Korea; Lao People's Democratic Republic; Malaysia; Myanmar; Philippines; Singapore; Thailand; and Viet Nam.
[2]Includes US and eurozone.
Source: Processed from *Direction of Trade Statistics*, International Monetary Fund.

Fig. 9.2 Increasing Intra-Asian exports

Armstrong 1995).[4] While economic factors are important, it is political integration that appears to drive convergence. Institutional forces outweigh market forces in drawing national economies closer together (Beckfield 2009). Economic arguments show freer trade and factor mobility from integration allow less-developed members to grow faster than more-developed ones. Factor price equalization further supports the convergence hypothesis (Stolper and Samuelson 1941). In a two-country resource-rich/resource-poor model, lowering tariffs has a negative effect on real wages in the resource-rich country (most gains accrue to resource rent), while the resource-poor country benefits through terms-of-trade. This also supports the convergence hypothesis.

However, an institutionalist economic explanation emphasizes more the formal structure and role actors play in integration initiatives. It suggests that as economic actors follow common rules in a more integrated system, and markets increase in size and complexity, convergence will likely result. It also stresses the importance of politically established institutions. Thus, to analyze convergence, political relations matter more than regional markets or economic development. Convergence can come from the diffusion of common development policies and the diffusion of common rules and market regulations.

[4] Evidence from Cross-Sectional Analysis of the Regional Growth Process within the European Union is shown in Armstrong (1995); and Ben-David (2001). Some, however, found a pattern of divergence; see Slaughter (2001); Arestis and Paliginis (1995). Part of the explanation rests on the interpretation of σ- and β-convergence (Martin 1996), where σ-convergence is a decrease in GDP dispersion, hence showing how the distribution of income evolves, and β-convergence points to a negative relationship between growth and initial level of GDP.

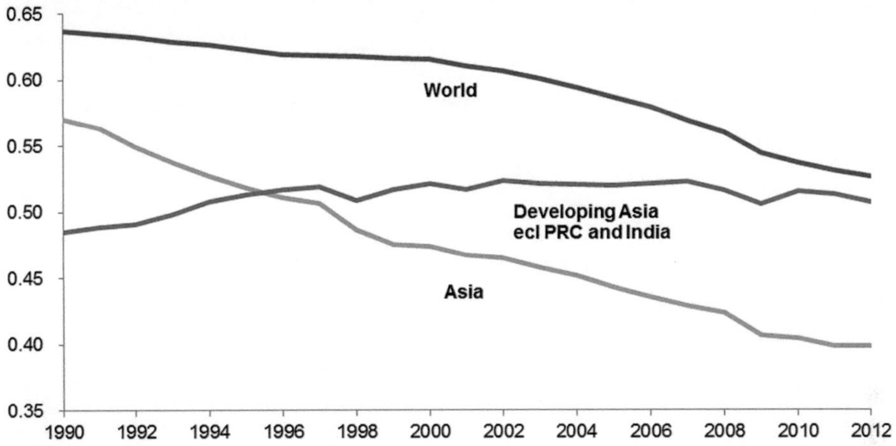

PRC = People's Republic of China.
Notes: Developing Asia is Asia excluding Japan and Oceania. Data for Afghanistan, Cambodia, Cook Islands, Maldives, Marshall Islands, Myanmar, Nauru, Palau, Timor-Leste, and Tuvalu are unavailable. Estimates are based on GDP per capita, PPP terms of a balanced panel of 161 countries, including 38 Asian countries.
[1]Gini Coefficient is computed as follows:

$$Gini = \frac{-(n+1)}{n} + \frac{2}{n^2 \mu_x} \sum_{i=1}^{n} ix_i$$

where x_i is the income of individual i, μ_x is the average income of the population, and n is the total number of individuals in the population. The Gini coefficient ranges from 0 (all individuals have equal income) to 1 (all income held by one person).
Source: *World Development Indicators*, accessed 23 January 2014; PWT 7.1 and government statistics for Taipei,China.

Fig. 9.3 Gini coefficient index

In Asia's case, convergence is also detected; inequality between countries has been declining (Fig. 9.3). Whether this is due to regional integration or other factors—or both—remains to be studied. Regardless, the theoretical arguments above are likely part of the reason. This has an important meaning since the goal of integration is fundamentally different than in Europe. Regional integration in Asia has not been based on a singular fixed set of beliefs. It is more market-driven, facilitated by unilateral investment and trade-enabling policies that change in different times (flexible), and with awareness of different level of development in different countries. While achieving "United States of Europe" is the goal in the European integration, the implicit goal of Asian integration is to manage the diversity.[5] The deliberate act of forging a common platform including forming regional institutions and inter-governmental organizations after the 1997/1998 Asian Financial Crisis does not negate such a goal.

[5] Recent comment by Vice President of the European Commission, Vivane Reding, who is the longest serving Brussels commissioner, is simply a re-confirmation. She called for the EU (European Union) to become a "United States of Europe as the best weapon against the Euro-skeptics." She further argued that "We need to build a United States of Europe with the Commission as government and two chambers. . .. .," implying that individual sovereignty would be a thing of the past.

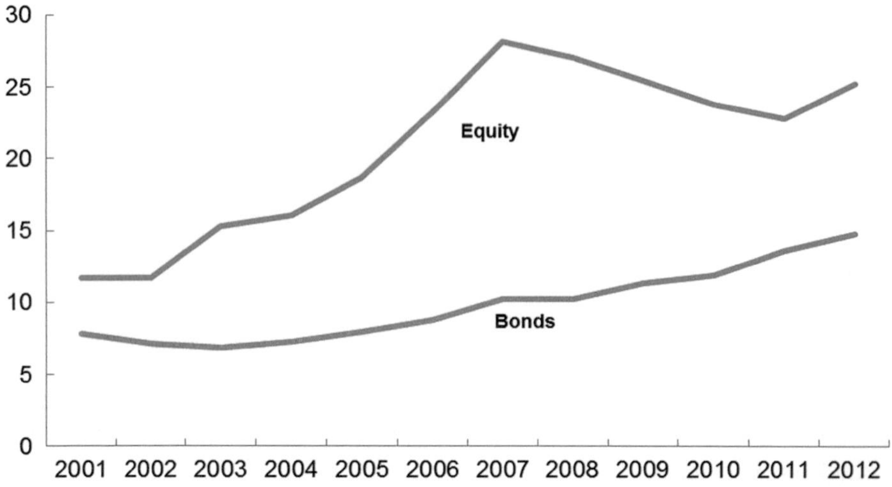

Notes: Asia refers to all 48 ADB member economies plus Australia, Japan, and New Zealand. Countries included as recipient differ from those included as source due to data availability. In particular, data for the People's Republic of China as source is not available.
Source: Processed from *Coordinated Portfolio Investment Survey*, International Monetary Fund. Accessed 5 December 2013.

Fig. 9.4 Intra-Asian portfolio investment

Proposition 3 Risk sharing is another possible benefit of integration; unfortunately, there is little empirical evidence that it happens.

Intuitively, risk sharing through integration makes sense. But many empirical studies show integration has a limited effect on the degree of risk sharing. Since the work of Backus et al. (1992), there have been several studies examining the presence of full risk sharing using cross-country income and consumption correlations. Most of them found that perfect risk sharing does not happen. Asia is no exception. Given an idiosyncratic shock, risk sharing was neither strong, nor did it improve.

What causes this mismatch? Based on numerous cross-country studies, it could stem from several factors, ranging from using domestic equity markets as a major source of finance (French and Poterba 1991), to time horizon and measurement errors (Canova and Ravn 1996), to consumption endowment uncertainty (Obstfeld 1994; Mendoza 1995), to the limited size of capital flows and higher sovereign default (Bai and Zhang 2005).

The effect of financial integration on economic growth has been well documented—more so than the effect of integration on risk sharing (Levine 2001). Theoretically, the consumption growth rate in integrating countries will be cross-sectionally independent of idiosyncratic variables as financial integration increases (Cochrane 1991). The key factor is greater insurance. If inter-regional or international capital markets are well integrated, countries can insure against idiosyncratic shocks. Individuals will invest more in high-risk and high-return assets if the risk can be shared or diversified (Obstfeld 1994).

Asia's financial integration is increasing, especially since the 2008/2009 global financial crisis, but remains limited (Fig 9.4). Using several welfare measures and alternative risk sharing scenarios, Azis (2007) concludes that "....while the level of East Asian financial integration may have increased, its benefits in terms of consumption and investment risk sharing have been limited. Even the advantage of having greater resilience to [an] external shock, that could be potentially reaped from greater synchronization of business cycles, has not been evident." In an IMF Working Paper, Borensztein and Loungani (2011) also conclude that intraregional risk sharing in Asia is low. For a given degree of contagion risk exposure, the US stands out as the one that reaps the most benefit from sharing risks with Asia. The study suggests that the region should work toward increasing risk sharing without exposing countries to greater contagion risks: "pursuing these regional policy avenues should receive a priority over a push for further overall financial integration whose welfare effect may be ambiguous."

All in all, while the level of Asia's financial integration may have increased, its benefits in terms of consumption and investment risk sharing have been limited. Although the concept of integration-driven risk sharing is ideal and conceptually sound,[6] the impact of regional integration must be predicated not on an ideal world, but on the world as it is.

9.2.2 Costs and Risks of Integration

When regional initiatives are launched to strengthen integration—for example, those that boost infrastructure connectivity, risk sharing, and market liberalization—we hear more of its benefits. Far less is heard on the risks of integration.

Proposition 4 The cascading effect of the Eurozone crisis was a vivid reminder of the contagion risk of highly integrated systems.

The main argument against excessive integration is that it exacerbates contagion in times of crisis. Examples abound of financial crises rapidly spreading from one country to another, especially when integration is deeper due to either geographical proximity or a regional arrangement.

While a shock may originate in the financial sector of one country, it can rapidly infect others across a region—affecting entire economies and damaging people's welfare. For Asia, the damage caused by the 1997/1998 Asian financial crisis is a

[6] Under certain circumstances, however, Stiglitz (2010) shows that risk sharing can be unfavorable. While the more integrated the regional economy the better risks can be dispersed, risk sharing can lower expected utility when the standard assumption of convexity and concave utility function does not hold. In particular, this is true when technologies are not convex. Following this dictum, and given the fact that information, externalities, and learning processes may give rise to a natural set of non-convexities, the intuition that integration should be desirable is not always accurate.

powerful reminder of the danger of contagion. An idiosyncratic shock, an exchange rate collapse followed by widespread financial crisis occurring in Thailand, leaped across boundaries, devastating economies of Indonesia, Malaysia, and the Republic of Korea. And yet the scale of integration in Asia at the time was more limited than now, despite some policy convergence. One can only imagine how much worse the crisis would have been had intra-Asian cross-border financial holdings had been larger than they were.

In adopting a single currency, the risks of integration cannot be overemphasized. Many studies prior to the formation of euro emphasized the benefits and opportunities of a single currency. This could be true for Asia as well. But when critical preconditions are not in place, and a desirable sequence is not followed (for example, political before economic integration), forcing a single currency on a group of countries can be risky—and costly as in the Eurozone today. Taking account of these risks and costs—some of which are intangible—a single currency remains a long-term prospect for Asia. Even after running some sensitivity tests, the outcome is the same (Azis 2009). Clearly, it could be counterproductive to focus on potential benefits while neglecting the risks and costs of a single currency in promoting regional integration.

Proposition 5 Trade diversion is another potential risk from regional integration, potentially detrimental to welfare.

Trade diversion is a classic integration risk debated among academics and policy makers alike.[7] It often occurs under free trade agreements (FTAs), when reduced tariffs give a less efficient producer under the FTA comparative advantage over a more efficient competitor outside. In Asia, the South Asia Free Trade Area is a notable example. Given relatively high levels of protection in the region, many predicted the risk of trade-diversion would be high (Baysan et al. 2006). This could be minimized, however, if regional integration was driven by unilateral and multilateral liberalization.[8] When geographical agglomeration effects are also at work, regional integration produces unequal net benefits; development takes place in a few rather than in all.

Proposition 6 Integration tends to increase inequality within countries

In a report by the Commission on the Measurement of Economic Performance and Social Progress, Stiglitz et al. (2010) viewed inequalities as the first cross-cutting challenge for quality-of-life indicators. They argued that inequalities should be assessed comprehensively by examining differences in the quality of life—across people, groups, and generations.

[7] A customs union is a form of regional integration likely to cause the largest trade diversion, where effects are distributed unequally.

[8] The trade-off between trade creation and trade diversion is often used to back north-south FTAs. South-south FTAs are more prone to trade diversion.

Unlike the relation between regional integration and income inequality *between* member countries, the relation between regional integration and income inequality *within* countries is based on the idea that market competition and the labor/capital balance of power is a key determinant of income inequality. Unfortunately, there are few empirical studies on this; and most relating to European integration. They argue that economic integration tends to create a larger labor market and increase wage competition between workers (Western 1997). With workers exposed to competition beyond national boundaries, their bargaining power weakens—either through unions losing influence or by other means. In this case, further integration is expected to increase internal inequality (Alderson and Nielsen 2002).

So what is the difference between the impact of globalization and regional integration, as both give rise to increased market competition? Labor markets expand more readily and labor is more competitive within regions than between regions. Consequently, firms can more easily exercise control over subsidiaries within than between regions. Also, political institutions are more similar within than between regions. So one can hypothesize that regional integration would likely reduce the power of labor unions, and thus have a more pronounced effect on income inequality.

In some cases, more developed institutions (like in Western Europe) can insulate workers from the pressures of international competition (Cameron 1978). Strong welfare states with generous unemployment benefits and training programs can help stabilize the national economy against the vicissitudes of international markets, such that worsening inequality can be averted when regional integration increases.

Most empirical evidence on this is based on Europe's integration. The welfare state shapes stratification directly through income transfers, and can reduce inequality and poverty. But European integration also limited individual government intervention. In addition, there was some retrenchment of Western European welfare states through spending limits imposed by the "convergence criteria" of the 1992 Maastricht treaty (Brady 2003). Thus, more limited national autonomy due to regional integration contributes to the shrinking of the welfare state, one consequence of which is greater income inequality.

In Asia, inequality within most countries has been rising (ADB 2012) alongside economic integration. This does not imply causality, however. With limited integration compared with Europe, it is hard to positively link the two. Deepening regional cooperation to remove barriers to trade and finance, and to further deregulate markets ("negative integration") can have a stronger impact than those from regulations designed to correct market failures ("positive integration"). This has happened in Europe,[9] and there is no reason it cannot happen in Asia as well. When it does, inequality and polarization within countries may worsen.

[9] The convergence effect of regionalization on between-country income inequality in Europe outweighs the polarizing effect of regionalization on within-country inequality, such that the net total income inequality has declined. In other words, regional integration has a positive net effect on reducing total income inequality (Scharpf 1997).

Unlike in the past, it is now widely acknowledged that income and wealth inequality clearly inhibits future growth [see, among others, Nissanke and Thorbecke (2010)]. Inequality is often associated with inadequate property rights, lowering investment. This is common knowledge. But the uncertainty created by the diffusion of political and social instability caused by inequality also tends to raise rent-seeking and dampens investment. All this challenges the standard argument for Kuznets' U-hypothesis.[10] Thus, if regional integration does lead to greater inequality within a country, growth and the prospect of improved welfare is diminished.

9.2.3 Welfare as the Goal and the Importance of Unilateral Policies

Like any policy and strategy, the goal of integration must be an improvement in welfare and quality-of-life—especially for the largest segment of society. Indeed, welfare measures must go beyond just consumption-based utility. Take trade integration. The volume and composition of trade are standard indicators in evaluating whether or not an FTA advances regional integration. But this is just part of the story. How those indicators change will either improve or weaken some socio-economic indicators matters as well. While these may not be on the FTA agenda, they must be taken into account from the perspective of overall development. Ignoring them could lead to misguided policies.

The policy response to a crisis caused by integration-driven contagion can damage welfare, especially when governments are belt-tightening. According to one Organisation for Economic Co-operation and Development (OECD) report, some 20 million jobs in both developed and developing countries disappeared since the 2008/2009 global financial crisis and 21 million jobs must be generated in G20 countries just to match the pre-crisis employment rate.[11] The report also says this is impossible in the near term. If anything, there is a risk the unemployment rate could increase. In a crisis, a crippled financial sector is bad enough; but nothing is worse than the true crisis costs to welfare when speaking about the risk of integration.

The environmental impact of a contagion-driven crisis poses another serious welfare risk. While a crisis can reduce pollution and resource consumption through reduced economic activity, a weakened economy also tends to lower environmental priorities.

While collective regional policies have their merit, unilateral policies can benefit both individual countries and the region. It may be true that some collective regional policies are superior to unilateral national policies. But that does not

[10] Inequality can also affect poverty by way of determining the growth-poverty elasticity. Added by the well-known effect of growth on poverty, a triangular relation between growth, inequality, and poverty is established [see (Bourguignon 2004)].

[11] OECD and International Labour Organization (ILO) 2012. Joint statement by ILO Director-General Juan Somavia and OECD Secretary-General Angel Gurría on the occasion of the G20 Labour and Employment Ministers Meeting. Guadalajara, Mexico, 17 May.

mean unilateral policies cannot, in fact, do the job better. The East Asia Miracle of the 1980s and early 1990s is testament to the value of unilateral liberalization. To say that without integration something bad will happen is farfetched. To argue the entire region will only benefit by joining an integration initiative or agreeing on some regional agenda is equally erroneous. Even without the risks of integration discussed earlier, this is the wrong way to think.

Countries commit to a regional agenda because it is to their national advantage. It expands opportunities, and allows them to allocate their own resources more efficiently. If they fail to see this and decide not to participate, nothing disastrous will happen. This is of course very different from a global commons like climate change.

If unilateral policies improve a country's economic performance, it is not difficult to imagine some positive spillover effects on the regional economy. In trade and financial integration, for example, if countries adopt policies that are good for themselves even without signing up for a regional initiative, their economic growth could become more robust and stable, which by itself also helps the region.

Thus, national policies are key. But they are also important in maintaining the integrity of those domestic institutions required for effective regional initiatives. Even in today's more globalized world, nations remain dominant, and democratic deliberation is largely organized around them. Each country has the right to create and protect its own regulations and institutions. For regional integration to work, there must be sufficient national or domestic policy space to maintain the integrity of domestic institutions.[12] Policy space, when filled in with the right measures, can advance the regional economy. The key principle is clarity and transparency—that the unilateral policy and national deliberation are based on facts and evidence for improving welfare. The cooperation agenda for regional integration can then focus on the rules and monitoring that ensure more effective implementation, while minimizing negative spillovers (as a safeguard). This approach can also improve the quality of national deliberations, making them more effective in boosting welfare.

9.3 Integration-Driven Capital Flows and Income Inequality

9.3.1 Liberalization and Capital Flows in Post-GFC

Financial sector liberalization (FSL) has been widely promoted as a way to better allocate capital and widens opportunities for savers and investors. It also creates an environment conducive to financial innovation. Some argue that FSL helps build discipline among policy makers in securing macroeconomic stability.

This has been the predominant thinking until recently. One of its most important—yet controversial—components is capital account liberalization

[12] A similar principle can also be applied to the concept of globalization; see Rodrik (2011).

(KAL). Capital flows resulting from KAL are channeled through domestic intermediaries—either by banks or firms—allowing greater competition and thus more efficiency. Indeed, countries adopting KAL often see a sudden jump in economic growth as they move away from financial repression. Yet, many of them, developing and developed countries alike, subsequently face instability, with some falling into financial crisis.[13]

Most analysts essentially defended KAL by citing the lack of preconditions for liberalization to explain why crises emerge. They blamed institutional factors such as corruption, weak enforcement, and limited understanding about how a liberalized financial sector operates. Policy recommendations thus centered on fixing those institutional factors, never questioning the virtue of KAL itself. Then came the 2007/2008 shock in the US, followed by the Eurozone crisis. Institutional factors in both economies were supposedly strong, well above those in many emerging markets. Obviously, the early analytical findings went out the window.

Only recently have analysts and scholars admitted early preaching on FSL and KAL was flawed [see, for example, CIEPR (2012)]. They now admit that the "First Best" approach of financial liberalization—where frictionless outcomes are emphasized—is faulty and should be replaced by a "Second-Best" approach in which financial regulation is given far greater importance, and where capital controls are no longer taboo. After decades of preaching the virtues of cross-border capital flows, the IMF finally admitted that some restrictions on capital flows can help protect an economy from financial turmoil. Central to the analysis is the need to maintain financial stability and macroprudential policy (IMF 2012).

The post-crisis low interest rate policy in advanced economies led to massive capital flows, most of which wound up in emerging markets, threatening their financial stability.[14] The resulting exchange rate pressure forced frequent market intervention to maintain trade competitiveness. But the problem does not end here. New financial vulnerability was created through bank-led inflows.

With additional funds flowing in from inexpensive sources (non-core or non-deposits), banks were more willing to take risks. Without foreign exchange market intervention, inflow-induced currency appreciation bolsters the balance-sheet of borrowers. And if used to augment bank loans, they can stimulate the economy to offset the fall in exports.[15] The problem arises when deleveraging or a

[13] In the 1990s alone, financial crisis struck Europe (1992/1993), Mexico/Latin America (1994), Asia (1997), Eastern Europe and Russia (1998). The recent global financial crisis began in the US and Europe. Crisis contagion has also become less regional and more global. Technology and information enable financial spillovers by reducing structural distance.

[14] Most analysis suggests the easy money policy in advanced economies was less effective than originally thought. The policy not only failed to strengthen recovery in the US and other advanced economies, but also provoked global monetary instability through capital flows, including those led by "carry traders," who exploit interest rate differentials across countries. See Mckinnon (2012).

[15] The amplified effect of cross-border flows on the supply of credit due to the changing risk behavior of banks is shown in Valentina and Shin (2012).

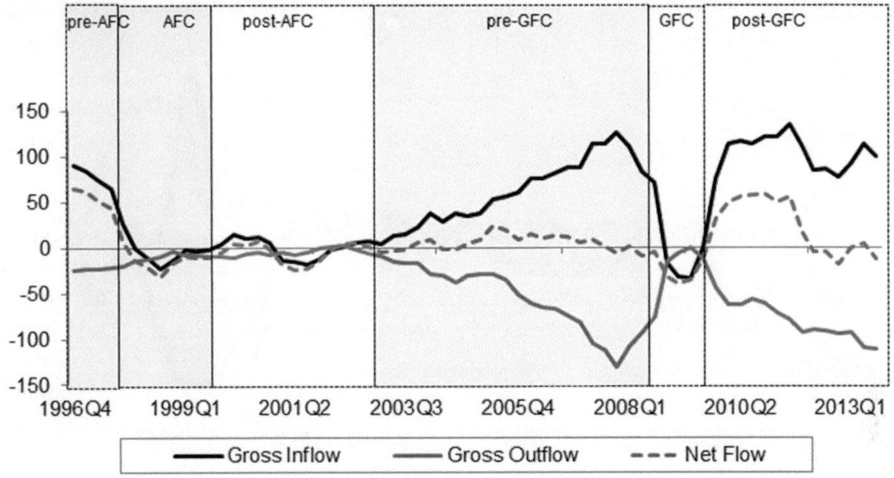

Fig. 9.5 Capital Inflows and Outflows. *Note*: Based on a 4-quarter moving sum. For inflows, refers to bank flows from other investments in the liabilities side (assigned a positive value); for outflows, in the asset side (assigned a negative value). *Source*: Processed from *Balance of Payments Statistics (both BPM5 and BPM6)*, International Monetary Fund

change in investor sentiment, for example, reverses the direction of capital flows. Loans can be disrupted, and the economy suffers a credit crunch. This occurred recently when European banks deleveraged by bringing back funds to strengthen their capital position.[16] On the other hand, a rising share in non-core liabilities can also alter bank behavior toward more risk-taking—for example by investing in securities and other risky financial assets. Indeed, data show this has already happened in some countries. Either way, increased bank-led inflows creates greater vulnerability. Only when recipient banks prudently manage and use new funds will the overall outcome be favorable.

The trend of gross capital flows in selected emerging Asian economies point to one common feature: the size and volatility of flows have increased since 2007/2008, more than that prior to the 1997/1998 Asian financial crisis (Figs. 9.5 and 9.6). While rising capital flows can be beneficial to recipient countries, their volatility and procyclicality can increase financial risks and imbalances.

Not all flows pose the same risks. It is useful to break them down into categories: (1) "equities" flows consisting of direct investment and equity portfolios; (2) "debt" flows comprising debt securities and others including derivatives; and (3) "bank"

[16] For example, in the Republic of Korea, each 1 % decline in external funding due to European bank deleveraging following the global financial crisis led to a 0.01 % decline in domestic credit by domestic banks [see Jain-Chandra et al. (2013)]. This occurred despite the country's relatively healthy foreign reserves, government efforts to provide foreign currency liquidity through bilateral and multilateral currency swap arrangements, and macroprudential measures that lowered domestic banks' reliance on short-term wholesale funding.

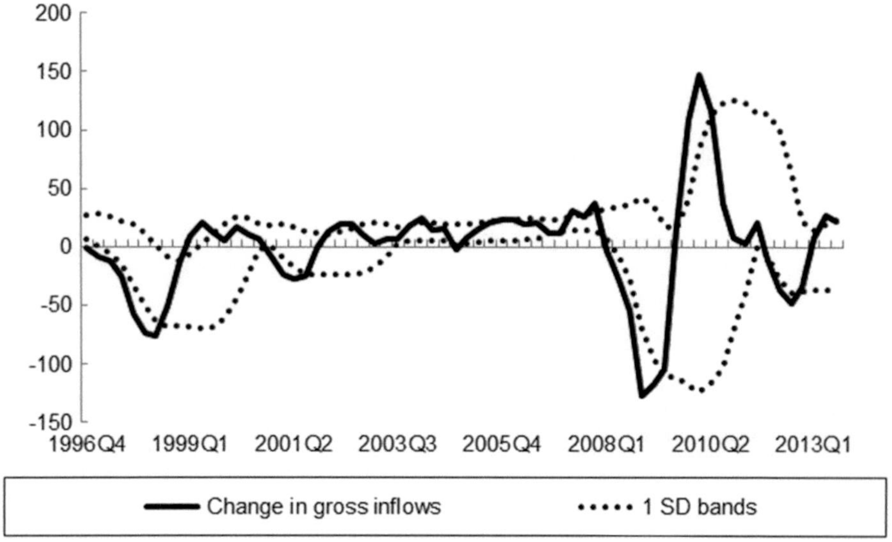

Fig. 9.6 Growth of gross inflows. *Source*: Author's construction based on the sources in Fig. 9.5

flows.[17] Inflows are equities-led, debt-led, and bank-led if the increase is mainly through equities, debt, and banks, respectively.

To better understand how each fluctuates, inflows are classified as "surges" if there is a sharp increase, and "stops" for a sharp decrease. For gross outflows, the terms "flight" and "retrenchment" are used. While "flight" refers to investors moving large amounts of capital abroad, "retrenchment" occurs when domestic investors liquidate foreign investments. Based on a one-standard deviation of the change in the mean capital flows as the limit (for example, in the case of inflows it is shown by dash lines in Fig. 9.6) beyond which they are labeled according to the above classifications, the following episodes are observed:

Surge Episodes
Equity-led: 2009Q4–2010Q1
Debt-led (excl. banking flows): 2002Q1–Q3; 2007Q2; 2007Q4
Bank-led: 1999Q1–Q3; 2004Q1; 2009Q3; 2010Q2; 2012Q4

Stop Episodes
Equity-led: 2000Q4; 2004Q4; 2006Q4–2007Q1; 2008Q1–Q3
Debt-led (excl. banking flows): 1997Q1–Q3; 2001Q1–Q3
Bank-led: 1996Q4; 1997Q4–1998Q2; 2008Q4–2009Q1; 2011Q3–2012Q1

[17] See Forbes and Warnock (2012). However, unlike their analysis, I distinguish "debt" from "bank" because banks are more prone to deleveraging and procyclicality, thus having more direct impact on the real sector.

Flight Episodes
Equity-led: 2007Q2–Q4
Debt-led (excl. banking flows): 1999Q2, 2005Q4; 2009Q4–2010Q2
Bank-led: 1999Q2–Q3; 2002Q4–2003Q2; 2004Q3; 2006Q1–Q2

Retrenchment Episodes
Debt-led (excl. banking flows): 1998Q1–Q2; 2008Q1–Q2; 2012Q2
Bank-led: 1996Q4–1997Q1; 1998Q3–Q4; 2002Q1–Q2; 2004Q4–2005Q2; 2008Q3–2009Q1; 2012Q3

Thus, the rise in capital flow volatility has not been uniform; debt and bank-led flows occurring most frequently. This poses challenges for financial stability. Yet, as long as ultra-easy monetary policy continues in advanced economies, capital flows are unstoppable and the threat to financial vulnerability remains. How this affects welfare, particularly when income inequality is measured, is examined next.

9.4 Financial Development and Income Inequality: A Model Framework

9.4.1 Model Framework

The links between financial development and income inequality is depicted in Fig. 9.7; the left part represents the dynamics in goods and factor markets (real sector) including exports and imports, with the right side the workings of financial markets. How the two interconnect determines the nature of the link between financial sector development and income inequality.

The real sector establishes the income generation from output production (X), with a portion covering the domestic market (D) and exports (E). Together with imports (M), those sold in the domestic market generate the total supply of goods and services (Q). In both allocations, the substitution is imperfect (not cost-less).[18] The process that generates output production (X) follows a standard input-output framework, where value added (VA) and intermediate inputs (INTM) jointly determines the level of output production (X). Expanding production networks and supply chains—where the location of production is different from the country where the intermediate inputs (INTM) are produced—suggests the need to distinguish between imported intermediate inputs (FINTM) and domestically produced intermediate inputs (DINTM). This distinction is important particularly for trade

[18] In a standard computable general equilibrium (CGE) model, for example, the allocation between the domestic market (D) and imports (M) follows Armington's constant elasticity of substitution (CES), while the allocation between domestic market (D) and exports (E) follows a constant elasticity of transformation (CET).

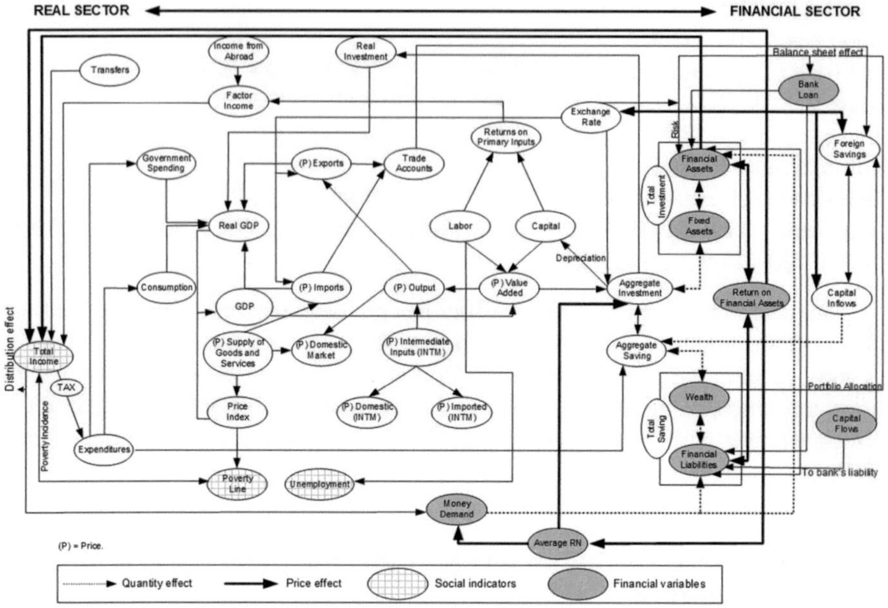

Fig. 9.7 Model framework—real-financial sector and income distribution. *Source*: Azis & Yarcia (2014) and Min (2014)

analysis in many emerging market economies where the import-content of many export products is large.[19]

The income generation is derived from the value added (VA), where returns on primary inputs of labor (L) and capital (K) are denoted by WF. In turn, these returns (WF) generate factor incomes (YF) including those from abroad (YFROW). However, total income (INC) consists of more than just factor income; it also includes transfers between agents/institutions (ITRAN). Tax payments that subtract and subsidies that add income are examples of these transfers, where size depends on the prevailing fiscal policy. Thus, income of different agents, including households, is influenced by both the level of economic activity and this non-factor income.[20] The way subsidies are allocated can have a significant impact on actual household income; typically, most subsidies go to low-income households.

[19] The dynamics of the use of imported inputs to produce exported goods, known as vertical specialization (VS), is analyzed in Hummels et al. (2001). Amador and Cabral (2009) show that vertical specialization (VS) in high-tech products has increased dramatically since the 1980s, especially in emerging Asia. Some even label it a new paradigm in the organization of world production, representing an important element of international trade.

[20] The effect of income level on macro variables works through the expenditure side. Together with government expenditure (GD) and net exports (E-M), real consumption (CD) reflects the size of agents' expenditure (EXP) out of their disposable income (YCON). The latter is determined by the income level (INC).

Yet, these two sources are still incomplete. The actual income received by rich and urban-based households holding financial assets can be well above income accrued by those who do not hold financial assets. With financial sector liberalization (FSL) and capital account liberalization (KAL), the former are far better able to reap benefits from a growing financial sector. During the 'bubble' period following capital inflows (FSAV), they benefit from the increased value of their financial assets (FIN ASSETS) as well as the income stream generated from those assets (RN), regardless of what is happening in the real economy. In many cases, this portion is larger than that generated from factor income. To the extent the financial sector often grows much faster than the real sector during a boom, the impact on income distribution can be predicted—the rich earn far more than the poor, and urban household income grows faster than rural income, both exacerbating income inequality.

After the global financial crisis, increased bank-led flows (see again Fig. 9.5) can be charted beginning with capital flows (CFLOW) in the bottom right of Fig. 9.7. Together with loans (BANKLOAN), these flows directly augment banks' financial liabilities (FIN LIABS).[21] This alters the rate of return on financial assets (RN) and financial returns received by asset holders (RN and INC are linked). Financial assets (RN) also have a two-way relationship with the size and composition of different agents' assets. Fixed assets (FIXAS) will be used directly for real sector investment (INVEST), such as in buildings, machinery, etc., while the rest—including financial assets (FIN ASSETS)—may move indirectly via financial markets, as for example, funds from equity issuance are used for business investment. Along with government spending (GD), consumption expenditure (EXP or CD), exports (E), and imports (M), this investment (ID) in real terms generates gross domestic product (RGDP).[22]

Increased capital flows (CFLOW) [captured through foreign savings SAV(fr)] also have macro-financial impacts: pressuring the currency via the exchange rate (EXR) to appreciate. The resulting trade account (TA) may thus worsen due to falling exports (E) and increased imports (M). In reality, however, almost all emerging market economies with large capital inflows respond by imposing some sort of capital controls—either directly (through taxes or levies for example) or indirectly (sterilized market intervention). This explains why net-exports in some countries continue to grow despite increased capital inflows. When net-exports shrink, the growth of consumption (CD) and investment (ID) can also offset the decline.

The resulting higher real gross domestic product (RGDP) fuels further financial sector growth either from strong fundamentals or simply market expectations. This

---

[21] Note that lending (BANKLOAN) is not only determined by the size of a bank's available funds, but also by changes in net worth and external finance premia of both borrowers and lenders; this "credit channel" hypothesis was elaborated in Bernanke et al. (1996), Adrian and Shin (2009), Stiglitz and Greenwald (2003), and Stiglitz (2001).

[22] Other financial variables can also affect aggregate economic activity by way of the money market.

further enhances rich household income along with savings (SAVING) or wealth (WEALTH), providing them with an additional income stream from financial returns.[23] If, through the portfolio allocation, the increased wealth is reinvested in financial instruments with lucrative returns, the financial assets and earnings of rich households increase yet again. In this way the magnitude of the growth-inequality nexus is amplified through this feedback cycle.[24]

The above framework clearly shows that in addition to standard factors like technology, globalization, education, and domestic institutions, rising inequality can be exacerbated by the non-inclusive nature of financial sector growth.

9.4.2 Model Verification

To verify the model framework, I use a financial general equilibrium model (FCGE) for one emerging Asian economy, Indonesia.[25] The model is the evolution of FCGE developed by the author since late 1990s (Azis 1997), subsequently applied to the dynamics of manufacturing sector in Azis (2000), and for impact analysis of financial crisis on socio-economic conditions in Azis (2003). With improved monetary block, the model was used to explain the disconnect between financial and real sector in Azis (2004). In Azis (2008) the author used the model to analyze debt sustainability by specifying debt maturity and debt re-profiling. More detailed household income distribution and poverty module were added in Azis (2009). Most parameters are calibrated on the actual data using the (non-linear) model specifications, while others are econometrically estimated. The validity and the predictive power of the model are tested by plotting actual data on some exogenous variables.

The model version used in the current study is based on the most recent available Financial Social Accounting Matrix (FSAM) and a more detailed specification of capital flows. It incorporates a credit channel component that includes the financial structure of lenders and borrowers in determining bank willingness to lend, and the amplified effect due to currency appreciation. The channel connecting financial

[23] Changes in the exchange rate also cause some valuation effects: the local currency value of any assets denominated in foreign currencies will increase (decrease) when local currency appreciates (depreciates).

[24] Aside from income inequality, poverty and unemployment (UEMPR) are two other social indicators endogenously determined in the model. While unemployment is derived from the difference between labor demand (L) and fixed labor supply, the aggregate variables in the real sector (X, D, E, M, and Q) are all determined along with their respective prices (PX, PD, PE, PM, and PQ). It is PQ that sets the overall price index (PINDEX). The poverty line (PL) can be derived from this. When PL is matched with the endogenously determined household income, the poverty level can be estimated.

[25] Indonesia is selected because like most emerging markets, its financial sector has been growing rapidly since financial liberalization began in the 1980s, and income inequality has worsened. Indonesia also became a primary destination for capital flows once advanced economies set ultra-easy monetary policy.

flows and income distribution is specified in detail by dissecting the flows as they appear on agents' balance sheets based on the type of income generated. Scrutinizing the role and detailed transmission within the financial sector allows us to analyze the dynamics of income earned from returns on financial assets held mostly by urban-based rich households.[26]

Given the shock from increased bank-led flows, two scenarios are constructed: one where recipient banks increase risk by investing in financial assets, particularly securities and equity markets; and the other where recipient banks spend the additional funds more prudently, by using them to strengthen credit and safe assets. In the first scenario (Figs. 9.8a, b, c), real GDP is only slightly higher than the baseline. So are investment and consumption. Inflation and unemployment rates are lower. However, the trade sector suffers: exports decrease and imports increase due to currency appreciation. Looking more closely, appreciation derives from higher interest rates, their level influenced by returns on financial assets. Because the issuance of financial assets increases under this scenario, prices will fall and yields rise, with interest rates also increasing. In the search for higher returns and yields, banks actively invest in these new assets instead of issuing more credit. This explains why the economy grows only slightly.

The effect on income distribution is far more obvious: it gets worse. Whether measured by disparity between rich and poor, or in terms of the rural/urban gap, inequality grows (Fig. 9.9a, b). Although the poverty line drops 1.2 % below the baseline, incomes for all household categories fall, despite growing GDP. Two factors are behind this: (1) wages fall due to lower prices, and (2) economic growth is mostly driven by expanding financial sector-related activity. These tend to benefit only urban-rich households whose depend far less on wages (factor income) than the rural-based poor. With more access to financial markets, the urban-rich accrue extra income from returns on the financial assets they hold. This is why increased bank-led flows under the risky behavior scenario worsen income inequality. And as expected, the largest deterioration is in financial income.

How does this compared with a scenario of prudent bank behavior? While real GDP in both cases is higher than the baseline, the increase is larger when banks act more prudently. There is higher growth in investment and consumption due to lower interest rates, and higher exports due to nominal and real exchange rate depreciation (see Fig. 9.8a, b). Unemployment rate is also much lower, though the price index is higher (see Fig. 9.8c). Thus, the macroeconomic impact is better when banks behave prudently.

Unlike the previous case, income in all household categories increases. More importantly, all factor incomes are higher. As increased liquidity from bank-led flows is not spent on financial assets, the urban rich do not receive extra income

[26] The model version used in the analysis here follows Min (2014). Due to space constraint, the detailed explanations of the model and simulation results are not shown here (they are available upon request).

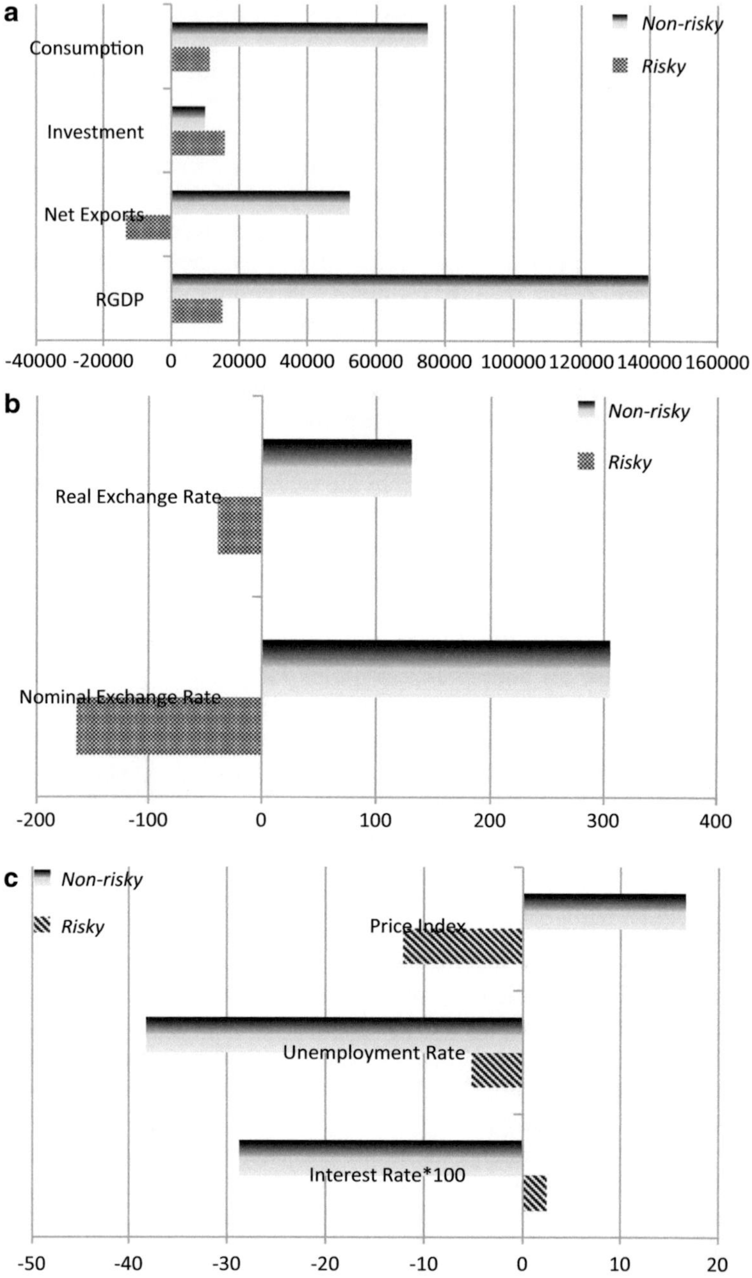

Fig. 9.8 (**a**) Aggregate impact of increased bank-led flows. *Source*: Results of model simulations. (**b**) Impact of increased bank-led flows on the exchange rate. *Source*: Results of model simulations. (**c**) Impact of increased bank-led flows on prices, interest rates, and unemployment rate. *Source*: Results of model simulations

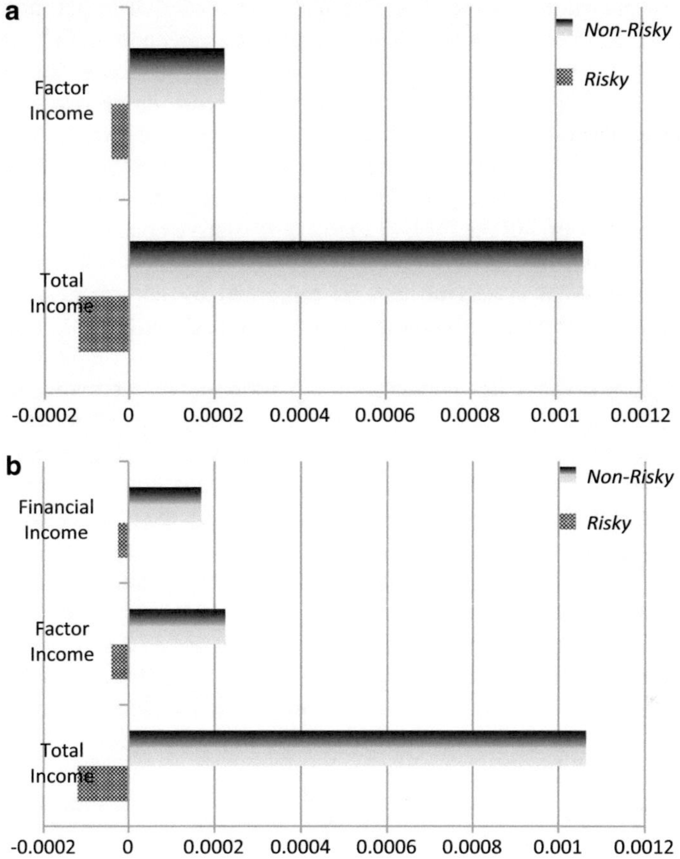

Fig. 9.9 (**a**) Impact of increased bank-led flows on poor/rich income rati. *Source*: Results of model simulations. (**b**) Impact of Increased Bank-Led Flows on Rural/Urban Income Ratio. *Source*: Results of model simulations

from these assets. As a result, overall income inequality between rich and poor, and between rural and urban households, narrows (see Fig. 9.9a, b).

It is abundantly clear that the repercussions of increased bank-led flows depend on how banks react. It is much better when banks act prudently and do not take on increased risk. The problem is, there is no guarantee banks will behave that way. Given the high frequency of financial crises worldwide, especially since the 1990s, it implies that most financial institutions on the receiving end of capital inflows tend to take on more risk. This all suggests policies need to adjust.

9.4.3 What's Next

Despite the risks and potential damage capital flows can cause, no country can stop them—especially when the spread of returns (interest rates) is large and push factors (slow growth in advanced economies) are substantial. What most countries can do and have done is to refocus their policy on the asset and liability side of bank balance sheet.[27] When external shocks strike, as in the case of European bank deleveraging during the Eurozone crisis, bank credit flows can be disrupted. Theoretically, outflows can be matched by retrenchment—returning foreign assets owned by domestic investors. The size of these assets, however, can only rise if capital outflows are encouraged. On the preventive side, some sort of capital controls can also help—in the form of direct quantitative controls such as imposing a levy on bank-led flows. In the context of regional integration and cooperation, the risk of integration-led contagion can be mitigated by complementing domestic financial safety nets with collective safety nets regionally.[28] The question is, which policy works best?

Three policy options are considered: (1) promote direct investment abroad, labeled "Encourage Outflows" (Fig. 9.10); (2) "Assign Levy" to non-core bank liabilities; and (3) strengthen regional financial safety nets, "Reg Safety Nets." The logic of regional financial safety nets is to support domestic safety nets, as these are far too inadequate given the potential damage caused by the unprecedented size and volatility of capital flows. The rationale for assigning a levy is to restrain rather than stop capital flows. Encouraging capital outflows helps maintain stability of net flows. In times of crisis, when capital tends to flow out during the boom-and-bust cycle, assets held abroad by domestic investors can act as a safeguard. They can provide a foreign asset buffer when markets become volatile. Indeed, the size of these ready-to-use foreign assets was important in some emerging market economies during the global financial crisis; the Republic of Korea is one example (Jain-Chandra et al. 2013).

I use the Analytic Network Process (ANP) to structure the model and quantify the weight of each model element [see Saaty and Vargas (2005)] for a detailed explanation about ANP). Achieving a balanced outcome of MACRO STABILITY, FINANCIAL STABILITY and improved SOCIAL ISSUES, depicted at the top of Fig. 9.10, is the strategic comparative goal. Each policy is weighted in terms of its relevance and contribution to BENEFIT, OPPORTUNITY, COST, and RISK (BOCR) that can be generated by increased bank-led flows. In the BENEFIT cluster, two sets of components are considered: (1) strengthen LIQUIDITY (the first box on the left of Fig. 9.10), through enhanced short-term securities and equity markets, along with boosted financial income; and (2) allow investment, consumption, financial income, and imported intermediate inputs to EXPAND (the second box on the left of Fig. 9.10). Some beneficial impacts of increased bank-led flows, such as improved CAPITAL MARKET, and enhanced RESILIENCE may emerge

[27] On the asset side, other than reducing loan-to-value ratio, efforts to contain excessive credit expansion and other risky investments are also made. On the liability side, mitigating the increase of non-core liabilities through bank-led flows is critical because they can heighten risky bank behavior and increase leverage. See Azis and Shin (2013) and Forbes and Warnock (2012).

[28] For the status of Asia's regional financial safety nets, see Azis (2012).

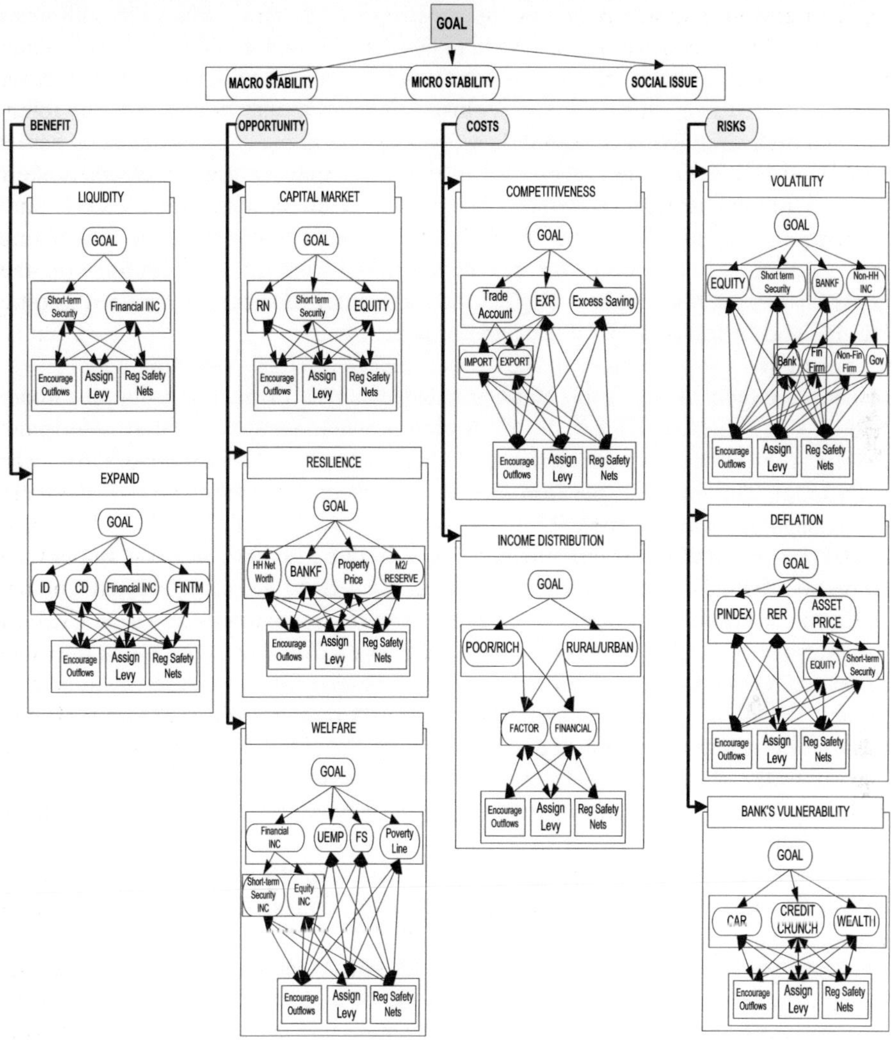

Fig. 9.10 Analytic network process (ANP) structure for policy options

only in the long-run. Recipient countries can also improve overall WELFARE, after a certain period, if they take advantage of the increased capital inflows. All these long-term benefits are grouped in the OPPORTUNITY cluster. The components connecting the GOAL and policies in Fig. 9.10 are most relevant and should be considered in prioritizing policies. For example, given an increase in bank-led flows, improvements in WELFARE can be fueled by a gradual increase in the financial income originating in short-term securities and equity earnings.

On the downside of bank-led flows, the short- and long-term costs (COST and RISK, respectively) are analyzed similarly, except that the priority ranking is based on the question: "which policy is the most costly or risky" when the following components are considered: COMPETITIVENESS and INCOME DISTRIBUTION under the COST cluster, and VOLATILITY, DEFLATION, and BANK VULNERABILITY under the RISK cluster. It is important to note that the reference for analyzing RISK is capital flow reversal—as in many crisis episodes with a boom-and-bust cycle, the biggest risk in massive capital inflows is precisely their reversal (procyclicality). However, one needs to distinguish this reversal from outflows by domestic investors. While useful in times of crisis, which is why one policy option is to encourage them, a flow reversal from investors pulling out will generate detrimental capital "flight" (see again the distinction between capital "flight" and "retrenchment" discussed earlier).

The policies at the bottom of the network in Fig. 9.10 are weighted with respect to each component and sub-component listed above them. For example, under BANK VULNERABILITY in the RISK cluster, where bank capital may deteriorate during a flow reversal, there is a risk that a bank's capital adequacy ratio (CAR) will deteriorate. The relevant question then is: Which of the three policies will likely create such a risk (most risky)?

All arrows under each component in Fig. 9.10 point in two directions, implying a feedback effect for every influence from an element to the other elements below it. Thus, the structure in each box under each cluster forms a network. Again, taking the example of BANK VULNERABILITY in the RISK cluster, a typical question to ask is: "Given a selected policy, which risk is least likely to be resolved by that policy?" Applying pairwise-comparisons, priority rankings for each feedback was made. The inputs used are a combination of the normalized quantitative data derived from the FCGE model simulations and analytical perceptions. The rankings based on the complex network structure are derived from the limiting super-matrix (see Appendix).

Table 9.1 shows the results of priority rankings for the three policies under the BOCR.[29] Thus, while to "Encourage Outflows" ranks highest in terms of its capacity to generate BENEFIT and OPPORTUNITY, the policy is also considered most costly and risky. For example, compared with "Assign Levy" and "Reg Safety Nets," "Encourage Outflows" will do the least in avoiding decreased competitiveness caused by the appreciation of real exchange rate (RER). On the RISK side, capital flow reversals may cause VOLATILITY in the EQUITY market. To "Encourage Outflows" will obviously make things worse.

Having calculated the above priorities, the next step is to apply them to some BOCR formula. Two types are used here: (1) the multiplicative approach (B $\times$ O)/

[29] For example, under the BENEFIT scenario in Table 9.1, three eigen-vectors are shown ("Ideals," "Normals," and "Raw"). While all three give the same ranking, i.e., encourage outflows being most preferred, followed by assigning levies, and regional financial safety nets (hence the ranking shown in the last column of Table 9.1), the normalized eigen-vector (0.4381; 0.4358; and 0.1261) under "Normal" with the sum equals unity is the most often used. All numbers under the column "Benefit," "Opportunity," "Cost," and "Risk" in Table 9.2 show the normalized eigen-vector.

Table 9.1 ANP results for benefit (B), opportunity (O), cost (C), and risk (R)

	Ideals	Normals	Raw	Ranking
Benefit				
1. Encourage Outflows	1	0.438129	0.858812	1
2. Assigning Levies	0.994591	0.435759	0.854167	2
3. Reg Safety Nets	0.287844	0.126113	0.247204	3
Opportunity				
1. Encourage Outflows	1	0.477338	0.826065	1
2. Assigning Levies	0.712725	0.340211	0.588757	2
3. Reg Safety Nets	0.382225	0.182451	0.315742	3
Cost				
1. Encourage Outflows	1	0.725513	1	1
2. Assigning Levies	0.08878	0.064411	0.08878	3
3. Reg Safety Nets	0.289555	0.210076	0.289555	2
Risk				
1. Encourage Outflows	1	0.488161	0.983803	1
2. Assigning Levies	0.313358	0.152969	0.308283	3
3. Reg Safety Nets	0.735147	0.35887	0.72324	2

Source: Results of ANP

(C × R) and (2) the additive approach (bB + oO-cC-rR); for the rationale of both, see Saaty and Vargas (2006).

Table 9.2 lists the results. The upper panel equally ranks BOCR, with the last column shows the superiority of "Assign Levy." The middle and lower panels display the results of sensitivity analyses; the middle reflects a more subdued option, where COST and RISK clusters are weighted more than BENEFIT and OPPORTUNITY, with the reverse case shown in the bottom panel, representing a 'buoyant' scenario. In either case, the highest preference for "Assign Levy" remains. Only the ranking of the other two policies is reversed when an additive approach is used. This suggests the superiority of placing a levy on bank-led flows is robust [see also Azis and Shin (2013)].

9.5 Summary

The global financial crisis and the eurozone crisis made many analysts question the virtue of financial liberalization and integration. These qualms coincide with dramatic changes in the global economy and economic thinking, particularly about the role of capital flows. Adding to the mix is growing income inequality in developed and developing countries alike. For a region where integration has been intensifying, where massive capital inflows came in response to easy money and low interest rates in advanced economies, and where income disparity is rising despite rapid recovery and strong growth, Asia is studied by taking a more balanced view of integration.

Table 9.2 Overall results based on multiplicative and additive BOCR

	Benefit	Opportunity	Cost	Risk				
ANP	b = 0.25	o = 0.25	c = 0.25	r = 0.25	(B × O)/(C × R)	Ranking	bB + oO − cC − rR	Ranking
Encourage outflows	0.438129	0.477338	0.725513	0.488161	0.590499767		−0.07455	3
Assign levy	0.435759	0.340211	0.064411	0.152969	15.04635304		0.1396	1
Reg safety nets	0.126113	0.182451	0.210076	0.35887	0.305205609		−0.0651	2
	b = 0.25	**o = 0.1**	**c = 0.35**	**r = 0.3**				
Encourage outflows	0.438129	0.477338	0.725513	0.488161	0.590499767		−0.2431	3
Assign levy	0.435759	0.340211	0.064411	0.152969	15.04635304		0.0745	1
Reg safety nets	0.126113	0.182451	0.210076	0.35887	0.305205609		−0.13141	2
	b = 0.35	**o = 0.3**	**c = 0.25**	**r = 0.1**				
Encourage outflows	0.438129	0.477338	0.725513	0.488161	0.590499767	3	0.06635	2
Assign levy	0.435759	0.340211	0.064411	0.152969	15.04635304	1	0.22318	1
Reg safety nets	0.126113	0.182451	0.210076	0.35887	0.305205609	2	0.010469	3

Source: Results of BOCR, based on ANP

Unlike in Europe, the process of integration in Asia is more market-driven, institution-light and bottom-up. Given the region's huge diversity of development level, the essence of Asian integration is to manage the diversity, not to seek for a "United Asia". To the extent greater integration poses risks—including financial contagion—both benefits and costs should be gauged carefully. The ultimate goal remains improved welfare. The market-driven process of integration needs to be carefully managed to reap its benefits while minimizing potential costs. It is argued that better and effective regional cooperation can indeed better manage integration and its potential risks. Strengthening financial safety nets is one important example of regional cooperation, given the uncertainty and volatility in financial markets.

On the seeming disconnect between financial sector growth and income inequality, I delve into the interconnections between financial and real sectors by simulating a general equilibrium model with financial module under a scenario of increased bank-led flows as experienced by many emerging market economies. When a large portion of the enhanced liquidity caused by capital inflows is invested by banks in financial assets and other risky loans, the impact on overall welfare is unfavourable. In particular, household income falls, most dramatically on the rural-poor. The only rising income derives from returns on financial assets, which is why the income gap between the rich and the poor widens, and the disparity between rural and urban households worsens. As investing in financial assets—as opposed to lending to firms—has a lower employment-generating capacity, the drop in unemployment is smaller than when banks act prudently. Indeed, prudent behavior is better from both macro-financial stability and income distribution perspectives.

The policy analysis suggests that during the tranquil period capital outflows should be encouraged to help stabilize net flows in times of market turmoil, at the same time strengthening competitiveness as the exchange rate weakens. But after taking into account the costs and the risks of such measure, imposing levy on bank-led flows is preferable. The resulting stable financial market feeds into the real economy, boosting factor income rather than returns on financial assets, suggesting that it will also reduce inequality. Through some sensitivity tests, the result is found to be robust. Clearly, taking one-sided approach in evaluating policy alternatives by neglecting the potential costs and risks of the policies may produce a sub-optimal result.

Appendix

The presence of feedback influences in a network model requires a large matrix—known as supermatrix—that contains a set of sub-matrixes. The supermatrix captures the influence of elements in a network on other elements in that network. Denoting a cluster by C_h, $h = 1, \ldots\ldots m$, and assuming that it has n_h elements $e_{h1}, e_{h2}, e_{h3} \ldots\ldots\ldots, e_{hmh}$, and laying out all the clusters and all the elements in each cluster both vertically on the left and horizontally at the top, we have the following supermatrix:

Supermatrix of a Network

$$
W = \begin{array}{cc} & \begin{array}{cccc} C_1 & C_2 & \cdots & C_N \\ e_{11}e_{12}\cdots e_{1n_1} & e_{21}e_{22}\cdots e_{2n_2} & & e_{N1}e_{N2}\cdots e_{Nn_N} \end{array} \\ \begin{array}{c} C_1 \begin{array}{c} e_{11} \\ e_{12} \\ \vdots \\ e_{1n_1} \end{array} \\ C_2 \begin{array}{c} e_{21} \\ e_{22} \\ \vdots \\ e_{2n_2} \end{array} \\ \vdots \\ C_N \begin{array}{c} e_{N1} \\ e_{N2} \\ \vdots \\ e_{NnN} \end{array} \end{array} & \begin{bmatrix} W_{11} & W_{12} & \cdots & W_{1N} \\ W_{21} & W_{22} & \cdots & W_{2N} \\ \vdots & \vdots & \cdots & \vdots \\ W_{N1} & W_{N2} & \cdots & W_{NN} \end{bmatrix} \end{array}
$$

The typical entry of this supermatrix is

Entries in Supermatrix of a Network

$$
W_{ij} = \begin{bmatrix} W_{i1}^{(j_1)} & W_{i1}^{(j_2)} & \cdots & W_{i1}^{(jn_j)} \\ W_{i2}^{(j_1)} & W_{i2}^{(j_2)} & \cdots & W_{i1}^{(jn_j)} \\ \vdots & \vdots & & \vdots \\ W_{in_i}^{(j_1)} & W_{in_i}^{(j_2)} & \cdots & W_{in_i}^{(jn_j)} \end{bmatrix}
$$

The entries of sub-matrixes in W_{ij} are the ratio scales derived from paired comparisons performed on the elements within the clusters themselves according to their influence on each element in another cluster (outer dependence) or elements in their own cluster (inner dependence). Judgments are elicited from which ratio scales are derived. The resulting unweighted supermatrix is then transformed into a matrix where each column sums to unity to generate a stochastic supermatrix. The derived weights are used to weight the elements of the corresponding column blocks (cluster) of the supermatrix, resulting in a weighted supermatrix which is also stochastic. The final ranking is derived from the limiting supermatrix, obtained by raising the stochastic supermatrix to large powers to read off final priorities, in which all matrix columns are identical. Each gives the relative priorities of the elements from which the priorities of the elements in each cluster are normalized to one (the powers of the supermatrix do not converge unless it is stochastic, ensuring that its largest eigenvalue is one). Using the example of the EXPAND component under the BENEFIT cluster in Fig. 9.10, the resulting limiting supermatrix is

Limiting super matrix

		Alternatives	Goal	EXPAND					
		Encourage Outflows	Assign Levy	Reg Safety Nets	EXPAND	CD	FIN INC	FINTM	ID
Alternatives	Encourage Outflows	0.18477	0.18477	0.18477	0.18477	0.18477	0.18477	0.18477	0.1848
	Assign Levy	0.23605	0.23605	0.23605	0.23605	0.23605	0.23605	0.23605	0.2361
	Reg Safety Nets	0.07918	0.07918	0.07918	0.07918	0.07918	0.07918	0.07918	0.0792
Goal	EXPAND	0	0	0	0	0	0	0	0
EXPAND	CD	0.0745	0.0745	0.0745	0.0745	0.0745	0.0745	0.0745	0.0745
	FIN INC	0.15758	0.15758	0.15758	0.15758	0.15758	0.15758	0.15758	0.1576
	FINTM	0.13292	0.13292	0.13292	0.13292	0.13292	0.13292	0.13292	0.1329
	ID	0.135	0.135	0.135	0.135	0.135	0.135	0.135	0.135

References

Adrian T, Shin HS (2009) Money, liquidity and monetary policy. Am Econ Rev 99(2):600–605

Alderson AS, Nielsen F (2002) Globalization and the great U-Turn: income inequality trends in 16 OECD countries. Am J Sociol 107:1244–1299

Amador J, Cabral S (2009) Vertical specialization across the world: a relative measure. North Am J Econ Financ 20:267–280

Arestis P, Paliginis E (1995) Divergence and peripheral fordism in the European Union. Rev Soc Econ 53:261–283

Armstrong HW (1995) An appraisal of the evidence from cross-sectional analysis of the regional growth process within the European Union. In: Armstrong HW, Vickerman RW (eds) Convergence and divergence among European regions. Pion, London

Asian Development Bank (2012) Asian development bank outlook 2012: confronting rising inequality in Asia. Manila

Azis IJ (1997) Impacts of Economic reform on rural-urban welfare: a general equilibrium framework. Rev Urban Reg Develop Stud 9(1):1–19

Azis IJ (2000) The non-linear general equilibrium impact of the financial crisis and manufacturing downfall. Develop Econ 38(4):518–546

Azis IJ (2003) Modeling the repercussions of financial shock on socio-economic indicators. In: Ananta A (ed) The Indonesian crisis: a human development perspective. Institute of Southeast Asian Studies, Singapore

Azis IJ (2004) Modeling the disconnect between financial policy and real sector using the financial social accounting matrix as a data system: the case of Indonesia. In: Charles J (ed) Banking disintermediation and its implication to monetary policy: a theoretical view and country experiences bank. Indonesia, Jakarta

Azis IJ (2007) Articulating the benefits and costs of regional financial arrangement in East Asia. In: Simorangkir I (ed) Global imbalances and their impact on emerging market economies: issues and challenges. Bank Indonesia, Jakarta

Azis IJ (2008) Fiscal policy and workouts from debt crises: the case of Indonesia's domestic debt. IPD Working Paper, Columbia University, New York, pp 1–25

Azis IJ (2009) Regional financial arrangement. In: Azis IJ (ed) Crisis, complexity and conflict. Emerald, London

Azis IJ (2012) Asian regional financial safety nets? Don't hold your breath. Pub Policy Rev 8(3): 321–340

Azis IJ, Shin HS (2013) How do global liquidity phases manifest themselves in Asia? Asian Development Bank, Manila

Azis IJ, Yarcia D (2014) How capital flows affect economy-wide vulnerability and inequality: flow-of-funds analysis of selected Asian economies. ADB Working Paper Series on Regional Economic Integration, No 136

Backus DK, Kehoe JP, Kydland FE (1992) International real business cycles. J Polit Econ 100(4): 745–775

Bai Y, Jin Z (2005) Financial integration and international risk sharing. Working Paper, Department of Economics, Arizona State University

Baysan T, Arvind P, Nihal P (2006) Preferential trading in South Asia. World Bank Policy Research Working Paper Series, 3813. World Bank, Washington, DC

Beckfield J (2009) Remapping inequality in Europe: the net effect of regional integration on total income inequality in the European Union. Int J Comp Sociol 50(5–6):486–509

Ben-David D (2001) Trade liberalization and income convergence: a comment. J Int Econ 55: 229–234

Bernanke BS, Gertler M, Gilchrist S (1996) The financial accelerator and the flight to quality. Rev Econ Stat 78(1):1–15

Borensztein E, Loungani P (2011) Asian financial integration: trends and interruptions. IMF Working Paper, No. 11/4, January, 2011

Bourguignon F (2004) The poverty-growth-inequality triangle. Paper Presented at the Indian Council for Research on International Economic Relations, New Delhi

Brady D (2003) The politics of poverty: left political institutions, the welfare state and poverty. Soc Forces 82(2):557–588

Cameron DR (1978) The expansion of the public economy: a comparative analysis. Am Polit Sci Rev 72:1243–1261

Canova F, Ravn MO (1996) International consumption risk sharing. Int Econ Rev 37(3)

CIEPR-Committee on International Economic Policy and Reform (2012) Banks and cross-border capital flows: policy challenges and regulatory responses. Washington, DC

Cochrane JH (1991) A simple test of consumption insurance. J Polit Econ 99(5):957–976

Forbes KJ, Warnock F (2012) Capital flow waves: surges, stops, flight and retrenchment. J Int Econ 88(2):235–251

French KR, Poterba JM (1991) Investor diversification and international equity markets. American Economic Review Papers and Proceedings, vol 81, pp 222–226

Hummels D, Ishii J, Yi K-M (2001) The nature and growth of vertical specialization in world trade. J Int Econ 54(1):75–96

International Monetary Fund (2012) The liberalization and management of capital flows: an institutional view. International Monetary Fund, Washington, DC

Jain-Chandra S, Kim MJ, Park SH, Shin J (2013) The impact of Foreign Bank deleveraging on Korea. IMF Working Paper, WP/13/101, Washington DC

Leonardi R (1995) Convergence, cohesion, and integration in the European Union. St. Martin's Press, New York

Levine R (2001) International financial liberalization and economic growth. Rev Int Econ 9(4): 668–702

Machiko N, Thorbecke E (2010) The poor under globalization in Asia, Latin America, and Africa. Oxford University Press, New York
Martin XS (1996) Regional cohesion: evidence and theories of regional growth and convergence. Eur Econ Rev 40:1325–1352
Mckinnon R (2012) Zero interest rates in the United States provoke World Monetary instability and constrict the US economy. SIEPR Policy Brief, Stanford University
Mendoza EG (1995) The terms of trade, the real exchange rate and economic fluctuations. Int Econ Rev 36:101–137
Min HH (2014) The analysis of the link between capital flows and macroeconomic-financial-spatial income distribution indicators: combining financial CGE and perception models. PhD Dissertation in Regional Science. Cornell University
Obstfeld M (1994) Risk-taking, global diversification, and growth. Am Econ Rev 84: 1310–1329
Rodrik D (2011) The globalization paradox. Making it. 24 August
Saaty TL, Vargas L (2005) Theory and applications of the analytic network process: decision making with benefits, opportunities, costs, and risks. RWS Publication, Pittsburgh, PA
Saaty TL, Vargas L (2006) Decision making with the analytic network process. Springer's International Series
Scharpf FW (1997) Economic integration, democracy, and the welfare state. J Eur Pub Policy 4: 18–36
Slaughter MJ (2001) Trade liberalization and per capita income convergence: a difference-in-differences analysis. J Int Econ 55:203–228
Stiglitz JE (2001) Information and the change in the paradigm in economics. Prize Lecture, Nobel Foundation, Stockholm
Stiglitz JE (2010) Risk and global economic architecture: why full financial integration may be undesirable. NBER Working Paper Series. No 15718. NBER, Cambridge, MA)
Stiglitz JE, Greenwald BC (2003) Towards a new paradigm in monetary economics. Cambridge University Press, Cambridge, MA
Stiglitz J, Amartya Sen E, Fitoussi J-P (2010) Mis-measuring our lives: why GDP doesn't add up/the report by the commission on the measurement of economic performance and social progress. The New Press, New York
Stolper W, Samuelson PA (1941) Protection and real wages. Rev Econ Stud 9(1):58–73
Valentina B, Shin HS (2012) Capital flows and the risk-taking channel of monetary policy. BIS working papers, No 400, December
Venables AJ (2009) Economic integration in remote resource-rich regions. OxCarre Working Papers No 022. Oxford Centre for the Analysis of Resource Rich Economies, Oxford
Western B (1997) Between class and market: postwar unionization in capitalist democracies. Princeton University Press, Princeton, NJ

Spatial Impact of Transportation Infrastructure: A Spatial Econometric CGE Approach

10

Zhenhua Chen and Kingsley E. Haynes

Abstract

This study introduces a new method called Spatial Econometric Computable General Equilibrium (SECGE), which integrates both spatial econometrics with computable general equilibrium modeling to improve the effectiveness of impact analysis on transportation infrastructure. Elasticities of factor substitution for the Constant Elasticity Substitution (CES) production function are estimated in a spatial econometric model with consideration of spatial dependence. CGE simulations adopting different substitution elasticities show a difference between spatial econometric estimation and traditional OLS estimation. Although the differences are relatively small in this aggregate case study, implications for more sensitive disaggregated regional models are clear.

10.1 Introduction

Transportation infrastructure plays an important role in regional economic development both in the stimulation of growth and as a response to output expansion (Haynes 2006). However, measuring these effects has proved difficult due to the complicated transmission mechanisms of transportation infrastructure. This complication is due to two reasons: first, regional impacts of transportation infrastructure are achieved through a mechanism that involves both a demand influence through the variation of transportation price and a supply influence implemented through the variation of transportation cost; second, impacts of transportation are usually evaluated in a regional context where the presence of unobserved local or regional variables may give rise to spatial autocorrelation. As a result, if not taking these interactions into account, impact analysis may become biased and spurious. This study introduces a new method called Spatial Econometric Computable

Z. Chen (✉) • K.E. Haynes
School of Public Policy, George Mason University, Arlington, VA 22201, USA
e-mail: zchen7@gmu.edu; khaynes@gmu.edu

P. Nijkamp et al. (eds.), *Regional Science Matters*, DOI 10.1007/978-3-319-07305-7_10,

General Equilibrium (SECGE), which integrates both spatial econometrics with equilibrium modeling to improve the effectiveness of impact analysis. We apply this new method to transportation infrastructure.

This study differs from previous studies in the following three aspects: First, through a spatial autocorrelation test, the presence of spatial dependence is observed and confirmed among the elasticities of factor substitution in the US. To deal with spatial dependence, spatial panel econometric techniques are introduced to estimate the elasticity of factor substitution of different sectors for the Constant Elasticity of Substitution (CES) production function with consideration of spatial direct and indirect effects.

Second, transportation impact analysis is conducted under different scenarios in a general equilibrium framework. Unlike partial equilibrium analysis, general equilibrium analysis allows researchers to obtain a more comprehensive understanding of transportation infrastructure's impacts given its consideration of interactions between demand and supply. The study validates the method by comparing traditional equilibrium simulation without controlling for spatial dependence and the new equilibrium simulation with consideration of spatial dependence. The comparison allows researchers to appreciate the spatial impacts of transportation infrastructure.

Third, the study is conducted with a focus on multimodal transportation systems that includes: road, rail, air, public transit, pipeline and water. Unlike a unimodal perspective, this multimodal perspective is essential to achieve a comprehensive understanding of the investment impacts in the transportation infrastructure system as a whole. It also enables us to compare impacts and their spillovers between different types of infrastructure and understand the relative importance of transportation investment by mode.

The rest of this paper is organized as follow. Section 10.2 lays a theoretical foundation for the study from relevant literature. Section 10.3 discusses the CGE structure. The modeling procedure is discussed in Sect. 10.4. Section 10.5 introduces the data and Sect. 10.6 presents simulation results, which is followed by a conclusion in Sect. 10.7.

10.2 Literature Review

The literature regarding economic impacts of transportation infrastructure can be reviewed from three fields based on the different analytical approaches: the traditional econometric approach, a spatial perspective and the general equilibrium approach. Further, the necessity for a multimodal investigation is also reviewed.

10.2.1 Traditional Econometric Approach

Since a series of papers by Aschauer (1989, 1990, 1994) argued that enhancing public infrastructure expenditure facilitated region achieving its economic potential, a large number of studies on public infrastructure have been conducted following a neo-classical approach using various forms of aggregate production functions (Gramlich 1994, 2001; Harmatuck 1996; Nadiri and Mamuneas 1996; Fernald 1999; Bhatta and Drennan 2003; Boarnet 1997; Boarnet and Haughwout 2000; Mattoon 2002; Duffy-Deno and Eberts 1991). These studies have been subjected to a variety of criticisms (Gramlich 1994). One of the criticisms is that these studies did not consider spatial interactions among units across geographic locations. Instead these studies assumed spatial independence among their research units.

10.2.2 Spatial Perspective

Spatial econometric theory indicates that estimation outcomes vary significantly if spatial dependence is considered in regional analysis (LeSage and Pace 2009). This is because regional performance is influenced by a region's own as well as other regions' performance. Munnell (1992) indicated that the estimated impact of public capital becomes smaller as the geographic focus narrows. She believed that this is because of the effects of leakages from an infrastructure investment could not be captured at a small geographic scale. Although this hypothesis may not be entirely accurate, as indicated by Boarnet (1998), it does suggest that the spatial dimension has influence on estimation and should not be neglected.

LeSage (1999) emphasized that traditional econometrics has largely ignored the spatial dimension of sample data. When data has geographic information, the issue of spatial dependence between observations may violate the Gauss-Markov theorem. Without considering this issue, estimation results may be statistically biased.

Thanks to the development of spatial econometric techniques by Paelinck and Klaassen (1979), Cliff and Ord (1981), Anselin (1988), LeSage and Pace (2009), Elhorst (2012) and many others, methodologies for empirical spatial analysis were developed. One of the dominant functions is to allow for the measuring of spatial spillover effects. These effects refer to the situation in which the input in one sector or region influences changes in neighboring local economies through trade linkages and market relationships (Bo et al. 2010). Transportation infrastructure may have spillover effects on regional economic growth because the impacts generated from infrastructure are not confined to that specific region (Moreno and López-Bazo 2007). To test the hypothesis empirically, different spatial models were adopted (Holtz-Eakin and Schwartz 1995; Kelejian and Robinson 1997; Cohen and Morrison Paul 2003, 2004).

Reviews of the existing literature regarding the spatial impacts of transportation infrastructure (Boarnet 1998; Holtz-Eakin and Schwartz 1995; Kelejian and Robinson 1997; Ozbay et al. 2007; Moreno and López-Bazo 2007; Cohen and Morrison Paul 2003, 2004; Cohen 2007; Mohammad 2009; Chen and Haynes 2013) show that the conclusions are not consistent given the fact that different data, methods, regions and periods were used. Despite the development of spatial econometric techniques enabling scholars to investigate spillover effects of infrastructures, most studies only considered certain types of spatial dependence in isolation as either spatial lag or spatial error. Without an adequate interpretation of the reasons for selecting a specific spatial model, it is possible that results of even these spatial studies may have estimation bias due to the neglect of a specific kind of spatial dependence.

10.2.3 General Equilibrium Approach

Furthermore, most of the aforementioned impact analyses are conducted under the partial equilibrium framework. The relationships between economic output and infrastructure are normally evaluated from the supply side by assuming a constant demand for infrastructure during the research period. As a matter of fact, the outcome of impact evaluation may be incomplete since impacts caused from the change in demand are not considered. For instance, transportation's impact on travelers' welfare measured by levels of utility cannot be measured under the partial equilibrium analysis. As a result, to obtain a comprehensive evaluation of infrastructure, a general equilibrium framework is required.

The CGE model enables impact analysis with consideration of both demand and supply. The theoretical framework relies on the Walras-Arrow-Debreu theory of general equilibrium, with modern modifications and extensions allowing for imperfect markets (Bröcker 2004). Because CGE provides a clear linkage between the microeconomic structure and the macroeconomic environment, the model can be used to describe the interrelationship among different industrial sectors and markets. More importantly, it can be used to assess both direct and indirect effects from the change of public policy on various economic variables such as output, employment, prices, income and welfare.

A CGE model usually consists of producer, consumer, government, and foreign economy blocks. The fundamental assumptions on producers and consumers in CGE are that producers seek profit maximization while consumers seek utility maximization both within constraints of their resources. The process of production can be illustrated either by a Cobb-Douglas form or a CES form. Government plays dual roles in CGE. On the one hand as a policy maker, the relative policy variable is introduced in CGE as an exogenous factor impacting the economy. On the other hand as a consumer, government revenue that comes from tax and tariff is spent on a variety of public expenditure such as public affairs, intergovernmental transfers and subsidies. As far as international trade is concerned, the distributional process between the domestic market and exports is illustrated by a constant elasticity transformation (CET) while the process between the domestic market and imports is normally illustrated by an Armington function (Bröcker 2004).

The applications of CGE in evaluating impacts of transportation infrastructure investment vary substantially. Impacts can be evaluated differently depending on specific research needs. Due to the fact that most transportation infrastructure achieves economic benefits through increasing accessibility and reducing transport costs, CGE analysis in transportation are usually constructed in a multi-regional structure. Miyagi (2006) evaluated economic impact in relation to the accessibility change using a spatial CGE (SCGE). In his model, economic impact was measured through reduction of congestion due to the specialized infrastructure investment (Miyagi 2006). Haddad and Hewings (2005) assessed economic effects of changes in Brazilian road transportation policy by applying a multiregional CGE model. By introducing non-constant returns and non-iceberg transportation costs, their model found asymmetric impacts of transportation investment on a spatial economy in Brazil. *CGEurope* is another SCGE model developed by Bröcker (1998). The model is primarily used for spatial analysis on the distribution of welfare effects linked to changes in accessibility within and between regions (Bröcker et al. 2001).

Pingo is a static CGE model used to forecast regional and interregional freight transportation (Ivanova 2004). The model contains 19 regions with 10 economic sectors. The *MONASH* model is a widely used multiregional and multi-sectoral dynamic CGE model (Dixon and Rimmer 2000). It allows for different choices in the level of sectoral and regional disaggregation. Transportation sectors in this model are treated as marginal sectors where the costs are imposed on the purchase price of goods and tradables.

Unlike the multiregional CGE model, *IFPRI* (International Food Policy Research Institute) is single regional CGE model which treats transportation cost as a type of transaction costs in trade (Löfgren et al. 2002). The model allows for assessing impacts through transaction cost variation. Transportation costs are treated as a part of trade in this type of CGE model. Some model transportation costs without an explicit representation of transportation sectors, such as *CGEurope*. In other models such as *Pingo*, *MONASH* and *IFPRI*, transportation costs are explicitly included in the price of final goods and services. None of these CGE models consider the issues of spatial dependence, despite most of them doing analysis in a regional context.

10.2.4 Unimodal vs. Multimodal

Another common feature of infrastructure impact studies is that many of them investigate transportation from a unimodal perspective. Some focus on public capital or transportation infrastructure in general (Duffy-Deno and Eberts 1991; Berndt and Hansson 1992; Kelejian and Robinson 1997) while others only focus on a specific mode such as highways, airports or ports (Holtz-Eakin and Schwartz 1995; Cohen and Morrison Paul 2003, 2004; Cohen 2007; Ozbay et al. 2007). Very few studies investigate the issue from a comparative and multimodal perspective (Adersson et al. 1990; Blum 1982; Cantos et al. 2005). It should be noted that these multimodal assessments are conducted under partial equilibrium framework, none of them were conducted from a general equilibrium perspective.

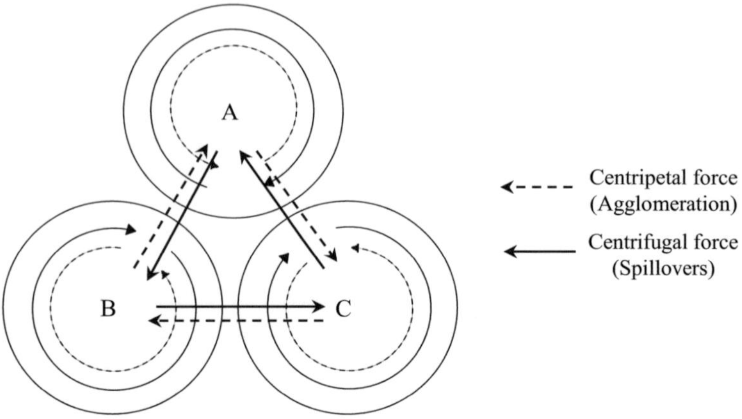

Fig. 10.1 Regional agglomeration and spillovers

10.3 Theoretical Motivation

Literature review indicates that despite CGE analysis provides a comprehensive way to assess impacts of transport with considerations of both the demand and the supply of an economy, the spatial aspects are only considered under a multiregional framework with a capacity to evaluate interregional flows. Issues such as the reliability of exogenous parameters, spatial dependence among elasticity of substitution have never been discussed or examined. As a result, these CGE models may suffer from estimation bias due to the lack of awareness of spatial dependence in the process of regional factor substitution.

In the view of the new economic geography, regional economic activities interact under two types of forces: centripetal and centrifugal (Krugman 1991). When the centripetal force outweighs the centrifugal force, regional agglomeration occurs, otherwise, regional spillovers dominate. Krugman also indicates the elasticity of substitution has a close linkage to economies of scale, which indirectly determines regional convergence or divergence. A higher elasticity of substitution implies smaller economies of scale under general equilibrium conditions, which then works against regional divergence. On the contrary, a lower elasticity of substitution implies higher economies of scale which is more likely to cause regional divergence. The elasticity of factor substitution matters both intraregionally and interregionally due to the effects of both agglomeration and spillovers.

As illustrated in Fig. 10.1, agglomeration and spillovers occur both intra-regionally and interregionally. Depending on the levels of transportation cost, changes of factor prices such as rental and wage rates may lead to the substitution of factors happening within regions A, B and C respectively and/or among them. As a result, unless the interregional elasticity of factor substitution equals zero, the aggregate values of the elasticity of factor substitution should be different from the value that is only driven by the two forces within each region.

Because most of the elasticities of substitution for general equilibrium analysis are exogenously provided, many CGE models retrieve the values from relevant literature that specifically focuses on elasticity estimation (Balistreri et al. 2003; Chirinko et al. 2004). However, a careful search of their sources show that all the elasticities of substitution are obtained through the method of econometric estimation, which usually requires substantial time series data or panel data to achieve robust estimation. However, spatial dependence in the process of estimating the elasticities of substitution has never been addressed, even though the data used for these estimations has a spatial perspective and may imply the existence of such an issue.

Anselin and Griffith (1988) indicate that spatial dependence matters in econometric estimation because ignoring such an issue may lead to serious estimation error. Most of the existing CGE analyses only rely on the elasticities of substitution from non-spatial econometric estimation, in other words, only the intraregional elasticity of substitution is considered. However, the interpretation of the new economic geography theory suggests that interregional activities on the elasticity of substitution may also exist. Limitations that occur due to omitting potential spatial dependence are likely to lead to negative consequences on policy impact analyses.

To address the issue of spatial dependence under general equilibrium framework, a new approach that integrates both spatial econometric estimation and CGE is needed. In addition, the lack of a multimodal perspective limits our understandings of the spatial impacts of transportation infrastructure, particularly in counties like the US where multiple modes of transportation infrastructure are comprehensively, competitively and maturely established. To fill the gaps in the literature, this study is conducted to answer the following questions:

- How does public transportation infrastructure contribute to economic outputs in the US under general equilibrium analysis?
- How do impacts vary among different modes of transportation including road, air, water and transit?
- Does the impact differ when comparing the estimation with and without consideration of spatial dependence in a CGE context?

10.4 CGE Structure

For demonstration purpose, this study adopts an edited version of a single country and static CGE model in the tradition of the *IFPRI* standard model, or the Dervis-DeMelo-Robinson tradition developed by McDonald (2005). An earlier application of this model to transportation infrastructure can be found in Chen and Haynes (2014). The model is an open economy including 13 commodities, 13 activities, 9 factors, 1 household and 1 rest of world account (ROW). Trade is modeled under the Armington assumption of imperfect substitution between domestically produced and imported goods, represented by a one level CES function (Armington 1969). In addition, exports are assumed to be imperfect substitutes for domestically produced goods, represented by a one level Constant Elasticity of Transformation (CET) function. The small country assumption is relaxed with the export demand function. The model allows for non-traded, non-produced and non-consumed domestic goods. The main model structures are discussed for different institutional blocks as described below.

10.4.1 Consumer

Consumer maximizes utility in a Cobb-Douglas form with constant returns to scale subject to a budget constraint. The household commodity consumption can be represented as:

$$\mathrm{PQD_i} \cdot \mathrm{QCD_i} = \mathrm{Comhav_i} \cdot \mathrm{HEXP} \tag{10.1}$$

where

$\mathrm{PQD_i}$: The purchase price of composite commodity i;
$\mathrm{QCD_i}$: Household consumption by commodity i;
$\mathrm{Comhav_i}$: Household consumption shares of commodity i in household;
HEXP: Household consumption expenditure in household.
Household income and household expenditure are denoted respectively as:

$$\mathrm{YH} = \sum\nolimits_{k=1}^{f} \mathrm{hvash_k} \cdot \mathrm{YF_k} + \mathrm{hwor} \cdot \mathrm{ER} \tag{10.2}$$

$$\mathrm{HEXP} = \mathrm{YH} \cdot (1 - \mathrm{tyh}) \cdot (1 - \mathrm{SADJ} \cdot \mathrm{kaphsh}) \tag{10.3}$$

where

YH: Household income;
$\mathrm{hvash_f}$: Share of income from factor f to household;
$\mathrm{YF_k}$: Income to factor f;
hwor: Transfers to household from ROW (constant in foreign currency);
ER: Exchange rate (domestic currency per world unit);
tyh: Direct tax rate on household;
SADJ: Savings rate scaling factor. The value assumes 1 in this study;
kaphsh: Shares of household income saved after taxes of household.

10.4.2 Producer

There are 13 firms that produce one commodity each, maximize their profits and face a nested production function, with capital, labor and inter-industry flows as factors of production. A two-stage production structure applies for producers in all sectors (See Fig. 10.2). The top level assumes Leontief technology with value added and intermediate inputs as factors of production The second level assumes value added CES technology with capital, labor and other endowments as factors of production, and intermediate inputs as a Leontief technology with the commodities

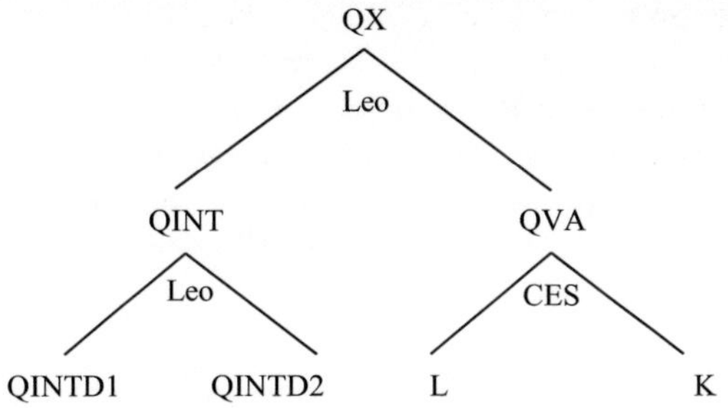

Fig. 10.2 Nested production structure

of all firms as factors of production. The CES multi-factor production function for activity is represented as:

$$QX_a = adx_a \cdot \left(\sum\nolimits_{k=1}^{f} deltax_{k,a} \cdot FD_{k,a}{}^{-rhox_a}\right)^{-\frac{1}{rhox_a}} \quad (10.4)$$

Where
QX_a: Domestic production by activity a;
adx_a: Shift parameter for CES production functions for QX;
$deltax_{f,a}$: Share parameters for CES production functions for QX;
$FD_{f,a}$: Demand for factor f by activity a;
$rhox_a$: Elasticity parameter for CES production functions for QX.

The Intermediate input demand by commodity function and the domestic commodity production can be denoted respectively as:

$$QINTD_c = \sum\nolimits_{k=1}^{a} ioqx_{c,k} \cdot QX_k \quad (10.5)$$

$$COMOUT_c = \sum\nolimits_{k=1}^{a} ioqxcqx_{c,k} \cdot QX_k \quad (10.6)$$

Where
$QINTD_c$: Demand for intermediate inputs by commodity;
$COMOUT_c$: Domestic commodity production;
$ioqx_{c,k}$: Use matrix coefficients;
$ioqxcqx:_{a,c}$: Share of commodity c in output by activity a.

Transportation services provided by transportation sectors are treated as intermediates in non-transportation sectors through the Leontief technology function. The value is added to the final product together with inputs from the CES production function. In transportation sectors of truck, air, transit and water, the factor inputs of the CES production function includes labor, private and public capital. The public transportation capital accounts are set to zero for the non-transportation sectors and for the two private transportation sectors rail and pipeline. The elasticities of factor substitution are assumed consistent for all factors substitution. It is exogenously estimated through both ordinary least squares (OLS) regression model and spatial econometric model.

10.4.3 Government

The government block includes functions representing government taxes and government income and expenditure. Five types of taxes are included in the model: tariff, export tax, sales tax, indirect tax and income tax from non-government institutions. The functions of different taxes revenue are denoted respectively as below:

$$\mathrm{MTAX} = \sum_{k=1}^{c} \mathrm{tm}_k \cdot \mathrm{pwm}_k \cdot \mathrm{ER} \cdot \mathrm{QM}_k \tag{10.7}$$

$$\mathrm{ETAX} = \sum_{k=1}^{c} \mathrm{te}_k \cdot \mathrm{PWE}_k \cdot \mathrm{ER} \cdot \mathrm{QE}_k \tag{10.8}$$

$$\mathrm{STAX} = \sum_{k=1}^{c} \mathrm{ts}_k \cdot \mathrm{PQS}_k \cdot (\mathrm{QINTD}_k + \mathrm{QCD}_k + \mathrm{QGD}_k + \mathrm{QINVD}_k) \tag{10.9}$$

$$\mathrm{ITAX} = \sum_{k=1}^{a} \mathrm{tx}_k \cdot \mathrm{PX}_k \cdot \mathrm{QX}_k \tag{10.10}$$

$$\mathrm{DTAX} = \mathrm{tyh} \cdot \mathrm{YH} \tag{10.11}$$

Where
MTAX: Tariff revenue;
ETAX: Export tax revenue;
STAX: Sales tax revenue;
ITAX: Indirect tax revenue;
DTAX: Income tax revenue from non-government institutions;
tm_c: Tariff rates on commodity c;
te_c: Export tax rate by commodity c;
pwm_c: World price of imports in dollars on commodity c;
PWE_c: World price of exports in dollars;
QM_c: Imports of commodity c;
QE_c: Domestic output exported by commodity c;
ts_c: Sales tax rates;
PQS_c: Supply price of composite commodity c;
QGD_c: Government consumption demand by commodity c;
The functions of government income, consumption and expenditure are denoted as:

$$\mathrm{YG} = \mathrm{MTAX} + \mathrm{ETAX} + \mathrm{STAX} + \mathrm{ITAX} + \mathrm{DTAX} + (\mathrm{govwor} \cdot \mathrm{E}) \tag{10.12}$$

$$\mathrm{QGD}_c = \mathrm{QGDADJ} \cdot \mathrm{qgdconst}_c \tag{10.13}$$

$$\mathrm{EG} = \sum_{k=1}^{c} \mathrm{QGD}_k \cdot \mathrm{PQD}_k \tag{10.14}$$

where

YG, QGD and EG denote government income, government commodity consumption and government expenditure respectively;

govwor: Transfers to government from world (constant in foreign currency);

QGDADJ: Government consumption demand scaling factor. The value assumes 1 in this study;

$qgdconst_c$: Government demand volume of commodity c;

10.4.4 Investment and Saving

Investment and saving block includes the following three equations:

$$\begin{aligned} TOTSAV = & \sum_{k=1}^{h} YH_k \cdot (1 - tyh_k) \cdot SADJ \cdot kaphsh_k \\ & + KAPGOV + (KAPWOR \cdot ER) \end{aligned} \tag{10.15}$$

$$QINVD_c = IADJ \cdot qinvdconst_c \tag{10.16}$$

$$INVEST = \sum_{k=1}^{c} PQD_k \cdot QINVD_k \tag{10.17}$$

where

TOTSAV: Total savings;

KAPGOV: Government Savings;

KAPWOR: Current account balance;

IADJ: Investment scaling factor. The value assumes 1 in this study;

$qinvdconst_c$: Investment demand volume.

10.4.5 Market Clearing Conditions

Market clearing conditions include equilibrium in factor markets, commodity markets, government, foreign trade, and savings and investment. These conditions can be represented in the following equations:

$$FS_f = \sum_{k=1}^{a} FD_{f,k} \tag{10.18}$$

$$QQ_c = QINTD_c + QCD_c + QGD_c + QINVD_c \tag{10.19}$$

$$KAPGOV = YG - EG \tag{10.20}$$

$$\begin{aligned}\text{KAPWOR} = & \sum_{k=1}^{cm} \text{PWM}_k \cdot \text{QM}_{cm} + \frac{\sum_{j=1}^{f} \text{YFWOR}_j}{\text{ER}} \\ & - \sum_{i=1}^{ce} \text{PWE}_i \cdot \text{QE}_i - \text{hwor} - \text{govwor} - \\ & \sum_{m=1}^{f} \text{factwor}_m\end{aligned} \tag{10.21}$$

$$\text{TOTSAV} = \text{INVEST} + \text{WALRAS} \tag{10.22}$$

where

FS_f: Supply of factor f;
QQ_c: Supply of composite commodity c;
$YFWOR_f$: Foreign factor income;
$factwor_m$: Factor payments from ROW (constant in foreign currency);
INVEST: Total investment expenditure;
WALRAS: Slack variable for Walras's Law.

10.5 Estimation Procedure

The estimation procedure of the study is carried out sequentially in the following four steps.

10.5.1 Step 1 Spatial Autocorrelation Test

The spatial autocorrelation, which is measured by values of Moran's I, is tested for in the capital-labor ratio variable and in wage-rental ratio variable using *GeoDa*, developed by the Spatial Analysis Laboratory at the University of Illinois at Urbana-Champaign (Anselin et al. 2006). The universal global Moran's I is defined as (Moran 1950; Cliff and Ord 1981):

$$I = \frac{n}{\sum_{i=1}^{n}\sum_{j=1}^{n} w_{ij}} \cdot \frac{\sum_{i=1}^{n} w_{ij}(x_i - \overline{x})(x_j - \overline{x})}{\sum_{i=1}^{n} w_{ij}(x_i - \overline{x})^2} \tag{10.23}$$

where n is the number of regions which includes 48 contiguous states and the District of Columbia for most sectors except the sectors of pipeline and water transportation, which only contain 48 regions and 36 regions respectively. x and $\overline{x}$ denote the specific state and the mean of x respectively. w_{ij} is the spatial weight matrix, representing the spatial relationship between region i and j. The spatial relationship in this study is defined as being contiguous to each other. Thus the spatial weight matrix is generated using the Queen Contiguity method.

Because Moran's I can only be tested on a yearly basis, Moran's I for each year from 1997 to 2011 is calculated. The results are similar for each variable in each

Table 10.1 Moran's I value of the selected years

	1997		2004		2011		
Sector	Ln(KL)	Ln(wr)	Ln(KL)	Ln(wr)	Ln(KL)	Ln(wr)	No. of states
Agriculture	0.14*	0.45***	0.20*	0.33**	0.25**	0.36***	49
Manufacture	0.40***	0.42***	0.31**	0.34***	0.23**	0.25**	49
Utility& Construction	0.43**	0.38***	0.35***	0.36***	0.29**	0.36***	49
Trade	0.21*	0.24**	0.22**	0.26**	0.07	0.12	49
Truck	0.48***	0.46***	0.58***	0.49***	0.22**	0.01	49
Rail	0.59***	0.59***	0.26**	0.27***	−0.51***	−0.50***	49
Air	0.01	0.01	0.07	0.07	0.10	0.10	49
Transit	−0.10	0.06	0.19**	0.17*	0.20**	0.14*	49
Water	0.13	0.07	−0.12	−0.06	−0.05	−0.07	36
Pipeline	−0.06	0.07	−0.12	−0.16*	−0.09	0.00	48
Warehouse	0.36***	0.37***	0.33***	0.32**	0.29***	0.31***	49
Information	0.23***	0.25**	0.21**	0.26**	0.28**	0.21**	49
Service	0.44***	0.46***	0.44***	0.47***	0.52***	0.53***	49

Note: ***, **, * denote coefficients are significant at 1 %, 5 % and 10 % statistical level, respectively
Source: Authors' calculation

year. Due to the space and page limits, Table 10.1 only shows the global Moran's I of capital-labor ratio (*KL*) and wage-rental ratio (*wr*) in the three selected years covering the beginning, the middle and the end of the investigation period. The Moran's I values of ratio variables of several sectors in most years are significant, which indicates spatial autocorrelation exists across different regions and years. Most of the values are positive indicating a tendency toward clustering, although some values such as the ratios of rail sectors in 2011are negative, which indicates a tendency toward dispersion. The existence of spatial dependence among both the dependent variable and independent variable implies that complicated spatial autocorrelation is an issue for this analysis.

10.5.2 Step 2 Non-Spatial Assessment

The second step is to obtain the basic values of elasticity of factor substitution for the CGE analysis. This study follows the classical CES production function estimation of elasticity of factor substitution. The basic equation can be written as:

$$Q = \left[\alpha_{kl} K^{\frac{\sigma_{kl}-1}{\sigma_{kl}}} + (1-\alpha_{kl}) L^{\frac{\sigma_{kl}-1}{\sigma_{kl}}}\right]^{\frac{\sigma_{kl}}{\sigma_{kl}-1}} \tag{10.24}$$

$$\ln\left(\frac{K}{L}\right) = \sigma_{kl}\ln\left(\frac{1-\alpha_{kl}}{\alpha_{kl}}\right) + \sigma_{kl}\ln\left(\frac{w}{r}\right) \quad (10.25)$$

Where Q is the composite goods of capital and labor, w and r represent wage and rental rates, respectively. σ_{kl} and α_{kl} are the substitution elasticity and distribution parameter of K and L. The equation can be simplified to a linear regression statement:

$$lny = \beta_0 + \beta_1 lnx + \varepsilon \quad (10.26)$$

where y is the capital-labor ratio, x is the wage-rental ratio, and ε is the independent and identically distributed (iid) error. The elasticity of substitution between capital and labor is represented by β_1.

A panel data includes the 48 states and the District of Columbia for 15 years from 1997 to 2011and are constructed following Balistreri et al. (2003)'s approach, which collected similar data from Bureau of Economic Analysis (BEA) but used it only at an aggregate level of analysis. Four data series are collected to operationalize equation 26: employment, total employee compensation, private fixed asset and property income. In the non-spatial assessment, the elasticity of factor substitution for different sectors is estimated using OLS. Panel regressions including both fixed effects and random effects estimations are also implemented. However, since some substitution elasticities have negative estimates that have no economic meanings, the estimates of panel regressions are not adopted for CGE integration.

10.5.3 Step 3 Spatial Econometric Analysis

The third step is to estimate the elasticity of factor substitution for different sectors using spatial econometric estimation to control for spatial dependence. Given the potential complexity of this issue, a generalized spatial model, "Spatial Durbin Model" (SDM) is adopted as the initial model for the assessment. The general form of substitution elasticity under SDM is written as:

$$\left(\frac{K}{L}\right)_{i,t} = \rho W\left(\frac{K}{L}\right)_{i,t} + \beta\left(\frac{w}{r}\right)_{i,t} + \theta W\left(\frac{w}{r}\right)_{i,t} + \varepsilon_{i,t} \quad (10.27)$$

$$\varepsilon_{i,t} \sim N\left(0, \sigma_{i,t}^2 I_n\right)$$

where $\frac{K}{L}$ and $\frac{w}{r}$ denote capital-labor ratio variable and wage-rental ratio variable, and $W\left(\frac{K}{L}\right)_{i,t}$ and $W\left(\frac{w}{r}\right)_{i,t}$ denote the spatial lag terms of capital-labor ratio variable and wage-rental ratio variable, respectively. i and t represent different regions and time periods. ρ, β and θ denote coefficients that need estimation. The analysis is conducted based on the same panel data as used in step 2.

To help identify the appropriate spatial panel model in a systematic way, Elhorst (2012) developed a spatial panel model selection routine, which can be executed in Matlab. The key process is illustrated in Fig. 10.3:

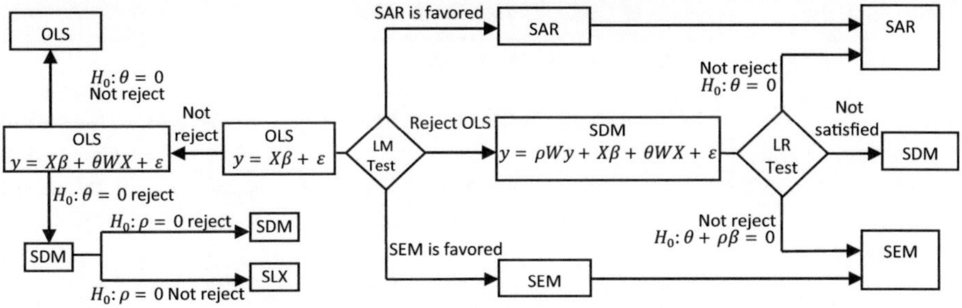

Fig. 10.3 Elhorst spatial model testing procedure. *Note*: SAR, SEM and SLX represent Spatial autoregressive model, Spatial error model and Spatial lagged X model. LM test and LR test represent Lagrange multiplier test and Likelihood ratio test

In this study, we follow the Elhorst spatial model procedure to test which spatial model is preferred technically. Table 10.2 illustrates the specific spatial model form and effect for each industry from Elhorst's routine. Hausman test suggests that spatial time fixed effect needs to be considered for most of the economic sectors. The spatial estimations for trade and warehousing do not include any effect. The pipeline sector is identified as needing to be controlled for both spatial fixed and time fixed effects.

A key function of spatial analysis is to investigate the spatial effects of factor substitution among different states. Because the spatial information of neighboring regions is added in the form of a spatial weight matrix, SDM is endowed with the capacity to investigate spillover effects from total effects (LeSage and Pace 2009). As a result, three types of impacts can be estimated through the spatial model: average direct impact, average indirect impact and average total impact (LeSage and Pace 2009). The first impact measures the influences of the explanatory variables that come from the same geographic unit as the dependent variable. The second impact, the indirect effect or spillover effect, measures the influence of explanatory variables that come from different geographic units. The third impact, total effect, consists of both the direct impact and indirect impact.

10.5.4 Step 4 SECGE

During step 4, a CGE model with an integration of spatial econometric estimates is established. The structure of CGE model has been discussed in Sect. 10.3. The elasticity of factor substitution was estimated under both non-spatial and spatial econometric models in step 2 and 3. These are the estimates utilized respectively for the CES production function in the CGE. The spatial econometric CGE (SECGE) is the second type of integration. Given the fact that the elasticity of factor substitution is not assumed or calibrated in this equilibrium model, the estimates based on historical data under the spatial econometric approach is expected to be more realistic for policy simulation. In addition, compared to the non-spatial econometric

Table 10.2 Estimation results of CES elasticity of substitution

	OLS	Spatial econometric (SE)				
Sector	Coef.	Total Coef.	Direct	Indirect	Model type	Spatial or time effect
Agriculture	0.62***	0.62***			SEM	TF
Manufacture	0.53***	0.36***	0.39***	−0.03***	SDM	TF
Utility& Construction	0.55***	0.53***	0.50***	0.03***	SDM	TF
Trade	0.84***	0.84***	0.84***	0.00	SDM	N/A
Truck	0.61***	0.43***	058***	−0.15***	SAR	TF
Rail	0.33***	0.24***	0.23***	0.01	SDM	TF
Air	0.77***	0.61***	0.72***	−0.11***	SAR	TF
Transit	0.72***	0.68***			SEM	TF
Water	0.70***	0.72***	0.70***	0.02	SDM	TF
Pipeline	0.51***	0.70***	0.47***	0.23***	SDM	SF+TF
Warehouse	0.93***	0.94***	0.93***	0.01	SDM	N/A
Information	1.03***	0.96***	1.02***	−0.06***	SDM	TF
Service	0.91***	0.92***	0.91***	0.001	SDM	TF

Model type indicates the specific spatial model used for Spatial Econometric (SE) estimation. Spatial or time effect indicates the types of effects being used based on the Elhorst (2012) testing routine. *TF* denotes time fixed effect, *SF* denotes spatial fixed effect, and *N/A* denotes no effect. OLS indicates result is estimated through OLS analysis while SE indicates result is estimated through spatial econometric analysis

Note: ***, **, * denote coefficients are significant at 1 %, 5 % and 10 % statistical level, respectively

Source: Authors' calculation

estimation, the spillover effects of factor substitution elasticity can be adequately estimated under the spatial econometric estimation procedure.

10.6 Data

Data used for this analysis includes two components: the first is panel data including quantity and price of capital and labor for the 13 economic sectors covering the 48 contiguous states in the US and the District of Columbia for the period from 1997 to 2011. The data is used to estimate elasticities of factor substitution for the 13 sectors. The second is a US national social accounting matrix (SAM), which is constructed from the GTAP 8 data base.

The GTAP 8 data base is developed by the Center of Global Trade Analysis at Purdue University. The latest version of GTAP data contains dual reference years of 2004 and 2007 as well as 129 world countries and regions for all 57 commodity types. Since our research interest is on multiple modes of transportation, non-transportation sectors are grouped into seven industrial sectors including agriculture, manufacture, utility and construction, trade, information, warehouse and service. Transportation sectors are originally divided into three commodity types in

GTAP: water transport (WTP), air transport (ATP) and other transport (OTP). Surface modes of transportation such as road, rail, pipelines and auxiliary transport activities are all combined in the sector OTP. For this analysis it becomes necessary to separate them.

Since 2007 is the latest reference year for the input-output tables and macroeconomic data in GTAP 8 (Narayanan et al. 2012), we use the BEA 2007 annual I-O table after redefinition as the complement information to further disaggregate the combined surface transportation sectors. Truck, rail, transit and other ground transportation, pipeline and warehousing and storage and others are separated out of the OTP based on their industrial shares in both make and use tables. The rest of OTP which includes auxiliary transport activities and travel agencies are combined with the service sector. Ultimately, six modal specific transportation sectors and seven non-transportation sectors are established.

Another challenge for the study was to add public transportation capital accounts in the US national SAM. Public transportation capital stock has important relationships to public transportation investment. The variation of public transportation capital is primarily influenced by level of investment,[1] thus a shock of public transportation capital in the CGE drives the social and economic variations that result from the level of transportation investment by mode.

Another important note is that public transportation investment in the US is highly modal biased. Highway and streets receive the most public investment while airport, transit and water transportation receives relatively less public investment. The pipeline and rail sectors in the US are primarily privately owned. So these massive infrastructure investments rely on the private sector. Public investment in pipeline and freight rail sectors is primarily used for safety and regulation related purposes and the amount is negligible compared to other modes of investment. Given this background, it is understandable that public transportation capital accounts can be added only for the road, air, transit and water related sectors.

In our CGE model, the four transportation sectors are considered differently to other sectors. The factor endowments consumed by truck, air, transit and water include not only labor and private capital, but also public capital. The ratios of public capital for road, air, transit and water are calculated based on the information of the national fixed assets from BEA.[2] Since the original capital account in GTAP 8 Data Base includes the entire capital stock (both public and private) of the economy, values of public capital for road, air, transit and water can be calculated using the public capital ratio times the total capital stock for each specific transportation sector.

To separate the public capital accounts from the original capital accounts for the four transportation sectors, two assumptions need to be made. The first is that

---

[1] Public capital stock is normally estimated through the Perpetual Inventory Method based on the level of depreciation rate and level of investment. The linkage can be written as $K_t = (1 - \delta) K_{t-1} + I_t$. Given the predetermined ratio of depreciation rate δ, capital stock is naturally primarily influenced by the investment level.

[2] Because there is no specific information on public transportation capital by mode except the highway and streets, the public capital shares for air, transit and water transportation have to be estimated based on their activity share.

non-transportation sectors do not have transportation capital. They solely depend on transportation sectors for transport services. The second is that the original capital account for truck transportation includes not only public capital in truck sectors, but also highway and street public capital. Similarly the original capital accounts of air transportation, transit and water transportation include not only public capital in each sector, but also include public capital for all the relevant infrastructures of each respective mode.

The assumptions are made based on the unique characteristics of the transportation sectors. Capital inputs for truck transportation include not only privately owned vehicles, trailers and relevant facilities, it also demands public capital such as the road networks to produce a road transport service. Air and water transportation sectors are similar. Capital stocks such as aircrafts and watercrafts are primarily privately owned while airports, air traffic control, ports and seaport terminals are mostly publicly owned. In other words, public transportation stocks are treated as factors for these transportation sectors to produce transportation services.

10.7 Results

Policy simulations are conducted under two scenarios of general equilibrium. The first scenario adopts CES elasticity of factor substitution from OLS estimation while the second scenario adopts estimates from spatial econometric models. The results of CES elasticity of factor substitution from the two estimations are displayed in Table 10.2.

The results show that the elasticities of factor substitution vary significantly across different sectors. The rail sector has the lowest value of substitution elasticity while the information sector has the highest. The comparison of OLS estimation and spatial panel estimation indicates differences of substitution elasticities exist among different sectors. For instance, the values for sectors of manufacture, utility and construction, truck, air, transit, water, pipeline, warehouse and information from the spatial econometric estimation are relatively higher than from the OLS estimation. This may imply the existence of positive spillover effects of factor substitution.

The spatial interactions of substitution elasticities are observed in the direct effects and indirect effects (See Table 10.2). Significant and positive indirect effects are found in sectors of manufacture, utility and construction, truck, transit and pipeline, which indicate that wage-rental ratios from adjacent regions have positive impacts on the local region itself. The results further confirm the existence of spatial dependence among these sectors.

To understand the total impacts of public transportation infrastructure as well as the impacts of each mode, public capital of different modes are shocked sequentially and respectively at the same level of a 10 % change. The impacts on welfare, GDP value added, household income, domestic production of each sector are simulated and compared, *ceteris paribus*.

The spatial impacts of different transportation modes are summarized in Table 10.3. The result shows that a 10 % increase shock of total public transportation capital is associated with a 0.04 % increase in both GDP value added and

Table 10.3 Economic impact of transportation infrastructure by mode

	Truck		Air		Water		Transit		All modes	
	OLS	SE	OLS	SE	OLS	SE	OLS	SE	OLS	SE
Agriculture	0.019	0.019	−0.044	−0.045	−0.007	−0.007	0.001	0.000	−0.031	−0.033
Manufacture	0.024	0.024	−0.044	−0.045	−0.005	−0.005	0.002	0.001	−0.024	−0.025
Utility&Cons	0.024	0.024	0.005	0.005	0.004	0.004	0.002	0.002	0.036	0.035
Trade	0.018	0.017	0.001	0.001	0.002	0.002	0.001	0.001	0.022	0.021
Truck	0.411	0.460	−0.022	−0.022	−0.001	−0.001	0.001	0.001	0.389	0.437
Rail	0.035	0.036	−0.032	−0.032	−0.004	−0.003	0.002	0.002	0.002	0.003
Air	−0.015	−0.021	2.708	2.753	−0.033	−0.032	−0.002	−0.003	2.655	2.694
Transit	0.011	0.010	−0.020	−0.021	−0.003	−0.003	0.736	0.819	0.725	0.806
Water	0.004	0.001	−0.088	−0.090	2.267	2.243	−0.001	−0.001	2.176	2.147
Pipeline	0.024	0.024	0.004	0.004	0.006	0.006	0.001	0.001	0.036	0.036
Warehouse	0.051	0.054	0.214	0.217	0.015	0.015	0.001	0.001	0.280	0.287
Information	0.013	0.012	0.009	0.009	0.001	0.001	0.002	0.002	0.025	0.024
Service	0.011	0.010	0.003	0.003	0.001	0.001	0.001	0.001	0.016	0.015
Household	0.020	0.020	0.014	0.014	0.005	0.005	0.002	0.002	0.042	0.042
GDPVA	0.020	0.020	0.014	0.014	0.005	0.005	0.002	0.002	0.042	0.042
Welfare	1603	1598	1137	1140	414	414	164	164	3311	3309

Note: Numbers indicate percentage change. Each column represents a CGE simulation result due to a 10 % increase of the corresponding transportation capital. OLS represents the results simulated using CES estimates from OLS models. SE indicates the results simulated using CES estimates from spatial econometric models

Source: Authors' calculation

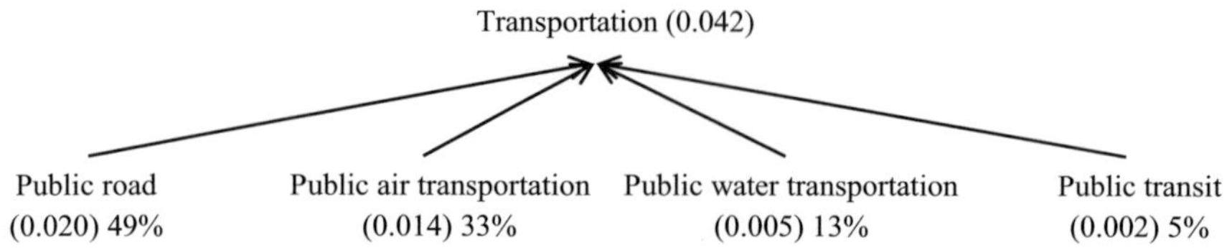

Fig. 10.4 Economic contributions of transportation infrastructure by mode in the US. *Note*: The values are obtained from separate simulations based on the condition that each mode of transportation capital increases by 10 %. The percentage indicates the share of contribution from the individual mode. Economic outputs are measured by the variations in value added GDP and/or household income

household income. In terms of the modal influences, the economic impacts vary significantly by modes. For instance, a 10 % increase of public capital in the truck sector, in other words, a 10 % increase of highway and street capital is associated with a 0.02 % increase in both the value added to GDP and to household income in 2007, *ceteris paribus*. Assuming a 10 % increase of public capital in air transportation sector, the US value added to GDP and household income are likely to increase by 0.014 %, *ceteris paribus*. Compared to truck and air sectors, the economic impacts of public capital in transit and water transportation sectors are much smaller. A 10 % increase of public capital in transit and water transportation sectors are associated with only a 0.002 and a 0.005 % increase in value added to GDP and to household income respectively, *ceteris paribus*. The comparison clearly indicates that public capital in highway and streets has the over whelming impact on growth of GDP and household income among the four public transportation modes. The relative economic contributions among the four modes of public transportation are illustrated in Fig. 10.4.

The results of welfare impact of public transportation infrastructure are similar to its economic impacts. Welfare effect in the study is measured by equivalent variation (EV), which is defined as "the income change at current prices that would be equivalent to the proposed change in the new equilibrium in terms of its impact on utility" (Varian 1992, 161). A positive value of the EV indicates a welfare gain and vice versa. Table 10.3 indicates that the increase of public capital in highways and streets generates the highest welfare gain. Public capital of air transportation has the second largest impact on welfare generation. Public water transportation and public transit still rank third and fourth in terms of their impact on welfare, respectively.

To assess whether spatial dependence has influence on the result of impact analysis is another task of the study. To achieve this goal, CGE simulation results of the two scenarios that adopt substitution elasticities from OLS and spatial econometric estimations are compared. Table 10.4 displays the estimation ratios of SECGE and traditional CGE. The ratio is calculated using the difference of CGE results based on the spatial econometric estimations and OLS estimation divided by the result of corresponding OLS estimation. The ratio indicates the magnitude of difference between traditional CGE and SECGE. For instance, a ratio of zero

Table 10.4 Estimation ratio of SECGE and traditional CGE

	Truck	Air	Water	Transit	All modes
Agriculture	−4 %	2 %	−2 %	**−34 %**	6 %
Manufacture	−1 %	2 %	−2 %	−7 %	6 %
Utility & Construction	−2 %	−4 %	1 %	−4 %	−2 %
Trade	−4 %	**−16 %**	2 %	−8 %	−5 %
Truck	**12 %**	2 %	−6 %	**−12 %**	**12 %**
Rail	3 %	2 %	−3 %	−2 %	**23 %**
Air	**35 %**	2 %	−2 %	25 %	1 %
Transit	**−11 %**	2 %	−4 %	**11 %**	**11 %**
Water	**−75 %**	2 %	−1 %	**57 %**	−1 %
Pipeline	−2 %	−5 %	0 %	**−10 %**	−2 %
Warehouse	7 %	2 %	−1 %	**−13 %**	2 %
Information	−7 %	0 %	4 %	2 %	−3 %
Service	−8 %	−3 %	3 %	−2 %	−5 %
GDPVA	0 %	0 %	0 %	0 %	0 %
Household	0 %	0 %	0 %	0 %	0 %
Welfare	0 %	0 %	0 %	0 %	0 %

Note: Variation over 10 percent is highlighted in bold
Source: Authors' calculation

indicates there is no difference of simulation results; a positive value indicates that SECGE provides higher values of simulation than traditional CGE based on OLS estimation of substitution elasticities. In spite of the fact that the two estimation routines do not show any differences in economic outputs at the aggregate level, the ratio does reveal that the results vary among different sectors and by different modes.

We indicate large difference (over ±10 %) from the estimation ratio of SECGE and CGE based on OLS in bold. Most changed impacts are among the transportation sectors which indicate the spatial interdependence across these sectors and the importance of a multimodal assessment. Next the largest changes outside the transportation sectors are between air transportation and trade and between transit and agriculture. The former is clear but the latter may reflect agglomeration or urbanization.

Conclusion

This study develops a new method that integrates both spatial econometrics and equilibrium modeling to improve the effectiveness of impact analysis on transportation infrastructure. Findings of the study have three implications:

First, the economic impacts of public transportation infrastructure in the US are confirmed to be positive under the general equilibrium framework. However, the magnitude of impact is much smaller than that have been found in many previous studies (Boarnet 1998; Holtz-Eakin and Schwartz 1995; Kelejian and Robinson 1997; Ozbay et al. 2007; Cohen and Morrison Paul 2003, 2004; Cohen 2007). There are two possible causes of this. First of all, the study differs from previous studies in that the evaluation focuses on the most recent period. Since the massive construction and expansions of transportation infrastructure in the US is mostly complete, it is reasonable to believe that the general impacts of the matured US transportation infrastructures are no longer as significant as they used to be during their evolving stages. Second, general equilibrium analysis may find a smaller effects than partial equilibrium analysis because of its consideration of the whole economy.

Second, the study identifies the relative importance of spatial impacts of different transportation modes in the US from a multimodal and comparative perspective. Under the same percentage of increase of public transportation capital, contribution from highways and streets takes about 49 % of total impacts of transportation, while the modes of air, water and transit only take 33, 13 and 5 %, respectively (see Fig. 10.4). The assessment confirms that the US highway and streets plays a dominant role among all transportation infrastructure systems in economic development while public transit and passenger transportation play the least important role among the systems.

Third, the study develops a SECGE model for transportation impact analysis. The method integrates spatial econometric estimation with general equilibrium analysis, which enables researchers to control for the issue of spatial dependence under equilibrium. This integration is important as spatial dependence has been observed among some economic sectors through these spatial autocorrelation tests. Without considering this issue, the elasticity of factor substitution will be biased in traditional OLS estimation, which then may impair validity of CGE assessments. This has been confirmed in our comparative analysis using both OLS estimation and spatial econometric estimation.

However, the differences are only found to exist among the sectoral productions especially among those sectors where spatial dependence is explicitly identified, but not among the aggregate economic outputs. The impacts of domestic production of different sectors become relative high when the substitution elasticities estimated from spatial econometric models. In particular, a higher percentage increase of outputs is observed among transportation sectors including truck, air, transit and water when a 10 % increase of overall transportation capital is implemented. The increase of output could possibly be explained by the inclusion of spillover effects of transportation infrastructure under SECGE, which could not be measured in traditional CGE models.

Given the fact that the study focuses on the national level, the spatial impacts do not vary significantly as spatial dependence may not be a serious issue at such a highly aggregate level of analysis. Although the differences are relatively small in this aggregate case study, implications for more sensitive disaggregated regional models are clear, and will be investigated in the future.

References

Andersson AE, Anderstig C, Harsman B (1990) Knowledge and communications infrastructure and regional economic change. Reg Sci Urban Econ 20(3):359–376

Anselin L (1988) Spatial econometrics: methods and models. Springer, Berlin

Anselin L, Griffith DA (1988) Do spatial effects really matter in regression analysis? Papers in Regional Science 65(1):11–34

Anselin L, IbnuSyabri, Youngihn Kho (2006) GeoDa: an introduction to spatial data analysis. Geogr Anal 38(1):5–22

Armington PS (1969) A theory of demand for products distinguished by place of production. Staff Papers-International Monetary Fund 16(1):159–178

Aschauer DA (1989) Is public expenditure productive? J Monetary Econ 23(2):177–200

Aschauer DA (1990) Why is infrastructure important? (Conference Series; [Proceedings]). Federal Reserve Bank of Boston, pp 21–68

Aschauer DA (1994) Infrastructure and macroeconomic performance: direct and indirect effects. Presented at the Paper to OECD Conference on Capital Formation and Employment, Amsterdam

Balistreri EJ, McDaniel CA, Wong EV (2003) An estimation of U.S. industry-level capital-labor substitution. EconWPA. Computational Economics

Berndt ER, Hansson B (1992) Measuring the contribution of public infrastructure capital in Sweden. Scand J Econ 94:S151–S168

Bhatta SD, Drennan MP (2003) The economic benefits of public investment in transportation. J Plan Educ Res 22(3):288–296

Blum U (1982) Effects of transportation investments on regional growth: a theoretical and empirical investigation. Pap Reg Sci Assoc 49(1):169–184

Bo CD, Florio M, Manzi G (2010) Regional infrastructure and convergence: growth implications in a spatial framework. Transit Stud Rev 17(3):475–493

Boarnet MG (1997) Infrastructure services and the productivity of public capital: the case of streets and highways. Nat Tax J 50(1):39–57

Boarnet MG (1998) Spillovers and the locational effects of public infrastructure. J Reg Sci 38(3): 381–400

Boarnet MG, Haughwout AF (2000) Do highways matter? Evidence and policy implications of highways' influence on metropolitan development. A discussion paper prepared for the brookings institution center on urban and metropolitan policy. Washington DC

Bröcker J (1998) Operational spatial computable general equilibrium modeling. Ann Reg Sci 32(3): 367–387

Bröcker J (2004) Computable general equilibrium analysis in transportation economics. Handbook 5 of handbooks in transport. Pergamon/Elsevier Science, Kidlington, pp 269–289

Bröcker J, Kancs A, Schürmann C, Wegener M, Spiekermann K (2001) Methodology for the assessment of spatial economic impacts of transport projects and policies. EERI Research Paper Series, (3)

Cantos P, Gumbau-Albert M, Maudos J (2005) Transport infrastructures, spillover effects and regional growth: evidence of the Spanish case. Transport Rev 25(1):25–50

Chen, Z., & Haynes, K. E. (2013). Transportation Capital in the United States: A Multimodal General Equilibrium Analysis. Public Works Management & Policy

Chen, Z., & Haynes, K. E. (2014). Public surface transportation and regional output: A spatial panel approach. Papers in Regional Science

Chirinko RS, Fazzari SM, Meyer AP (2004) That elusive elasticity: a long-panel approach to estimating the capital-labor substitution elasticity

Cliff AD, Ord JK (1981) Spatial processes: models & applications. Pion

Cohen JP (2007) Economic benefits of investments in transport infrastructure. Discussion paper no 2007–13 December 2007. OECD/ITF

Cohen JP, Morrison Paul CJ (2003) Spatial and supply/demand agglomeration economies: state- and industry-linkages in the U.S. food system. Empir Econ 28:733–751

Cohen JP, Morrison Paul CJ (2004) Public infrastructure investment, interstate spatial spillovers, and manufacturing costs. Rev Econ Stat 86(2):551–560

Dixon PB, Rimmer MT (2000) MONASH: a dynamic, computable general equilibrium model of the Australian economy. Centre of Policy Studies, Monash University

Duffy-Deno KT, Eberts R (1991) Public infrastructure and regional economic development: a simultaneous equations approach. J Urban Econ 30(3):329–343

Elhorst JP (2012) Matlab software for spatial panels. Int Reg Sci Rev. doi:10.1177/0160017612452429

Fernald JG (1999) Roads to prosperity? Assessing the link between public capital and productivity. Am Econ Rev 89(3):619–638

Gramlich EM (1994) Infrastructure investment: a review essay. J Econ Lit 32(3):1176–1196

Gramlich E (2001) Infrastructure and economic development. Remarks by Governor Edward M. Gramlich at the Texas trade corridors new economy conference. San Antonio, TX

Haddad EA, Hewings GJD (2005) Market imperfections in a spatial economy: some experimental results. Q Rev Econ Finance 45(2–3):476–496

Harmatuck DJ (1996) The influence of transportation infrastructure on economic development. Logist Transport Rev 32(1)

Haynes KE (2006) Infrastructure: the glue of megacities, vol 9. Kenniscentrum Grote Steden

Holtz-Eakin D, Schwartz AE (1995) Spatial productivity spillovers from public infrastructure: evidence from state highways (working paper no 5004). National Bureau of Economic Research

Ivanova O (2004) Evaluation of infrastructure welfare benefits in the Spatial Computable General Equilibrium (SCGE) framework. Department of Economics, University of Oslo

Kelejian HH, Robinson DP (1997) Infrastructure productivity estimation and its underlying econometric specifications: a sensitivity analysis. Pap Reg Sci 76(1):115–131

Krugman P (1991) Increasing returns and economic geography. J Polit Econ 99:484–99

LeSage JP (1999) The theory and practice of spatial econometrics. A manual to accompany the spatial econometrics toolbox. Available at: http://www.spatial-econometrics.com

LeSage J, Pace RK (2009) Introduction to spatial econometrics, 1st edn. Chapman and Hall/CRC, London

Löfgren H, Harris RL, Robinson S (2002) A standard computable general equilibrium (CGE) model in GAMS (vol. 5). Intl Food Policy Res Inst

Mamuneas TP, Nadiri MI (2006) Production, consumption and the rates of return to highway infrastructure capital (unpublished manuscript). University of Cyprus, and New York University and National Bureau of Economic Research

Mattoon R (2002) Midwest infrastructure: assessing the contribution of basic infrastructure to economic growth. Chicago Fed Letter, 184b (Special Issue December)

McDonald S (2005) The provide project standard computable general equilibrium model: version 2. Technical paper series 15625, PROVIDE project

Miyagi T (2006) Evaluation of economic impacts from the accessibility-change by transportation investment: a SCGE modeling approach

Mohammad M (2009) Infrastructure investment in Indonesia: process and impact (Dissertation). University of Groningen, Netherlands

Moran PAP (1950) Notes on continuous stochastic phenomena. Biometrika 37(1–2):17–23

Moreno R, López-Bazo E (2007) Returns to local and transport infrastructure under regional spillovers. Int Reg Sci Rev 30(1):47–71

Munnell AH (1992) Policy watch: infrastructure investment and economic growth. J Econ Perspect 6(4):189–198

Nadiri MI, Mamuneas TP (1996) Contribution of highway capital to industry and national productivity growth (No BAT-94-008). Report prepared for Apogee Research, Inc., for the federal highway administration office of policy development

Narayanan GB, Aguiar A, McDougall R (eds) (2012) Global trade, assistance, and production: the GTAP 8 data base. Center for Global Trade Analysis, Purdue University

Ozbay K, Ozmen-Ertekin D, Berechman J (2007) Contribution of transportation investments to county output. Transport Policy 14(4):317–329

Paelinck JHP, Klaassen LH (1979) Spatial econometrics. Saxon House, Farnborough

Varian HR (1992) Microeconomic analysis, 3rd edn. Norton & Company, New York

Demographic Challenges to Regional Development

11

Geoffrey J.D. Hewings and Euijune Kim

Abstract

While the field of regional science has focused considerable attention on the role of structural change in affecting regional development, relatively modest attention has been directed to the influence of demographic-induced changes with the exception of migration and the impact of the demographic transition in developing countries. Further, most models of the regional economy portray significant disaggregation of the industrial system yet often capture the impacts of households in one sector. Increasing attention to household heterogeneity is justified on the basis of different consumption behavior by age, income and location. Since households account for 40–70 % of gross domestic product on the expenditure side, addressing the role of household heterogeneity would appear to be essential. In this paper, a review of recent work that captures this heterogeneity will be reviewed and some illustrative results from applications in Korea and the US will be provided.

11.1 Introduction

In his provocatively titled Presidential address to the Regional Science Association, Torsten Hagerstrand (1970) asked "What about people in Regional Science?" He was commenting on the tendency of researchers in the nascent field of regional science to assign the role of individuals or households as afterthoughts in the development of

G.J.D. Hewings (✉)
Regional Economics Applications Laboratory, University of Illinois, 607, S. Mathews, #318, Urbana, IL 61801-3671, USA
e-mail: hewings@illinois.edu

E. Kim
Department of Agricultural Economics and Rural Development, Seoul National University, Seoul, South Korea
e-mail: euijune@snu.ac.kr

P. Nijkamp et al. (eds.), *Regional Science Matters*, DOI 10.1007/978-3-319-07305-7_11,

economic models. Over 40 years later, has much changed? The answer is mixed; transportation models have been significantly expanded to embrace choice models; migration has moved beyond the "net migrant" [see Rogers (1990)] and microsimulation (Birkin and Clarke 2011; van Leeuwen 2010) offers a way to combine individual decision-making within a macro context. However, when one looks at the structure of most regional macroeconomic models, once again "people" have been consigned (aggregated) to one vector of consumption and one row of receivers of wages and salaries; the heterogeneity of household skill levels (and thus their ability to command differential factor payments) and Engels' well known laws about the relationship between consumption preferences and income are all but ignored. What is troubling is that household consumption accounts for 70 % of expenditures on the expenditure side of GDP in countries like the US yet the heterogeneity of tastes and preferences, the effects of differences in household size, income and age are all collapsed. These variations could be generated by changes in:

- The age composition of households since consumption patterns change with age
- Income distribution, since there are important differences in the way income is allocated depending on the level of income
- In- and out-migration, not only in terms of volume but also in terms of composition
- In retirement patterns and especially the propensity for retirees to remain in a region
- In social security costs and the way these are allocated across households over time

Figure 11.1 shows the way in which a more elaborated set of interactions between the demographic and economic parts of the economy can be handled. In essence, demographic changes—whether from the perspective of the individual or household—have the capacity to exert significant influences on what and where commodities and services are produced and consumed. In turn, these market signals affect significant sources of income from wages and salaries and the profitability of enterprises affects returns to capital and savings that ultimately affect disposable income for retiree households.

In this paper, a series of issues surrounding household behavior will be explored in terms of their economic impacts. The analysis will draw on both quantity adjusted and price adjusted general equilibrium models. The final set of analyses will describe some alternative approaches to role of education investment in economies with ageing populations drawing on work completed in the US and Korea using general equilibrium models with slightly different assumptions about household behavior.

11.2 Unraveling Household Heterogeneity[1]

With many national and regional economies experiencing significant demographic changes—ageing of the population, differential (in terms of income and occupational characteristics) out- and in-migration and deepened income disparities—there is a

[1] This section draws on Yoon and Hewings (2006).

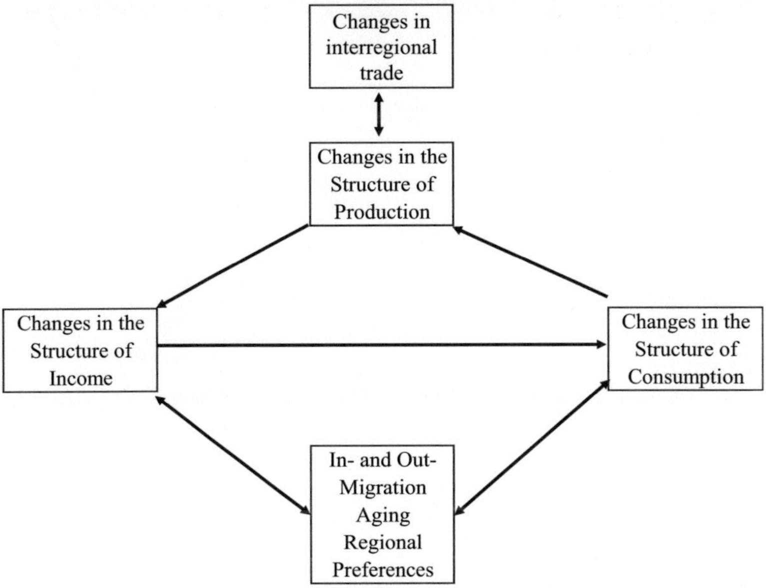

Fig. 11.1 The demographic-economic connection

need to consider ways in which some of these demographically-induced changes can be handled. In a sense, the duality between production structure and the structure of income distribution advanced in the context of social accounting systems (see Fig. 11.1) can be enhanced by a broader vision of the demographic influences on consumption, income distribution and thus production.

Some progress has been made on this topic. Hewings (1982) and Hewings et al. (1989) emphasized the role of the household sector and the importance of consumption patterns in the analysis using extended input-output and social accounting models. When the input-output table is embedded in a social accounting systems, almost all the analytically important entries concentrate in the parts of system capturing economy ↔ household interactions. Li et al. (1998) and Rose and Li (1998) constructed an income distribution matrix to explore various facets of income distribution at the national and regional levels. Rose and Beaumont (1988, 1989), and Rose and Li (1998) calculated estimates of interrelational income multipliers using the method proposed by Miyazawa (1976). More recently, Wakabayashi and Hewings (2007) found some implications of life cycle changes on Japanese consumption behavior using a modified AIDS estimation system in the interregional context.

Initially, to explore some implications of demographic changes on consumption behavior in the Chicago region, an extended Chicago Region Econometric Input-Output Model (CREIM) was developed, into which a modified Almost Ideal Demand System (AIDS) was integrated. A recent prediction shows that the proportion of the ageing population will accelerate after 2010 in both the US and Chicago (see Fig. 11.2). The US experience is not unique; the proportion of the elderly population (65+) to those of working age (15–64) in Korea is expected to rise from

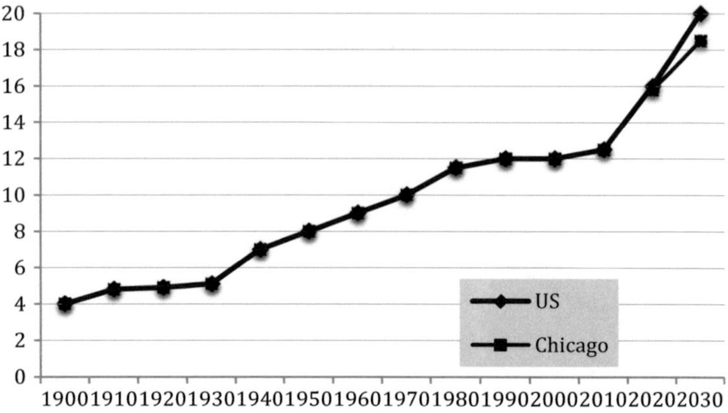

Fig. 11.2 Percentage of the population age 65 and over in the US and Chicago, 1900–2030

11 % in 2008 to an estimated 23 % in 2030. Dealing with this reality presents a unique challenge for the government, as the welfare programs typically used to alleviate socio-economic problems—such as public pension payments, medical insurance benefits, and public assistance for the low-income classes—are in need of rapid expansion, and yet tax revenue is on the decline (Chun 2006).

Figure 11.3 reveals that there are likely to be significant changes in the consumption volume (and composition of goods and services) by age groups between 2003 and 2030. Essentially, the lower age groups will have flat or declining growth with much more rapid growth forecast for the upper age groups.

One of the major issues to be addressed is the degree to which it is analytically important to estimate consumption by age group or income quintile; in this sense, the analysis takes the next step from the Wakabayashi and Hewings (2007) analysis by incorporating disaggregated consumption functions into an impact and forecasting model by integrating an Almost Ideal Demand System (AIDS), proposed by Deaton and Muellbauer (1980), into an econometric input-output model. In this way, it is possible to explore the importance of household consumption disaggregation in contrast to results generated with a single representative household.

11.2.1 The Modified Almost Ideal Demand System[2]

The AIDS system is derived from the PIGLOG (price-independent log)-class expenditure function defined as follows.

[2] This section draws on Wakabayashi and Hewings (2007).

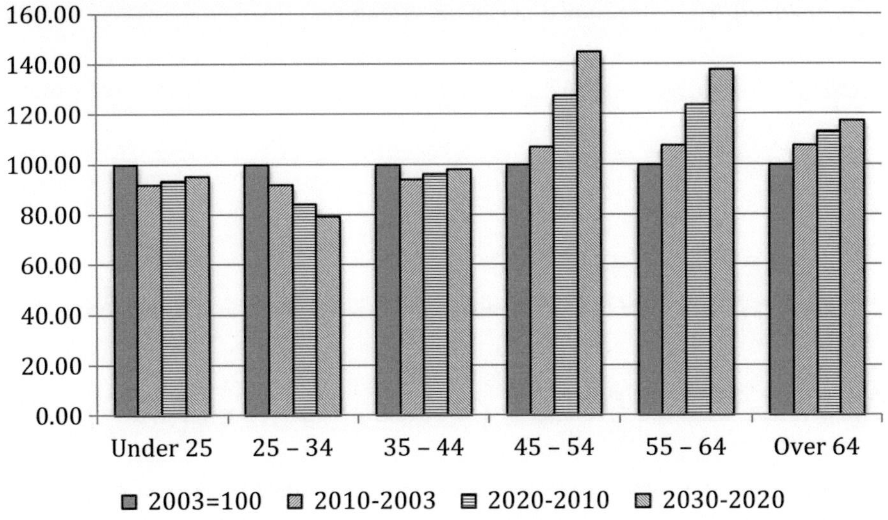

Fig. 11.3 Consumption growth by households of different ages (2003 = 100)

$$\ln C(U,P) = (1-U)\ln A(P) + U\ln B(P) \tag{11.1}$$

where $\ln A(P) = \alpha_0 + \sum_k \alpha_k \ln P_k + \frac{1}{2}\sum_k \sum_j \gamma^*{}_{kj} \ln P_k \ln P_j$ and

$\ln B(P) = \ln A(P) + \beta_0 \prod_k P_k^{\beta_k}$

To derive price elasticities that reflect both own price effects and cross price effects, the K-good expenditure function is expressed in terms of a "two-good economy" producing ι and $-i$.[3]

$$\ln C(U,P) = (1-U)\ln A(P) + U\ln b(P) \tag{11.2}$$

where

$$\ln A(P) = \alpha_0 + \alpha_i \ln P_i + \alpha_{-i} \ln P_{-i} + \ln A \frac{1}{2}\left(\gamma^*_{ii} \ln P_i \ln P_i + \gamma^*_{i-i} \ln P_i \ln P_{-i}\right.$$

$$\left. + \gamma^*_{-ii} \ln P_{-i} \ln P_i + \gamma^*_{-i-i} ln P_{-i} \ln P_{-i}\right.$$

and $\ln B(P) = \ln A(P) + \ln B(P) = \ln A(P) + \beta_0 \left(P_i^{\beta_i} P_{-i}^{\beta_{-i}}\right)$

[3] This approach is also helpful to preserve degree of freedom in a small sample, since it can save k-2 degrees of freedom in the estimation. And the expenditure function changes from $C = \sum_k P_k Q_k$ for k goods to $C = P_i Q_i + P_{-i} Q_{-i}$ for good i and $-i$.

A modified AIDS cost function can be written as:

$$\ln C(U,P) = \alpha_0 + \alpha_i \ln P_i + \alpha_{-i} \ln P_{-i} + \frac{1}{2}(\gamma^*_{ii} \ln P_i \ln P_i + \gamma^*_{i-i} \ln P_i \ln P_{-i} + \gamma^*_{-ii} \ln P_{-i} \ln P_i +$$

$$\gamma^*_{-i-i} \ln P_{-i} \ln P_{-i}) + \beta_0 \left(P^{\beta_i}_{-i} P^{\beta_{-i}}_{-i}\right) \tag{11.3}$$

where $\gamma_{ii} = \gamma^*_{ii}, \gamma_{i-i} = \gamma^*_{i-i} = \gamma^*_{-ii}$

Applying Shepherd's lemma to this expenditure function, the function can be described as:

$$w_i = \frac{\partial \ln C(U,P)}{\partial \ln P_i} = \frac{P_i Q_i}{C(U,P)} = \alpha_i + \gamma_{ii} \ln P_i + \gamma_{i-i} \ln P_{-i+} \beta_i U \beta_0 \left(P_i^{\beta_i} P_{-i}^{\beta_{-i}}\right) \tag{11.4}$$

For a utility-maximizing consumer, total expenditure X is equal to $C(U, P)$ and this equality can be inverted to give U as a function of P and X, the indirect utility. Equation (11.4) can be rewritten as the AIDS demand functions in budget share form:

$$w_i = \alpha_i + \gamma_{ii} \ln P_i + \gamma_{i-i} \ln P_{-i} + \beta_i \ln(X/P) \tag{11.5}$$

where w_i is the budget share of the good i for the household, P_i is the price of good i, and (X/P) is the total expenditure on all goods and services in real terms. A price index P is defined by

$$\ln P = \alpha_0 + \alpha_i \ln P_i + \alpha_{-i} \ln P_{-i} + \frac{1}{2}(\gamma_{ii} \ln P_i \ln P_i + \gamma_{i-i} \ln P_i \ln P_{-i} + \gamma_{-ii} \ln P_{-i} \ln P_i +$$

$$+\gamma_{-i-i} \ln P_{-i} \ln P_{-i}) \tag{11.6}$$

where $\gamma_{i-i} = \frac{1}{2}(\gamma^*_{i-i} + \gamma^*_{-ii})$

Since the price index P is defined as (Eq. 11.6), the AIDS model is non-linear. Deaton and Muellbauer (1980) suggested the real price index P can be replaced by the Stone price index $\hat{P}$ to transform the AIDS into a linear one.

$$w_i = \alpha_i + \gamma_{ii} \ln P_i + \gamma_{i-i} \ln P_{-i} + \beta_i \ln(X/\hat{P}) \tag{11.7}$$

$$\ln \hat{P} = w_i \ln P_i + w_{-i} \ln P_{-i} \tag{11.8}$$

The AIDS with the Stone price index is called 'Linear Approximate Almost Ideal Demand' (LA/AIDS), employed by many researchers because of it is more convenient to estimate. To be consistent with consumption theory, the model should require the following conditions.

$$\sum_i \alpha_i = 1, \sum_i \gamma_{ij} = 0, \sum_i \beta_i = 0 \text{ (adding up)} \tag{11.9}$$

$$\sum_j \gamma_{ij} = 0 \text{ (homogeneity)} \tag{11.10}$$

$$\gamma_{ij} = \gamma_{ji} \text{ (symmetry)} \tag{11.11}$$

Since homogeneity implies $\gamma_{ii} = -\gamma_{i-i}$, (Eq. 11.7) can be re-written as

$$w_i = \alpha_i + \gamma_{ii}\ln(P_i/P_{-i}) + \beta_i\ln(X/\hat{P}) \tag{11.12}$$

Since the modified AIDS is flexible, it is not guaranteed to satisfy the homogeneity and symmetry conditions. So these conditions are introduced as parameter restrictions in the estimation process. The modified AIDS assumes that the size of the family affects budget share and N, the number of household members, was introduced as a shift parameter of α_i. Then the model can be expressed as:

$$w_{it}^k = \alpha_i + e_i N_t^k + \gamma_{ii}\ln(P_{it}/P_{-it}) + \beta_i\ln(X_t^k/\hat{P}_t^k) + e_i^k \tag{11.13}$$

where e_i^k is the error term and i and k refer to ten consumption types and six ages of reference person.

$$w_{it}^r = \alpha_i + e_i N_t^r + \gamma_{ii}\ln(P_{it}/P_{-it}) + \beta_i\ln(X_t^r/\hat{P}_t^r) + e_i^r \tag{11.14}$$

where e_i^r is the error term and i and r refer to ten consumption types and five quintiles of income.

The empirical results of (Eq. 11.13) show life cycle changes in consumption behavior (a sample is provided in Fig. 11.4). From the estimation results (Eq. 11.13) and (Eq. 11.14), the price elasticity ε_{it}^j and expenditure elasticity η_{it}^j are:

$$\begin{aligned} \varepsilon_{it}^j &= \frac{\gamma_{ii}}{w_{it}^j} + w_{it}^j - 1 \text{ for } j \\ &= k(\text{age of reference person}), r(\text{quintiles of income}) \end{aligned} \tag{11.15}$$

$$\eta_{it}^j = \frac{\beta_i}{w_{it}^j} + 1 \text{ for } j = k(\text{age of reference person}), r(\text{quintiles of income}) \tag{11.16}$$

11.2.2 The AIDS-Type Chicago Region Econometric Input-Output Model

To predict the demographic changes in the Chicago region up to 2030, this research extends the Chicago Region Econometric Input-Output Model (CREIM). The regional econometric input-output model (REIM), initially designed by Conway

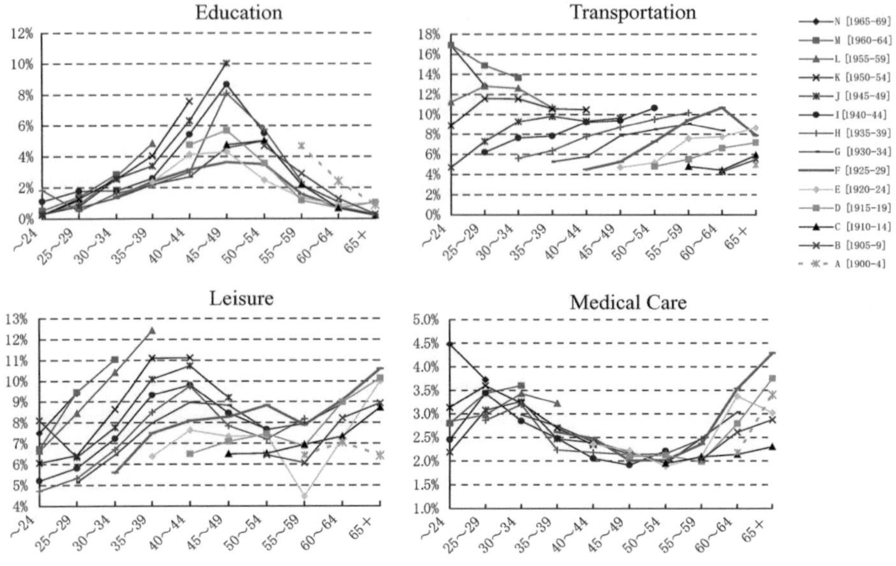

Fig. 11.4 Consumption shares by age, 1980s to 2020s [share of total income spent on selected goods and services]

(1990, 1991), and further developed by Israilevich et al. (1997), provides an alternative perspective for impact analyses or forecasts. The CREIM, which generates forecasts of the Chicago economy on an annual basis, with the forecast horizon extending up to 30 years, is a computable regional general equilibrium model, based on the Marshallian equilibrium of outputs. The model combines traditional input-output analysis with time-series analysis. The input-output component enables a detailed analysis of purchases and sales between industries, while the time-series component allows for the analysis of inter-temporal change in the transaction flows of goods and services. Together, these two components yield a detailed analysis of structural change over time at the sectoral level.

The CREIM model is a system of linear and nonlinear equations formulated to predict 264 endogenous variables. The CREIM identifies 53 industries and three government sectors. For each industry, there are projections of output, employment, and earnings. Out of the 253 equations, only 53 relate to the linear input-output components. Many of the non input-output equations are nonlinear and estimated in a recursive fashion. As a result, the relationships of one sector to another include the formal input-output link as well as a set of complex links through a chain of actions and reactions that could potentially involve the whole economy. Further details may be found in Israilevich et al. (1997).

To predict the impacts of demographic changes in the Chicago region, a modified AIDS system was integrated into the CREIM. An AIDS-type CREIM can be constructed by including a modified AIDS system into equations of final demand by households. The AIDS system equations that are to be included are derived from estimates using a 20-year time series (1984–2003) in the final demand equations (see Yoon and Hewings (2006) for details of the integration).

Although international (legal and illegal) immigration is an increasingly important component of national population change, the region's demographic structure is determined by the combination of natural increase (births–deaths), and two types of migration, international and interregional.

The next sections of this paper describe some of the analyses that have been conducted to explore various aspects of the synergies presented in Fig. 11.1. The outcomes provide a mix of results that meet a priori expectations, produce some surprises and also create outcomes whose impacts depend on the time period chosen.

11.2.3 Applications: Population Composition and Changes Time

Given, the potential changes generated by the ageing phenomenon noted in Fig. 11.2, how does this translate into demonstrable economic impacts? Figure 11.3 revealed the expected consumption by six age groups in comparison to aggregating the effects into a single household type. The evidence suggests that it is important to pay attention to age if for no other reason than changes in the rate of growth by age are so different. However, it is not just the rate of growth but also differences in consumption patterns as revealed in Fig. 11.4 for a sample of goods and services.

Here, we find some important differences in the way households allocate income; the figures show how a typical dollar is divided among a sample of goods and services. For example, on average in 2003, households allocated almost 13 % of their income for food, 36 % for housing (including mortgage, other loans, maintenance expenditures etc.) and 17 % for all forms of transportation. The food expenditure allocation varied from 12.4 % (45–54 age group) to 14.5 % (under 25) while the transportation allocations varied from 18.1 (under 25) to 14.7 (over 65). Over time, many of these expenditures are forecast to change. For example, people over 65 will spend a declining share of their income on food but an increasing share on other goods and services that include restaurants.

The health care allocations generate some interesting outcomes; while all age groups will experience an increase in the share of income allocated to health care, the greatest increases occur not in the over 65 age group but in the other age groups, increasing from 3.9 to 5.9 % (35–44), from 4.4 to 5.9 % (44–54) and 6.2 to 8.1 % (55–64). Since income usually follows a growth path that peaks in middle to pre-retirement, the implication here is that not only will a larger share of income go towards health care but the volume of expenditures on health care will increase as well.

For the Chicago economy, these results have important implications. There is a strong synergy between the structure of production (the distribution of production across manufacturing and services for example), the structure of income distribution (what type of jobs and incomes are associated with the production system) and the structure of expenditures by households (the triangular relationship identified in Fig. 11.1). Changes in any one of these links will generate important changes on the other two. Different consumption patterns caused by demographic changes such as an ageing population will change the industrial production structure of the Chicago region in the future. In turn, these changes in production structure will have important implications on the profile of activities that remain competitive in the Chicago region, creating further feedback effects on the nature of local jobs and wage and salary income.

A final concern is this phase of the analysis was the implications for the distribution of income; in parallel to the division of consumption expenditures by age, we also explored differences due to levels of income. A combination of factors will see the income inequality rise in Chicago through 2030 although this result is modified by the effects of migration and consideration of assets (especially non wage and salary income).

11.2.4 Implications

In this research, a further important question can be raised. Is consumption becoming more important in the Chicago region in the future? To answer this question, the feedback effects of consumption on the production in the Chicago region can be evaluated. To accomplish this, the trends of consumption multipliers in the Chicago region up to 2030 were estimated. From 2003, the consumption multipliers are predicted to increase gradually until 2030 rising from 1.98 to 2.03. More critically, are the differences in consumption patterns by age group are they statistically significant? To check whether the differences in consumption behaviors between the one representative household and the disaggregated households (six age groups and five income quintiles) and those between disaggregate representative households are statistically significant, F tests are applied. The results revealed statistically significant differences between the one representative household and six age groups, and similarly between age groups. Compared with the one representative household, most age groups show statistically significant differences in housing consumption, but all age groups present no statistically significant differences in the consumption of personal care and pensions, and most age groups show no statistically significant differences in the consumption of beverages, transportation and entertainment. Between age groups, most age groups show significant differences in the consumption of housing, health care and education. However, most age groups present no significant differences in the consumption of clothing, entertainment and personal care. In summary, the combination of differences in the consumption by age group—in total and in terms of differences in types of goods and services consumed—and the changes in income distribution generated by continuing changes in economic structure will combine to create important impacts on the Chicago economy that could not be revealed through models using a single representative household.

11.3 Ageing and the Macro Economy[4]

The analysis presented thus far has taken a modified view of a classic perspective on how an economy functions, paying special attention to the role of the productive system. However, it still fails to capture the full effects of the role of households. A slightly different version of the model, a more traditional computable general equilibrium model in the Walrasian tradition integrated with an overlapping generations framework, was constructed using the same data base to explore changes in household behavior on the economy whereas the analysis presented thus far still explores a set of households that are reacting to changes in the economy rather than generating those changes (see Park and Hewings (2009) for more details of the model).

11.3.1 Ageing and Its Impact

To accomplish this task, behavior by households of different ages, 21–85 was identified; to simplify the analysis, it was assumed that individuals were forward looking (i.e., they considered the future in making decisions about whether to spend or save) that they had some uncertainty about how long they would live and that their income consisted of wage and salary (and dividends) while they were working and only dividends and pensions in retirement. Further, it was assumed that all individuals retired at age 65 and died at 85 (see Fig. 11.5).

Figure 11.5 also shows the various components of income over a typical household's lifetime; since it was assumed that individuals die at 85 (or unexpectedly earlier), their consumption patterns reflect a finite expectation for the calculation of expenditures from income (drawing down their non pension assets over the period from 65 to 85). Figure 11.5 assumes an economy in which there are no changes in the age distribution over time; assume that the expectations suggested by Fig. 11.2 come into play, how will this effect the expectations presented in Fig. 11.5? It turns out to be a lot more complicated!

Figure 11.6 presents the outcomes under an ageing population scenario. Not surprisingly, untaxed wages increase under an ageing population, reflecting the relatively scarcity of labor. Nonetheless, total income decreases over almost all age cohorts. For working age cohorts, this happens because the sharp increase in social security tax under ageing population reduces the net wage income from labor supply. For early retirees, though relatively smaller than workers, the fall in the interest rate caused by relatively abundant capital contributes to reducing the capital income from savings. With these different changes in income by age, the effect of ageing population on savings is also sensitive to the age cohorts. That is, before the retirement, the difference in saving is not large enough to generate major interest. The possible reason is that even an ageing population will motivate

[4] This section draws on Park and Hewings (2007a, b, c, 2009).

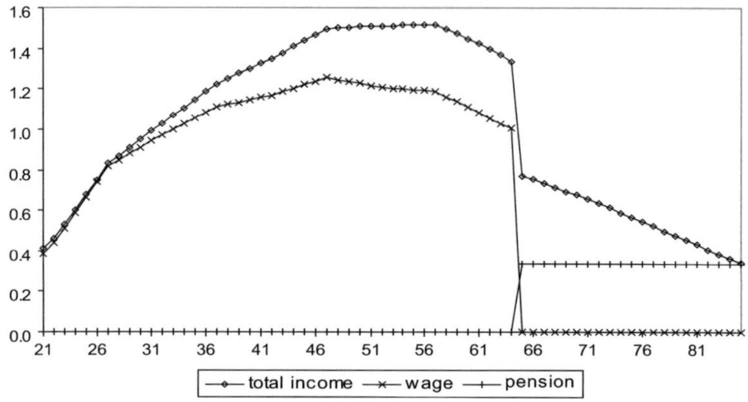

Fig. 11.5 Contributions to income over a lifetime (no change in population structure)

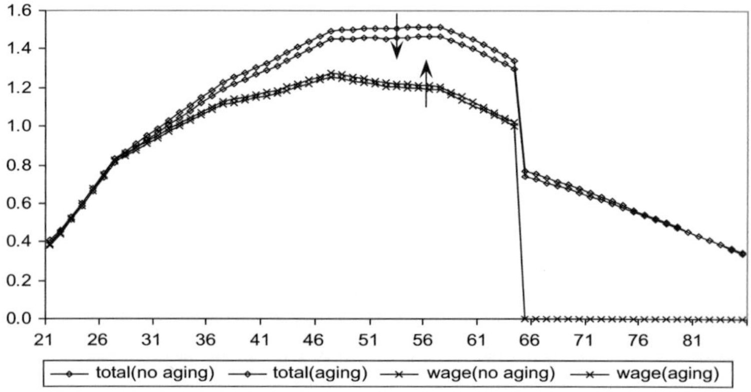

Fig. 11.6 Contributions to income over a lifetime (ageing population)

precautionary saving for the working age cohorts but they cannot afford to sufficiently increase savings due to the fall in total income. As a result, consumption under an ageing population drops significantly except for the oldest cohorts, reflecting a decline in total income and strong precautionary saving motives.

Figure 11.7 shows the transitional path of Gross Regional Product (GRP). The fall in aggregate savings accompanied by the smaller labor force eventually leads to the fall in the GRP compared to the before-ageing population. However, in the initial periods, the transition to an ageing population helps to increase the absolute level of effective labor and capital stock because baby boomers are still at work enjoying higher productivity and accumulating a larger amount of assets preparing for ageing. Both the increases in labor and capital necessarily drive the regional output above the level of GRP before the ageing population. However, in the subsequent period, GRP starts to decrease up to 2040s, and then converges at a level that is lower by approximately 9 % compared to the base year (2005). This

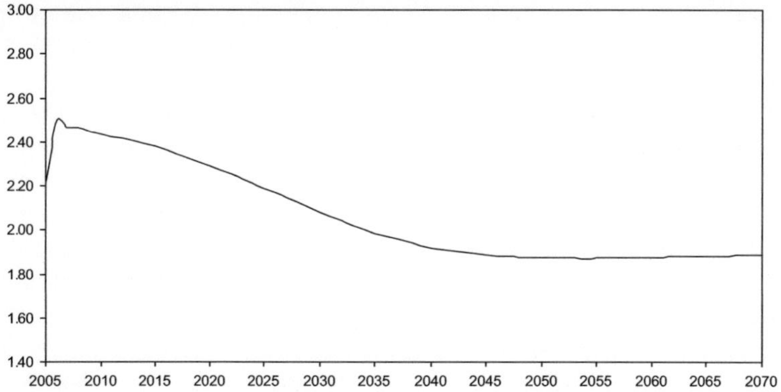

Fig. 11.7 Gross regional product (ageing population)

happens because after an initial overshoot, the capital stock starts to decrease gradually reflecting the fall in aggregate savings, and thus, two negative impacts, smaller capital stock and labor force, fuel the decline in GRP. The decreasing GRP leads, in turn, to a fall in the per capita GRP.

In contrast to the earlier finding, when an ageing population is considered in this more behavioral manner, the income inequality declines rather than increases. A major reason for this outcome may be traced to changes in social security payments by wealthier workers, increased returns from assets and with more forward-looking behavior, retirees will have more assets from which to draw income in retirement. The earlier analysis failed to include the effect of assets (non wage and salary income) and, increasingly, these will form a major part of the income base for retirees.

11.3.2 Immigration, Ageing and the Regional Economy

This part of the analysis explores changes in the impacts of immigration policies; it is assumed that the immigration policies between local and federal government are differentiated. This differentiation is not in terms of issues such as quotas, visa requirements, or guest worker programs but more in terms of a region's ability to compete more effectively for the pool of in-migrants. Hence, it is assumed that the local governments in the Chicago region implement a more favorable set of incentives to attract more immigrants than the federal government (as a share of total population). These might include housing subsidies, enhanced social and health care programs, pro-active recruiting policies (through public-private partnerships) and general enhancement of the current process of channelization of immigrants flows (regions with high existing levels of immigrants have a higher probability to compete more effectively for new immigrants using family and community (Chicago)-to-community (source of immigrants in their home country) ties.

The issues surrounding international immigration has become one of the most debated topics because it has both positive and negative impacts on the host economy. One of the biggest costs that immigration may create might generate would be through adverse effects on the local labor market by crowding out; increased immigration could reduce wages and exhaust employment opportunities for native workers, especially for those who are young and have low skills. Also, higher income disparities could be generated due to the large decline in the income of low-skilled workers. On the other side, however, immigration fundamentally changes the age structure, and may very helpful in contributing to a solution to the demographic imbalance caused by ageing population. Also, one of the most common arguments in favor of immigration is that it will significantly alleviate the solvency problem of the social security program because immigrants pay social security tax, and usually have no parents who are currently drawing on the system. Of course this assumes that the immigrants participate in the formal economy (whether they are legal or not) and thus contribute through direct and indirect taxes.

Among U.S. states, Illinois has long been a major immigrant settlement place as the fifth leading immigrant-receiving state. It has admitted the nearly 0.4 million legal immigrants in the last decade, an average of 40,000 immigrants per year. The cumulative total of legal immigrants in Illinois between 1965 and 2002 was estimated to be 1.3 million. In addition, according to the Immigration and Naturalization Service (INS), over 0.4 million illegal immigrants reside in Illinois, and most of them are concentrated in Chicago region. Among these immigrants, more than three-fifths (64.7 %) of all immigrants since 1993 came from Mexico, Poland, India, Philippines, former Soviet Union, and China. Mexico alone accounted for nearly one-quarter of all new immigrants (24.8 %). This continuing influx of new immigrants will account for a much more significant share of the Chicago's population; now, the Latino population of Chicago slightly exceeds that of the African-American population and is growing more rapidly as a result of higher rates of natural increase as well as through in-migration (including both interregional and international contributions).

Simulations were conducted for the following three scenarios that are differentiated by the size of immigrants for both regions; Chicago and rest of the U.S. Scenario 1 assumes that each region admits new immigrants amounting to 0.6 % of the regional population every year, which is equivalent to the historical average of immigrants admitted in the Chicago region between 1993 through 2002. In contrast, scenario 2 assumes that only the Chicago region admits more immigrants (1.2 % of the population, or about 0.1 million per year) while rest of the US fixes the share of immigrants at 0.6 %. Scenario 3 assumes that the Chicago local government is more successful and attracts 1.5 % of its population, or about 0.12 million per year. According to these scenarios, the dependency ratio [the percentage of the dependent old age populations (those $\geq$65) to the population in the working age groups (between 15 and 64)] in the Chicago region is expected to be substantially reduced over the next several decades. For example, without immigration, the model projects a significant increase in the dependency ratio from 19 to 32 % over the next 30 years, whereas new immigrants admitted

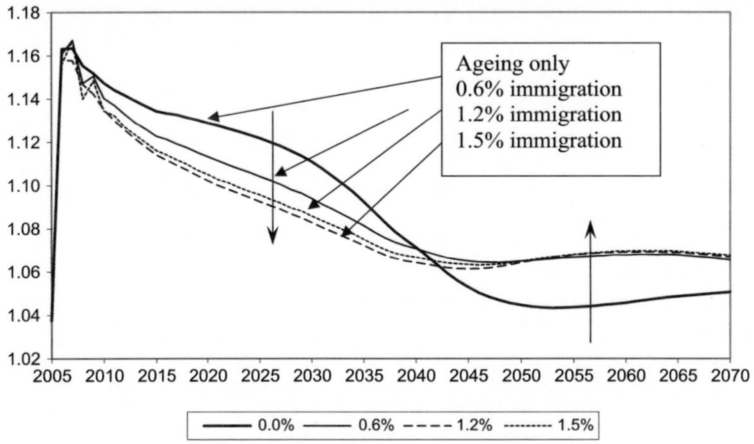

Fig. 11.8 Impacts of immigration on wages

following the Scenario 3 contribute to dropping the dependency ratio in the 2030s to 19 %, which is the same level (in 2005) before the impacts of an ageing population. Taking into account the characteristics of immigrants, who are usually younger and lower-skilled than the resident population, newly admitted immigrants are assumed to be equally distributed between the ages of 21 and 35, and whose average productivity is about 60 % of the peak at 47 years of age. The baseline Scenario, whose results are compared with Scenario 1 through 3, assumes an ageing population with no immigration. This is the scenario that was introduced in the previous section.

Figure 11.8 examines the impacts on wages. The inflow of young immigrants, initially, lowers the capital/labor ratio, which in turn contributes to a decrease in wages. However, after the initial period, the fall in the capital/labor ratio corresponding to accumulating immigrants decreases and ceases its downward trend around 2040, about 5 years earlier than the case of baseline (no ageing or immigration). After 2040, the wages under favorable immigration remain higher than the baseline case. This result is somewhat counter intuitive because large immigration should be expected to exert a strong downward impact on wages. One possible reason for this result is because the first immigrants start to retire in the early 2040s, resulting in an increase in the capital/labor ratio. However, there are two more important factors at work for this result to happen. The first factor is that the more immigrants that are admitted, the more native workers can save since immigrants will significantly reduce the social security tax burden (by increasing the after-tax income of native workers). Further, at the time of immigration, it is assumed that the capital does not flow into the host country with immigration, but once immigrants start to work and acquire the higher levels of productivity, they can accumulate more savings, thereby increasing aggregate capital stock. This is a critical assumption, especially as it pertains to the second and succeeding generations of offspring from the original immigrants.

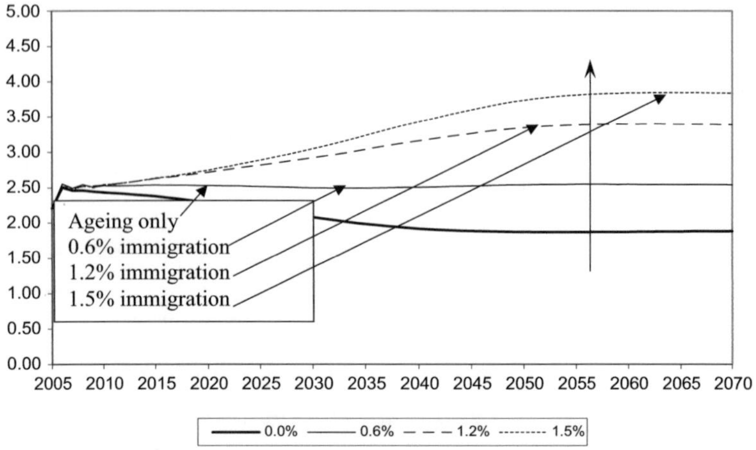

Fig. 11.9 Impact of immigration on Chicago gross regional product

Figure 11.9 shows how the gross regional product would be changed by different immigration streams over time. An increase in immigrants appears to have more positive impacts on regional output growth. For example, in the case of the maximum contribution by the most favorable policy (Scenario 3), the Chicago region appears to grow annually by 0.9 % between 2005 and 2070, while without immigration it will face negative growth (−0.2 % per year) over the same period due to ageing population. This result can be fully expected because immigration provides a positive labor supply shock to the local economy.

However, the transitional profile of per capita GRP (Fig. 11.10) is not similar to that of aggregate GRP as shown in Fig. 11.1. During the initial period, relatively larger immigration, in Scenarios 2 and 3, keeps the per capita GRP remaining at a lower level than that of the baseline case because the immigration increases (by assumption) only the supply of low skilled workers. However, after the 2030s, when the first immigrants really begin to acquire higher levels of productivity, per capita GRP has an upward trend and grows faster than the baseline case. This positive trend also substantially contributes to reducing the decline of per capita GRP under ageing population. For example, between 2005 and 2070, the negative 5.5 % per capita GRP growth under an ageing population is reduced to, ranging from negative 2.6 % in scenario 1 to negative 1.9 and negative 1.2 % in scenario 2 and 3, respectively. The national GRP share of the Chicago region noticeably increases from 3.0 % to around 3.5 ~ 4.0 % in Scenario 2 and 3 because both scenarios assume relatively higher share of immigrants are admitted only in the Chicago region.

The projected effect on the social security tax rate is shown in Fig. 11.11. Not surprisingly, a larger number of working-age immigrants appears to have a significant downward impact on the social security tax rate. Thanks to this downward pressure, in 2050, the social security tax rate is projected to return to the level established before the impacts of an ageing population. This is one of the most

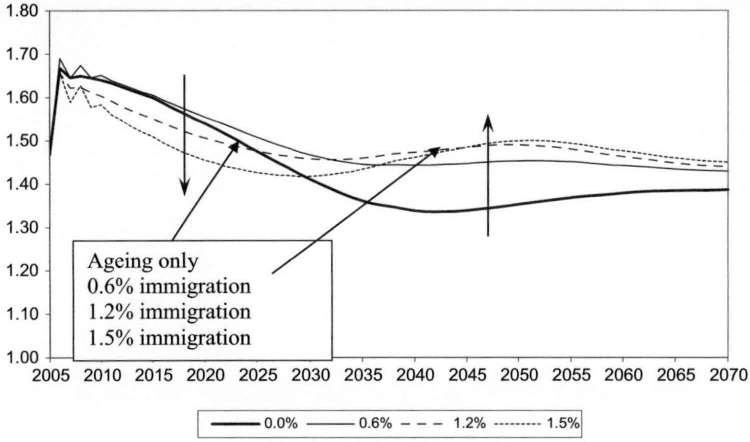

Fig. 11.10 Per Capita Chicago gross regional product

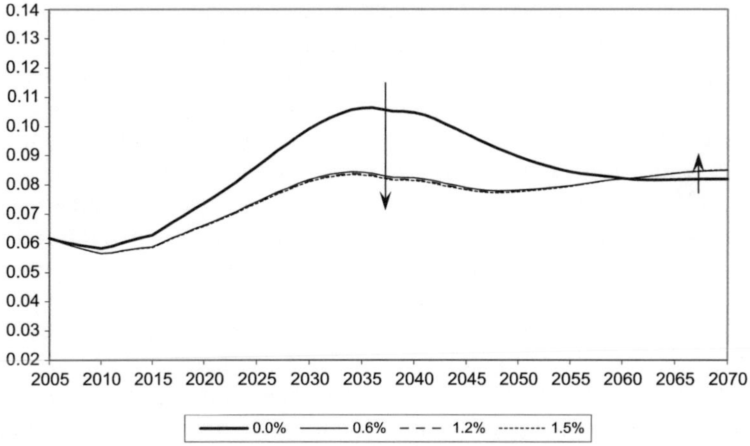

Fig. 11.11 Impact of immigration on the social security tax rate

significant benefits generated from immigration. However, on closer look, the benefit for social security system is reversed when the immigrants start to retire. After 2050, the social security tax rate starts to increase and eventually converges to around 9 %, higher than the rate expected under no immigration. This result reveals that in the longer run, immigration could generate a different impact; as immigrants age, like everyone else, a sustained policy of immigration has little long-run impact on the age structure of the population, and thus its benefit declines. Another important policy implication, especially for local government, arises from the different stance on immigration between federal and local governments. In Scenarios 2 and 3, only the Chicago local government optimistically attracts more immigrants than the national average. However, the social security tax rate

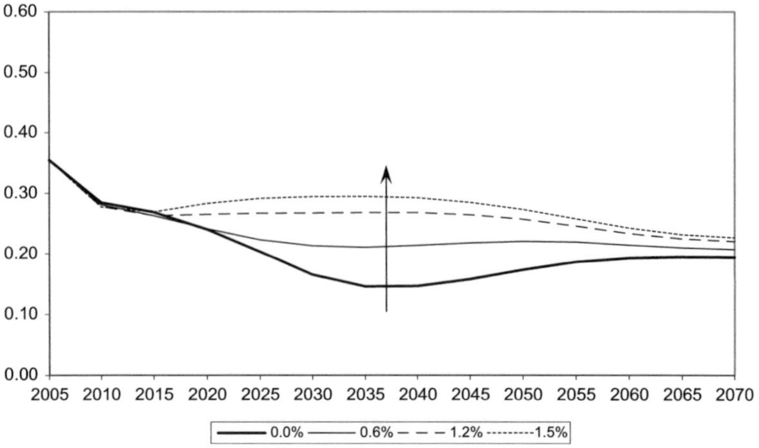

Fig. 11.12 Immigration impacts on income distribution: Gini coefficients for Chicago

changes insignificantly because the additional working-age immigrants in Chicago region are not of a large enough size to decrease the tax rate that is influenced by changes in the national population. Therefore, locally increased immigration may only hurt the local labor market without generating additional tax benefits. This is an important point; local autonomy in the case of a small region has limited impact of national policy that in turn could affect the outcome in Chicago.

Figure 11.12 present the effects of immigration on income distribution; immigration turns out to have a negative impact on equality in terms of income distribution, i.e., the income Gini coefficient becomes larger as more immigrants are admitted. There are two reasons for this. First, younger, lower income groups substantially rely on labor income, while middle-aged populations earn larger incomes from both asset holdings and labor earnings. Thus, the younger populations become relatively poorer as more immigrants decrease wage income, whereas rich middle-aged populations are not much affected by the immigration because they earn large capital income thanks to the increases in the interest rate. The second reason is closely related to the change in the demographic structure associated with immigration. Before the first immigrants start to retire around 2040s, the share of the population with larger income increases relatively faster than the younger and older poor populations because more immigrants acquire higher skills and become richer. This structural change in population increases the aggregate income gap between middle-aged richer population and poor young and old populations. However, after 2040s, since wages start to increase and immigrants start to retire, the Gini coefficients in all immigration scenarios starts to fall.

In contrast to the income distribution effect, immigration improves the equality of asset distribution until the mid 2030s, i.e., the asset Gini coefficient falls. However, the effect of immigration on asset distribution is reversed during the subsequent period. Basically, immigration has an upward pressure on asset Gini coefficient since it increases the asset holdings of the wealthiest group without significant changes in asset holdings of younger generations who face liquidity constraints. However, in the initial period, the increasing number of younger populations associated with new immigrants drives the asset Gini coefficient down, reflecting the reduced gap of the aggregate asset between middle-aged wealthy population and the younger, poor population.

The welfare effects of the immigration derived from the simulation results reveal that the current young populations appear to be big gainers of the favorable immigration policy[5]. The rationale for this is that even with the wage declines in the initial period, the prospect of higher disposable income for the rest of their lives obtained by both increased interest rates and reduced social security tax outweighs the negative effect from the wage loss. This is good news for current young generations. However, unlike the assumption of this model, if more immigrants fail to adapt to conditions in the host region's labor market, and thus, remain lower skilled workers, then immigration cannot make a sufficient contribution to increasing tax contributions.

11.3.3 How Does a Change in Retirement Age Affect a Regional Economy?

The final part of the analysis considers the impact of changes in the retirement age. Recall that it was fixed at 65 but the flexibility afforded by an absence of requirements to retire at this age is generating longer attachments to the labor force. Does this have much of an impact on a regional economy?

If the worker learns that he/she will live longer than previously expected, he/she would consume less or work longer before retirement to finance the additional consumption expenditure during their extended lifetime. In this model, even the maximum lifetime is limited to the age of 85, the average expected lifetime is assumed to increase due to the lower probability of death under an ageing population. Thus, the optimal behavior of each individual under an ageing population should be similar to that of the situation where an individual lives longer. However, an individual is not likely to choose smaller consumption because it could hurt his/her welfare; instead, he/she will continue to work, if possible, beyond the initial retirement age, so that he/she could adjust income in proportion to increased life expectancy, which would then allow the same amount of consumption as before. In this respect, increasing the retirement age can be considered as an alternative policy measures to compensate for the loss of labor supply under an ageing population. Moreover, since increasing the time spent working is equivalent to a delay in the age of initial social security benefit receipts, it might contribute to lessen the government's fiscal burden from public social security pension system. To some extent, the increase in the age at which recipients can receive the full social security benefits reflects a policy decision that derives from longer labor force attachment and the fact that people are living longer post retirement.

[5] The welfare benefit is measured by a consumption equivalent variation (EV), which computes the consumption change required to keep the expected utility in the initial condition equal to that achieved in the new condition under immigration policies [see Park and Hewings (2009)].

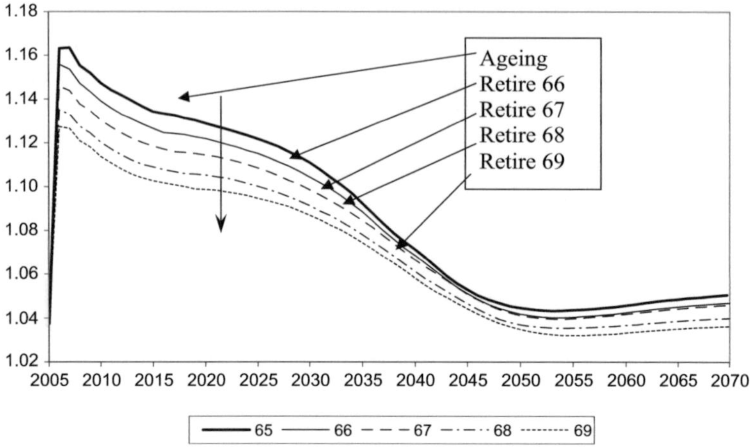

Fig. 11.13 Extending retirement age and the impact on wages

Simulations for impact analysis are conducted through the four scenarios, which are differentiated by retirement age. It is assumed that the retirement age is delayed by 1 year for each Scenario, i.e., for Scenario 1 through 4, individual is supposed to retire at 66, 67, 68, and 69, respectively. Once again, the baseline scenario is one in which the population ages as before. Increasing the retirement age generates a smaller capital/labor ratio compared to Baseline Scenario since the labor force increases as much as the working age is expanded. The lower capital/labor ratio leads to a fall in wages as shown in Fig. 11.13. According to the simulation results, if the retirement age is delayed by 4 years, i.e., retirement at the age 69, then wages fall by 7 ~ 8 % until 2030s compared to the baseline. Figure 11.14 shows what happens to the per capita GRP. Basically, the rise in the retirement age contributes to an increase the output, and thus the per capita GRP also increases since there is no change in the size of population. In particular, if individuals could continue working beyond the age 65 by at least 2 or 3 years longer, then the per capita GRP around 2050s starts to rise above the level before the ageing population occurs. However, the additional gain in per capita GRP corresponding to a 1-year increase in retirement age becomes smaller, reflecting the fact that the productivity of population decreases dramatically from age 65.

There is also an impact on social security tax rates. Not surprisingly, there is a marked decline in social security tax rate over the transition period. For example, the maximum tax rate around 2030s decreases from 11 % in the Baseline Scenario to below 6 % in Scenario 4, which is even lower than before the ageing population. The significant fall in tax rate becomes possible thanks to both increases in pension contribution by increased working-age populations and delay in the payment of pension benefits.

There is a small effect from increasing retirement age on income and asset distribution. The Gini coefficients for both cases appear to increase, though not noticeably, over the entire transition period as the retirement is delayed to the later age. This result is consistent with expectations suggested by transitional paths of

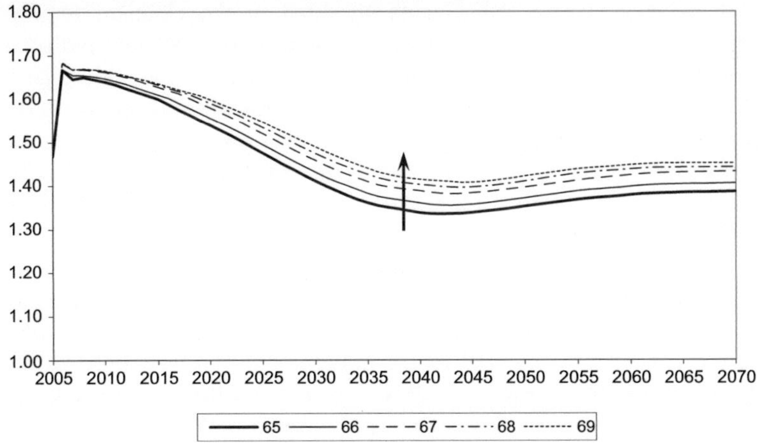

Fig. 11.14 Retirement age changes (65 through 69) and Chicago Per Capita gross regional product

wage and interest rates since wages fall and interest rates increase. Smaller wages under the increasing retirement age scenario reduces the income of young poor generations who significantly rely on labor income. On the contrary, higher interest rate increases the capital income of the middle-aged rich populations who holds large accumulated assets thanks to the reduced social security tax payment.

By affecting the social security tax rate, the increasing retirement age influences the allocation of consumption over the lifetime, and this reallocation may cause either increase or decrease in welfare. The welfare benefits changes depending on an increase in retirement ages. All individuals over the whole age cohorts appear to favor the increasing retirement age. Furthermore, younger generations gain more than old generations who have already retired. For younger generations, they benefit from the longer payrolls with smaller taxes until far in the future, whereas for the older generations welfare gains are limited since all the benefits are generated from increasing capital income arising from the increases in the interest rate.

What happens when immigration is also considered? According to the simulation, the optimal immigration occurs at the share of immigrants in the neighborhood of 0.6 %. However, beyond this point, like pension reforms, an increase in immigrants generates welfare cost. Now, it is a question why the optimal share of immigration does not occur at zero percent, since immigration distorts the local labor market by reducing the wage. The main reason is because immigrants significantly contribute to reducing the social security payroll tax. However, this benefit does not imply that an increase in the share of immigrants necessarily result in more welfare benefits. The main reason for this prediction is attributable to the difference of the immigration policies between federal and local government. Increasing the immigration share from zero percent to 0.6 %, both regions (Chicago and the Rest of the US) admit the same share of immigrants, so that the social security tax rate is substantially lowered since it is determined at the national level. Beyond the 0.6 %, only Chicago region adopts more new immigrants. However, the additional number of new immigrants admitted in the Chicago region is not sufficient to lower the tax rate, meaning that the increase in the number of local immigrants beyond the national average does not generate the additional benefits arising from tax cut that might be possible from a nation-wide increase in immigration. Instead, the welfare cost arising from smaller wages increase as more local immigrants are newly admitted. Of course, there still exits the welfare benefits for local residents because more immigrants generally stimulate higher economic growth. However, as long as the share of local immigration is higher than national average, this benefit is necessarily dominated by welfare cost.

However, the policy implications become complicated when immigration, pension reform, changing retirement age and skill acquisitions of the immigrant children are considered—as well as the effects of differential in-migration rates for Chicago and the Rest of the US. This is an area of research that needs far more attention—especially for the development of optimal policies.

11.4 Endogenous Investment in Human Capital[6]

The previous sections have highlighted an economy in which endowments of human capital are assumed to remain fixed. In the Chicago case studies, one of the strong assumptions was that the immigrants, while contributing to ameliorating the negative impacts of an ageing population in the short and medium term present problems in the longer term absent significant investment in human capital. A complementary perspective might ask whether enhanced investment in the human capital of the resident (non-immigrant) population might also contribute positively to reducing the negative impacts of population ageing. The idea here is that a more productive, albeit smaller workforce, could stem some of the anticipated decline in gross product (see Fig. 11.7 for example) by increasing per capita and potentially total output. In this section, some alternative strategies focused on Korea and the Midwest of the US will be used to illustrate some of the options available, their anticipated outcomes and the types of policy initiatives that might be required.

11.4.1 The Interregional CGE Model of Korea

An Interregional CGE Model of Korea (ICGE) was developed to analyze three industrial sectors of two major Korean regions: the Seoul Metropolitan Area (SMA) and the rest of Korea (ROK). The industrial sectors under consideration have been classified as primary (i.e., agriculture, forestry, fishing, and mining), manufacturing, and service industries. As to the regions, the SMA accounted for roughly 47 % of the national population and over 80 % of all major enterprise headquarters, while only accounting for 12 % of Korea's territory. The ICGE model accounts for the behaviors of the economic agents of six producers, two regional

[6] This section draws on Kim et al. (2013) and Kim and Hewings (2013a, b, c).

households, two regional governments, a central government, and the rest of the world. The policy simulations from the ICGE model identify how an ageing population affects regional economic growth and disparity over 15 periods. In addition, this paper examines how these economic losses can be compensated by educational investments in the 20–29 age cohort.

As noted earlier, the proportion of the elderly population (65+) to those of working age (15–64) in Korea is expected to rise from 11 % in 2008 to an estimated 23 % in 2030. Dealing with this reality presents a unique challenge for the government, as the welfare programs typically used to alleviate socio-economic problems—such as public pension payments, medical insurance benefits, and public assistance for the low-income classes—are in need of rapid expansion, and yet tax revenue is on the decline (Chun 2006). The main objective here is to estimate how much of an increase in labor productivity is required to offset the effect of a shrinking labor force. The increase will be assumed to be derived from educational investments on economic losses incurred from the population aging using an Interregional CGE Model (ICGE) of Korea. The ICGE model identifies three industrial sectors: primary (i.e., agriculture, forestry, fishing, and mining), manufacturing, and service industries. As to the regions, the SMA accounted for roughly 47 % of the national population and over 80 % of all major enterprise headquarters, while only accounting for 12 % of Korea's territory. The ICGE model accounts for the behaviors of the economic agents of six producers, two regional households, two regional governments, a central government, and the rest of the world. The policy simulations from the ICGE model identify how an aging population affects regional economic growth and disparity for 15 periods.

Analysis of the structural changes in population composition, age-earning profile, and income sources found that as the growing aging population becomes less economically active, they are likely to become increasingly economically dependent. This potential outcome is supported by the fact that the 60+ age cohort depends significantly more on outside income (e.g., social security) than other age groups, while simultaneously possessing considerably lower overall income levels. Non-wage income sources, such as grown-up children, subsidies from state or local governments, and social security benefits, account for over 50 % of the total income of the 60+ population. Hence, it is important to understand how the elderly population functions before the combined effects of their dependency and increased longevity (due to higher life expectancy rates) acts as a damper on contemporary economic growth.

In contrast to the earlier work reviewed here (Kim and Hewings 2013a, b, c) that integrated a perfect foresight assumption for households with an optimization behavior for consumers and producers. More attention need to be directed to the specification of interregional economic interactions and the roles of financial sectors in the market. For example, the ICGE model follows an adaptive expectation for price inflation, whereas Kim and Hewings' (2013a, b) model employs Auerbach and Kotlikoff's OLG model to accommodate uncertainty and idiosyncratic risk. This means that household consumption in Kim and Hewings (2013a, b) is accounted for in terms of a mathematical optimization process, whereas in the ICGE model, household consumption is determined by the saving rate.

Details of the model may be found in Kim et al. (2013); on the demographic side, the most relevant characteristics are the disaggregation of population into eight age cohorts: 0–9, 10–19, 20–29, 30–39, 40–49, 50–59, 60–69, and an 70+ age group. Each age cohort carries different parameters and values for labor productivity, mortality rates, and participation rates in the labor market (i.e., share of labor supply relative to total population size) on the supply side, and saving rates and consumption behaviors on the demand side. Each regional labor input by age cohort is assumed to be homogeneous and to possess intersectoral mobility, whereas it is assumed that capital stock cannot move from one region to another. The labor demand by various regions and industry is derived from the producers' value-added maximization of the first order conditions, whereas labor supply relies on the participation rates of the various age cohorts and the total population size of the region overall. The labor productivity by region and working age group is estimated through a form of Mincerian earning regression, in which the determinants of the productivity are gender, education level, job experience, type of occupation and industrial sector, and possession (or lack thereof) of a professional license. Under the neoclassical closure rule for the labor market, the labor participation rate is derived by balancing out total labor demand against total labor supply. If the population flows among the regions are not exogenous to the model, then in–migration is assumed to be in response to interregional differences between origin and destination regions in terms of wage per capita and unemployment rate, as well as the physical distance between the regions. Hence, the population of a given region is the sum of the natural growth of the native population combined with the net gain (or loss) of migrant populations.

Two tiers of government structure are specified in the model: two regional governments and one national government. Government expenditures consist of consumption expenditures, subsidies to producers and households, and savings. Revenue sources include taxation of household income, value-added, and foreign imports. With regard to the macroeconomic closure rule for the capital market, aggregate savings determines investments. There is one consolidated capital market, consisting of household savings, corporate savings of regional production sectors, private borrowings from abroad, and government savings. There are no financial assets in the model, so overall consistency requires equating total domestic investment to net national savings plus net capital inflows. The sectoral allocation of total investment by destination is endogenously determined by the capital price from each sector and the allocation coefficient of investment. The investment allocation by destination is transformed into the sectoral investment by origin through a capital coefficient matrix. This price adjustment is required for the Walrasian equilibrium condition, and every price is measured on a relative scale.

The ICGE model is a recursive and adaptive dynamic model, composed of a within-period model and a between-period model. The within-period model determines equilibrium quantities and prices under objectives and constraints for each economic agent, in which the balance between supply and demand is achieved in a perfectly competitive market. The between-period model finds a sequential equilibrium path for the within-period model over the multiple periods by updating the values of all exogenous variables, such as government expenditures, from one period to another. For example, the current capital stock is expanded with new

Table 11.1 Major equations in the ICGE model

Output	Output = Leontief (Value added, Intermediate demand)
Value added	Value added = Total Factor Productivity × CD (Capital stock, Labor by age cohort)
Supply	Output = CET (Foreign exports, Domestic supply)
Domestic supply	Domestic supply = CET (Regional exports, Intraregional supply)
Demand	Demand = Armington (Foreign imports, Domestic demand)
Domestic demand	Domestic demand = Armington (Regional imports, Intraregional supply)
Labor demand	Labor demand = LD (Wage by age cohort, Value added, Net price)
Wage by age cohort	Wage by age cohort = WA (Education, Type of employment, Job experience)
Labor supply	Labor supply = LS (Labor market participation rate, Population)
Population	Population = Natural growth of previous year's population + Net population inflows
Regional incomes	Regional incomes = Wage + Capital returns + Government subsidies
Migration	Migration = TODARO (Incomes and Employment opportunities of origin and destination, Distance between origin and destination)
Consumption by commodity	Consumption by commodity = CC (Price, Population size by age cohort, Incomes)
Private savings	Private savings = PS (Saving rate by age cohort, Population size by age cohort)
Government revenues	Government revenues = Indirect tax + Direct tax + Tariff
Government expenditures	Government expenditures = Government current expenditure + Government savings + Government investment expenditure + Government subsidies
Labor market equilibrium	Labor demand = Labor supply
Capital market equilibrium	Private savings = Total investments
Commodity market equilibrium	Supply of commodities = Demand for commodities
Government	Government expenditures = Government revenues
Capital stock	Capital stock = Depreciated lagged capital stock + New investments

investment but also reduced with a constant depreciation rate. The within-period model is a square system of equations with 196 equations and 231 variables; a unique solution can be found because the number of endogenous variables is the same as the number of the equations under convexity. The exogenous variables include world market prices and government expenditures. The numeraire of the model is set as the consumer price index. In addition, we calibrated a Social Accounting Matrix (SAM) as a benchmark for the development of the ICGE model. The SAM consists of six accounts—factors, households, production activities, government, capital, and the rest of the world—and is treated as an initial equilibrium for the ICGE model. The values of some parameters are adjusted to replicate the equilibrium conditions for the base year, 2005. Table 11.1 summarizes major equations in ICGE model.

11.4.2 Simulation with the ICGE Model for Korea

In general, an aging population carries with it both positive and negative effects for economic development. Negatively, it results in an intergenerational imbalance between benefits, costs of pensions, and overall healthiness, thereby reducing the total labor supply and lowering saving and investment rates (Canning 2007; Horioka 2007; Bloom et al. 2009). Positively, however, it induces higher rates of labor market participation from older workers, thereby leading to longer working lives. Moreover, the declining fertility rate associated with an aging population may lead to a more general increase in female labor force participation, which may help to offset the foregoing negative effects arising from lower male labor force participation and savings rates. The impacts an aging population has on changes in the consumption expenditures, however, can be rather ambiguous in that the effect depends on life expectancy rates and expected retirement age. For example, an individual may consume less and work more in order to finance additional consumption expenditures after retirement (Park and Hewings 2007c). Yet, if one's expected retirement age were to be deferred in accordance with prolonged life expectancy rates, future welfare needs may be deemed too distant; hence, consumption patterns might decline drastically as one nears or enters retirement age.

In this section, the ICGE model was applied to estimate the effects of an aging population on the regional economies of residents for the Seoul Metropolitan Area and the rest of Korea. The baseline assumed that population aging had not yet gone into effect. The population share by cohort in this baseline is projected using the trend from 2000–2005; likewise, all parameters are kept at their 2005 levels, so as to simulate a scenario in which all population-related policies remain unchanged from that same year. The base year and the snapshot year for this analysis are 2006 and 2020 respectively; that is, the baseline accounts for what would happen without any significant changes for a 15-year period. In this paper, the aging population and the composition of population by age cohort are key variables, and their changes based on the scenario are considered 'shocks' to the ICGE model.

Regarding simulations, there are at least three shocks including (1) population effects, such as changes in population growth rate, number of households, working age population, and labor supply available, all effect levels of private consumption; (2) technological innovations, which enhance productivity; and, (3) public policies, such as reforms in pension benefits, education and training programs, and insurance systems, which may have a variety of welfare related effects. However, only the third shock is not considered here due to the model structure designed for the real-side economy. The two other shocks can be taken into account in the model simulation if the ICGE model is expanded to the financial side economies. The shock from the population variables is modeled and injected into the ICGE, and then a new set of equilibrium values can be generated for regional production and prices for a 15-year span, thereby satisfying the price normalization rule subject to the exogenous consumer price inflation rate without operation of the interregional migration module. Long-term effects come into play during the second period in the form of the stock-accumulation effects of changes in capital stock and population

Table 11.2 Differences in share of population by cohort between baseline and population aging Scenario (unit: %)

	2006	2010	2015	2020
Seoul metropolitan area				
0–9	−0.1	0.0	1.5	3.1
10–19	0.2	0.7	−0.4	−1.1
20–29	0.0	0.0	0.8	2.3
30–39	−0.1	−0.6	−0.9	−1.2
40–49	−0.3	−2.1	−5.0	−8.3
50–59	0.3	1.7	3.0	2.1
60+	0.0	0.3	1.0	3.1
Rest of Korea				
0–9	−0.2	−0.4	0.7	2.0
10–19	0.3	1.3	0.3	−0.4
20–29	0.1	0.3	1.6	3.2
30–39	−0.1	−0.6	−1.0	−1.4
40–49	−0.5	−2.1	−4.9	−8.1
50–59	0.3	1.2	2.2	1.1
60+	0.1	0.3	1.1	3.6

Note: Difference share = (Population aging scenario—baseline)

demographics. In the counterfactual scenario that population aging continues, composition of the population by age cohort continues along the projections provided by the National Statistical Office of Korea so as to reflect the demographic effect of population aging for future generations (see Table 11.2). Roughly, the scenario has a lower population share of the 20–59 age cohort by 0.5 % point on average and a higher share of the 60+ age cohort by 1 % point compared to the baseline.

If the shares of the age cohort 20–49 decrease for both regions, according to the population aging for 15 years, there will be a sharp downturn in the economy, an outcome that is similar to traditional intuition and work. The direct impact of the population aging on the economies is a reduction in the labor supply. Total labor supply would decline by 3.01 % on a 15-year average compared with the baseline, as shown in Table 11.3. Government revenues are expected to drop by 1.98 %. In addition, the population aging has a negative effect on the GDP by −2.37 % on average; −0.69 % in the period 2006–2010, −1.90 % in the period 2011–2015, and −3.55 % in the period 2016–2020. The negative effects on the GDP tend to become more severe over the period of 15 years at an increasing rate. The population aging could generate negative impacts on the economic growth of both the SMA and ROK regions, but it seems to pose a greater threat to the SMA economy. The growth rate of the Gross Regional Product (GRP) of the SMA is −2.78 % on average, which is lower than that of ROK by 0.74 % point. The reason for this difference is that the ROK has already attained higher population shares of over 60s than the SMA under the baseline: the level of the ROK is larger than that of the SMA by 0.1 ~ 0.6 % point.

Table 11.3 Impacts of population aging on value added, government revenues and labor supply (unit: %)

	GDP	GRP of SMA	GRP of ROK	Government revenue	Labor supply
2006	−0.21	−0.14	−0.26	−0.19	−0.24
2007	−0.43	−0.45	−0.42	−0.38	−0.56
2008	−0.65	−0.64	−0.66	−0.57	−0.85
2009	−0.95	−0.96	−0.94	−0.81	−1.24
2010	−1.22	−1.23	−1.21	−1.01	−1.56
2011	−1.43	−1.50	−1.38	−1.16	−1.77
2012	−1.71	−1.89	−1.57	−1.36	−2.12
2013	−1.81	−1.96	−1.68	−1.39	−2.12
2014	−2.14	−2.40	−1.94	−1.65	−2.69
2015	−2.40	−2.79	−2.09	−1.83	−3.20
2016	−2.73	−3.20	−2.35	−2.08	−3.92
2017	−3.11	−3.69	−2.64	−2.40	−4.82
2018	−3.31	−4.11	−2.66	−2.56	−5.44
2019	−3.75	−4.62	−3.04	−3.01	−6.43
2020	−4.84	−5.72	−4.13	−4.35	−7.33
Average	−2.37	−2.78	−2.04	−1.98	−3.01

SMA Seoul metropolitan area, *ROK* Rest of Korea

How can these economic losses be compensated for in the long run? A few policy options have been proposed, e.g., securing new labor forces such as retirees, immigrants, and female workers through incentive tools and improving the technological skills through education and training programs. This paper focuses on investment policies concerning the university education of the 2020s groups to cope with the labor supply shortage created by the population aging. Seven alternatives for the investment simulation are set up depending on the annual growth rates of the educational investment expenditures from 0 % to 12 % compared with the baseline (see Table 11.4 where the results for selected years are shown). The increase in the education investments with the annual growth rate of 3 % could result in the GDP decreasing by −1.92 %, which is higher than the case without the investment (0 %) by 0.45 % point. Even though the investment growth rate is expanded to 11 % for 15 years, the GDP would not be larger than the result of the baseline with the population aging. Finally, the simulation with the annual growth rate of 12 % could lead to the GDP increasing by 0.26 % on the 15-year average, offsetting the negative economic effects caused by the population aging in Korea. Under this scenario, the government is required to increase the expenditure by 12 % every year for at least 11 years to generate a net growth of the GDP in the sense that the GDP growth rate increases from −0.34 % in the 11th period (2016) to 0.15 % in the 12th period (2017).

Contrary to traditional intuition, the results in Table 11.4 show that the increase in the educational investment expenditures would not reduce GDP losses caused by

Table 11.4 Impacts of education investments on GDP growth rate and regional income ratio (unit: %)

	0 %	3 %	5 %	8 %	10 %	11 %	12 %
(1) GDP growth rate							
2006	−0.21	−0.28	−0.34	−0.42	−0.48	−0.51	−0.54
2010	−1.22	−1.18	−1.16	−1.17	−1.19	−1.21	−1.23
2015	−2.40	−1.92	−1.60	−1.16	−0.89	−0.76	−0.64
2020	−4.84	−3.37	−2.87	−0.87	0.93	2.15	3.48
Average	−2.37	−1.92	−1.67	−1.11	−0.60	−0.27	0.06
(2) Ratio of regional domestic products of ROK to SMA							
2006	1.389	1.388	1.388	1.388	1.387	1.387	1.387
2010	1.302	1.292	1.285	1.276	1.269	1.266	1.264
2015	1.254	1.217	1.194	1.162	1.141	1.131	1.122
2020	1.255	1.179	1.131	1.064	1.022	1.001	0.980
Average	1.277	1.240	1.217	1.183	1.161	1.151	1.140

SMA Seoul metropolitan area, *ROK* Rest of Korea

the population aging; rather, it would worsen the problem in the early stage of investment periods (2006–2010). It would take 6 years to see a net positive change in the GDP through investments in the universities. This consequence is attributed to a pronounced mismatch or timing discrepancy between the cost input and the output generation, where the slope of cost curve is constant over all simulation periods and the GDP increases in a logistic trend. So the GDP gap (=increase in GDP − expenditure costs) could be negative from the first period (2006) to the fifth period (2010), and positive after the sixth period (2011). On the other hand, the improvement in labor productivity through educational investments is expected to reduce the economic losses in terms of national economies but also increase the income difference between the two regions. For example, the ratio of the GRP of the ROK to the SMA decreases from 1.277 in the simulation with null growth rate to 1.140 in the simulation with a 12 % annual growth rates on the investments. This disparity is caused by different elasticity values of education by age cohort with respect to the labor productivity for two regions: 1.062 for the SMA and 0.819 for the ROK. More investments in the ROK could be a solution to tackle this regional problem, but the consequence is unclear due to the chronic brain drain from the ROK into the SMA, according to Shim and Kim (2012), who argue that the latter region had a pulling effect on the college graduates of the former.

11.4.3 US Midwest Analysis

Drawing on some innovative proposals by Sadahiro and Shimasawa (2002) and Ludwig et al. (2007), Kim and Hewings (2013a, b, c) explored the role of options for enhancing productivity through investment in additional training. However, the timing of the training and its frequency were not considered. A multiregional model involving six regions (Illinois, Indiana, Michigan, Ohio, Wisconsin and the rest of US) was developed. The economy was assumed to be closed to the rest of the world. There are two types of economic agents in each region: a representative firm and households and in each year, there are 65 overlapped generations in the household sector. The federal government operates the social security system that operates in all regions. Households now have three decisions for disposable income: (1) allocation between consumption and saving (inter-temporal); (2) allocation between goods produced in any region (inter-regional) and (3) allocation between education (investment in human capital) and working.

The results revealed significant differences across regions in existing levels of productivity and growth rates over time. In contrast to the findings from Korea, there were some significant positive impacts from investment in human capital, although the time period for them to occur was extensive (greater than 10 years). However, the interregional CGE model used in the US analysis was not as comprehensive as the one for Korea and one suspects that these differences together with the way in which initial endowments from one generation influence future generations [see Sadahiro and Shimasawa (2002)] account for the variations in outcomes. Clearly, the interactions with financial considerations and assumptions about government behavior need to be addressed more effectively. Further, adoption of an assumption about inter-temporal optimization will need to be modified to account for idiosyncratic behavior; in addition, the institutional responses to almost continual skill upgrading will need to be considered to ensure that there is a greater degree of flexibility and responsiveness to changing market conditions.

11.5 Limitations and Further Research Agenda

One of the important considerations that has not been fully embraced in all the models is the impact on equity of mitigation problems. Consider the Korean case: while the aging trend could cause a reduction in the GDP by 1.94 % on average, the damage could be compensated for through an increase in the educational investments in the 20–29 age cohort by 12 % for at least 15 consecutive years. However, the investment policy has a negative impact on the regional income distribution due to different elasticity values of education by age cohort with respect to labor productivity. However, the issue becomes more complicated when other factors and policy options are considered—as revealed in Sect. 11.2. As Borsch-Supan (2004) noted, regional policies for coping with the challenges of shifting demographic structures can be practically formulated if the responses of economic agents are specified according to economic incentives and motivations for labor participation by recognizing age-specific productivities. For example, which populations are willing to migrate from labor-abundant to labor-scarce regions? What policy measures lead to greater participation in training and education programs? What barriers exist that discourage women from participating in the labor market? What are the positive and negative effects of raising the mandatory retirement age? Should national or local governments encourage increased immigration flows to deal with labor shortage issues? Are less restrictive capital flows more capable of overcoming the various political barriers regarding cross-border labor flows?

Additional considerations, especially related to equity concerns, need to be addressed. While Yoon and Hewings (2006) found that ageing generated a negative impact on the Gini coefficient of income equality, Park and Hewings (2009) found that when assets (i.e. non wage and salary income) were included, the impact was reversed—although not continuously over time. However, the choices of non-wage and salary assets was not specified. Hence, it would seem appropriate to extend the current real-side models of the ICGE type to the real-financial side so as to measure the effects of allocating financial assets on the economies. The agent could choose a set of financial assets including not only simple savings with commercial banks but also bonds and equities, depending on their own capital costs and returns. In addition, another possible extension would be to focus more on the role of migration particularly specified by age and, for the working population, levels of human capital endowment, that is likely to be heterogeneous by age group. Further, retiree migrants will exert a more significant role as they shift the locus of their expenditures from SMA to ROK in Korea or from the US Midwest to locations in the southeast or the west.

In summarizing the demographic challenges to economic development, there are several important conclusions that can be drawn from the nascent literature. Household consumption varies by age and income level; as the composition (age structure or income structure) of households change, there are likely to be important changes in the type of goods and services demanded. Absent immigration and investment in human capital—particularly in the prime working age populations—there is likely to be a negative effect on gross product. Immigration, investment in human capital and removal of retirement age restrictions may help offset the negative effects of ageing populations. However, internal migration may serve to complicate the interregional impacts. Further, the impacts of immigration on regional competitiveness have not been explored especially the potential for differential impacts over time [see Park and Hewings (2007b)]. Finally, the financial implications—particularly the fiscal stress likely to be placed on social security systems by increases dependency ratios (number on non working age people dependent on each worker) create considerable uncertainties for effective planning.

Better health care has also created another issue—extending the anticipated lifetime for many retirees. Assets planned for a 10-year retirement period may now need to stretch to 20 or 30 years; as a result, many retirees have re-attached to the labor force. Do they reclaim prior occupations or move into other fields? Do they remain where they have retired or re-locate, perhaps back to communities in which they worked previously? Retiree cohorts are not homogeneous and thus there is a need to focus on further disaggregation of those post 65 years of age. Finally, there is the issue of the housing stock; retiree households are usually smaller in size and thus housing demands—in terms of size and location—may be very different from the current supply, much of which is oriented to working-age families. The implications for the housing market, transportation demands and the provision of public goods (such as schools) will all require significant modifications.

The challenges are significant, requiring innovative adjustments to the set of existing models; impact analyses and forecasts that ignore these demographic dynamics are likely to be seriously in error. The issues are most decidedly spatial in character, providing the possibilities for the next generation of scholars to expand the regional science toolbox and thus continue to build on the foundation that Walter Isard provided.

Acknowledgments Parts of this paper draw on joint work of SangGyoo Yoon, Seryoung Park and Tae-Jeong Kim whose contributions are gratefully appreciated. The comments of an anonymous referee and the support of the National Science Foundation Grant 0818575 is gratefully acknowledged.

References

Birkin M, Clarke GP (2011) The enhancement of spatial microsimulation models using geodemographics. Ann Reg Sci 49:513–532

Bloom DE, Canning D, Fink G (2009) The graying of global population and its macroeconomic consequences. Harvard School of Public Health

Borsch-Supan, AH (2004) Global ageing: issues, answers, more questions. Michigan Retirement Research Center. Research Paper No. WP 2004-084

Canning D (2007) The impact of ageing on Asian development. Paper presented at seminar on ageing Asia: a new challenge for the region. ADB annual meeting, Kyoto, Japan

ChunYJ (2006) Population aging, fiscal policies, and national saving: predictions for Korean economy. NBER Working Paper No. 12265, National Bureau of Economic Research

Conway RS (1990) The Washington projection and simulation model: ten years of experience with a regional interindustry econometric model. Int Reg Sci Rev 13:141–165

Conway RS (1991) An empirical comparison of regional multipliers. In: Dewhurst JJLL, Hewings GJD, Jensen RC (eds) Regional input-output modeling: new developments and interpretations. Avebury, Aldershot, pp 178–195

Deaton A, Muellbauer J (1980) An almost ideal demand system. Am Econ Rev 70(3):312–326

Hägerstrand T (1970) What about people in regional science? Pap Reg Sci Assoc 24:7–21

Hewings GJD (1982) Trade, structure and linkages in developing and regional economics. J Develop Econ 11:91–96

Hewings GJD, Fonseca MAR, Guilhoto JJM, Sonis M (1989) Key sectors and structural change in the Brazilian economy: a comparison of alternative approaches and their policy implications. J Polcy Model 11:67–90

Israilevich PR, Hewings GJD, Sonis M, Schindler GR (1997) Forecasting structural change with a regional econometric input-output model. J Reg Sci 37:565–590

Kim T-J, Hewings GJD (2013a) Endogenous growth in an aging economy: evidence and policy measures. Ann Reg Sci 50:705–730

Kim T-J, Hewings GJD (2013b) Inter-regional endogenous growth under the impacts of demographic changes. Appl Econ 45:3430–3448

Kim, T-J, Hewings GJD (2013c) Aging population in a regional economy: addressing household heterogeneity with a focus on migration status and investment in human capital. Int Reg Sci Rev (forthcoming). doi: 10.1177/0160017613484930

Kim E, Hewings GJD, Lee C (2013) "Impact of Educational Investments on Economic Loss from Population Aging Using an Interregional CGE Model of Korea", Presented at the 52nd European Congress of the RSAI. Bratislava, Slovakia, August 21–25:2012

Li P-C, Rose A, Eduardo B (1998) Construction of an input-output income distribution matrix for the US. In: Hewings G (ed) Advances in spatial analysis: essays in honor of Kenichi Miyazawa. Springer, Berlin, pp 347–365
Ludwig A, Schelkle T, Vogel E (2007) Demographic change, human capital and endogenous growth. Working Paper, MEA, University of Manheim
Miyazawa K (1976) Input-output analysis and the structure of income distribution. Lecture Notes in Economics and Mathematical Systems, vol 116. Springer, Heidelberg
Park S, Hewings GJD (2007a) Aging and the regional economy: simulation results from the Chicago CGE model. Discussion Paper, 07-T-4, Regional economics applications laboratory, University of Illinois, Urbana, IL
Park S, Hewings GJD (2007b) Immigration, aging and the regional economy. Discussion Paper, 07-T-5, Regional Economics Applications Laboratory, University of Illinois, Urbana, IL
Park S, Hewings GJD (2007c) Does a change in retirement age affect a regional economy? evidence from the Chicago economy. Discussion Paper 07-T-6, Regional economics applications laboratory, University of Illinois, Urbana, IL
Park S, Hewings GJD (2009) Immigration, ageing and the regional economy. Cityscape: A J Policy Develop Res 11:59–80
Rogers A (1990) Requiem for a net migrant. Geogr Anal 22:283–300
Rose A, Beaumont P (1988) Interrelational income distribution multipliers for West Virginia. J Reg Sci 28:461–475
Rose A, Beaumont P (1989) Interrelational income distribution multipliers for the U.S. Economy. In: Miller R, Polenske K, Rose A (eds) Frontiers of Input-Output Analysis. Oxford University Press, New York
Rose A, Li P-C (1998) Interrelational Multipliers for the US Economy: An Application to Welfare Reform". In: Hewings G (ed) Advances in spatial analysis: essays in honor of Kenichi Miyazawa. Springer, Berlin, pp 347–365
Sadahiro A, Shimasawa M (2002) The computable overlapping generations model with an endogenous growth mechanism. Econ Model 20:1–24
Shim J, Kim E (2012) Analysis of occupational mobility of college graduates. Korea Spatial Plan Rev 75:37–51 (in Korean)
van Leeuwen ES (2010) The effects of future retail developments on the local economy; combining micro and macro approaches. Pap Reg Sci 89:691–710
Wakabayashi M, Hewings JDG (2007) Life cycle changes in consumption behavior: age-specific and regional variations. J Reg Sci 47:315–337
Yoon SG, Hewings GJD (2006) Impacts of demographic changes in the Chicago Region. Discussion paper 06-T-7, Regional economics applications laboratory, University of Illinois, http://www.real.illinois.edu/d-paper/06/06-t-7.pdf

Part III

Regional Policy and Uncertainty

Cluster Life-Cycles, Entrepreneurship and Regional Economic Development with a Case Study of the Korean Shipbuilding Cluster

12

Roger R. Stough

Abstract

This paper views industrial clusters as complex systems composed of multiple interdependent dimensions, firms and enterprises, organizations. A literature review is used to construct a life-cycle framework for assessing cluster dynamics and guiding related policy development. The framework is erected on seven dimensions along with measures and development related patterns associated with the different life-cycle stages. Examples of policy guidelines are provided across all seven dimensions and by stage of cluster development. Finally, a case study of the South Korea shipbuilding cluster is provided in an effort to illustrate how the framework can be used to analyze cluster dynamics and to guide related policy responses. Conclusions and future research opportunities are presented in the final part of the paper.

12.1 Introduction

The purpose of this paper is to develop and apply a framework for assessing cluster dynamics and providing policy guidelines. The cluster concept is well known and often the basis for contemporary policy and practice in economic development following the work of Michael Porter (1990). Clusters are organized around one or more industries and attract investment and related companies and organizations because they enable benefit capture from Marshallian positive externalities such as reduced procurement costs, strong knowledge spillovers, and lower transaction costs. These and other advantages such as improved market knowledge and information are amplified as strong internal and later external networks evolve (Porter 1998 and Rocha 2004). Further, empirical studies by Rocha and Sternberg (2005) and Delgado et al. (2010) provide convincing evidence that clusters also motivate

R.R. Stough (✉)
School of Public Policy, George Mason University, Arlington, VA 22201, USA
e-mail: rstough@gmu.edu

P. Nijkamp et al. (eds.), *Regional Science Matters*, DOI 10.1007/978-3-319-07305-7_12,

and support increased new firm formation. More general discussions of the relationship between clusters and entrepreneurship appear in Stough et al. (2014).

12.2 Cluster Dimensions

Despite evidence that clusters contribute to strong regional economic growth and development as above, Martin and Sunley (2011) recently observe that it is the increasing returns part of clustering that has tended to be recognized by earlier cluster investigators. Consequently, serious attention to all parts of the full pattern of cluster development are lacking if not missing. To the extent that cluster dynamics have been considered it is mostly with respect to life-cycle theory (Bergman 2008; Bode and Alig 2011; Martin and Sunley 2011; Lefebvre 2012; and Tichy 1998) and largely in the recent past.[1] In this paper the dynamics of clusters are viewed from the perspective of merging the life-cycle approach with concepts from the fields of social complexity and general systems theory.

Dynamic industry clusters are defined in this paper as complex systems composed of multiple interdependent dimensions and multiple firms and enterprises. Thus clusters are understood as integrated ensembles of interdependent dimensions whose values or weights of relevance change over time. The dimensional element of this conception was first envisioned in a study of the shipbuilding cluster in Northern Ireland (Klink and De Langen 2001). The dimensions used in the present study are derived in part from the Klink and De Langen (2001) study but also have been merged with others that have appeared repeatedly in the cluster literature (Baptista and Mendonca 2010; Bergman 2008; Brocker and Soltwedel 2010; Karlsson 2008; Martin and Sunley 2011; Menzel and Fornahl 2009; Stough et al. 2014) to create a seven dimension manifold for analyzing cluster dynamics. These dimensions are important because life-cycle theory implies that certain states or stages are reached as a system (cluster) evolves through the life cycle.[2] At each stage these dimensions are envisioned to have specific values or qualities thus providing the basis for estimating how well a cluster follows the theoretical development path. The dimensional profiles also provide targets or guides for policy intervention. Based on a review of the cluster literature the dimensions are:

1. Spatial concentration—greater in early and more dispersed in later stages
2. Industrial cluster strength—scale and scope increase; decrease in later stages

[1] However, some alternative views consider cluster dynamics from an adaptive cycle and nested systems theory (Martin and Sunley 2011) perspective and the dynamics of cross cluster interaction (Engel and Palacio 2009).

[2] This statement is not to imply that clusters evolve in a linear progression to some end state. Rather clusters seem to evolve in a highly noisy event that often shows trial and err "learning" that may truncate at any point. For example, there are many mining clusters that have had initial take offs in economic growth only to play out and revert to pre-cluster status as global prices decline or the ore lode plays out prematurely.

3. Knowledge—heterogeneous in early stages; more homogeneous in later stages
4. Entrepreneurship—greater within clusters than outside them
5. Convergence—convergence around best practices and standards increases after the early &Mid-cycle is reached; is correlated with a change in the nature of knowledge creation and information from heterogeneous to more homogeneous
6. Network linkages—strengthen over the life-cycle; decrease as clusters lock-in on decline
7. Cooperation—minimal to modest initially and in the take-off and early growth stages, increase as scale and scope increase; remain strong until lock-in on decline progresses

12.3 Cluster Life Cycles

Figure 12.1 depicts a staged life-cycle model for clusters largely derived from Bergman (2008) based on his synthesis of the work of others including Bode and Alig (2011), Hassink and Shin (2005), Ingstrup and Damgaard (2012), Knop and Olko (2011), Lefebvre (2012), Lorenzen (2005), Martin and Sunley (2011), Menzel and Fornahl (2009), Sonderegger and Taube (2010), Swann (2002), and Tichy (1998). This body of work evidences the recent yet sustained scholarly interest in life-cycle theory as a framework for modeling cluster development and evolution.

General life cycle theory gained recognition from its earlier application in business, industry and technology development (Klepper 2007; and Utterback and Abernathy 1975). It assumes that a growth process has an origin, a take-off leading to or initiating extended growth that begins at the first inflection point of the model after which growth occurs at an increasing rate (Fig. 12.1) but later slows after the middle part of the cycle and rapidly declines to zero thus reaching an asymptotic plateau. At exhaustion, cluster growth may remain in stasis but often experiences a period of decline (locking in on decline) or reinvention whereby a new cycle of growth is initiated. The state of each of the seven-cluster dimensions varies as the cluster moves through this model framework. There appears to be some dependency among the cluster dimensions as their behavior appears to change differentially.

While the life-cycle model implies a deterministic cluster process, it is not when one recognizes clusters may evolve in a quite erratic fashion, may never complete the process or the full growth cycle and/or may reach an asymptote prematurely with respect to the model. The model thus provides a benchmark against which to assess cluster development. Factors that may 'cause' the cluster to deviate include physical environment events such as hurricanes, earthquakes, droughts, tornadoes, climate change and floods. Societal or manmade factors such as business cycles, new technology (especially radical), change in political leadership, war, post war recovery, and conditions that eliminate competition, e.g., a region dominated by mafia rule may induce deviation from the cluster life-cycle model.

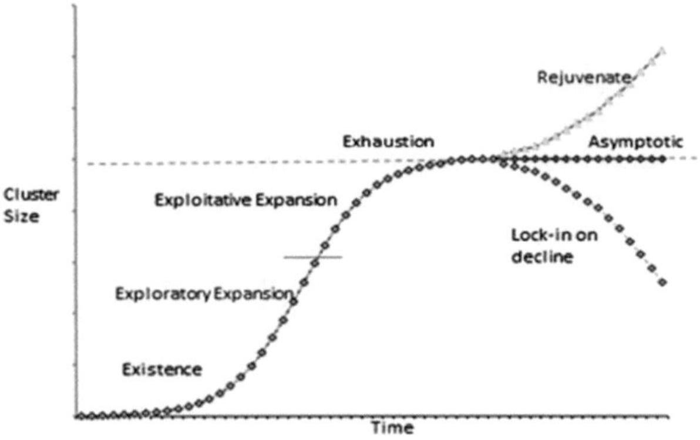

Fig. 12.1 Cluster Life Cycle Development Stages (*Source*: The author, designed after bergman 2008)

12.4 Cluster Dimensions and Stage Analysis

The dimensional analysis frame is a 7 × 4 × 5 matrix where the rows represent the seven cluster dimensions; and, the four columns represent: (1) the dimensions, (2) related indicators or measures, (3) patterns, and (4) policy guidelines. The third dimension arrays the five primary stages of the life cycle which are presented below with five different Tables 12.1, 12.2, 12.3, 12.4 and 12.5. Thus, each of the tables represents one of the stages of the life-cycle with rows and columns as described above. See Table 12.1 for an illustration.

12.4.1 The Pre-cluster Stage (See Table 12.1)

Even relatively larger numbers of firms that might form a cluster at this stage are not concentrated geographically and are nearly randomly dispersed or relatively small (often branch plants) thus there are few if any industries that might support a meaningful concentration. The number of firms and establishments in even larger industries is small and their relative strength is low with location quotients (LQs) often below 1.0, indicating industry presence in the study area is less intensive than the average of the industry intensity in other locations at more aggregate levels. Knowledge is dispersed and highly heterogeneous. The level of entrepreneurship is low and focused on non-productive entrepreneurship. There is no strategic convergence around a cluster concept at this stage as there is no cluster. There is limited business or industry networking at this stage. Finally, a low level of intra-or inter industry dependence (buying and selling to each other) exists. In short at the pre-cluster stage the economic ecosystem of a region is under developed as there is no basis for clustering.

Local regional public policy can do little more with respect to cluster formation than trying to improve workforce capability and quality, maintaining the infrastructure or supporting the provision of business assistance, and promoting economic

Table 12.1 Cluster analysis and policy pre-cluster stage

Cluster dimensions	Indicators and measures	Patterns	Policy guidelines
Spatial concentration	Geographic spread (boundary) density of core industry firms	Random; difficult to bound; firms highly dispersed	Local regional level 1. Workforce improvement 2. Maintain infrastructure 3. Business attraction 4. Business assistance centers 5. Promote economic cooperation; e.g., buying and selling cooperatives Macro level (EU or National) 1.Provide grant assistance for lagging regions 2. Workforce development and infrastructure maintenance 3. Grant support for business assistance Notes and caveats 1. Scale and scope issues make it difficult to promote/envision a cluster 2. Pre-cluster conditions imply need to improve workforce and infrastructure 3. Attract new and retain old firms 4. Hope a part of the economy will attract enough firms for a take-off 5. Non-productive entrepreneurship will dominate given scale and scope, networking and knowledge and information etc. deficiencies
Industry/Cluster strength	Number of firms in core industry; employment, income; I–O analysis; location quotient (LQ)	No core industry so cannot specify	
Knowledge and Information	Patents; university STEM research funding; industry survey; expert interview; focus groups	Very few patents; overall university presence small; S&E and enrollments per 100,000 population small	
Entrepreneurship	Number of start-ups and churn rate	Productive entrepreneurship low; non-productive may be high; churn low	
Convergence	Agreed upon strategic plan; business assn. membership; survey; expert interviews; focus groups	No convergence as no core industry exists	
Networking	Density and complexity of industry/cluster networks; number of foreign firms and FDI (inbound and outbound FDI)	Social networks may be extensive but business networks largely focused on business attraction and retention	
Cooperation	I–O Analysis; cluster industry survey; expert interviews; focus groups	Low level of inter-industry dependence; cooperation may exist but functional cooperation, e.g., minimal purchasing from each other locally	

Source: The author

Table 12.2 Cluster analysis and policy take-off stage

Cluster dimensions	Indicators and measures	Patterns	Policy guidelines
Spatial concentration	Geographic spread (boundary); density of core industry firms	Geographical concentration emerging—core industry boundary discernible; firm density is weak	Local regional level 1. Workforce improvement 2. Maintain infrastructure 3. Business attraction and retention 4. Business assistance centers 5. Promote economic cooperation – for example, buying and selling cooperatives 6. Recognize and communicate (information and data) that pre- cluster conditions are emerging Macro level (EU or National) 1. Assistance for lagging regions: workforce development and infrastructure maintenance 2. Support for business assistance 3. Planning assistance support for regions with emerging clusters Notes and caveats 1. There are usually scale and scope issues in regions with emerging clusters so may be difficult to envision a cluster being formed out of the local economic ecosystem 2. Take-off conditions mean that some part of the economy is emerging as a leading industry. It is critical to recognize this and to promote further development of the emerging core industry at the minimum 3. Promotion of entrepreneurship should begin to heavy focus on productive entrepreneurship and the creation of growth companies 4. Support incubation and workshops on productive entrepreneurship
Industry/Cluster strength	Number of firms in core industry; employment, income; I–O analysis; Location Quotient (LQ)	At least one industry is growing and has LQ 1; other core industry indicators are growing; increasing flows among core industry and supplier industries (I–O Table)	
Knowledge and information	Patents; presence of university STEM research funding; industry survey; expert interviews; focus groups	Patents increasing; university S&E sponsored research increasing; knowledge intensive industries are growing	
Entrepreneurship	Number of start-ups and churn rate	Start-ups increasing but modestly; non-productive entrepreneurship still dominate; modest increase in churn rate	
Convergence	Agreed upon strategic plan; membership in business assns.; survey; expert interviews; focus groups	No apparent convergence; economic development focused on business attraction and retention; planning not focused on growing the dominant industry and related emerging cluster	
Networking	Density and complexity of business/ industry networks; number of foreign firms and FDI (inbound & outbound); social/business network analysis	Not well developed with small nodes of firms beginning to emerge; modest level of internal sourcing	

Cooperation	I–O Analysis; Cluster industry Survey; expert interviews; focus groups	Low cross industry flows (regional I–O Table); region not focused yet on building the cluster

Source: The author

Table 12.3 Cluster analysis and policy exploratory expansion

Cluster dimensions	Indicators and measures	Patterns	Policy guidelines
Spatial concentration	Geographic spread (boundary); density core industry firms	Rapid geographical concentration; discernible core industry boundary; number of firms in the core industry increasing along with density; supporting industry firms increasing rapidly.	Local regional level 1. Market and promote the cluster 2. Expand and maintain hard and soft infrastructure 3. Accommodate new businesses (start-ups and in-migrating)—create one stop assistance center 4. Promote development of incubation and accelerators 5. Facilitate networking among firms in the cluster; sponsor or co-sponsor informative workshops and networking events 6. Streamline regulatory environment to support cluster development 7. Maintain and disseminate information on workforce needs and promote training by local education institutions Macro level (EU or National) 1.Provide assistance for cluster development after take-off; 2. Assistance for lagging regions: workforce development, infrastructure Notes and caveats 1. The Exploratory Expansion stage means that the economy is experiencing increasing returns and high growth. The role of government is to recognize this and promote further development of the core industry and networking 2. Promotion of entrepreneurship should focus on creation of growth companies and Incubation, and workshops on growth company formation. Non-productive entrepreneurship should be continued but as a secondary function
Industry/Cluster strength	Number of firms in core industry; employment; income; I–O analysis; location quotient analysis (LQ)	At least one industry is growing rapidly with LQ increasing rapidly from the take-off level; rapidly increasing internal flows between core industry and supplier industries (I–O analysis)	
Knowledge and Information	Patents; number of universities and STEM research funding; industry survey; expert interviews; focus groups	Rapidly increasing patents and university STEM sponsored research; university-industry linkages increasing growing; but still considerable knowledge & information heterogeneity	
Entrepreneurship	Number of start-ups and churn rate	Rapidly increasing new start-ups including company spin offs; focus on non-productive entrepreneurship secondary; churn rate rapidly increasing	
Convergence	Agreed upon strategic plan; membership in business assns.; cluster industry survey; expert interviews; focus groups	Signs of convergence begin to appear in latter part of exploratory stage; business associations focused on core industry and cluster firm growth (and new associations forming, e.g., technology councils); focus is not on standardization or building agreement on best practices; marketing the cluster region has begun	
Networking	Density and complexity of business/industry networks; number of foreign firms and FDI (inbound and outbound); social network analysis	Business/industry local networks evolving rapidly with expanding internal sourcing (I–O analysis)	

Cooperation	I–O Analysis; cluster industry survey; expert opinion; focus groups	Cross industry flows increasing rapidly (regional I–O Table); regional organizations recognize a cluster exists and focused on building and marketing a vibrant cluster (regional I–O analysis); new business/cluster related organizations may emerge, e.g., technology councils.

Source: The author

Table 12.4 Cluster analysis and policy exploitive expansion stage

Cluster dimensions	Indicators and measures	Patterns	Policy guidelines
Spatial concentration	Geographic spread (boundary); density of core industry firms	Geographical concentration occurring at a decelerating rate; core industry firms boundary observable and expanding geographically; density of firms in the core industry stabilizing; increasing for supporting firms.	Local regional level 1. Provide information about cluster dynamics to the industry and supporting businesses and associations: during early part of this stage to promote agreement on standards and best practices and strategic plan development; but during latter part of the stage convey that cluster dynamics are slowing and strategic planning is necessary to determine how to regenerate or sustain the cluster
Industry/Cluster strength	Number of firms in core industry; employment; income; I–O analysis; location quotient (LQ) analysis	Core industry still growing but at a decreasing rate with LQ stabilizing; increasing flows among core industry and supplier industries (I–O Table).	2. Work with cluster representatives to streamline regulatory environment in latter part of the stage to thwart attempts to seek regulatory protection from external competitors 3.Maintain, renew (and create new) hard and soft infrastructure 5.Encourage industry groups to support incubation and accelerators by outsourcing spinoff companies
Knowledge and information	Number of universities and STEM research funding; industry survey; expert interviews; focus groups	Patents increasing at decreasing rate; STEM sponsored research increasing at decreasing rate; university-industry linkages increasing but at decreasing rate; knowledge and information shifting to homogeneous	6. Maintain and disseminate information on workforce/training needs and promote training by local education institutions Macro level (EU or National) 1.Provide planning assistance (technical and grants) and relevant information for clusters during the final parts of the Exploitative stage to support regeneration 2. Create special fund to support planning for sustaining clusters
Entrepreneurship	Number of start-ups; churn rate	Rate of new start-ups decreasing as cluster matures; the churn rate decreases and by end of this stage there are few new start-ups and failure rates increase in the few cases	Notes and caveats 1. The Exploitative Expansion stage means that the cluster economy is maturing and returns are increasing but at decreasing rates, nearing zero at end of this stage; Increasing returns growth of the Exploratory stage has declined as broad agreement has been reached on best practices and overall strategic plan that may be flawed because of a failure to anticipate homogenizing trends.
Convergence	Agreed on strategic plan; membership in business assns;	Convergence occurs rapidly from mid-stage; networks and	2. The role of government is to recognize that the cluster is maturing and if growth slows showing signs of moving toward locking in on

	survey; expert interviews; focus groups	business assns. focus on protecting cluster industries not growth. Early part of stage focused on standardization; building best practices agreement; marketing remains strong	decline it should ensure this information is shared with cluster organizations and that they are incentivized to plan for regeneration. 3. Promotion of entrepreneurship should continue and focus on creation of growth companies, incubation, & workshops. Heavy incentives should be provided to induce larger companies to form spinoffs and outsource them to regional incubators/accelerators. Maintain non-productive entrepreneurship support
Networking Cooperation	Density and complexity of business/industry networks; number of foreign firms and FDI (inbound and outbound); social network analysis I–O Analysis; cluster/industry survey; expert interviews; focus groups	Social-business networks evolved into highly complex supporting structures; internal sourcing is highly evolved; cluster companies and organizations increasingly involved in external networks. Cross industry flows well developed and highly integrated across multiple industry segments (regional I–O Table); in latter part of this stage the region is focused on maintaining the cluster and its markets; business/cluster related organizations involved in maintaining the cluster and its markets	

Table 12.5 Cluster analysis and policy exhaustion stage

Cluster dimensions	Indicators and measures	Patterns	Policy guidelines
Spatial concentration	Geographic spread (boundary); density of core industry firms	Cluster is bounded but becoming increasingly dispersed; density decreasing because of decrease of companies and large spread of cluster geography	Local regional level 1. This situation is similar to the pre-cluster stage in some ways; however, it is different at least in the early stages of exhaustion because the infrastructure and companies that define the clusters are still located in the cluster region; they are just loosing competitiveness 2. Leadership for rejuvenation is needed such as information and data, analytical support, and public relations. 3. Strategic planning grants are needed for rejuvenation and marketing 4. Redesign of regulatory environment to support rejuvenation is needed 5. Important to maintain and renew hard and soft infrastructure 6. Encourage industry groups to support technological innovation and the conversion of that into new products and services (the D part of R&D critical here) 7. Facilitate the transfer of new technology from elsewhere and use to support incremental innovation (if not radical) in the core industry; support incubation and accelerators by encouraging outsourcing of spinoff companies 6. Workforce development for staffing new products and services of the envisioned rejuvenated cluster Macro level (EU or National) 1.Provide planning assistance and relevant
Industry/Cluster strength	Number of firms in core industry; employment; income; I–O analysis; location quotient (LQ) analysis	All strength measures are declining or at best static; employment, income, cross industry flows (I–O analysis); LQs	
Knowledge and information	Patents; number of universities and STEM research funding; industry survey; expert interviews; focus groups	Patents decreasing; STEM university research funding stable but may be decreasing.	
Entrepreneurship	Number of start-ups churn rate	Little or no start-ups; recently formed firms disappearing due to lack of competitiveness or out migration to more favorable locations	
Convergence	Agreed on strategic plan; membership in business assns.; survey; expert interviews; focus groups	Convergence around agreed upon best practices and processes is stable or static; homogenizing trends impact competitiveness negatively; membership in associations waning as decline sets in	
Networking	Density and complexity of business/ industry networks; number of foreign firms and FDI (inbound and outbound); social network analysis	Network density and complexity are declining; foreign firms as well as local firms out migrating at increasing rates or going out of business; inward bound FDI decreasing or dried up	
Cooperation	I–O Analysis; cluster/industry survey; expert interviews; focus groups	Cooperation may be declining as firms are trying to stay competitive especially when previous attempts to stave off	

	decline have failed; association membership declining and loosing effectiveness to support the cluster	information for cluster rejuvenation 2. Assistance for lagging regions (which the exhausted cluster now is): workforce development, infrastructure maintenance and planning for promotion of cluster development in the local exhausted cluster and for clusters in other regions <u>Notes and caveats</u> 1. The exhaustion stage means cluster will lock-in on decline if it is not rejuvenated. The role of government and community (associations) is to provide cluster renewal assistance for cluster renewal as firms will either not have resources to do it or will moving operations out of the region 2. The role of government is to recognize that the cluster is exhausted and thus ensure relevant information and data is available and shared with cluster organizations in an effort to incentivize planning for regeneration. 3. Promotion of entrepreneurship should focus on creation of growth companies, incubation, workshops. Heavy incentives to induce core industry companies to form spinoffs and outsource their development

cooperation via, for example, promotion of buying or selling cooperatives. At the macro level assistance for workforce improvement, business assistance, development planning and infrastructure, and support for lagging regions via a regional policy regime is a possibility. At best the region can implement business attraction and retention policies and hope a segment of the economy will, via planning or serendipitously, attract enough firms for a take-off.

12.4.2 The Take-Off Stage and Policy Guidelines (See Table 12.2)

In the take-off stage one or more industries begins to emerge and exhibit signs of clustering (see Table 12.2). An area within which most of the companies in the industry are contained can be identified. Yet, the density of firms is modest. The strength of the core industry and the emerging cluster is growing and characterized by a LQ that rises above 1.0 demonstrating that the core industry is evolving into a lead cluster sector. It is during this stage that cross industry cooperation begins as represented by increasing cross industry flows (from a regional I–O table).

Knowledge and information heterogeneity may be expanding but modestly as cohesion of the core industry(s) is still in a nascent stage. Yet, that which does exist is highly heterogeneous. It is not likely that knowledge indicators would experience much change as take-off occurs.

Entrepreneurship indicators such as start-ups may show modest increases and there may be efforts to form a business incubator. Yet, most of the start-ups will continue to be of the non-productive form. Convergence around a strategy that envisions a cluster is not usually of major interest in the form of action steps.

Networking may be evolving at this stage but will not be well developed around the concept of an emerging cluster. There may be a few new firms formed with the goal of supplying inputs or marketing assistance to businesses in the core industry (s) but this will be modest. There may be isolated instances of firms in the core industry(s) cooperating on matters such as joint bidding on projects or supplying goods or services jointly in cooperation to a market segment. Cooperation in general will tend to be modest as the region is not yet focused on building the cluster.

The role of local regional government at the take-off stage includes providing information about the economy regarding the emergence of a core industry(s) and providing assistance for workforce development and infrastructure maintenance and development. It also has the opportunity to provide support for business assistance, for example, training and promotion of buyer-seller cooperatives. A major local government leadership role is to communicate to all stakeholders that cluster conditions are evolving and that there is a need and opportunity for productive entrepreneurship. A role for national government is to provide planning and advisory assistance for regions that exhibit the potential for cluster development. Continued support for lagging regions is important as regions with emerging clusters may also be lagging regions.

12.4.3 Dimensional Patterns in the Exploratory Expansion Stage (See Table 12.3)

The exploratory expansion Stage emerges shortly after successful take-off and extends until the middle part of the steep growth segment of the S-shaped life-cycle (Fig. 12.1) model is reached. From the middle of this segment growth continues to accelerate but at a decreasing rate. The upper boundary of the exploratory expansion phase is the lower boundary of the exploitive expansion phase (see Fig. 12.1). This phase is usually the most dramatic part of the life-cycle (along with perhaps rejuvenation) as it is where trial and error business approaches are tried and are at a maximum and thus start-ups and the churn of start-ups and failures is most pronounced. The churn of ideas and knowledge during this period is why this stage is called exploratory as it is a time of testing to discover the best trajectory for the cluster.

The cluster becomes defined around one or more core industries as this stage unfolds and the cluster boundary with some creative approaches can be specified using GIS techniques however this is made more complicated when one is attempting to map flows of purchases or goods among core firms and members of the supply chain. Throughout this phase more and more companies are formed or attracted to the cluster so that density is increasing and proximal relations become more pronounced. The strength of the cluster also increases dramatically with considerable growth in employment, income and wealth creation. Core industry location quotients often increase to concentration levels three or more times the provincial or national average.

Considerable new knowledge and information is being created both in the industry and supporting firms or supply chain members. This industry churn spills over into other organizations leading to more patenting not only among the companies but also via growing involvement of research institutions which find a growing market for research. This interrelated activity produces considerable entrepreneurial activity so start-ups and spinoffs are increasing and focused on the productive variety while emphasis on non-productive business formation becomes secondary.

The notion of convergence around a strategy tends to be conceptual during the early and middle parts of the exploratory expansion stage. However, in the latter part a formal strategy may emerge due to an increasing focus of business associations on the further growth of the core industry and its supporting businesses. Also, new facilitating and supporting organizations may arise such as technology councils and venture capital or lobbying associations. In this part of the exploratory stage the focus is mostly on growing the cluster by internal processes and by company attraction including foreign firms that would bring inbound foreign direct investment (FDI). This creates a highly positive environment for new knowledge and information, and heterogeneity of knowledge and information is growing and spilling over into the local regional community.

Networking during this stage is evolving rapidly with the expansion of business associations that are promoting cluster development and thus bringing cluster industry and business representatives together in a variety of contexts, e.g., training, workshops, award ceremonies, webinars, social and speaker events. The growing cluster "ecosystem" includes an increasing variety of firms both in the core industry and the evolving supply chain and there are increasing flows among these as joint bidding for projects and joint delivery of products and services become increasingly commonplace. Business and related associations tend to provide forums to promote further networking among members of the cluster and as flows among buyers and suppliers become increasingly complex the level of cooperation increases considerably.

Given considerable dynamics in the cluster during the exploratory phase it may seem that there is little role for government. But leadership from government at the local regional level is important to continue to promote the cluster and its development. In the early parts of this phase the evolving cluster is potentially quite fragile and can sometimes leap from a seemingly rosy future to exhaustion due to unanticipated forces. Government at the national level has standing programs for facilitating recovery from disequillibrating events, however, local regional government organizations need to be prepared to quickly lead initiatives for recovery resources from higher levels of government to adjust.

At the local regional government level there are other roles to be played. In the early part of the exploratory phase there is need for government supported grants and assurances for supporting business assistance, workforce training, entrepreneurship, smoothing out regulatory processes by facilitating the creation of one stop services for applications, registrations and licensing. At the national or EU level grant programs to assist cluster development during the early parts of the exploratory phase are important because this is a fragile period for the cluster.

12.4.4 Dimensional Patterns in the Exploitive Stage (See Table 12.4)

The gains that have occurred during the expansion stage tend to become consolidated into a more routine portfolio of functions and processes during the exploitive expansion stage. During this stage the geography stabilizes and the bounded cluster area tends to experience a reduction in the rate of spatial expansion or stop as infilling becomes the dominant method of sitting for expansion of existing or in-migrating companies. The density of firms tends to increase especially in the early and middle parts of the exploitation stage and stabilize in the latter parts.

Industry and cluster strength continue to grow as the cluster region stakeholders build deeper network relations but the increasing trend toward interdependence evolves at a decreasing rate. A high level of interconnectedness among core industry and the supply chain is retained. However, stability will tend to occur in the latter parts of this stage.

Knowledge and information will be produced at a very high rate during the early part of the exploitive stage. A high rate of patenting and cluster related sponsored research at universities and other research organizations will occur during the early parts of this stage but this will decrease later. The character of knowledge and information will become more homogenous. Consequently, entrepreneurship will tend to stabilize and decrease. These patterns tend to occur because the likelihood of convergence and agreement around a strategic plan during this phase is high as the core industry and related suppliers solidify agreement on best practices and standards. Once this happens, the need or motivation for R&D and experimentation by entrepreneurs for enhancing practices, processes and widgets, is of less interest. Historically this has been a critical development in the process leading to cluster decline. It is thus important during the middle to latter parts of the phase that the strategic approach include sustainability concepts and thus the regeneration of the cluster when or if maturation tends toward non-sustainable processes.

The slowing rate of growth creates an environment that begs for increased growth! One avenue in the view of cluster stakeholders may be to gain regulatory protection and tax breaks to help the bottom line of firms in the cluster and competitiveness. In an era where internal and external networking, especially in the latter stages of the exploitive stage, become national and global due in part to increasing outbound FDI, it is difficult to protect a cluster from external competition. Strategic plans need to focus as much on sustainability and rejuvenation as upon establishing agreed on best practices. In short, it is important that such plans include processes and procedures to ensure that there is a continued flow of knowledge and information into the regional cluster ecosystem and thus that the renewal dynamic provided by entrepreneurship is maintained at a high level. Protection from external competition is the opposite of what is needed for ensuring rejuvenation and sustaining the cluster rather than letting it lock-in on decline.

Local regional government has several important roles during the exploitive expansion stage. First, is providing information and data on cluster dynamics to cluster stakeholders. Second, is to interpret this information and data for stakeholders by conveying in the early stages the need for finding general agreement on cluster strategy and best practices. This includes explaining that strategy needs to address the issue of economic sustainability as the cluster matures. Third, there is a continued need to streamline the regulatory environment and to facilitate compliance with measures like one stop process facilities (on-line or physical). Fourth, is to ensure that public infrastructure is provided and/or maintained at levels that facilitate low transaction costs for the movement of goods, services and information (bytes). There is also a continuing need to maintain workforce training.

At the national or EU level the role of government is to provide flexible grant programs and information to facilitate managing the cluster during the latter parts of the exploitative stage. One major focus should be on sustaining the cluster and planning for its rejuvenation before it falls into exhaustion. Maintenance of resources for productive entrepreneurship and related technological change for undergird successful innovative clusters and thus sustainability is of central importance.

12.4.5 Dimensional Patterns in the Exhaustion Phase (See Table 12.5)

If efforts to rejuvenate and sustain the cluster and new growth during the exploitive expansion stage fail, cluster growth will decline to zero and eventually into absolute decline. Given that all stakeholders of the cluster including firms, employees, governments and non-profit associations want to avert a long period of decline this discussion focuses mostly on rejuvenation of the cluster.

The spatial concentration of a cluster that has reached exhaustion gradually becomes less dense as companies fold, are acquired or relocate. The spatial structure of a cluster will tend to hollow out and industry and cluster strength will decline due to lock-in where unemployment increases and income and wealth decrease, and effective cluster networking and cooperation suffer. Knowledge and information will narrow as they become homogenous and entrepreneurship will revert toward non-productive entrepreneurship. While a high level of cooperation may continue it will tend to be difficult to develop a successful rebounding plan without external help. But the infrastructure and some of the industry members and associations still exist (especially in the early part of the exhaustion stage). So there are still resources to stage a comeback if organized or focused through strategic leadership.

The roles of local government and higher levels of government are major and critical to achieving rejuvenation once exhaustion sets in. That is why in the middle and latter parts of the exploitive expansion stage it is so important so seed sustained cluster development. It is easier to achieve rejuvenation when the cluster is still relatively strong. So what should governments do? At the local regional level it is important to continue to provide information, data and analytical support to understand the condition of the cluster. Local government partnered with industry and cluster associations should undertake planning to create a strategy for renewal. It is important to restore, maintain and possibly build new infrastructure. There is a need to encourage industry groups to invest in and facilitate the identification and transfer of technological innovation to the region as these will be critical elements of a plan to renew the cluster and productive entrepreneurship regardless of other elements of the plan. Finally, there will be major workforce development needs as the plan is formed and implemented. Local government should be promoting and facilitating the creation of the infrastructure to deliver this through secondary and post-secondary educational facilities, and industry associations.

The national government should provide grants for planning assistance and infrastructure renewal for clusters in the exhaustion stage to help support cluster renewal or to help the design and implementation of a new economy. Existing programs for lagging regions may become accessible for clusters in the exhaustion phase. Also, if the cluster region can claim that disaster level disequillibrating forces have contributed to exhaustion then disaster recovery grants and awards may be available.

12.4.6 Testing the Cluster Dynamics Model

The above discussion lays out a life-cycle model for cluster dynamics. This model is conceptual and rests upon interpretive analyses and hypotheses derived from the literature. In this regard it begs for careful evaluation and testing of not only the interpretations of how the seven dimensions of clusters vary with the different stages of the model but also how they are linked to one another. Carrying out this work is a large challenge and for sure not one that can be addressed fully in this paper. A case study of the Korean Shipbuilding Industry is now presented and its development is interpreted from the life-cycle perspective that has been offered in this paper.

12.5 The Korean Shipbuilding Cluster: Assessing the Dynamics of this Cluster

12.5.1 Background: The Korean Shipbuilding Cluster

The shipbuilding industry in Korea is the largest and most advanced in the world (Shin and Hassink 2011).[3] Its roots can be traced to the nineteenth century where small and medium size wooden boats were built along the southeast coast of Korea in the areas of Geoje, Changwon, Busan and Ulsan. A Japanese company established a shipyard near Busan in the 1880s which the authors view as the start of the modern Korean shipbuilding industry. The Japanese brought new steel shipbuilding knowhow to the region. Subsequently and throughout the colonial period (1910–1945) there was a near continuous flow of Japanese shipbuilders to the Southeast coastal part of the Korean peninsula but the associated expanded investment in infrastructure still focused on small and a few medium sized ships. From 1880 to 1945 the industry which mainly built small and medium sized ships and boats and was maintained in this way. The industry was sustained but exhibited no major changes until about 1970.

In the 1970s as part of a series of national 5 year economic plans emphasizing machinery and heavy industry focused investment from the private sector(s) in the automobile and shipbuilding industries was motivated in part by public sector investment in not only infrastructure but also in private companies. This policy initiative in collaboration with Hyundai, Daewoo and Samsung *chaebol* companies laid plans and began to produce large ships such as oil and LNG carriers. Technologies were obtained from other places for the start-up phase of the newly invigorated industry. These were soon internalized and with subsequent

[3] Much of the following description of the evolution of the Korean shipbuilding industry is taken from the paper by Shin and Hassink (2011) and from discussions with senior Korean informed decision makers and leaders such as Dr. Kunmo Chung (2013 and 2014, multiple personal communications) who was the founding director of KAIST and the founder of the Korean Nuclear power industry during several visits to Korea in 2013.

incremental technology innovations led to the rapid advancement of Korea's shipbuilding industry.

Encouraged with government policies that supported the goal to produce larger ships and in part because of rapidly expanding global demand, several large shipbuilding docks were constructed at multiple sites in the Southeast. The Korean government also provided focused political, financial, educational and research support including government and industry financed shipbuilding and shipping related ocean studies. It also facilitated the formation of the Association of a Korean Shipbuilders public-private partnership (PPP) to provide promotional information and services for the industry. The industry grew by 30 % every year between 1976 and 1985 and employment tripled. Due to oil supply shocks and other economic factors shipbuilding faced a significant declining international market. While Korean shipbuilders suffered along with others this was temporary; the big three *chaebols* later enjoyed considerable success from their earlier aggressive investment and marketing. Thus in the late 1980s and 1990s Korea's shipbuilding industry and cluster became the largest in the world. During this period the three *chaebo*ls remained dominant and stabilized their areas of specialization and to some extent expanded the scale of production; and diversified their product mix, and managerial and technological knowhow. At the same time Hyundai, Daewoo and Samsung developed supply chain linkages with smaller companies and/or helped create spinoffs often led by entrepreneurial management.

In the late 1990s the Asian economic crisis became a major threat and obstacle to not only growth but also the maintenance of the superior position of Korea's shipbuilding industry. Restructuring occurred resulting in the case of Daewoo becoming independent and forcing less competitive companies out and creating new, medium sized companies often led by innovative entrepreneurs and thus more competitive shipbuilders (e.g., STX Group) and component producers. The industry emerged once again from this short period of decline as a dominant playing in the global economy. Restructuring motivated by the Asian economic crisis was the event that led to restructuring and increased competiveness.

In conclusion, it may be claimed that the Korea Shipbuilding cluster was not created independent from its historical context or even by historical accident as it evolved out of a previous existing traditional non-steel ship and boat building industry upon which Japanese investors amplified the scale. Korean government leadership in collaboration with a few large private companies propelled the development of the ship building cluster with a strong commitment to the growth of this industry in the early 1970s. Entrepreneurial and innovative leadership in the 1970s and 1980s took advantage of the country's supportive policies and the global environment by investing aggressively in the development of the industry.

The cluster dynamics background described above along with the life-cycle model dimensions described above are now used to examine the Korean shipbuilding cluster and to determine its current life cycle. The first dimension is network linkages.

12.5.2 Network Linkages

The case summary above provides evidence that network linkages became progressively deeper and broader over the cluster history beginning at or shortly after take-off *circa* 1970. Today there are deep vertical linkages to robust supply chains within each of the *chaebols* with recent foreign investments in shipbuilding infrastructure and cooperative relations in China, Viet Nam, Brazil, Norway and Romania (Shin and Hassink 2011). However, horizontal networking across the major *chaebol* companies of Hyundai, Daewoo and Samsung has remained minimal with most cooperation focused on addressing problems that face the whole industry such as the impact of heavy regulation in some countries constraining Korean market entry. To the extent *chaebol* cooperation has occurred it has been indirect and through membership in non-profit trade and promotion organizations. Very little direct cooperation has occurred. The status of evolution and recent network linkages in general appear to place the shipbuilding cluster in middle of the Exploitive Expansion stage.

The strategy from the beginning has been one organized around the three *chaebols*. This has been the general strategy of South Korean development planning since the 1960s which has had a heavy industry focus that included shipbuilding. This strategy worked well until the late 1980s and 1990s when market conditions for large but relatively simple ships (oil tankers, LNG carriers) faded. In the late 1990s the large conglomerates as well as the whole Korean economy was negatively impacted by the Asian economic crisis. The *chaebols* were forced to restructure and in the case of Daewoo to uncouple its shipbuilding division to achieve debt reduction and in turn the acquisition of many new projects, and a demonstration that it could manage multiple large orders as well (Shin and Hassink 2011). In short, the crisis created many financial challenges which led to restructuring and the development of several new companies or the rolling up of medium sized but relatively inefficient companies into new more efficient and entrepreneurial ventures, e.g., STX Group. The resulting strategy was to create more competition and to restructure the former lead companies into models that were more efficient and innovative. Restructuring resulted in reinvigorating a cluster that was trending toward Exhaustion. As such it resulted in a reengineering of the cluster and moving it back into the earlier Exploratory Expansion Stage. At the same time the network linkages that grew in the earlier stages still existed and could also be redesigned and redirected to facilitate more rapid movement toward agreement on new best practices and standardization. Consequently, by the late 2000s the shipbuilding cluster can also be viewed as in the Exploitive Expansion stage.

As described above the shipbuilding cluster in Korea is dominated by several *chaebols* which dominate the structure of the industry. While there are several medium sized companies that to some extent build ships or major parts of ships, they generally provide inputs to the larger shipbuilding companies. Beneath this "top heavy" structure are many small companies that are for the most part tied or linked to one of the *chaebols*. From the take-off until the late 1990s the value chain remained top heavy. In the late 1990s renewal of the cluster's formation of new companies decelerated, an indication that the cluster was moving into the Exploitive Expansion stage of the life-cycle. However, renewal of the cluster from an increase and restructuring among the middle sized companies meant that the level of new companies increased due to increased entrepreneurship and related startups, and the associated appearance of new and more diverse information. In short, the cluster today based on the status of network relationships appears to be in the Exploratory Expansion stage of development.

12.5.3 Entrepreneurship

While there is a strong and deep value chain around each of the major *chaebols* this supply chain has mostly arisen around *chaebol* incentivized start-ups (spinoffs), selected companies and/or subsidiaries. Grass roots company formation and development into ventures that later find their way into the supply chain of one of the three conglomerates is not the primary way entrepreneurship seems to work in Korea. This is due largely to the prevailing *chaebol* dominated industry structure and policy. However, major entrepreneurial behavior at a high level was observed in the late 1990s that led to major changes in the previous shipbuilding industrial structure. New or transformed older companies at the mid-size level emerged and brought new and more efficient enterprises into being such as the STX Group (Shin and Hassink 2011). This led to creation of new best practices and increased scale and scope of the cluster. Nonetheless, *chaebol* dominated structure appears to have constrained the development of entrepreneurship in terms of the emergence of independent growth companies and thus while the Korean shipbuilding cluster has grown considerably it has not been associated with the pattern of entrepreneurship that theory suggests at the Take-off, Exploratory or Exploitive stages. This is despite the fact that one of the key motivating factors of entrepreneurship is heterogeneity of information and knowledge which as noted below follows the trajectory indicated by the life-cycle model. Given the constrained level of entrepreneurship while preconditions for its expansion existed, a qualification of how this dimension of cluster dynamics evolves may be needed in future study.

Like the competitiveness dimension, entry and exit behavior of firms was slowing in the middle and late 1990s. The renewal of the cluster at the end of the 1990s motivated a reversal in the entry and exit declining trend. As the cluster moved into the 2000s, entries and exits once again increased signifying that the cluster was mostly in the Exploratory Expansion stage of the cycle.

12.5.4 Geographic/Spatial Concentration

Company concentration was high in the pre-cluster and early cluster stages and was primarily centered in the southeastern part of Korea. As the cluster grew in the Exploratory Expansion era (1980s and 1990s) the cluster remained geographically concentrated in this part of Korea with the exceptions that some high level inputs such as research and human capital development (education) occurred with greater intensity in the northwest in the vicinity of Seoul and south central in Daejeon. From the late 1990s the entrepreneurially driven industry transformation of the shipbuilding industry and cluster experienced geographically diversified investments not only further in Korea but internationally as explained above in the network linkages discussion. So the shipbuilding cluster in Korea follows the path that the life-cycle model asserts. The recent geographic expansion suggests that the cluster which in the 1990s appears to have been approaching stasis may, due to renewal, now be in the latter part of the Exploratory Expansion stage or in the earlier part of the Exploitive Expansion stage.

12.5.5 Industry/Cluster Strength

As noted in the discussion on strategic relations there is evidence that the shipbuilding cluster evolved through the exploratory expansion stage and almost to the end of the exploitive expansion stage by the late 1990s. In fact, one could argue that there were indications that the cluster was proceeding rapidly toward exhaustion and decline. However, due to the restructuring that occurred on the part of the major companies which in all cases was motivated by government leadership, the original strategic organization of the cluster around three large conglomerates was restructured which once again increased the heterogeneity of knowledge and entrepreneurial behavior and internal competitiveness and new external market knowhow. While it could be argued that the Cluster in terms of dynamics was approaching Exhaustion, restructuring helped move it toward an earlier life-cycle stage, the Exploration Expansion stage.

Scale and scope of the shipbuilding cluster was narrow in terms of both size and diversity in the early and take-off stages of the life-cycle. After take-off both scale and scope grew with increasing size and product complexity up to the present. Along the way there have been changes in the rate of scale and scope development. The major one was in the late 1990s when the shipbuilding industry was re-organized thus evidencing a renewal trend away from what then appeared to be an approach nearing Exhaustion. Great increases in scale and scope occurred as this reforming process unfolded. Today scale and scope continue to grow at a high rate which suggests in terms of this dimension the shipbuilding cluster has slipped back to the later part of the Exploratory Expansion stage or the early part of the Exploitive Expansion stage.

The late 1990s restructuring served to not only renew growth rates and related attributes associated with the Exploratory Expansion stage of the cluster life-cycle but also to accelerate the growth in scale and scope of the industry and once again drive knowledge toward greater heterogeneity. It also induced de-concentration of the cluster members' product mix and markets as operations and infrastructure in other parts of the world were established thus leading to a major decentralization and diversification of the cluster.

Interpreting the stage or status of competitiveness of the shipbuilding industry is complicated. If one focuses on restructuring then competitiveness would be expected to be strong and increasing, following the late 1990s restructuring. The data as presented by Shin and Hassink (2011) suggest that indeed this cluster became more competitive after the late 1990 restructuring. If this is the focus then competitiveness would indicate the cluster is in the latter part of the Exploratory Expansion stage. At the same time however, before the late 1990s restructuring, growth of the industry and other dimensions suggested the cluster was converging (slowing down). This suggests a decrease in competitiveness and thus places the cluster in the later part of the Exploitive Expansion stage. Given that the strongest indicator is one of continued growth and expansion of the cluster it is viewed as being in the later part of the Exploratory Expansion stage of the life-cycle.

12.5.6 Knowledge and Information

Knowledge and information was highly dispersed early in the shipbuilding cluster's development (take-off and Exploratory Expansion stages). Yet in the early stages of development the focus was on building much larger ships and boats but ones that required minimal new knowledge and knowhow such as oil tankers and LPG carriers (Shin and Hassink 2011). Increasingly in the 1980s and especially after the 1990s when knowhow and demand awareness where at a peak, market exploitation led to broadening the product mix and creating increasingly more diverse and complex products such as ocean liners, drilling platforms and military ships (submarines, battle ships, and destroyers). Thus, as the cluster companies move toward a comprehensive shipbuilding portfolio the range and diversity of knowledge appears to have been moving toward standardization and thus toward greater homogeneity. Given this it appears that the state of the knowledge dimension is at the middle to latter part of the Exploitive Expansion life-cycle stage.

Early in the history of the shipbuilding cluster development of new and best practice technology was acquired from other countries. After this technology was mastered and incorporated by Korean cluster companies incremental innovation occurred in an effort to transform the basic technology into brand name products and services. Production process was continually renewed from internal company innovation and the creation of public-private research facilities for the creation and testing of prototypical new technologies. Like for production process, market analysis and research expanded as the cluster moved into the early stages of the Exploratory Expansion stage continuing on into what increasingly appears to be early entry into the Exploitive Expansion stage. Beyond investment in new technology, Korean companies also invested in increasingly large ship building infrastructure well into the 1990s and then again after the restructuring in the late 1990s to accommodate expanding the cluster market into shipbuilding areas requiring more sophisticated inputs such as ocean liners, military ships and boats and drilling platforms (Shin and Hassink 2011). These developments included new infrastructure techniques such as the MegaBlock concept where modular units of a ship are built on land and then combined in or on the water to create a new ship. Many of these technologies came from others and thus may be viewed mostly as imitation innovation. However, unlike many other countries subsequent to absorption of the newly foreign acquired technologies, additional Korean innovations were erected incrementally on top of the original best practices. The ability to innovate on top of best practice is a capability that is not well developed in many of the emerging economies of the world. Korea has demonstrated that it is somewhat unique in this regard.

12.5.7 Cooperation

Cooperation was at all times strong between the cluster industry leaders and the central Korean government. Initially the government set a strategic direction and incentivized the small and to some extent middle sized companies to expand the infrastructure and operations needed to support shipbuilding. Later government helped to incentivize development of non-profit organizations and associations of business, industry and government officials in an effort to increase cross organization networking, knowledge sharing, and learning. When the shipbuilding cluster began to experience declining or perceived conditions of decline in the late 1990s the government provided leadership in the form of incentives and information resulting in company, industry and cluster restructuring. During the stage of exploratory expansion the role of government seems to have receded in importance. Government activity appears to have been strongest during the take-off, exploratory and exploitive stages. Instead of adopting a strong protectionist role when cluster growth began to slow rapidly the Korean government provided leadership and incentives (e.g., loans and grants) for restructuring in order for the industry to renew its competitiveness. The way the government has behaved with respect to the shipbuilding industry cluster provides support for an interpretation that the cluster was nearing the Exhaustion stage in the late 1990s. But because of government intervention and accommodating industry leadership the cluster soon was reengineered into renewal and the rebirth of an Exploratory Expansion stage cluster.

The role of government as a leader and incentive provider has been significant at and before take-off, during the Exploratory Expansion and Exploitive Expansion stages and particularly important during the pre-exhaustion stage that the Korean shipbuilding cluster experienced in the late 1990s. Given the current status of the shipbuilding cluster in Korea it seems that the cluster from the perspective of the role of government is now in the latter part of the 2nd Exploratory Expansion stage. It is important to recognize that government has exercised its role at nearly all stages of the shipbuilding cluster's development.

12.5.8 Convergence

It may be argued that the shipbuilding cluster was approaching deep and pervasive convergence among its industrial organizations in the late 1990s as argued above that the cluster was moving toward Exhaustion at this time. There were many indicators that the cluster was at best in the latter part of the Exploitive Expansion stage if not already in the Exhaustion stage at that time However, government and business cooperation then led to restructuring of the cluster's industry organizations into more entrepreneurial and efficient agents. As such the convergence tendency of the late 1990s was reversed and rejuvenation leading to Exploratory Expansion was achieved.

12.6 Discussion

The dimensional analysis of the Korean shipbuilding cluster provides several confirming if not new insights into cluster dynamics. It illustrates that such an analysis can surface in an analytical way deviations from the basic cluster life-cycle model. The deceleration of cluster growth identified in the late 1990s was a significant structural deviation that can either be interpreted as a temporary wrinkle in the progression of the life-cycle or as prematurely reaching the end of the cycle or stasis. If the latter interpretation is accepted then the subsequent renewal may be viewed as moving or recycling the cluster back to the Exploratory Expansion stage. If the former interpretation is accepted then the renewal may be viewed as simply putting the cluster back on the development path and thus in the Exploitive Expansion stage. That nearly all of the cluster dimensions indicated that the cluster was nearing Exhaustion in the late 1990s suggests the cluster was nearing the end of its cycle. If only several of the dimensions exhibited behavior indicative of stasis then the crisis may be interpreted as an aberration away from the maturation process that the life-cycle model offers.

In addition to providing evidence that the cluster apparently approached the end of the life-cycle and apparently was rejuvenated there are other useful elements of the case that life-cycle dimensional analysis helps to identify. One of these is that it shows that the Korean governance style and culture results in the ever present role of the national government throughout the cluster life cycle. Yet, the presence of government in cluster development seems to have been moderate during the initial Exploratory Expansion stage. Also, the analysis provides insight into how a cluster comes into being based on the Shin and Hassink (2011) historical analysis which showed the early infrastructure preconditions that propelled subsequent Japanese investment and later the adoption of shipbuilding as a national strategic development priority. In this case the coastal location of early shipbuilding may be viewed as a critical initial factor leading to the later development of the industry. However, it does not explain why it began in the southeast coastal region rather than other parts of coastal Korea. Additional analysis is needed to determine if the initial conditions also included some environmental factors (for example, favorable sea conditions or good fishing) or more specific cultural factors. The analysis also provides insight into the nature of entrepreneurship in this cluster and provides a hypothesis about Korean entrepreneurship as being heavily dominated by *chaebol* influence and thus resulting in a high level of spinoff type of company formation and a lower level of organic or independent entrepreneurship. Finally, the dimensional approach enabled a structural analysis of what is here interpreted as a cluster nearing stasis or the end of its life-cycle and the dynamics of its regeneration and renewal. The various aspects of the life-cycle analysis of the Korean shipbuilding cluster suggest this methodology could be used to identify the historical evolution of clusters and provide insight into qualitative phase changes in the development path of clusters. The resulting insight may also provide the basis for policy guidance to help facilitate cluster renewal and therefore increasing returns economic growth and development benefits.

12.7 Conclusions of the Paper

This chapter set out to address the concern that earlier cluster investigators voiced regarding a lack of understanding about cluster dynamics given a tendency for cluster enthusiasts to focus on the increasing returns part of cluster dynamics, that is, the Exploratory and Exploitive Expansion stages. Work prior to the preparation of this chapter was focused on life cycle theory as a framework for investigation and thus understanding cluster dynamics. This previous work considering the life cycle frame focused almost completely on a case study approach. Yet the case studies adopted a methodology for the most part that was historical in character. In this chapter a dimensional methodological approach was introduced and applied (tested) in an application to the analysis of the Korean shipbuilding cluster. This cluster was selected because a recent detailed historical case study analysis had been published that provided considerable detailed data and insight into its evolution (Shin and Hassink 2011). This enabled examining the Korean shipbuilding case at various times in its evolution in terms of each of the seven cluster dimensions proposed in this chapter. The major contribution of this paper is the creation of this dimensional approach. That said the seven dimensions while not selected arbitrarily require greater foundation and explanation than offered here. This is a task for future research. That said each dimension has been identified in the literature as dimensional correlates of cluster dynamics.

Future research is needed as noted to develop a full rationale for the elements of the dimensional analysis. Additional applications of the method to other case studies are needed to better determine its value and usefulness. Importantly such analyses should be used to provide policy guidance or at least to suggest what appropriate policies are for different cluster stage situations. Finally, there is a need to also provide a way to represent the status of a cluster at any point in time on the various dimensions. For example, seven-point star diagrams could be used to represent the status of clusters at various times in a space that represents the life-cycle stages: existence, take-off, exploratory expansion, exploitive expansion, exhaustion, lock-in decline or rejuvenation. Of course a measurement methodology would need to be created to enable obtaining estimates of the seven dimensions in specific cases and at specific times.

Finally, it is important to recognize that the life-cycle approach developed in this paper is presented as is not nested in a broader theoretical framework. Mulligan (2014, personal communication, February 26) suggests that the life-cycle approach "could benefit by re-examining the literature on city sizes" and that a "cluster could be viewed as a single unit where the individual firms are all stochastically shifting, much like a city system where some places are growing and others are falling; so there is some kind of steady state distribution that perhaps should not be violated by too much or harm falls to many." He further notes that Foster and Wild (1996) considered logistic diffusion models and carrying capacities where the survival of individual firms were examined. Expansion of this type of thinking could lead to placing the concept of cluster dynamics in a broader theoretical context. For example, in such a framework events leading to the exhaustion stage might be viewed as explanations of cases where a lack of rejuvenation leads to a lock-in on decline. Alternatively, cases where rejuvenation does occur may not in fact be rejuvenation of the cluster but more of a metamorphosis resulting in a transformation that leads to a cluster of quite a different nature than the original cluster (Jackson 2014, personal communication, February 24) such as in the case of Pittsburgh, PA in the U.S. where rejuvenation of its steel making cluster resulted in the emergence of a technology services cluster. The observations by Mulligan and Jackson suggest some exciting new directions that consider cluster dynamics that go beyond the focus in this paper on the life-cycle model and thus offer a broader frame of reference for future work.

12.8 Discussion and Application of the Life-Cycle Model for Policy

A dimensional model has been developed for the assessment of cluster dynamics and used to examine cluster development stages including policy suggestions for local regional and national or EU policy makers and practitioners. The paper has focused on industrial clusters in general and more specifically on innovative clusters (sometimes called technology clusters). Here some discussion is offered on how to identify cluster stages and the most productive cases for policy intervention.

The core industry or industries for innovative clusters are technology intense in terms of business model and workforce, and sustainability. These are achieved through continuous renewal and improved technology. This implies a high rate of entrepreneurship as this is the major vehicle for achieving continuous innovation. Innovative clusters thrive in regions where knowledge and information production and spillover is high which in turn are the optimal conditions for entrepreneurship.

Patents and patent applications offer one of several measures that correlate positively with performance of innovative clusters. Several maps are presented to

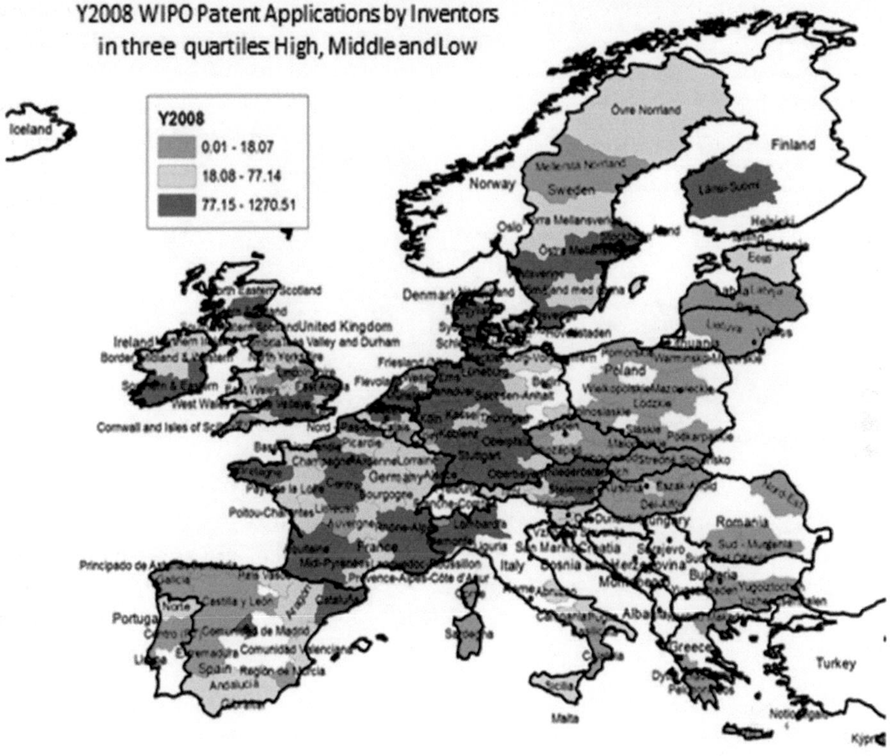

Map 12.1 Y2008 WIPO patent applications by inventors in three quartiles high, middle and low (*Source*: The author)

show the distribution of patent intensity, (a major indicator of innovativeness) for NUTS 2 regions in Europe (2000–2008). Regions with high levels of patents and patent applications would be expected to have high levels of innovation[4] and a strong correlation with innovativeness. While NUTS 2 regions do not necessarily contain clusters, they are the regional ecosystems within which geographic clusters exist.

Map 12.1 displays patent applications by region for 2008 where the NUTS 2 regions have been sorted into three groups (high, medium and low patent production). Those with high levels of patents are in red, low levels in green and middle in yellow. Regions that lie mostly in the former area called Western Europe (South East England, France, Belgium, the Netherlands, Germany and Austria, Northern Italy and Denmark and the South of Sweden and West Central Finland in

[4] Patents are not an ideal measure of innovativeness and it would be ideal to have measures for which there are not so many caveats. However, all that is readily available from archival sources is the patent data.

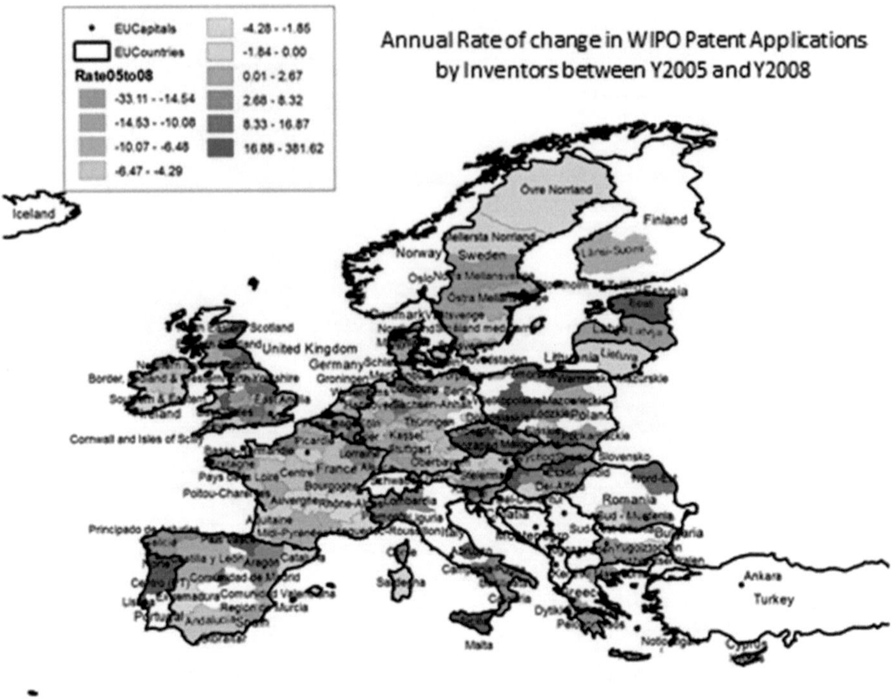

Map 12.2 Annual rate of change in WIPO patent applications by inventors between Y2005 and Y2008 (*Source*: The author)

the vicinity of Tampere) evidence the highest levels of patent production in 2000. In short the annual rate of patenting applications between 2005 and 2008 (Map 12.2) illustrates that the highest average rates of change occurred outside the highest patent producing regions of Europe observed in 2008 (Map 12.1). Data for 2008 in Map 12.1 enables identification of regions where patenting is high and thus ones where the potential for innovative clusters is strong. For example, the South of Sweden evidences a high rate of patenting. However, its rate of patenting from 2005–2008 (see Map 12.2) is quite modest. A working hypothesis based on this data could be formulated that this region is either in the latter part of the exploitive expansion stage of the life-cycle or may be prematurely moving toward exhaustion. This suggests the need for additional analysis to further document these possibilities, for example from data for the seven dimensions of the cluster life-cycle methodology. If additional analysis were to support either of the above possibilities (exhaustion or pre-mature exhaustion) leadership to help reinvigorate the cluster including a plan would be implied.

Map 12.2 data also points to other possibilities, for example, the Pais Vasco region in the Basque Country in Spain. Pais Vasco exhibits a modest level of patenting performance in 2008 but the rate of growth from 2005–2008 is one of the highest in Europe. This suggests that Pais Vasco has passed through the take-off stage and is in the early part of the exploratory expansion stage of cluster development. Of course additional analysis using the seven-dimenison cluster life-cycle approach is needed to validate this hypothesis.

These two examples have been selected because they appear to represent one of the three most defensible situations for which support from higher levels of government (nation and/or EU) is warranted: approaching exhaustion or reaching it prematurely (South of Sweden); and, experiencing take-off or the early part of the exploratory expansion stage (Pais Vasco). These are the situations where the development and/or sustainability of innovative clusters are most vulnerable and at the same time may only require modest investment to either restore an approach to sustaining the cluster or growing it. While there are many reasons why investment should be made at all stages of cluster development as noted above in the policy suggestions relevant to the different stages of the cluster life-cycle the high vulnerability of the three cluster situations noted above, as noted, offer the best return on public sector investment.

Acknowledgments An earlier version of this contribution was presented 53rd Meetings of the Western Regional Science Association, San Diego, California—USA in February 2014. The author recognizes the support he received from the Directorate General for Research and Innovation: Economic Analysis and Indicators or the European Union in the conduct of research leading to this and other papers regarding clusters and entrepreneurship, and cluster dynamics.

References

Baptista R, Mendonca J (2010) Proximity to knowledge sources and the location of knowledge based start-ups. Ann Reg Sci 45(1):5–29

Bergman EM (2008) Cluster life-cycles: an emerging synthesis. In: Karlsson C (ed) Handbook of research on cluster theory. Edward Elgar, Cheltenham, 316 pp

Bode A, Alig S (2011) Cluster development and cluster life cycle, Technische Universitat Darmstadt [http://www.tci2011.com/portals/147/files/Workshop-2011/01/simon/]

Brocker J, Soltwedel R (2010) Agglomeration and growth in knowledge-based societies: an introduction to this special issue. Ann Reg Sci 45(1):1–4

Delgado M, Porter ME, Stern S (2010) Clusters and entrepreneurship. J Econ Geogr 10:495–518

Engel JS, del-Palacio T (2009) Global networks of clusters of innovation: accelerating the innovation process. Bus Horizons 52:493–503

Foster J, Wild P (1996) Detecting evolutionary change in an augmented logistic diffusion econometric model using time-varying spectral methods. Economics discussion paper no 192, University of Queensland

Hassink R, Shin D-H (2005) The restructuring of old industrial areas in Europe and Asia. Environ Plan A 37:571–580

Ingstrup MB, Damgaard T (2012) Cluster facilitation from a cluster life cycle perspective. Eur Plan Stud 2012:1–19

Karlsson C (ed) (2008) Handbook of research on cluster theory. Cheltenham, Edward Elgar, 316 pp

Klepper S (2007) The evolution of geographic structures in new industries. In: Franken K (ed) Applied evolutionary economics and economic geography. Edward Elgar, Cheltenham, pp 69–92

Klink VA, De Langen P (2001) Cycles in industrial clusters: the case of the shipbuilding industry in the Northern Netherlands. Tijdschrift voor Economische en Sociale Geografie 92:449

Knop L, Olko S (2011) Crises in the cluster life-cycle. Management knowledge and learning, international conference, Silesian University of Technology, Gliwice, Poland

Lefebvre P (2012) Organized clusters evolutions: from cluster life-cycle approaches to the action trajectories approach, Paper presented at the 7th international seminar on regional innovation polices, cluster life-cycle sessions, Porto (Portugal) October 11–12, 2012, 41 pp

Lorenzen M (2005) Why do clusters change? Eur Urban Reg Stud 12(3):203–208

Martin R, Sunley P (2011) Conceptualizing cluster evolution: beyond the life-cycle model? Utrecht University, Urban and Regional Research Centre Utrecht. [http://econ.geo.uu.nl/peeg/peeg.html]

Menzel PE, Fornahl D (2009) Cluster life cycles: dimensions and rationales of cluster development. Jena Economic Research Papers no. 2007-076. Max Plank Institute in Economics, Jena, Germany

Porter ME (1990) The competitive advantage of nations. Free Press, New York

Porter ME (1998) Clusters and the new economics of competition. Harvard Business Review. November-December, pp 77–90

Rocha HO (2004) Entrepreneurship and development: the role of clusters. Small Bus Econ 23:363–400

Rocha HO, Sternberg R (2005) Entrepreneurship: the role of clusters theoretical perspectives and empirical evidence from Germany. Small Bus Econ 24:267–292

Shin DH, Hassink R (2011) Cluster life cycles: the case of the shipbuilding industry cluster in South Korea. Reg Stud 45(10):1387–1402.

Sonderegger R, Taube F (2010) Cluster life cycle and diaspora effects: evidence from the Indian IT cluster in Bangalore. J Int Manage 16:383–397

Stough RR, Karlsson C, Johansson B (2014) Agglomeration, clusters and entrepreneurship. In: Karlsson C, Johansson R, Stough RR (eds) Regional economic development: agglomerations, clusters and entrepreneurship. Edward Elgar, Cheltenham, 298 pp

Swann GMP (2002) Towards a model of clustering in high-technology industries. In: Swan GMP, Prevezer M, Stout D (eds) The dynamics of industrial clustering. Oxford University Press, Oxford

Tichy G (1998) Clusters: less dispensable and more risky than ever. In: Steiner M (ed) Clusters and regional specialization: on geography, technology and networks. Pion, London

Utterback JM, Abernathy WJ (1975) A dynamic model of process and product innovation. Omega 3(6):639–656

Toward a Public Policy Agenda for Regional Science: Planning Versus Measuring Impacts

13

Peter D. Blair

Abstract

Much of Walter Isard's lifelong devotion to advancing the emerging disciplines of regional science and peace science focused on a better understanding of the forces shaping the social and economic development of geographic regions and the role that analysis can play formulating public policy. This paper explores generally the use of regional science tools applied to public policy questions in two fundamentally different ways: (1) to fashion steps directed at the most desirable outcome, however desirable is defined—an explicit *planning objective* or (2) to articulate the consequences of possible alternative courses of action—a perhaps more modest *impact analysis objective*. The paper illustrates the circumstances suggesting one approach versus the other with an example involving the use of optimization tools and input–output analysis.

13.1 Introduction

Walter Isard dedicated much of his life to advancing the emerging disciplines of regional science and peace science, which both focused on a better understanding of the forces shaping the social and economic development of geographic regions. He also was deeply interested in the role that analysis can play in understanding and peacefully resolving human conflict, perhaps especially social conflict. Inherently

---

Executive Director, Division on Engineering and Physical Sciences, National Research Council, National Academy of Sciences. Views expressed in this paper are the author's and not necessarily those of the National Academy of Sciences. The principal examples used in this paper are drawn from Ronald E. Miller and Peter D. Blair, *input–output Analysis: Foundations and Extensions*, London: Cambridge University Press, 2009, and the updated materials located on the publisher's website accompanying the text: http://www.cambridge.org/aus/catalogue/catalogue.asp?isbn=9780521517133

P.D. Blair (✉)
National Research Council, National Academy of Sciences, Washington, DC, 20001, USA
e-mail: pblair@nas.edi

P. Nijkamp et al. (eds.), *Regional Science Matters*, DOI 10.1007/978-3-319-07305-7_13,

both these quests involve dimensions of shaping public policy to improve the human condition. In regional science in particular, the role of analysis is to improve our ability to understand the implications of alternative policy choices affecting regions. The tools that have been developed in the field over the past half century in the course of this still emerging field—input–output analysis, location theory, regional econometrics, geographical information systems, to name a few—are perhaps utilized for this purpose more than any other. Many, if not most of these areas had beginnings in Isard's work.

In applying these frameworks today to public policy, it is useful distinguish between two fundamentally different conceptual approaches to addressing public policy questions: (1) to fashion steps directed at the most desirable outcome (however desirable is defined)—an explicit *planning objective* or (2) to articulate the consequences of possible alternative courses of action—a perhaps more modest *impact analysis objective*. This paper first explores conceptually these two approaches and the circumstances suggesting one approach versus the other and illustrates the differences with an example involving the use of optimization tools and input–output analysis.

13.2 The Role of Analysis in Public Policy: An Illustration

I recall my arrival in Washington, DC in the early 1980s, on leave from academia eager to apply my engineering and economic analysis skills to solving national policy problems. As a visiting senior analyst at the former Congressional Office of Technology Assessment (OTA)[1] my initial charge was to examine the cost and performance of new electric power technologies in order to assess their relative prospects for increased deployment in the United States over the next two decades relative to conventional alternatives of the time, and to consider the strengths and weaknesses of alternative public policy measures for accelerating deployment of different types of new power technologies in various regions of the country.[2] It was

[1] In 1972 the United States Congress established the Office of Technology Assessment (OTA) as a small analytical agency to become better informed about implications of new and emerging technologies. The agency's architects intended the reports and associated information it produced to be tuned specifically to the language and context of Congress. OTA's principal products—technology assessments—were designed to inform Congressional deliberations and debates about issues that involved science and technology dimensions but without recommending specific policy actions. The agency's unique governance by a bicameral and bipartisan board of House and Senate Members helped ensure that issues OTA addressed were relevant to the Congressional agenda and that assessments were undertaken with partisan and other stakeholder bias minimized. Over a span of 23 years OTA completed 755 reports on a wide range of topics including health, energy, defense, space, information technology, environment, and many others until Congress terminated the agency's annual appropriation of funds to operate in 1995 (see Blair 2013).

[2] Office of Technology Assessment, *New Electric Power Technologies: Problems and Prospects for the 1990s*, Washington, DC: U.S. Congress, Office of Technology Assessment, OTA-E-246, July 1985.

an interesting time for analysts in the electric utility industry and in the public policy environment in which the industry operated as outlined in the introduction of the OTA assessment report:

> During the 1970s, the environment within which utilities made investment decisions changed from a relatively predictable continuation of past trends to a highly uncertain and complicated maze of interrelated financial, regulatory, and technology considerations. As electric utilities face the 1990s, the experiences of the 1970s have made them much more wary of the financial risk of guessing wrong and overcommitting to large central station coal and nuclear plants. At the same time, the possibility of being unable to meet electricity demand exists, causing growing concern among utilities as the next decade approaches.
>
> As a result, utilities are now taking steps to enhance their flexibility in accommodating future uncertainties. In addition to continued and primary reliance on conventional technologies, supplemented by coal combustion technology enhancements to reduce pollution emissions and increase efficiency, utilities are considering a variety of less traditional options. These include life extension and rehabilitation of existing generating facilities, increased purchases from and shared construction programs with other utilities, diversification to nontraditional lines of business, increased reliance on less capital-expensive options such as load management and conservation, and smaller scale power production from a variety of conventional and alternative energy sources. Such options offer utilities the prospects of more rapid response to demand fluctuations than traditional, central station power plants. (Office of Technology Assessment 1985, p. 1)

A key feature of the time was that uncertainty dominated the electric utility decision-making environment of 1980s. Historical projections of electricity demand were shown to be consistently wrong for the previous decade, consistently underestimating the impacts of energy efficiency improvements brought about by higher electricity prices—growth projections fell from nearly 8 % annually in 1974 to 2 % in 1985. The renewable power industry was in its infancy following passage of the Public Utility Regulatory Policies Act of 1978 (PURPA) and implementation of a federal renewable energy tax credit and similar credits in a number of states. At that time natural gas had been declared a premium fuel reserved for home heating and precluded from use in electric power generation with the Powerplant and Industrial Fuel Use Act of 1978 (PIFUA), which seems particularly ironic now given the shale gas boom in the United States in the last several years. Electric utility stock values and bond ratings had fallen precipitously as rate increases proposed by utilities around the country were being challenged by state public utility commissions as based on imprudent investments in unneeded power generating capacity—mostly large central station nuclear power and coal-fired generation facilities.

Major amendments in 1977 to the 1970 Clean Air Act were also especially influential in shaping policy at the time. They added provisions for the Prevention of Significant Deterioration (PSD) of air quality in areas attaining the National Ambient Air Quality Standards (NAAQS) in addition to new requirements on pollution sources (including especially power plants) in so-called non-attainment areas for NAAQS—geographic areas that did not meet one or more of the federal air quality standards. These new provisions established major new permit review requirements that injected considerable uncertainty in the future of coal-fired power generation and no new nuclear power plants had been constructed in the United States since 1972.

As all these forces came together to shape the energy and environmental public policy environment in the 1980s, there were many major urgent technical, economic, and policy issues to be addressed regarding the future of the electric utility industry in the United States. Of particular significance at the time and even today are the substantial differences in the circumstances of regions in the United States regarding resources, regulation, supply and demand market structures for electricity and fuels, and many other features. For the public policy questions of the time, in retrospect, my somewhat naïve academic approach to the issues was to simply examine what was known about the cost and performance of new electric power generating technologies, to model the impacts of various regulatory actions, incentives or other policy measures on the possible cost and performance of those technologies relative to conventional alternatives, and recommend the most cost effective measures. However, taking this path required making assumptions about the decision-making context well beyond the much narrower framework of analysis with which we had been charged for the OTA assessment.

The essential assumptions at issue are not just technical ones about such things as inflation rates, costs of capital, regulatory assumptions, etc., but also the basic political assumptions regarding the receptivity within Congress to various public policy measures, economic assumptions such as how the regulatory structure of the utility industry itself was evolving, and even ideological assumptions about the rationale for public policy incentives such as incorporating environmental externalities or federal budget priorities. I learned that in this situation the most effective advice for the Congress is to estimate, to the extent possible, the consequences of alternative courses of actions without necessarily providing a recommended course, particularly if a candidate recommendation hinged on an assumption that was subject to debate by stakeholder interests. Indeed, providing a specific recommendation under these circumstances might undermine the effectiveness of the analysis in informing the debate, which was the principal objective of the analysis in the first place.[3]

13.3 Policy Decisions

My initial experience in Washington illustrates the challenges of informing a policy debate in a productive way. Making all the assumptions necessary to come to a definitive policy recommendation, even with complete transparency about those assumptions, often runs the risk of relegating the advice more to serving as the basis for an existing advocacy position than informing the debate about making choices, particularly if heroic assumptions must be made in the course of the analysis,

[3] See Blair (2013) for additional details about the purpose, structure, and history of OTA technology assessments.

e.g., about federal budget priorities, tax policy, discount rates, available data, modeling assumptions, etc.

As my experience illustrates, informing policy decisions free of unintended bias may be most productive in principle by articulating the consequences of alternative actions without suggesting the policy decision. For example, if an objective is to accelerate the role of clean energy technologies to mitigate climate change impacts or other objectives, there are many policy pathways, and analyzing the implications of a range of them has considerable value. If a particular pathway is recommended, however, it is important to gauge the degree to which that recommendation hinges on assumptions about the policy context or other factors.

As an example, extending illustrations in energy policy, if government is considering policy mechanisms for reducing the nation's dependence on imported oil, economists will argue generally that market-based mechanisms, such as a tax, comprise by far the most efficient means in terms of internalizing non-market costs to the economy such as national security or environmental benefits of reduced oil use. Historically, however, public aversion to tax increases of any kind in the United States has often rendered such pathways problematic. For example, at various times over the past three decades the prospects of increasing the federal gasoline tax have been put forward in the Congress in an effort to better capture environmental, national security, and other economic costs not captured in fuel markets, but such efforts have virtually always been shouted down in a public backlash.

Even if mechanisms to better capture market externalities in the form of increasing prices become feasible politically, many other issues remain, such as regressivity of the proposed tax, regional differences in patterns of gasoline demand or vehicle miles travelled or other economic impacts, and other factors that are often well outside the scope of an analysis designed to measure success in terms of reduced oil use. For example, in recent debates about the prospects of a carbon tax to internalize the long term costs of carbon emissions from burning fossil fuels, the issue of regressivity has seldom become a significant factor in the debate.[4]

13.4 Regional Science Applications: Illustration Using Generalized Input–Output Analysis

A commonly posed public policy analysis problem familiar to regional scientists is to analyze the implications of a new spending program (usually government, but certainly not exclusively so) on an economy, which may involve comprehensive examination of a wide variety of factors associated with that spending program, such as impacts on employment, pollution, infrastructure utilization, or capital expenditures. In the following a general framework expanded from Leontief's basic input–output analysis formulation is applied for tracing these impacts

[4] While the discussion of regressivity has been often absent from this policy debate, analysis of the topic is not, such as in Parry et al. (2005) or Oladosu and Rose (2007).

associated with interindustry production generated in response to a spending program interpreted in traditional Leontief parlance as a new vector of final demands presented to the economy.[5]

A straightforward and commonly applied[6] approach to accounting for quantifiable impacts associated with interindustry activity, such as pollution generation, is to first define a matrix of direct impact coefficients, $\mathbf{D}^p = [d_{kj}^p]$. Each element, d_{kj}^p, in the case of pollution impacts as an illustration, is the amount of pollutant type k, e.g., sulfur dioxide, generated per dollar's worth of industry j's output. Hence, the level of pollution associated with a given vector of total outputs can be expressed as $\mathbf{x}^{p*} = \mathbf{D}^p\mathbf{x}$ where $\mathbf{x}^{p*}$ is the vector of pollution levels. By adding the traditional Leontief model, $\mathbf{x} = \mathbf{L}\mathbf{f}$, where $\mathbf{A}$ is the matrix of technical coefficients, $\mathbf{f}$ is the vector of final demands, and $\mathbf{x}$ is the vector of total outputs, and $\mathbf{L} = (\mathbf{I} - \mathbf{A})^{-1}$, we can compute $\mathbf{x}^{p*}$ as a function of final demand, that is, the total pollution of each type generated by the economy directly and indirectly in supporting that final demand, $\mathbf{x}^{p*} = \mathbf{D}^p\mathbf{L}\mathbf{f}$.

The matrix product, $\mathbf{D}^p\mathbf{L}$, is a matrix of total environmental impact coefficients; that is, an element of this matrix is the total pollution impact generated per dollar's worth of final demand presented to the economy.[7] Similarly, we can expand the pollution coefficients matrix with similar coefficients for other factors associated with interindustry activity that we assume vary linearly with output, such as employment or energy consumption.

For this illustrative example we restrict the generalized framework to energy use, environmental pollution, and employment as illustrative of the more general case in which we define three direct-impact coefficient matrices relating energy requirements, pollution generation and employment to total output: $\mathbf{D}^e$, $\mathbf{D}^p$ and $\mathbf{D}^l$, respectively, in units such as British thermal units (Btus) of energy of energy used (coal, oil, electricity, etc.), pounds of pollution emissions (sulfur dioxide, particulates, carbon dioxide, etc.), and person-years of employment all expressed per dollar's worth of output. For convenience, we concatenate these matrices to yield a direct-impact coefficient matrix $\mathbf{D} = \begin{bmatrix} \mathbf{D}^e \\ \hline \mathbf{D}^p \\ \hline \mathbf{D}^l \end{bmatrix}$ and similarly define a vector of total impacts, $\mathbf{x}^*$, by concatenating $\mathbf{x}^{e*} = \mathbf{D}^e\mathbf{x}$, $\mathbf{x}^{p*} = \mathbf{D}^p\mathbf{x}$, and $\mathbf{x}^{l*} = \mathbf{D}^l\mathbf{x}$ to yield $\mathbf{x}^* = \begin{bmatrix} \mathbf{x}^{e*} \\ \hline \mathbf{x}^{p*} \\ \hline \mathbf{x}^{l*} \end{bmatrix}$ or $\mathbf{x}^* = \mathbf{D}\mathbf{x}$. For accounting convenience, we can include $\mathbf{x}$ along with the corresponding final demands associated with the generation of a particular vector of

---

[5] See Blair (1979) and Miller and Blair (2009), Chap. 10.

[6] Beginning with Victor (1972) and many others over the past four decades; see Miller and Blair (2009), Chap. 10.

[7] Measuring energy and environmental activities in input–output models in monetary units can create accounting inconsistencies as described in Miller and Blair (2009), but adopting hybrid units for such calculations, e.g., energy units for energy transactions, have limitations as well as described in Dietzenbacher and Sage (2006).

total impacts, $\mathbf{x}^*$, by defining a new vector of total impacts and concatenating the vector of final demands with $\mathbf{x}^*$ to obtain this expanded vector of *total* impact, $\tilde{\mathbf{x}} = \begin{bmatrix} \mathbf{x}^* \\ \hline \mathbf{f} \end{bmatrix}$. We can similarly expand the matrix of direct-impact coefficients by concatenating $\mathbf{D}$ and $(\mathbf{I} - \mathbf{A})$ to yield a new expanded direct-impact coefficients matrix: $\mathbf{G} = \begin{bmatrix} \mathbf{D} \\ \hline (\mathbf{I} - \mathbf{A}) \end{bmatrix}$. This formulation is particularly well suited to using input–output with mathematical programming models.[8]

As an alternative to the preceding formulation, we may wish to express total impacts as a function of final demands in much the same way we traditionally use the Leontief inverse. That is, we may find the total impacts in terms of energy, pollution generation and employment associated with some given level of final demand. In this case we can write our earlier expression for total impacts, $\mathbf{x}^* = \mathbf{D}\mathbf{x}$, equivalently as $\mathbf{x}^* = \mathbf{D}^*\mathbf{f}$, where $\mathbf{D}^* = \mathbf{D}\mathbf{L}$ is the matrix of total-impact coefficients.[9] In this case we compute $\mathbf{x}^*$ as a function of $\mathbf{f}$ while in $\mathbf{x}^* = \mathbf{D}\mathbf{x}$ we computed $\mathbf{x}^*$ as a function of $\mathbf{x}$. Finally, once again for convenience, we may wish to include $\mathbf{x}$ itself in our vector of total impacts. This can be accomplished easily by concatenating $\mathbf{x}$ with the vector of total impacts in the same manner we concatenated $\mathbf{x}^*$ with $\mathbf{f}$ in constructing $\tilde{\mathbf{x}}$, i.e., we define a new expanded vector of total impacts to be: $\bar{\mathbf{x}} = \begin{bmatrix} \mathbf{x}^* \\ \hline \mathbf{x} \end{bmatrix}$. As before with $\mathbf{G}$, we can similarly expand the total-impacts coefficients matrix by concatenating the Leontief inverse with the total-impact coefficients; we call the new expanded total-impacts coefficients matrix, $\mathbf{H}$, so that $\mathbf{H} = \begin{bmatrix} \mathbf{D}^* \\ \hline \mathbf{L} \end{bmatrix}$.

The expressions for $\tilde{\mathbf{x}}$ and $\bar{\mathbf{x}}$ are equivalent descriptions of the same situation, of course, since $\mathbf{x}$ and $\mathbf{f}$ uniquely define one another in a Leontief model—for every given $\mathbf{f}$, there is one and only one $\mathbf{x}$, and vice versa. Note also that we can create a matrix of impacts generated by each industry separately by $\mathbf{H}\hat{\mathbf{f}}$ where $\hat{\mathbf{f}}$ designates the elements of $\mathbf{f}$ placed along the diagonal of a square matrix.

13.5 Summary: Generalized input–output Formulations

Recall that the generalized input–output model becomes possible with a set of direct impact coefficients, $\mathbf{D} = [d_{kj}]$, each element of which is the amount of an impact variable k, for example, pollution or energy, generated per dollars' worth of industry j's output. Using $\mathbf{D}$ we can pose the generalized input–output model in

[8] As examples, see Thoss (1976) and Blair (1979).

[9] This formulation was originally applied by Just (1974) and Folk and Hannon (1974) to examine the impacts of new energy technologies. Other more recent applications are summarized in Forssell and Polenske (1998), including, in particular, Qayum (1991 and 1994), Schäfer and Stahmer (1989) and Lang (1998).

what we will refer to as either its *impact analysis* or *planning* forms, which we define as the following:

Impact Analysis Form: $\overline{\mathbf{x}} = \mathbf{Hf}$ where $\mathbf{H} = \begin{bmatrix} \mathbf{D}^* \\ \hdashline \mathbf{L} \end{bmatrix}$ and $\overline{\mathbf{x}} = \begin{bmatrix} \mathbf{x}^* \\ \hdashline \mathbf{x} \end{bmatrix}$ and $\mathbf{D}^* = \mathbf{DL}$

Planning Form: $\tilde{\mathbf{x}} = \mathbf{Gx}$ where $\mathbf{G} = \begin{bmatrix} \mathbf{D} \\ \hdashline (\mathbf{I} - \mathbf{A}) \end{bmatrix}$ and $\tilde{\mathbf{x}} = \begin{bmatrix} \mathbf{x}^* \\ \hdashline \mathbf{f} \end{bmatrix}$

The impact analysis form is most traditionally considered in input–output applications where the question is what industry outputs and factors associated with interindustry activity, such as energy use, environmental pollution levels and employment, result from a given schedule of final demands presented to the economy, as in Johnson and Bennett (1981), Hannon et al. (1983) and many others. However, the planning form has advantages in applications where one seeks to optimize an objective other than the objective implicit in a traditional input–output model.[10] In the following we explore an example of extending this framework to planning applications.

For the planning form, the basic input–output framework can be posed the solution of a system of n linear equations in n unknowns, which is certainly one of the most attractive features of the framework—a straightforward and unique solution. Equivalently, but useful as a starting point, is posing the generalized input–output model as an optimization problem. In general, this means relaxing fundamental assumptions in the framework in order to adapt it to specific situations, such as allowing technical coefficients to vary as a function of relative prices in the case of econometric extensions to the basic model, adding capital coefficients in dynamic input–output models, or adding trade coefficients in multi- or interregional models, as examples. Also, in using input–output in planning applications, where one seeks to optimize (maximize or minimize) some objective function related to interindustry activity, it is useful to begin by formulating the input–output analysis framework as a simple linear programming problem.

We start with the generalized input–output formulation presented in its planning form, $\tilde{\mathbf{x}} = \mathbf{Gx} = \begin{bmatrix} \mathbf{D} \\ \hdashline (\mathbf{I} - \mathbf{A}) \end{bmatrix} \mathbf{x}$, where $\mathbf{D}$ is the matrix of direct impact coefficients relating factors such as energy use, pollution emissions, and employment to industry output, i.e., $\tilde{\mathbf{x}} = \mathbf{Gx} = \begin{bmatrix} \mathbf{x}^* \\ \hdashline \mathbf{f} \end{bmatrix}$, where $\mathbf{x}^*$ represents the levels of total impact of energy use, pollution, and employment associated with output $\mathbf{x}$ and, of course, final demand $\mathbf{f}$. We use $\mathbf{G}$ to represent the more generalized structural relationships governing not only the Leontief production possibilities but also the levels of energy consumed, pollution discharged and employment generated that are associated with those production possibilities. Hence, if we define $\mathbf{v}$ as the vector of value added

[10] We will see later that the implicit objective function in an input–output model is to maximize the sum of all final demands or, equivalently, to minimize the sum of all value added inputs.

coefficients and, hence, $q = \mathbf{v}'\mathbf{x}$ as the total value added of an economy associated with a vector of total outputs—the economy's gross domestic product, we can rewrite the generalized input output formulation as a linear programming (LP) problem to determine the minimum value of q that satisfies deliveries to final demand or, equivalently, what values of the elements of $\mathbf{x}$ minimize q while satisfying final demand?[11]

$$\begin{array}{c} Min \quad q = \mathbf{v}'\mathbf{x} \\ subject \ \ to : \mathbf{Gx} \geq \widetilde{\mathbf{x}} \end{array}$$

Since the input–output model has a unique solution for a given level of final demand, and those conditions are part of the constraint equations, $\mathbf{Gx} \geq \widetilde{\mathbf{x}}$, then either of two situations apply: (1) the additional constraints beyond the Leontief conditions included in the constraint equations (e.g., energy, environmental and employment equations) over-constrain the problem, i.e., present conflicting constraints, so that there is no feasible region and, hence, no possible LP solution or (2) the additional conditions are not binding constraints, i.e., they are fully satisfied by the Leontief conditions. In our example so far the latter situation applies and the LP solution is identical to the impact analysis solution determined earlier. In the former case, however, when there is no feasible region, we will need to resort to other approaches in order to find a solution, e.g., multiobjective programming techniques.[12]

Even the basic LP formulation to the generalized input–output planning problem gives us the flexibility to include alternative or multiple objective functions in approaching planning problems. For example, one might be interested in minimizing the value added cost to meeting a target final demand, minimizing pollution emissions and energy conservation all as goals. However, the problem of conflicting constraints requires other methods of solution. As an illustration, linear goal programming (GP) is one commonly applied extension to LP that accommodates multiple objectives in a very straightforward manner and can be used to extend the Leontief framework to deal with environmental issues involving conflicting objectives and constraints.

[11] More extensive economic interpretations of the Leontief model as a linear programming problem are included in Dorfman et al. (1958) and Intriligator (1971). An important advantage of posing the input–output framework in this way is that we can include alternative objective functions and/or additional constraints as part of a planning problem.

[12] Decision making with multiple objectives is another well-developed area in operations research with many approaches available. Surveys of such approaches are found in Cohen (1978), Cochrane and Zeleny (1973), Nijkamp and Rietveld (1976), Trzaskalik and Michnik (2002), and Tanino et al. (2003).

13.6 Linear Goal Programming[13]

In a traditional linear programming problem the optimal solution must lie within the feasible region defined by the constraint equations. LP maximize or minimizes the objective function subject to having already satisfied what can be considered a preliminary objective of considering solutions that are within the feasible region. In goal programming the implicit objective of any solution lying within the feasible region in the linear programming approach is considered instead as an explicit rather than implicit objective. A traditional method of accomplishing this is to introduce so-called *deviational* variables d_k, for $k = 1, 2, \ldots m$, for m constraint equations. The values of d_k measure "deviation" from the right-hand side values of the constraint equations or deviation from an established goal for the explicit objective function, q. In all cases $d_k \geq 0$.

We can think of the deviational variables as either exhibiting *overachievement* or *underachievement* of the established goal. For overachievement of goal k we call the level of overachievement a *positive deviation*, which is indicated by a non-zero value for a *positive deviational variable,* d_k^+ (sometimes referred to as a *surplus* variable). Likewise, for underachievement of the established goal, we call the level of underachievement a *negative deviation* from the goal indicated by a non-zero value for a *negative deviational variable*, d_k^- (sometimes referred to as a *slack* variable). In GP if $d_k^+ > 0$ then $d_k^- = 0$ and vice versa, i.e., at most one of the paired positive and negative deviational variables, d_k^- and d_k^+, can be greater than 0; if both are 0 then, of course, the goal is achieved exactly.

In GP, to account for the relative order of priority on the objectives, we assign the objectives to *pre-emptive priority* classes, each of which is denoted by P_l, for $l = 1, \ldots, L$ where L is the total number of priority classes. The GP equivalent of an LP would be to two pre-emptive priority classes, P_1 and P_2, associated with first satisfying all LP constraint equations as the highest priority (the implicit objective of any solution falling within the feasible region) and then maximizing q as the second priority. Continuing the relationship with the LP case, in general, minimizing positive deviational variables is equivalent to satisfying a $\leq$ constraint or minimizing an objective function and minimizing negative deviational variables is equivalent to satisfying a $\geq$ constraint or maximizing an objective function. Hence, seeking to drive d_k^+ for the constraint equations to zero is equivalent to satisfying the constraints originally specified in the LP problem and seeking to drive d_k^- to zero for q is equivalent to maximizing the objective function specified in the LP problem. The key difference is that in GP we no longer really distinguish between what in the LP formulation were objective and constraint equations[14]

[13] Goal programming was first suggested by Charnes and Cooper (1961). Useful characterizations of the approach and further refinements are included in Lane (1970), Lee (1971, 1972, 1973), and Ignizio (1976).

[14] Constraints in LP are generally specified as inequalities, but by introduction of slack or surplus variables, they can be specified as equations.

and, rather, consider them all goal equations that are optimized in a specified preemptive order of priority.[15]

It is important to note that GP provides considerable flexibility in handling multiple, even conflicting objectives. In LP we are required to begin with a feasible region in order to find our way to an optimal solution. In GP the feasible region is essentially unspecified since any point can be completely specified in terms of the deviational variables and it is the priority order of objectives that determines the solution, because by specifying a priority order a solution can be found even if the objectives are conflicting. This places much more of the planning burden on determining the priority order, which in a policy context is often quite complex, especially with sometimes competing policy objectives, such as economic growth and environmental quality. Blair (1979) employs an approach called *analytic hierarchies* (Saaty 1980) for this purpose. Others use a wide range of multiobjective decision-making approaches, such as Nijkamp and van Delft (1977) and Cohen (1978).

Another commonly cited limitation of linear goal programming (and linear programming generally for that matter) is that tightly constrained problems are insensitive to how close one is to a given target because solutions, at least as we have developed the methodology so far, are developed by satisfying goals completely in pre-emptive order (ordinal) or lexicographic order. This means that one has to fully satisfy a higher order goal equation before moving on to the next. This can lead to illogical solutions, especially in tightly constrained problems. For example, if an employment goal has a higher priority than, say, a pollution goal, then the last unit of employment achieved could be at the expense of an enormous amount of pollution. The literature includes many approaches to address this problem, such as Lane (1970).

As with LP, when the number of variables and equations increases beyond two, solution procedures become much more complex. However, there are a variety of solution approaches to GP problems. In GP, as we have just seen with an example, through sequential imposition of constraints we arrive at a solution—GP is sometimes referred to as "weighting within constraints." Linear GP problems can be solved via a basic simplex algorithm similar to that commonly used in LP.[16]

13.7 Applications to the Generalized Input–Output Planning Problem

Let us return to our generalized input–output planning example noted earlier to illustrate the GP solution in this context. Recall the constraint equations, which in the GP context are no longer called constraints, but rather goal equations. We refer

[15] This applies to the linear version of GP; the features of alternative GP formulations, such as developed in Lane (1970) or Cohen (1978), address some of the limitations of the linear approach that will be apparent in what follows, e.g., when problems are very tightly constrained.

[16] As in Blair (1979) and Lee (1971, 1972, 1973). Other approaches are explored in Ijiri (1965), Cohen (1978), and Ignizio (1976).

to the relationships among the goal equations collectively as the set of *system process functions*:

$$\tilde{\mathbf{x}} = \mathbf{G}\mathbf{x} = \begin{bmatrix} \mathbf{D} \\ \hline (\mathbf{I}-\mathbf{A}) \end{bmatrix} \mathbf{x} = \begin{bmatrix} \mathbf{x}^{*} \\ \hline \mathbf{f} \end{bmatrix}$$

where the partitions of the matrix **G** are the coefficients for the energy, environment, employment and economic (Leontief) equations, respectively, as defined earlier.[17] If the values for $\mathbf{x}^{*}$ and **f** are planning targets then this can be formulated as a GP problem as above by introducing deviational variables which will measure under and over achievement of the established targets.

13.8 Policy Programming

Blair (1979) combines the impact analysis version of the generalized input–output model along with the planning form and goal programming in an integrated approach called *policy programming*.[18] In this approach, first, a number of alternative future regional energy development scenarios, or *future scenarios*, for short, are defined. These scenarios are all defined to satisfy a set of generalized input–output system process equations, as defined earlier. That is, each future scenario is defined in terms of values of industry output, energy consumption, pollution emissions and regional employment that comply with the basic Leontief identities and accompanying direct impact coefficients for energy use, environmental pollution emissions and regional employment. We define a collection of values of these variables that comply with these system process functions as a *consistent* scenario.

In policy programming the method of analytic hierarchies[19] is applied to define *preference scenarios*, which are simply linear combinations of future scenarios. A set of relative weights derived from the method of analytic hierarchies is applied to assemble the linear combination of future scenarios that reflect the relative desirability of the alternative future scenarios. Preference scenarios are assembled for each of a number of defined policy makers or other significant decision-makers in the planning process, e.g., in the case of energy-environment problems, electric utilities, government regulators or industrial consumers. Since these preference scenarios are linear combinations of consistent scenarios, they are also consistent by the above definition.[20]

[17] This is the framework adopted in Blair (1979) and Miller and Blair (2009).

[18] Illustrated and applied in Blair (1979) and Miller and Blair (2009) and summarized here.

[19] The method of analytic hierarchies, often referred to as the Analytic Hierarchy Process, is a theory and method of decision-making based on deriving priorities from a matrix of pairwise comparisons of alternatives; see Saaty (1980); there are, of course, many other ways of deriving priorities as well.

[20] Posed as a theorem in Blair (1979).

The variables defined in the system process functions are then separated into *clusters*, over which different policy makers (or others relevant to the planning problem) have differing degrees of relative influence. The methodology of analytic hierarchies is applied once again, this time to define weights of relative influence of policy makers over clusters of decision variables. A *composite scenario* is then assembled by bringing together all clusters of variables, with the values derived from the corresponding clusters of variable values drawn from preference scenarios weighted by the relative influence of policy makers over those clusters. This composite scenario, however, will not likely be consistent (by the earlier definition) with the set of system process functions since the relative weights over clusters are derived independently and the composite scenario is not a linear combination of consistent scenarios.

To show this analytically, recall that the *Impact Analysis Form* is specified as $\overline{\mathbf{x}} = \begin{bmatrix} \mathbf{x}^* \\ \hline \mathbf{x} \end{bmatrix} = \mathbf{Hf} = \begin{bmatrix} \mathbf{D}^* \\ \hline \mathbf{L} \end{bmatrix}\mathbf{f}$ and the *Planning Form* is specified as $\tilde{\mathbf{x}} = \begin{bmatrix} \mathbf{x}^* \\ \hline \mathbf{f} \end{bmatrix} = \mathbf{Gx} = \begin{bmatrix} \mathbf{D} \\ \hline (\mathbf{I}-\mathbf{A}) \end{bmatrix}\mathbf{x}$. We define three new final demand vectors $\mathbf{f}_1$, $\mathbf{f}_2$ and $\mathbf{f}_3$ that correspond to three different possible *future* scenarios. For each final demand vector, we can compute the generalized impact of each as $\overline{\mathbf{x}}_i = \mathbf{Hf}_i$ for $i = 1, 2$ and 3. For convenience, if we define $\mathbf{F} = [\mathbf{f}_1 \, \vdots \, \mathbf{f}_2 \, \vdots \, \mathbf{f}_3]$ as a matrix, the columns of which are the final demand vectors, then it is easy to define $\overline{\mathbf{X}} = [\overline{\mathbf{x}}_1 \, \vdots \, \overline{\mathbf{x}}_2 \, \vdots \, \overline{\mathbf{x}}_3]$ as the matrix of corresponding generalized impact vectors, which can be expressed as

$$\overline{\mathbf{X}} = \mathbf{HF} = [\overline{\mathbf{x}}_1 \mid \overline{\mathbf{x}}_2 \mid \overline{\mathbf{x}}_3] = \left[\begin{array}{c|c|c} \mathbf{x}_1^* & \mathbf{x}_2^* & \mathbf{x}_3^* \\ \hline \mathbf{x}_1 & \mathbf{x}_2 & \mathbf{x}_3 \end{array}\right]$$

These scenarios (each of which is specified in a column of $\overline{\mathbf{X}}$) are *consistent* as defined earlier, since, in each case, $\mathbf{x}_i^* = \mathbf{DLf}_i$.

We can define a *composite scenario* as a linear combination of these future scenarios. For example, consider the composite scenario $\mathbf{f}_c$ defined as a simple average of the future scenarios, i.e., $\mathbf{f}_c = \sum_{i=1}^{3} \beta_i \mathbf{f}_i$, where $\beta_i = 1/3$ for $i = 1, 2$ and 3 (note that $\sum_{i=1}^{3} \beta_i = 1$). It is easy to show that $\mathbf{x}_c^* = \mathbf{DLf}_c$, which confirms that the composite scenario is consistent as defined above. Blair (1979) shows that this is true for any linear combination of consistent future scenarios (not just a simple average as above). In the example, we can define the consistent composite scenario as $\overline{\mathbf{x}}_c = \mathbf{Hf}_c = \mathbf{H}[\beta_1\mathbf{f}_1 + \beta_2\mathbf{f}_2 + \beta_3\mathbf{f}_3]$.

Since $\overline{\mathbf{x}} = \begin{bmatrix} \mathbf{x}^* \\ \hline \mathbf{x} \end{bmatrix}$, we write $\overline{\mathbf{x}}_c = \begin{bmatrix} \mathbf{x}_c^* \\ \hline \mathbf{x}_c \end{bmatrix}$ and if $\mathbf{x}_c^* = \mathbf{Dx}_c$, then the composite scenario is consistent as well as defined above; this will be true for any linear combination of consistent future scenarios. We now define variable clusters for these scenarios as energy, pollution, and employment, necessary to distinguish between variables over which priorities can be assigned. If we assemble a different composite scenario in which the relative weights for combining the future scenarios are different for each cluster of variables, then it is easy to show that the resulting composite scenario is not necessarily consistent, and in general unlikely. To denote both different future scenarios to be combined and different weights to be applied to clusters of variables, instead of β_1, β_2 and β_3 we then define these weights for each cluster of variables as β_i^e, β_i^p and β_i^l for scenarios $i = 1$, 2 and 3, where e, p and l denote, as before, the energy, pollution and employment clusters of variables, respectively. To account for the variable clusters, we can express each future scenario i as $\overline{\mathbf{x}}_i = \begin{bmatrix} \mathbf{x}_i^e \\ \mathbf{x}_i^p \\ \mathbf{x}_i^l \\ \hdashline \mathbf{x}_i \end{bmatrix}$ or equivalently as $\tilde{\mathbf{x}}_i = \begin{bmatrix} \mathbf{x}_i^e \\ \mathbf{x}_i^p \\ \mathbf{x}_i^l \\ \hdashline \mathbf{f}_i \end{bmatrix}$. Hence, we can define a composite scenario, reflecting the different weights for each cluster of variables, as: $\overline{\mathbf{x}}_c = \begin{bmatrix} \sum_{i=1}^{3} \beta_i^e \mathbf{x}_i^e \\ \sum_{i=1}^{3} \beta_i^p \mathbf{x}_i^p \\ \sum_{i=1}^{3} \beta_i^l \mathbf{x}_i^l \\ \hdashline \sum_{i=1}^{3} \mathbf{x}_i \end{bmatrix}$.

This new composite scenario is not generally consistent, as defined earlier. The problem of developing a *consistent composite scenario* can be addressed via GP, by formulating the problem to find a consistent scenario that is "as close as possible" to the composite scenario but that complies fully with the defined system process functions. The literature is replete with attempts to extend the input–output framework applied to environmental problems using various kinds of multiobjective decision-making tools in addition to the simple approach illustrated here.[21]

13.9 Strengths and Weaknesses of the Illustrative Model Assumptions for Policy Applications

Note the contrast, conceptually, between the impact analysis and the planning formulations just illustrated. The analytical assumptions in the impact analysis formulation are already substantial from a policy making point of view, including all the assumptions of the Leontief model, the assumptions of linear relationships

[21] Also illustrated in Miller and Blair (2009) and Blair (1979). A similar example is given in Hipel (1992).

among energy consumption, pollution, employment, etc. associated with interindustry production, and many others. The planning formulation, however, requires yet additional normative assumptions of planning priorities and targets and of how to measure proximity to planning targets. Not included at all are considerations of uncertainty in the data, other analytical limitations of the analysis tools, scale and aggregation, geographic scope and many other considerations that may be significant in a policy making context. This is but one example of fashioning a policy analysis in either an impact analysis or planning form, but the significance of assumptions necessary will progress similarly with essentially any analysis framework.

There are many contemporary policy issues where these illustrative considerations apply. For example the application in Blair (1979) was to regional energy planning questions. Today, the United States faces substantial analogous questions such as the resilience and reliability of the nation's electric power transmission and distribution system, the implications of the recent domestic additions of oil and gas supply resulting from aggressive use of hydraulic fracturing of shale gas resources, the challenges associated with substantial adoption of renewable electric power generating technologies, and the nature and degree of environmental regulation of energy production. These questions are different in many ways from those faced in the late 1970s in the United States, but they still possess many of the same analytical characteristics. For example, the considerable differences among U.S. regions in terms of available energy supply resources and characteristics of energy use, regulatory context, economic conditions and patterns of economic structural change and the pace of that change would make any modeling framework more challenging, particularly if the objective was to acquire results at the national level.

Adapting the policy programming framework outlined earlier to these contemporary problems, utilizing either of the impact analysis or planning approach, would benefit from enhancements to deal more effectively with current circumstances. As an example, Miller and Blair (2009, Chap. 9) show that in order to construct the matrix of direct energy coefficients, posed earlier as $\mathbf{D}^e$ such that $\mathbf{x}^{e*} = \mathbf{D}^e\mathbf{x}$, presumes implicitly that for each energy type prices are uniform across all consuming sectors for that type of energy. This may be inconsequential in some circumstances, but in highly developed economies where different sectors use energy quite differently, actual interindustry energy prices will vary considerably. For example, Griffin (1976) shows that the condition of uniform interindustry prices across all energy-consuming sectors does no hold at all historically for the U.S. Economy. Weisz and Duchin (2006) show similar results. Miller and Blair (2009, Chap. 9) show that a so-called hybrid units formulation of the Leontief model accommodates non-uniform interindustry energy prices quite effectively, but introduces other analytical issues, such as those noted earlier raised by Dietzenbacher and Sage (2006).

Conclusions

The role of analysis in public policy questions has always been a challenging one. On the one hand, analysis provides a path to assembling evidence for alternative policy options in a transparent way that helps articulate the consequences of the alternatives. On the other hand, analysis can also be used to justify a particular policy option, but the credibility of the analysis for such a purpose rests even more on the credibility of the assumptions that have to be made to accept the findings and conclusions of the analysis.

Opponents of a selected policy option will be quick to point out the implausibility of assumptions required to accept the findings of any analysis that is not consistent with their position. Moreover, any ideological premise will be especially subject to challenge. For example, the oft cited conclusion that market-based solutions are more efficient than regulatory measures is certainly easy to support with ample evidence in economics, as in the case of energy taxes noted earlier. But if the context of the policy decision being addressed includes complications of regional differences, political constituencies, regressivity of impacts, and many other features, then alternatives often have to be considered.

The framework of analysis should be tuned to the policy context and, very often, if policy issues involve deep ideological differences or complications of the kind just noted, it can be counterproductive to advance a normative recommendation of a particular course of action and, instead, it may be much more productive to articulate the consequences of alternative courses of action in as transparent a manner as possible to enable policy makers to be informed by analysis rather than be obliged to accept it or reject in order so support a particular policy position.

Problems in regional science and public policy are especially subject to this tradeoff. Researchers carrying out analysis to help advance a policy debate should be aware of the tradeoffs from the outset of their work. Examining the implications of assumptions made that are relevant to the policy context is important beyond the reason of good analysis, although perhaps that alone should be sufficient justification. It is essential for policy makers to understand not only the implications of alternative policy options but also the limits of the analysis used to articulate those implications.

References

Blair PD (1979) Multiobjective regional energy planning. Martinus Nijhoff, Boston, MA

Blair PD (2013) Congress's own think tank: learning from the legacy of the congressional office of technology assessment (1972-1995). Palgrave Macmillan, New York

Charnes A, Cooper WW (1961) Management models and industrial applications of linear programming, Volumes 1 and 2. Wiley, New York

Cochrane J, Zeleny M (eds) (1973) Multicriteria decision-making. University of South Carolina Press, Columbia, SC

Cohen JL (1978) Multiobjective programming and planning. Elsevier Science & Technology Books, New York

Dietzenbacher E, Sage J (2006) Mixing oil and water? Using hybrid input–output tables in a structural decomposition analysis. Econ Syst Res 18:85

Dorfman R, Samuelson P, Solow R (1958) Linear programming and economic analysis. McGraw-Hill, New York

Folk H, Hannon B (1974) An energy, pollution, and employment policy model. In: Macrakis M (ed) Energy: demand, conservation and institutional problems. MIT, Cambridge, MA, pp 159–173

Forssell O, Polenske KR (1998) Introduction: input–output and the environment. Econ Syst Res 10:91–97

Griffin JM (1976) Energy input-output modeling: problems and prospects, EA-298. Electric Power Research Institute, Palo Alto

Hannon B, Costanza R, Herendeen R (1983) Measures of energy cost and value in ecosystems. University of Illinois at Urbana-Champaign, Center for Advanced Computation, September

Hipel KW (1992) Multiple objective decision making in water resources. J Am Water Resour Assoc 28:3–12

Ignzio JP (1976) Goal programming and extensions. Lexington Books, Lexington, MA

Ijiri Y (1965) Managerial goals and accounting for control. North-Holland, Amsterdam

Intriligator M (1971) Mathematical optimization and economic theory. Prentice-Hall, Englewood Cliffs, NJ

Johnson MH, Bennett JT (1981) Regional environmental and economic impact evaluation. Reg Sci Urban Econ 11:215–230

Just J (1974) Impacts of new energy technology using generalized input–output analysis. In: Macrakis M (ed) Energy: demand, conservation and institutional problems. MIT, Cambridge, MA, pp 113–128

Lane MN (1970) Goal programming and satisfying models in economic analysis. Ph.D. Dissertation, University of Texas, Austin

Lang G-M (1998) Applying an integrated natural resource accounts and input–output model to development planning in Indonesia. Econ Syst Res 10:113–134

Lee S (1971) Decision analysis through goal programming. Decis Sci 2:172–180

Lee S (1972) Goal programming for decision analysis. Auerbach, Philadelphia

Lee S (1973) Goal programming for decision analysis of multiple objectives. Sloan Manage Rev 14:11–24

Miller RE, Blair PD (2009) Input–output analysis: foundations and extensions. Cambridge University Press, London

Nijkamp P, Van Delft A (1977) Multicriteria analysis and regional decision-making. Martinus Nijhoff, Leiden

Nijkamp P, Rietveld P (1976) Multiobjective programming models: new ways in regional decision making. Reg Sci Urban Econ 6:253–274

Oladosu G, Rose A (2007) Income distribution impacts of climate change mitigation policy in the Susquehanna River Basin economy. Energy Econ 29(3):520–544

Office of Technology Assessment (1985) New electric power technologies: problems and prospects for the 1990s. U.S. Congress, Office of Technology Assessment, OTA-E-246, July, Washington, DC

Parry IWH, Sigman H, Walls M, Williams RC III (2005) The incidence of pollution control policies. In: Folmer H, Tietenberg T (eds) International yearbook of environmental and resource economics 2006/2007. Edward Elgar Publishers, Cheltenham, UK

Qayum A (1991) A reformulation of the Leontief pollution model. Econ Syst Res 3:428–430

Qayum A (1994) Inclusion of environmental goods in national income accounting. Econ Syst Res 6:159–169

Saaty TL (1980) The analytic hierarchy process. McGraw-Hill, New York

Schäfer D, Stahmer C (1989) Input–output model for the analysis for environmental protection activities. Econ Syst Res 1:203–228

Tanino T, Tanaka T, Inuiguchi M (eds) (2003) Multi-objective programming and goal programming: theory and applications. Springer, Heidelberg
Thoss R (1976) A generalized input–output model for residuals management. In: Karen R (ed) Advances in input–output analysis, proceedings of the sixth international conference on input–output techniques, Vienna, April 22–26, 1974. Ballinger, Cambridge, MA, pp 411–432
Trzaskalik T, Michnik J (eds) (2002) Multiple objective and goal programming: recent developments. Springer, Heidelberg
Victor PA (1972) Pollution: economy and environment. George Allen and Unwin, London
Weisz H, Duchin F (2006) Physical and Monetary input–output analysis: what makes the difference? Ecol Econ 57:534–541

Representing Negotiation Space

14

J.H.P. Paelinck

Abstract

The elements of negotiation theory, and relations between them, in short "negotiation space", s first investigated from the point of view of some of its properties. A two-agents case is taken up, both linear and non-linear variants being studied, and some analytical results with their properties are presented; the analysis is then generalized to multi-dimensional situations, both in terms of agents and of issues. Finally one can picture topographically the results of various negotiation stages; an example is presented.

14.1 Introduction

On the invitation of Walter Isard, we started analyzing conflicts by means of hypergraph analysis (Paelinck and Vossen 1983a).

Conflict and negotiation analysis is still being currently studied [see e.g. Varii auctores (2012)].

Formal specifications can give quite more hints at and insights in the workings of those processes.

In the next section of this paper negotiation space will be explored, space being defined in the strict mathematical sense of a set structured by a certain number of relations between its elements (Paelinck 1983b, pp. 39–52). Moreover, it will be argued again that concepts of economic analysis are able to play an important role in bringing to the fore implied properties of such spaces (Paelinck 1996). A purely linear case will be explored, after which non-linearities and multiple dimensions are taken up.

Another representation is more topographical, and concerns the picturing of various stages of a negotiation process; this aspect is taken up in Sect. 14.3.

J.H.P. Paelinck (✉)
School of Public Policy, George Mason University, Arlington, VA 22201, USA
e-mail: j.paelinck@planet.nl

P. Nijkamp et al. (eds.), *Regional Science Matters*, DOI 10.1007/978-3-319-07305-7_14,

General conclusions and references follow; but before proceeding to the core of the exposition, two main points have to be cleared.

The first one is about the heavy formal content of this paper; let it be said that such formalizing can lead to very practical insights; we refer for this aspect to the paper by de Koster and Paelinck (1985), where a conflict situation is studied by laying bare its weakest link. We love to mention here that, having had once the privilege to comment on one of Walter's expositions in the field, we could show the isomorphism of two approaches.

Second, that we choose a novel approach—hypergraph theory—explains that few (recent) reference could be quoted, together with the fact that another field of investigation—to wit spatial econometrics—took up much of the author's attention.

14.2 Negotiation Space

14.2.1 Linear Case

To start with, the following assumptions are put forward:

A_1: two agents, A and B, negotiate about two divisible issues represented on the 0-1 line;

A_2: the agents are each commanding one of the issues;

A_3: they are assumed to handle a linear preference function, which can include externalities (namely the result obtained by the opponent);

A_4: constraints in negotiation space are also linear.

The following representation will be explored (Fig. 14.1):

Figure 14.1 represents the starting positions and also two constraints on the willingness of concession; for instant agent *A* is willing to move to point α at a constant substitution rate *a*, and analogously for *B* (constraint substitution rate *b*), α and β being respectively right and down from the (0,1) and (1,0) points. *A* has a linear preference function with a substitution rate strictly larger than *a*, *B* one strictly smaller than *b,* those substitution rates possibly including externalities as defined previously. The feasible region is the polytope (simplex, later polyhedron) to the north-east of the relevant area.

The problem has however no feasible solution as logical constraints:

$$x_A + x_B = 1 \tag{14.1a}$$

$$y_A + y_B = 1 \tag{14.1b}$$

prevail, only *exchange* of all or part of the content of the litigious issues being possible.

Constraints (14.1a) and (14.1b) imply graphically that the solutions should lie on circle(s) (segments) centered around the unweighted center of gravity of the relevant area, *C*, so retention of a non-zero part of the initial "endowments" (α and β) precludes final agreement.

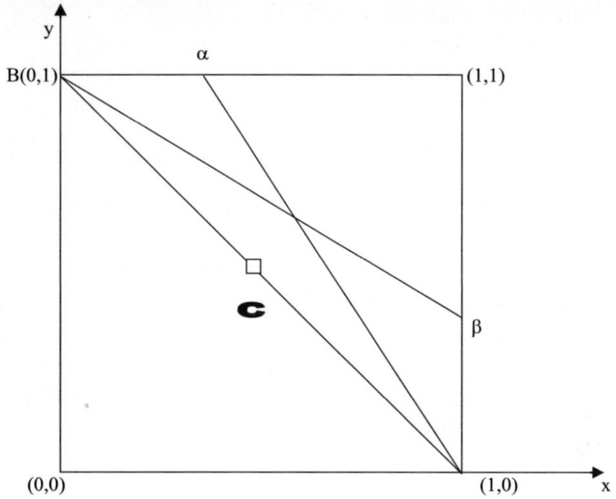

Fig. 14.1 Linear case

The problem becomes feasible from the moment *A* and *B* have in their respective constraints substitution rates equal or smaller, and equal or larger than *1*, the slope of the diagonal connecting the north-west and south-east corners of the relevant area. If moreover *A*'s preference substitution rate is defined as r_A and that of *B* as r_B, then under the condition:

$$r_B \geq 1 \geq r_A \tag{14.2}$$

the new (optimal) situations will be A(0,1) and B(1,0).

As a numerical example take the following preference functions:

$$\varphi_A = x_A + 2\, y_A \tag{14.3a}$$

$$\varphi_B = 2\, x_B + y_B \tag{14.3b}$$

Table 14.1 hereafter gives some "extreme" values of those functions:

Obviously, negotiations will exclude the (0,0) point, but if even if some of the feasible region would lie below the NW-SE diagonal, solutions would still be A(0,1) and B(1,0), given the assumptions on the rates of substitution of the respective preference functions [they are respectively 2 and ½ for functions (14.3a) and (14.3b)]. Analytically, the problem can be formulated as:

$$\max_{x,y} \varphi_A, \varphi_B \tag{14.4}$$

s.t.

$$y_A \geq a* - a\, x_A \tag{14.5a}$$

$$y_B \geq 1 - b\, y_B \tag{14.5b}$$

Table 14.1 Values of preference functions

	(0,0)	(1,0)	(0,1)	(1,1)
φ_A	0	1	2	3
φ_B	0	2	1	3

($a^* > 1$), plus the constraints (14.1a) and (14.1b), and in addition:

$$0 \leq x_A, x_B, y_A, y_B \leq 1 \qquad (14.5c)$$

The solution, on the pay-off frontier, is known to be a strong equilibrium (on this concept, see Friedman (1977), p. 168).

A question that could be raised is why not use a classical Edgeworth box-diagram as a representation? The answer is that the way the problem was set up here gives clearer insights into the workings of the model hypothezised, as the previous and following reasonings show.

14.2.2 Non-Linear Case

Suppose now first the constraints to be non-linear; a reasonable assumption (A_4*) is that the absolute values of their slopes decrease as a function of the importance of the issue (x, y) withheld. Figure 14.2 illustrates this, and also generates a possible interesting case:

There is now a perfectly feasible *area* for negotiation, despite the fact that α and β do not reach the extreme coordinates (*0,1*) and (*1,0*); indeed, all points belonging to the closed area around C satisfy Eqs. (14.1a) and (14.1b). Full remittance, however [solutions *A(0,1)* and *B(1,0)*], can once more only be attained if the non-linear constraints reach those points.

As to the preference functions, some of them exclude rhe attainability of those extreme points; such is the case of a Cobb-Douglas specification, as it has to "jump" from a given isoquant for strictly positive values of x and y to a point with a zero coordinate; quadratic functions—and as will be seen below, extended Cobb-Douglas functions—do not suffer from that drawback.

Nevertheless, fractional solutions might *reveal* the presence of Cobb-Douglas specifications, hence the following considerations on the case of Fig. 14.2. Be it first noticed that in the case of linear preference functions under the conditions of Sect. 14.2, A and B will move out from their initial endowments until they hit the limit of the feasible region, but nothing guarantees that conditions (14.1a) and (14.1b) are in fact satisfied, moreover the separate optima not being necessarily points on the diagonal; a simple graphical argument can show this. Under certain circumstance—to be studied hereafter, to wit the introduction of a "relative price"—the solution will lie on the diagonal, the point of shortest Euclidean distance to C determining the relative positions of A and B.

If both A and B are driven by Cobb-Douglas specifications, the same problem will arises, but the analytical computation of the (fractional) solutions is easier. Suppose A to pick a point below the diagonal; B will then be stuck in the

Fig. 14.2 Non-linear case

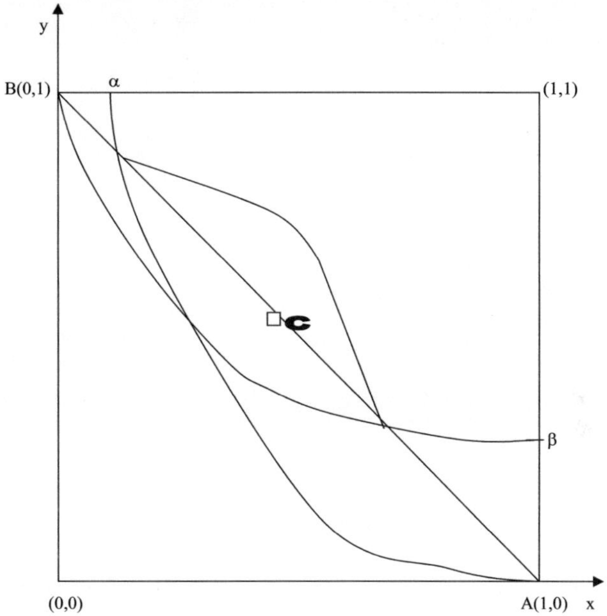

symmetrical point with respect to *C* [conditions (14.1a) and (14.1b)]. But A can be better off if he moves towards the diagonal (a simple isoquant argument underpins this), but *B* has then to retreat towards the diagonal, so a deal will be very problematic, and the same will occur in the reverse situation. So a first solution concept, taking into account externalities mentioned previously, is to try and end up on the diagonal.

A then maximises:

$$\varphi_A = x_A{}^{\rho}(1 - x_A)^{1-\rho} \tag{14.6a}$$

leading up to:

$$x_A{}^{o} = \rho \tag{14.6b}$$

leaving $1-\rho$ to *B*; similarly, *B* with characteristic parameter σ, ends up with:

$$x_B{}^{o} = \sigma \tag{14.7}$$

leaving *A* with $1-\sigma$.; the so-called "Pareto-circlets" denote optimal values. However, conditions (14.1a) and (14.1b) are clearly only satisfied if $1-\rho \equiv \sigma$, so individual maximising behaviour leads to non-compatible solutions. This compatibility can be restored—and this is a second solution concept—if a relative "price" of *y* with respect to *x* is introduced; this leads to the system:

$$x_A = \rho \tag{14.8}$$

$$x_B = p\sigma \tag{14.9}$$

$$x_A + p\, y_A = 1 \tag{14.10}$$

$$x_B + p\, y_B = p \tag{14.11}$$

$$x_A + x_B = 1 \tag{14.12}$$

$$y_A + y_B = 1 \tag{14.13}$$

one of the equations being redundant. The solutions are then:

$$x_A{}^o = \rho \tag{14.14}$$

$$x_B{}^o = 1 - \rho \tag{14.15}$$

$$y_A{}^o = \sigma \tag{14.16}$$

$$y_B{}^o = 1 - \sigma \tag{14.17}$$

$$p = \sigma^{-1}(1 - \rho) \equiv x_B{}^o / y_A{}^o >< 1 \tag{14.18}$$

provided they lie on the feasible domain; the "price" is clearly only equal to one if $\sigma = 1 - \rho$, which was already hinted at in the discussion of the first solution concept. The new variable p plays indeed the role of an exchange rate of y for x; for instance $p > 1$ implies that A exchanges more units of x than he would receive units of y. This reminiscence of a Walrasian "umpire' leads up to the possible role of a third "good offices" body, persuading (in the case studied, A) to accept the "price" given the relative preferences of A and B.

A mixed linear non-linear solution is also possible, but the solution, as in the double linear preference function case, is analytically less tractable.

It has already be said about the Cobb-Douglas specification that *(1,0)–(0,1)* solutions are excluded as a result of maximising behaviour, even asymptotically, as "inferior" isoquants are nearer to the axes than 'superior" ones, without any of them ever cutting those axes. A modified specification might include the extreme points, e.g.:

$$\varphi_A = (x_A + q)^\rho (y_A + \theta)^{1-\rho} \tag{14.19}$$

giving as "diagonal" optimal values:

$$\rho \le \theta(1 + 2\theta)^{-1} \rightarrow x_A{}^o = 0, y_A = 1 \tag{14.20a}$$

$$\theta(1 + 2\theta)^{-1} < \rho < (1 + \theta)(1 + 2\theta)^{-1} \rightarrow \begin{aligned} x_A{}^o &= (1 + 2\theta)\rho - \theta \\ y_A{}^o &= (1 + 2\theta)(1 - \rho) - \theta \end{aligned} \tag{14.20b}$$

$$\rho \ge (1 + \theta)(1 + 2\theta)^{-1} \rightarrow x_A{}^o = 1, y_A{}^o = 0 \tag{14.20c}$$

which matches the relative preference elasticities for x and y; anyhow, interior solutions remain possible. In fact, the resulting "demand equations" are of the well-known "linear expenditure" type.

Still another possible solution concept is of a Cournot-oligopoly nature. The preference equations then are e.g. of type (14.6), plus the conditions (14.1a) and (14.1b), leading up to the following solutions:

$$x_A{}^o = \rho(\rho + \sigma)^{-1} \tag{14.21a}$$

$$x_B{}^o = \sigma(\rho + \sigma)^{-1} \tag{14.21b}$$

$$y_A{}^o = (1 - \rho)\,[2 - (\rho + \sigma)]^{-1} \tag{14.21c}$$

$$y_B{}^o = (1 - \sigma)\,[2 - (\rho + \sigma)]^{-1} \tag{14.21d}$$

These solutions have this time an *implicit* "price":

$$p = x_B{}^o / y_A{}^o = \sigma[\,2 - (\rho + \sigma)\,(\rho + \sigma)\,(\,1 - \rho)\,]^{-1} \tag{14.22}$$

which is only equal to one for $\rho + \sigma = 1$; again the solution needs in general a feasible region exceeding the diagonal to the south-west, and using specification (14.19) allows of obtaining exact or constrained binary *(0-1)* solutions.

It should finally be mentioned that the interior equilibria are not of the strong type (again Friedman 1977, p.168), except possibly for the mixed linear/non-linear case and the extended Cobb-Douglas preference functions; a further remark is that in the interior cases the existence of different strategies leads to multiple possible equilibria, implying a variety of cases in terms of values of the outcomes for both parties concerned.

14.2.3 Multiple Dimensions

A first case is that of more than two issues, some of them being commanded by *A*, the others by *B*. Take the case of *A* handling only one, *B* two; Figs. 14.1 and 14.2 have then to be generalised to *3* dimensions, as illustrated by Fig. 14.3:

A possibility exists that, even if in the (*x*,*y*)-direction the program is feasible, *B* imposes a constraint on *z* (see the line ending up at γ) and *A* imposes $y = z$, the conjunction of both conditions excluding "symmetrical" solutions with respect to the centre of gravity, so even if the (*x*,*y*)-problem might have a solution, the generalised (*x*,*y*,*z*)-problem has (provisionally: see Sect. 14.2.4) none. This example shows, on the one hand that behind the scenes "implicit" constraints may lurk, and have to be laid bare, on the other that issues are to be approached globally (see Paelinck and van Gastel (1991), p.74).

The above multidimensional specification might also be used for the case of three (or more) agents, each handling one (or several) of the issues; the analysis is similar.

Fig. 14.3 Three-dimensional case

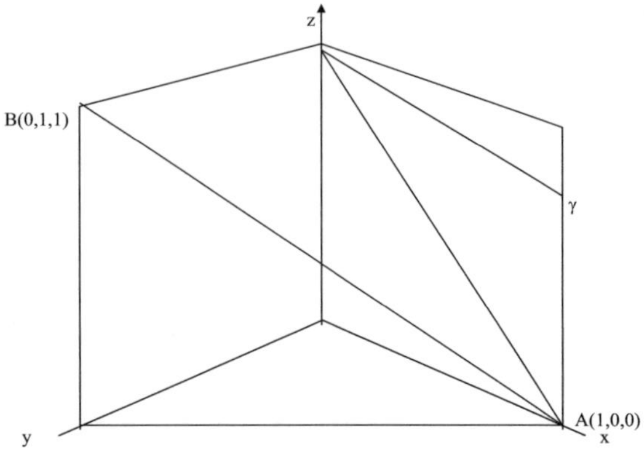

14.2.4 Extensions

The first conclusion is an analytical one: the specifications presented allow of describing clearly the positions taken up by actual or potential negotiators (or, equally, "good offices" people). Furthermore, some links can be laid with dynamic negotiation processes, for instance the speed of reaction in terms of proposals and counter-proposals (again Paelinck and van Gastel 1991, p.84).

The second conclusion is that elements of the specification need (optimal) modification to allow the negotiation to progress and succeed; in the cases presented above the extension of the feasible space is the most important one, and possibilities exist to formulate *that* optimizing problem (Paelinck 2000).

Finally Boulware ("all or nothing", "take it or leave it"; Paelinck and van Gastel 1991, p.75) strategies could be investigated, using for instance a function inspired of the *arg ch*-type:

$$\varphi \approx \ln\left[e + (z - 1)^{1/2}\right] \qquad (14.23)$$

which is clearly not defined for $z < 1$.

14.3 Monitoring Conflict Resolution: A Topographical Approach

This section takes up the representation problem again, but in a different vein, to wit the topographical representation of different phases of the process of negotiation; a topographical representation (or envelopment, or embedding) is proposed, an example being worked out; mathematics are consigned in an appendix.

Next section first recalls the main ideas of hypergraph conflict analysis; it can be shown indeed (Paelinck and van Gastel 1991, p.73) that this approach can be considered as a discrete version of the continuous negotiation space treated in

Sect. 14.2. To be complete, it should be mentioned that conflict and negotiation are complementary, the latter being one form of conflict resolution (Paelinck 2000).

14.3.1 Hypergraph Conflict Analysis

Hypergraph Conflict Analysis (HCA) was introduced by Paelinck and Vossen (1983a). Since then a certain number of further contributions have appeared, in particular three studies on hypergraph conflict resolution (or reduction: de Koster and Paelinck 1985; van Gastel and Paelinck 1992; Paelinck 2000).

Suppose there to be a number of agents, or groups of agents (A_i), confronted with a set of possible options (O_j), whatever the latter may be; in spatial analysis one might encounter infrastructural projects, regional development issues, and many other strategic choice problems. The agents could agree or disagree with some of the options, and this state of affairs can be set out in a table or matrix (**C**), Table 14.1 reproducing a 3×3 binary (only full agreement or disagreement is assumed here) case.

The hypergraph nature of Table 14.2 results from the fact that for each agent the agreeable options are a subset of the overall set of possible options.

Several measures can be proposed to show the "degree of conflict". One, which will be noted δ, divides the minimum number of zeros taken over the columns of C by the number of agents, n_i (for Table 14.1 this is 2/3). Another measure, noted τ, is the transversal number, defined as the cardinal of the minimal set of options on which all agents taken together agree; for Table 14.1 this number is 3. A relative transversal number, τ^*, would divide $\tau - 1$ by the cardinal of the set of potential options minus one, so in the case of Table 14.1, $\tau^* = 1$. It should be intuitively clear that δ and τ (or τ^*) are interrelated.

If $\delta = 0$, or alternatively, $\tau = 1$, (or $\tau^* = 0$) there would be no conflict, as all agents agree on at least one option. Hypergraph conflict resolution aims at computing an optimal way of "turning over" agents, so as to drive δ down to zero or τ up to 1.

Fuzzy generalisation of the above ideas is possible, but this topic will not be addressed here.

14.3.2 Topographical Representation

Starting point will be the initial situation, for which we assume a multidimensional framework (see Paelinck (2000), Sect. 14.3.4); in each dimension k, $\delta_k = n_i - 1$, or alternatively, $\tau_k = n_k$, where n_k is the corresponding number of options; one defines, as said in Sect. 14.2:

$$\tau_k * = (\tau_k - 1)/(n_k - 1) \tag{14.24}$$

and so obviously $0 \le \tau_k \le 1$.

Table 14.2 Matrix **C**

Agents/Options	O_1	O_2	O_3
A_1	1	0	0
A_2	0	1	0
A_3	0	0	1

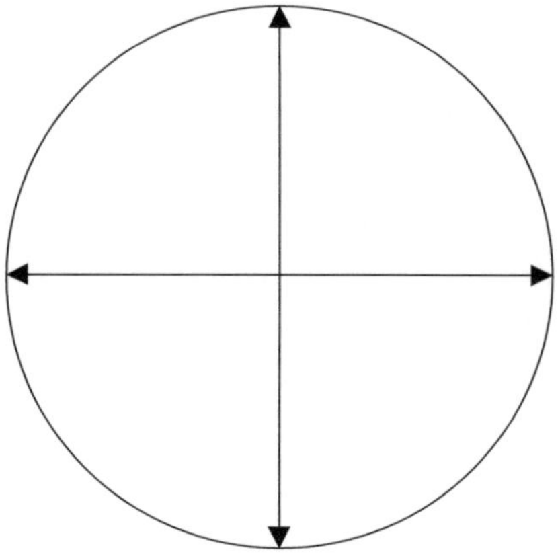

Fig. 14.4 Topographical representation

Figure 14.4 hereafter gives an example of representation of this situation for k = 1,. . .,4, the radius of the circle being equal to 1.

If the circle is centered on (0,0), the four end points *A, B, C, D* will have coordinates (0,1), (1,0), (0,−1), (−1,0).

The idea is to follow and measure the process of conflict resolution, in particular as a function of the τ_k*; two more figures follow to picture this (Figs. 14.5 and 14.6).

A first indicator of the "displacement' of the conflict would be the coordinates of the new centres of gravity, g_l, where l is the index of the conflict situation (here $l = 1, 2, 3$); in the example they moved from (0,0) to (−0.1875, −0.0625) and then to (−0.0625, 0); one notices in situation 2 the role of the weights of **C** and **D**, and in situation 3 that of the weight of **D**. If conflict resolution were to be completed, the coordinates would again be (0,0), but this could also be the case in intermediate situations, so g_l is not a sufficient statistic.

So an additional indicator, borrowed from Paelinck (2001), computes the quadratic envelopment of the relevant points and measure its area, s_l; for the initial situation, this area is obviously equal to π. The mathematics of the envelopment process are exposed in Appendix.

14.3.3 Computational Results

Table 14.3 hereafter presents the results of the envelopments computations for the three situations presented above; the method used is the following.

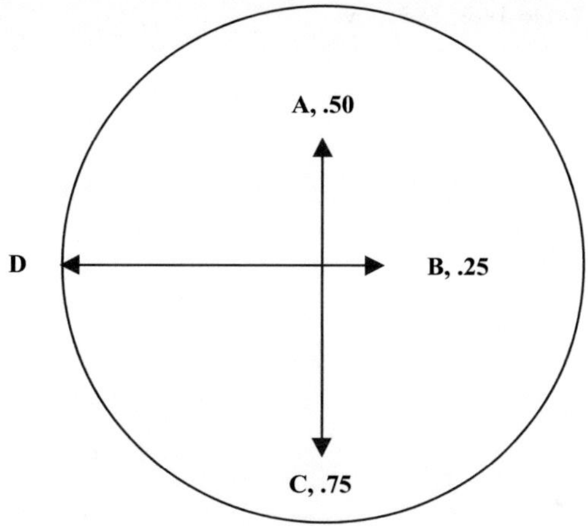

Fig. 14.5 Situation 2

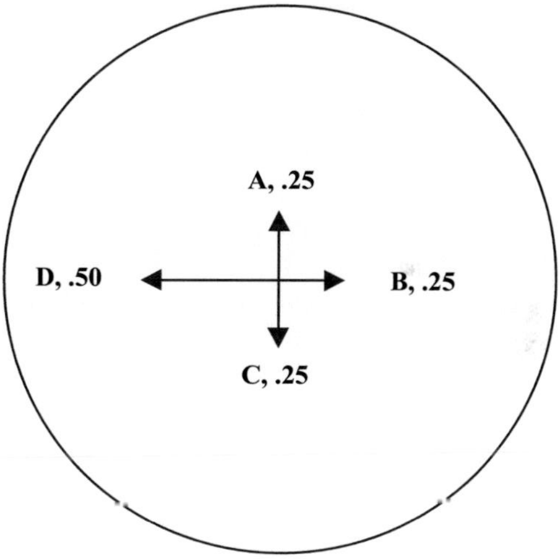

Fig. 14.6 Situation 3

Table 14.3 Computational results

Sit./Par.	a	b	c
1	1	1	0
2	1.6498	2.3324	1.9342
3	5.2245	15.6735	0

From the centre of gravity (coordinates x*, y*) one computes for each dimension k: $u_k = (x_k - x^*)^2, v_k = (y_k - y^*)^2, z_k = (x_k - x^*)\ (y_k - y^*)$; let **u, v, z,** be the column vectors of those magnitudes, and $\mathbf{X} = [\mathbf{u}, \mathbf{v}, \mathbf{z}]$ the resulting matrix. One then constructs **X'X**, a 3×3 matrix, and defines a column vector **e** of three coefficients *a, b, c*, from which the following program is derived:

Table 14.4 Synthesis

Conflict situations	g_l	s_l	σ_l/μ_l
S_1	(0,0)	3.1416	0
S_2	(−0.1875,−0.0625)	1.8408	0.4472
S_3	(−0.0625,0)	0.3472	0.3416
S_4	(0,0)	0	0

$$\underset{a,b,c}{\text{Min}}\ \phi = \mathbf{e'X'Xe} - \mathbf{2i'Xe} \tag{14.25}$$

s.t.:

$$\mathbf{au} + \mathbf{bv} + \mathbf{cz} - \mathbf{i} \leq \mathbf{0} \tag{14.26}$$

The respective areas, s_l, have been computed as π, 1.8408 and 0.3472; one notices the progressive decrease of the area of the "conflict surface".

Again, for complete conflict resolution, the area would be zero.

14.3.4 Extensions

Other measures could of course be added, such as, e.g., an extra indicator of "skewness" of a given conflict situation, in addition to the location of the center of gravity; an obvious choice would be the coefficient of variation. Table 14.4 summarizes all the findings, including S_4, total conflict resolution.

Hopefully practical applications will contribute to enrich the arsenal of useful describers of disagreement occurrences.

Conclusions

The two parts of this study may, at first sight, appear to be unrelated or disconnected; however, this is not the case.

Sure, the second part seems to be more linked to practical applications, the first one to pure theory. But the latter allows to analyze in depth the reasons for some observed negotiating behaviour, as already hinted at in Sect. 14.2.4, where examples were given.

The second point taken up there—extension of the feasible space—leads immediately to possible stages of the negotiation process, as illustrated by graphs 2 and 3 and the computational results of Tables 14.3 and 14.4.

The general point then is to underscore the importance of appropriate mathematical representation structures to investigate complex processes, of which conflict and negotiation are outstanding examples.

Mathematical Appendix

Elliptic Area

The ellipse has characteristic equation:

$$a\,x^2 + b\,y^2 = \rho^2 \tag{14.27}$$

where *x* and *y* are the Cartesian coordinates and ρ a given constant; *a* and *b*, inverses of the squares of the semi-axes, are here strictly positive parameters, other cases being however possible (hyperbolas and parabolas). Equation (14.27) is written as follows in polar coordinates:

$$r^2\left(a\ \sin^2\varphi + b\ \cos^2\varphi\right) = \rho^2 \tag{14.28}$$

and where *r* is now the radius, φ the angle expressed in radians.

From (14.28) one computes the elliptic area as:

$$s = \int_0^{2\pi}\int_0^{\rho(a\cos^2\varphi + b\sin^2\varphi)^{-1/2}} r\,dr\,d\varphi \tag{14.29}$$

$$= 1/2\,\rho^2 \int_0^{2\pi} (a\cos^2\varphi + b\sin^2\varphi)^{-1} d\varphi \tag{14.30}$$

The value of the definite integral (14.30) is known (C.R.C., p.313, No 419) and equal to:

$$s = \pi\rho^2(ab)^{-1/2} \tag{14.31}$$

Reduction of an Oblique Quadratic to its Canonical Form

Starting from:

$$a\,x^2 + b\,y^2 + c\,xy = r^2 \tag{14.32}$$

one applies the following transformation (Fisher and Ziebur 1967, p. 421, Theorem 95-1):

$$a* = a\ \cos^2\alpha + b\ \sin^2\alpha + c\ \sin\alpha\cos\alpha \tag{14.33}$$

$$b* = a\ \sin^2\alpha + b\ \cos^2\alpha - c\ \sin\alpha\cos\alpha \tag{14.34}$$

$$c* = (b - a)\sin 2\alpha + c\cos 2\alpha = 0 \tag{14.35}$$

from where it follows that:

$$\text{cotg}\,2\alpha = (a - b)/c, 0 < \alpha < \pi/2 \tag{14.36}$$

Under certain conditions (sign of the discriminant, i.e. negativity of a^* or b^*), Eq. (14.32) represents a hyperbola, as sometimes happens with the methodology exposed; a solution then is to switch to logarithms [see Paelinck (2001)].

References

CRC (1959) Standard mathematical tables, 12th edn. Chemical Rubber Publishing Company, Cleveland, OH

de Koster MBM, Paelinck JHP (1985) Reduction of conflict. Organ Conflict Peace Sci 4:1–17

Fisher RC, Ziebur AD (1967) Integrated algebra and trigonometry, with analytic geometry, 2nd edn. Prentice Hall, Englewood Cliffs, NJ

Friedman JW (1977) Oligopoly and the theory of games. North Holland, Amsterdam

Paelinck JHP (1996) Measuring the regional quality of life. In: Chatterji M (ed) Regional science. Perspectives for the future. Macmillan-St Martin Press, London, pp 80–87

Paelinck JHP (2000) Controlling complexity in spatial modelling. CESifo, Working paper series, working paper nr 362

Paelinck JHP (2001) Analyse topographique de complexes industriels, Paper presented at the 40th annual conference of the Association de Science Régionale de Langue Française, Bordeaux, September 2001

Paelinck JHP, van Gastel MAJJ (1991) Axiomatische Conflictvoorstelling (Axiomatic Conflict Representation). In: Kaplan M, Rinnooy Kan AHG (eds) Onderhandelen. Structuren en Toepassingen Academic Service, Schoonhoven, pp 67–88

Paelinck JHP, Vossen PH (1983) Axiomatics of conflict analysis. In: Isard W, Nagao Y (eds) International and regional conflict. Ballinger Publishing Company, New York, pp 33–52

Paelinck JHP (with the assistance of J.-P. Ancot and J.H. Kuiper) (1983b) Formal spatial economic analysis. Gower Press, Aldershot

van Gastel MAJJ, Paelinck JHP (1992) Computing the solution to a conflict situation by means of continuous multicriteria analysis. In: Chatterji M, Forcey LR (eds) Disarmament, economic conversion and the management of peace. Praeger Publishers, New York, pp 27–39

Varii auctores (2012) L'économie des conflits. Revue d'économie politique, N° 2

An Integrated Disaster Relief Supply Chain Network Model with Time Targets and Demand Uncertainty

15

Anna Nagurney, Amir H. Masoumi, and Min Yu

Abstract

As the number of natural disasters and their impacts increase across the globe, the need for effective preparedness against such events becomes more vital. In this paper, we construct a supply chain network optimization model for a disaster relief organization in charge of obtaining, storing, transporting, and distributing relief goods to certain disasterprone regions. Our system-optimization approach minimizes the total operational costs on the links of the supply chain network subject to the uncertain demand for aid at the demand points being satisfied as closely as possible. A goal programming approach is utilized to enforce the timely delivery of relief items with respect to the pre-specified time targets at the demand points. A solution algorithm for the model is also provided. A spectrum of numerical examples illustrates the modeling and computational framework, which integrates the two policies of pre-positioning relief supplies as well as their procurement once the disaster has occurred.

Keywords

Supply chains • Disaster relief • Humanitarian logistics • Network optimization • Cost minimization • Goal programming • Time constraints • Variational inequalities

A. Nagurney (✉)
Department of Operations and Information Management, Isenberg School of Management, University of Massachusetts, Amherst, MA 01003, USA

School of Business, Economics and Law, University of Gothenburg, Gothenburg, Sweden
e-mail: nagurney@isenberg.umass.edu

A.H. Masoumi
Department of Management and Marketing, School of Business, Manhattan College, Riverdale, NY 10471, USA
e-mail: amir.masoumi@manhattan.edu

M. Yu
Pamplin School of Business Administration, University of Portland, Portland, OR 97203, USA
e-mail: yu@up.edu

P. Nijkamp et al. (eds.), *Regional Science Matters*, DOI 10.1007/978-3-319-07305-7_15,

15.1 Introduction

A natural disaster, by definition, is *some rapid, instantaneous* or *profound impact of the natural environment upon the socio-economic system* (Alexander 1993). Cyclones, tornadoes, hurricanes, landslides, earthquakes, tsunamis, volcanic eruptions, and floods are among the natural phenomena ravaging our planet each year, causing immense loss of life as well as suffering. The number of natural disasters and the sizes of the populations affected by such events have been growing (Schultz et al. 1996; Nagurney and Qiang 2009). Scientists are warning that we can expect more frequent extreme weather events in the future. For instance, tropical cyclones—which include hurricanes in the US—are expected to be stronger as a result of global warming (Sheppard 2011; Borenstein 2012).

The amount of damage and loss following a disaster depends on the vulnerability of the affected region, and on its ability to respond (and recover) in a timely manner, also referred to as *resilience*. Disasters are believed to occur when hazards meet vulnerability (Blaikie et al. 1994). Hence, being prepared against potential disasters leads to reduced vulnerability and a lower number of fatalities. As the institutional experts say: "during a natural disaster, one has only two options: to become a victim, or to become a responder" (Alvendia-Quero 2012). Thus, viable resiliency against natural disasters has to be achieved and sustained not only by the cognizant organizations, from governmental to humanitarian ones, and private enterprises, but also by individuals.

Disasters necessarily affect regions and pose challenges in all phases of disaster management. Vivid examples of disasters such as Hurricane Katrina in August 2005, the Haiti earthquake in January 2010, Fukushima in March 2011, Superstorm Sandy in October 2012, and tropical cyclone Haiyan in November 2013 have challenged researchers, practitioners, as well as policy-makers and other decision-makers and have yielded multidisciplinary approaches to models, methods, and techniques with major contributors being from regional science (cf. West and Lenze 1994; Israelevich et al. 1997; Rose et al. 1997; Okuyama et al. 1999; Cho et al. 2001; Okuyama 2004; Rose and Liao 2005; Ham et al. 2005; Greenberg et al. 2007; Grubesic et al. 2008; Reggiani and Nijkamp 2009; Nagurney and Qiang 2009; Rose 2009, among others). Background on the quantification of the economic impacts of disasters under risk and uncertainty can be found in the book by Dacy and Kunreuther (1969) (see also Kunreuther (1967)) with an updated treatment in Kunreuther and Michel-Kerjan (2012) and with a critique of the former book by Okuyama (2003).

The complexity of disaster relief supply chains, in turn, originates from several inherent factors. The associated large demands for relief products pose challenges to the logistics planning authorities (Lin 2010) with the level of uncertainty adding to the complexity. According to Beamon and Kotleba (2006), there may exist irregularities in the size, the timing, and the location of relief product demand patterns. In addition, disaster-driven supply chains are, typically, formed as incident-responsive ones with temporary configurations of disparate resources. Commercial supply chains, on the other hand, often involve supplierbuyer relationships that have evolved over years, along with the refinement of policies. In contrast, disaster relief systems develop new networks of relationships within days or even hours, and have very short life-cycles (Oloruntoba and Gray 2006; Denning 2006). Thus, *time* plays a substantial role in the construction and operation of such networks. As noted by Tzeng et al. (2007), once a disaster such as an earthquake strikes, effective disaster efforts can mitigate the damage, reduce the number of fatalities, and bring relief to the survivors.

The fact that time is a critical element in disaster relief is also noted in the key benchmarks defined for the US Federal Emergency Management Agency (FEMA)'s response and recovery. The key benchmarks are: to meet the survivors' initial demands within 72 h, to restore basic community functionality within 60 days, and to return to as normal of a situation within 5 years (Fugate 2012). According to FEMA's first national preparedness report, states have developed fatality management plans, yet not all of them are adequate and practical. Specifically, it is challenging to measure the progress of preparedness activities, according to a US Department of Homeland Security (2012) report. Walton et al. (2011) further emphasize that the importance of speed is noted in leading emergency response guidelines with disaster relief operations intrinsically requiring the need for speed (see also USAID (2005) and UNHCR (2007)).

The timely and efficient delivery of relief goods to the affected population not only decreases the fatality rate but may also prevent chaotic situations. In the case of cyclone Haiyan, for example, the strongest typhoon ever recorded in terms of wind speed, which devastated areas of Southeast Asia, particularly the Philippines, where 11 million people were affected, slow relief delivery efforts forced people to seek any possible means to survive. A number of relief trucks were attacked and had food stolen, and some areas were reported to be on the brink of anarchy (Chicago Tribune 2013; CBS News 2013).

In this paper, we propose an integrated supply chain network model for disaster relief. Our mathematical framework is of system-optimization type where the organization aims to satisfy the uncertain demands subject to the minimization of total operational costs while the sequences of activities leading to the ultimate delivery of the relief good are targeted to be completed within a certain time. The first criterion in this model captures the total costs of all the activities in the supply chain as well as the expected shortage and surplus penalties at the demand points. The second criterion consists of the penalties associated with the time deviations on paths of the relief items to the demand points with respect to the pre-specified target times. Our model allows this time requirement to vary from one demand point to another. In addition, our model integrates the possibility of the pre-positioning of relief items before the occurrence of the disaster as well as the case in which the organization procures after the disaster hits the region. The solution to our model yields the optimal levels of activities associated with procurement, storage, and transportation of the relief items, whether purchased/procured in advance or after the disaster strikes—if need be. The solution also provides the optimal values of the over-the-target time deviations. This feature enables the organization to estimate the anticipated delays in the delivery of the relief items to the demand points.

We now review the related literature in the area of disaster relief supply chains. About half of the research has applied quantitative techniques (Charles and Lauras 2011). The other half utilizes case studies and empirical research to investigate past disasters and to present best practices. Time is frequently a criterion/objective aimed to be maintained along with the total cost being minimized (see, e.g., Zhenling 2009; Tzeng et al. 2007). Such a formulation, however, may result in excessive delivery times to the demand points if the distribution to certain locations is too costly. In our model, in contrast, rather than evenly minimizing all travel/delivery times, we seek to minimize the time deviations of activities on paths with respect to the pre-determined target times. Thus, one can prioritize the demand points based on their locations, demand patterns, number of elderly and children, etc.

Sheu (2010) constructed a dynamic fuzzy model of disaster relief response in large-scale problems. Nagurney et al. (2011) presented a supply chain network design model for critical needs with the possibility of outsourcing. Nagurney et al. (2012b) developed a multiproduct supply chain for the production and distribution of disaster relief items. Nagurney et al. (2013) presented supply chain network frameworks for various timesensitive, perishable healthcare products such as human blood, pharmaceuticals, medical nuclear products, etc. Nagurney and Yu (2014), in turn, constructed a game theory supply chain network model for the case of time–based oligopolistic competition.

Hale and Moberg (2005) proposed a set covering location model to identify secure sites for the storage of emergency supplies. Barbarosoglu and Arda (2004) and Falasca and Zobel (2011) developed two-stage stochastic models for the procurement and transportation of the vital disaster relief items. Also, Mete and Zabinsky (2010) proposed a two-stage stochastic model for the storage and distribution of medical supplies to be used in case of emergencies. Balcik and Beamon (2005) studied facility location in humanitarian relief. Huang et al. (2012) presented performance measures for the efficiency, efficacy, and equity of relief distribution.

Liu and Nagurney (2013), in turn, constructed a supply chain network model with quickresponse production and outsourcing under uncertain demand and cost. Nagurney and Qiang (2012) developed network robustness and performance measures in addition to synergy measurement of network integration in the case of humanitarian partnerships. Qiang and Nagurney (2012) proposed a bi-criteria indicator to evaluate the performance of supply chains of critical needs under capacity and demand disruptions. MacKenzie and Barker (2011) integrated a risk-management approach with a Multiregional Input-Output model using ideas from Isard et al. (1998) to quantify the regional economic impacts of a supply shortage. Simpson and Hancock (2009) applied simulation to the case of resource allocation in an emergency response system. Rottkemper et al. (2012) presented a bi-criteria mixed-integer programming model for the inventory relocation of relief items. Furthermore, Ortuño et al. (2011) and Vitoriano et al. (2011) developed goal programming frameworks for the distribution of relief goods while considering targets for attributes such as the cost and travel time. A recent edited volume on disaster management and emergencies is by Vitoriano et al. (2013) with a survey on decision aid models for humanitarian logistics therein by Ortuño et al. (2013).

Our model, in contrast to much of the above cited literature, does not consider targets for cost; instead, it minimizes the total operational costs of the activities in the supply chain network. This, we believe, makes more sense in that the relief operations budget largely depends on the intensity of damage and the fatality toll which may be difficult be determine a priori. In addition, our model—unlike the aforementioned ones—allows for the pre-disaster and the post-disaster procurement of relief items, and involves the time and the cost associated with each strategy or a combination of both. Moreover, we handle nonlinear costs, which capture congestion effects, a big issue in disaster relief, and an aspect that has been missing from much of the literature on the topic (cf. Haghani and Oh 1996).

The organization of this paper is as follows. In Sect. 15.2, we present the supply chain network topology of the disaster relief organization, construct the optimization problem, and establish its variational inequality formulation which provides us with nice computational features. We also present several illustrative numerical examples and conduct sensitivity analysis. In Sect. 15.3, we propose an algorithm that, when applied, computes the incurred optimal levels of disaster relief product flows in the supply chain network as well as the optimal path time deviations. The algorithm is then used to solve a large-scale numerical example in Sect. 15.4, along with a variant. In Sect. 15.5, we summarize the results and present our conclusions.

15.2 The Integrated Disaster Relief Supply Chain Network Model

In this section, we develop the integrated disaster relief supply chain network model. We mention that, depending on the structure of the specific disaster relief (humanitarian) organization, the scale/intensity of the disaster, and the region or the country that the disaster takes place in, the supply chain network may vary in specific details. Nevertheless, the general framework of the majority of non-profit disaster aid organizations follows a pattern similar to the one that we describe below.

15.2.1 Components of the Disaster Relief Supply Chain Network

The disaster relief supply chain network model, as we shall demonstrate, captures both the *preparedness* phase and the response phase of the disaster management cycle (with the other two phases being *mitigation* and *recovery*) (Tomasini and Van Wassenhove 2009). We take into account the pre-disaster preparations including the procurement, the pre-positioning, and the storage of disaster relief items given the estimated demand in disaster-prone areas. We also take into consideration the relevant issues surrounding the transportation and the ultimate distribution of the relief goods to the demand points once a potential disaster takes place. Furthermore, the case where an organization procures the humanitarian aid items after the occurrence of a disaster is also integrated into our model.

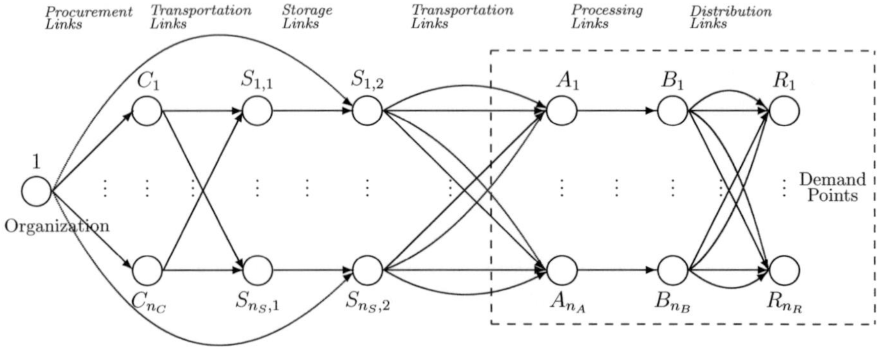

Fig. 15.1 Network topology of the integrated disaster relief supply chain

Regardless of the strategy that the relief/humanitarian organization will eventually adopt—the procurement of relief goods in advance or subsequent to the disaster or a combination of both—the ultimate delivery of the items to each demand point is aimed to be completed within a specific time target associated with that point as closely as possible. Such goal constraints enhance the reliability of a disaster relief system. All disaster relief organizations, whether owned and run by the public or the private sector, are non-profit. Thus, a centralized system-optimization approach is adopted as recommended by Nagurney et al. (2011).

Figure 15.1 depicts the network topology of a disaster relief supply chain associated with a humanitarian organization responsible for the procurement and the delivery of a disaster relief product. The progression of the required activities in such systems over time is graphically depicted. Depending on the location and the type of the relief product, be it water, food, medicines or vaccines, for example, the disaster relief organization can represent the US National Guard, the American Red Cross, the International Federation of Red Cross and Red Crescent Societies (IFRC), Doctors without Borders (MSF), etc. The organization, in Fig. 15.1, is denoted by the origin node. Every other node in the supply chain network represents a facility in the supply chain. A path connecting the origin node to a demand node (destination) consists of a sequence of directed links with each one corresponding to an activity in the supply chain. This sequence of activities ensures that the disaster relief item is procured, positioned, stored, and distributed to the people in need. We assume that there exists at least one path in the supply chain network connecting the origin (node 1) with each demand point located at the rightmost (final) tier of the supply chain (with additional assumptions stated later).

In the topology in Fig. 15.1, we assume that the organization is considering n_C procurement facilities, denoted by C_1, C_2,. . ., Cn_C. These are the locations where the purchased aid item is received from, or the item is collected from donors, such as people, companies, etc. Associated with each type of procurement is an individual cost function. Collection nodes constitute the second tier of the supply chain, and the links connecting the origin to these nodes correspond to the activities of "procurement."

The next two sets of nodes, i.e., those associated with tiers 3 and 4, represent the storage facilities. These are the facilities to which the collected goods are shipped, and then stored until needed. There are n_S of such nodes in the network, and the links connecting tiers 2 with 3, and 3 with 4 are referred to as "transportation" and "storage" links, respectively. The reason that the nodes in tiers 3 and 4 are sharing similar notation (e.g., $S_{1,1}$ and $S_{1,2}$) is that they correspond to the same geographical locations but reflect different points in time.

As mentioned earlier, our model not only allows for the pre-planned procurement and storage of relief goods, but also integrates the direct procurement of aid items. The case of immediate procurement is reflected in the network in Fig. 15.1 with the curved links connecting the origin node to the storage facilities. This is the case where the organization initiates procurement after the occurrence of the disaster. The organization may benefit from avoiding storage costs. Even if the organization decides to "store beforehand" as its dominant strategy rather than to "procure afterward," the latter can be expected to be available as a backup plan especially under severe circumstances.

The next set of links, connecting tiers 4 and 5 of nodes, are of "transportation" type, and correspond to the shipment of the product—either stored or freshly procured—to the affected region. These links are used once the disaster strikes a region. As seen in Fig. 15.1, there may exist multiple links connecting every pair of nodes in tiers 4 and 5. This is to allow for different modes of transportation: air, ground, and/or sea, as feasible, for the shipment. Thus, nodes $A_1, A_2, \ldots, A_{n_A}$ represent the arrival portals of disaster aid items into the affected region—which is represented by the dashed box in the network topology.

The next set of links represent the "processing" of the received items at the portals of the affected region. Typically, a variety of goods are sent to the areas hit by disasters—some of which may even be useless or inappropriate. As a result, the unloading, unpacking, sorting, and the loading of the goods have to be done before the ultimate distribution takes place. Processing is often done at the same location as the arrival portals (e.g., airports, ports, etc.), and the corresponding facilities are denoted by $B_1, B_2, \ldots, B_{n_B}$. Processing at international locations may include, in addition, customs inspections and payments, etc.

Finally, the last set of nodes reflects the actual demand points in the affected region, denoted by $R_1, R_2, \ldots, R_{n_R}$. These nodes can be the hospitals, community centers, or points of distribution of the relief product to the disaster victims. The links connecting the last two sets of nodes are "distribution" links, and are represented by multiple links to allow for the possibility of various modes of transportation within the affected region. This may be crucial when parts of the infrastructure have been severely damaged by the disaster.

The supply chain network topology is denoted by $G = [N, L]$, where N and L, respectively, denote the sets of nodes and links (arcs).

15.2.2 Formulation of the Disaster Relief Supply Chain Network Model

Let f_a denote the flow of the disaster relief product on link a, be it a procurement, storage, transportation, processing, or distribution link. Let $c_a(f_a)$ and $\hat{c}_a(f_a)$ denote the unit operational cost function and the total operational cost function on link a, respectively. The link total cost functions are assumed to be convex and continuously differentiable. We have:

$$\hat{c}_a(f_a) = f_a \times c_a(f_a), \quad \forall a \in L. \tag{15.1}$$

$\mathcal{P}_k$ denotes the set of paths connecting the origin (node 1) to demand point k with $\mathcal{P}$ denoting the set of all paths joining the origin node to the destination nodes. The total number of paths in the supply chain, i.e., the number of elements in $\mathcal{P}$ is given by n_p.

In the model, we assume that the demand is uncertain due to the unpredictability of the actual demand at the demand points. Similar examples of system-optimized models with uncertain demand and associated shortage and surplus penalties can be found in the literature (see, e.g., Dong et al. 2004; Nagurney et al. 2011, 2012a; Nagurney and Masoumi 2012).

The probability distribution of demand is assumed to be available. It may be derived using census data and/or information gleaned and obtained over the course of the preparedness phase. If d_k denotes the actual (uncertain) demand at destination point k, we have:

$$P_k(D_k) = P_k(d_k \leq D_k) = \int_0^{D_k} \mathcal{F}_k(t)dt, \quad k = 1, \ldots, n_R, \tag{15.2}$$

where P_k and $\mathcal{F}_k$ denote the probability distribution function, and the probability density function of demand at point k, respectively.

Let v_k be the "projected demand" for the disaster relief item at point k; $k = 1, \ldots, n_R$. The amounts of shortage and surplus of the aid item at destination node k are denoted by Δ_k^- and Δ_k^+, respectively, and are calculated as follows:

$$\Delta_k^- \equiv \max\{0, d_k - v_k\}, \quad k = 1, \ldots, n_R, \tag{15.3a}$$

$$\Delta_k^+ \equiv \max\{0, v_k - d_k\}, \quad k = 1, \ldots, n_R. \tag{15.3b}$$

Hence, based on the probability distribution of the demand, the expected values of shortage and surplus at each demand point are:

$$E(\Delta_k^-) = \int_{v_k}^{\infty} (t - v_k)\mathcal{F}_k(t)dt, \quad k = 1, \ldots, n_R, \tag{15.4a}$$

$$E(\Delta_k^+) = \int_0^{v_k} (v_k - t)\mathcal{F}_k(t)dt, \quad k = 1, \ldots, n_R. \tag{15.4b}$$

Therefore, the expected penalty assigned to the humanitarian organization due to the shortage and surplus of the relief item at each demand point is equal to:

$$E\left(\lambda_k^- \Delta_k^- + \lambda_k^+ \Delta_k^+\right) = \lambda_k^- E\left(\Delta_k^-\right) + \lambda_k^+ E\left(\Delta_k^+\right), \quad k = 1, \ldots, n_R, \tag{15.5}$$

where λ_k^- is the unit penalty corresponding to the shortage of the relief item at demand point k which can represent the social cost of a death due to the inefficiency of the relief distribution system. In order to avoid such shortages, a large penalty should be assigned by the authorities, which can vary across the demand points. Furthermore, λ_k^+ denotes the unit penalty associated with the surplus of the relief item. This penalty is taken into account so as to minimize the over-shipping of goods, which results in congestion and additional efforts. Depending on the criticality of the situation, the unit surplus penalty can be expected to be lower than that of shortage, for a given demand point, and can be equal to zero. Similar ideas have been applied in the case of critical needs (Nagurney et al. 2011) and human blood (Nagurney et al. 2012a; Nagurney and Masoumi 2012).

x_p represents the flow of the disaster relief goods on path p joining node 1 with a demand node which must be nonnegative, since the goods are procured, stored, and shipped in nonnegative quantities, that is,

$$x_p \geq 0, \quad \forall p \in \mathcal{P}. \tag{15.6}$$

The projected demand at destination node k, v_k, is equal to the sum of flows on all paths belonging to $\mathcal{P}_k$, that is:

$$v_k \equiv \sum_{p \in \mathcal{P}_k} x_p, \quad k = 1, \ldots, n_R. \tag{15.7}$$

The relationship between the flow on link a, f_a, and the path flows is as follows:

$$f_a - \sum_{p \in \mathcal{P}} x_p \delta_{ap}, \quad \forall a \in L. \tag{15.8}$$

Here, δ_{ap} is an indicator of link a's relation with path p, and is equal to 1 if link a is contained in path p and is 0, otherwise.

Next, we present the expressions that capture the *time* aspect of our integrated disaster relief supply chain model.

Let τ_a denote the completion time of the activity on link a, which is assumed to be a linear function of the flow of the product on that link. We have:

$$\tau_a(f_a) = g_a f_a + h_a, \quad \forall a \in L, \tag{15.9}$$

where $h_a \geq 0$, and $g_a \geq 0$. We allow (some of) these terms to take on zero values for modeling flexibility purposes, as we shall show in a forthcoming numerical example.

Considering the above link time functions, we note that, in the context of transportation science, g_a can be referred to as the *congestion factor* on link a while h_a is the *uncongested term*, also known as the *free-flow travel time* on link a (see Dafermos and Sparrow 1969; Beckmann et al. 1956). Here, the congestion factor addresses the impact of the volume of the product on the time associated with the link activity, whether it is a procurement link, a storage link, a processing link, or a transport/shipment/distribution link. The uncongested cost term on a link represents the best case scenario time associated with completing the activity on the link.

The completion time of the sequence of activities on path p is denoted by τ_p, and is derived as follows:

$$\tau_p = \sum_{a \in L} \tau_a(f_a)\delta_{ap} = \sum_{a \in L} (g_a f_a + h_a)\delta_{ap}, \quad \forall p \in \mathcal{P}. \tag{15.10}$$

The sum of the uncongested terms h_as on path p is constant and is denoted by h_p:

$$h_p = \sum_{a \in L} h_a \delta_{ap}, \quad \forall p \in \mathcal{P}. \tag{15.11}$$

Hence,

$$\tau_p = h_p + \sum_{a \in L} g_a f_a \delta_{ap}, \quad \forall p \in \mathcal{P}. \tag{15.12}$$

Note that, at this point in the model development, the completion time of each path in accordance with the demand points can be expressed as:

$$\tau_p \leq T_k, \quad \forall p \in \mathcal{P}_k; \quad k = 1, \ldots, n_R, \tag{15.13}$$

where T_k is the target for the completion time of the activities on paths corresponding to demand point k determined by the organization's decision-maker. In other words, all activities on paths connecting the origin node to a specific demand point are aimed to be completed by a certain time. The time targets at the demand points, T_ks, may vary from one demand node to another since the decision-maker may need to prioritize demand points based on their strategic locations, potential injury tolls, classes of populations (elderly, children, etc.). Hence, the choice of these time goals by the authorities has to be done carefully. Using (15.12), inequality (15.13) can be re-written as:

$$\sum_{a \in L} g_a f_a \delta_{ap} \leq T_k - h_p, \quad \forall p \in \mathcal{P}_k; \quad k = 1, \ldots, n_R. \tag{15.14}$$

We also define the target time T_{kp} for demand point k with respect to path p. This new target time is defined as the difference between target time T_k and h_p:

$$T_{kp} = T_k - h_p, \quad \forall p \in \mathcal{P}_k; \quad k = 1, \ldots, n_R. \tag{15.15}$$

We assume that the values of the T_{kp}'s are all nonnegative; otherwise, we remove path p from the network due to infeasibility. We further assume that there is at least one path $p \in \mathcal{P}_k$ for each k such that $T_{kp} > 0$; otherwise, the organization will have to relax the corresponding time target T_k (make it less restrictive).

Thus, inequality (15.14) can be re-written as:

$$\sum_{a \in L} g_a f_a \delta_{ap} \leq T_{kp}, \quad \forall p \in \mathcal{P}_k; \quad k = 1, \ldots, n_R. \tag{15.16}$$

Note that the goal constraints introduced here are not hard constraints meaning that, under certain circumstances, the organization might be forced to deviate from the goal. However, this deviation will be minimized, as we shall see, along with the minimization of total cost throughout the network subject to the uncertain demand. For example, depending on the actual completion time, the sequence of activities on path p leading to the delivery of the relief products to demand point k will be completed either at, before, or after the determined time goal target. Let z_p denote the amount of deviation with respect to target time T_{kp} corresponding to the "late" delivery of product to point k on path p, which was assumed to be nonnegative. Using (15.16), we now construct the following constraints:

$$\sum_{a \in L} g_a f_a \delta_{ap} - z_p \leq T_{kp}, \quad \forall p \in \mathcal{P}_k; \quad k = 1, \ldots, n_R. \tag{15.17}$$

The path time deviations must be nonnegative for all paths in the network; that is,

$$z_p \geq 0, \quad \forall p \in \mathcal{P}_k; \quad k = 1, \ldots, n_R. \tag{15.18}$$

Using (15.8), we can replace the link flows with the path flows in (15.17), so that

$$\sum_{q \in \mathcal{P}} \sum_{a \in L} g_a x_q \delta_{aq} \delta_{ap} - z_p \leq T_{kp}, \quad \forall p \in \mathcal{P}_k; \quad k = 1, \ldots, n_R. \tag{15.19}$$

Let $\gamma_k(z)$ denote the tardiness penalty function corresponding to demand point k which is a function of the time deviations on paths leading to that point. These functions are assumed to be convex and continuously differentiable.

Interestingly, Nagurney et al. (1996) utilized goal targets in the case of spatial economic markets, whereas Nagurney and Ramanujam (1996) considered penalties associated with transportation targets with associated penalty functions that could be nonlinear (as in the case above).

$\hat{C}_p(x)$ denotes the total operational cost function on path p and is constructed as:

$$\hat{C}_p(x) = x_p \times C_p(x) = x_p \times \sum_{a \in L} c_a(f_a)\delta_{ap}, \quad \forall p \in \mathcal{P}, \tag{15.20}$$

with notice to (15.8), where C_p denotes the unit operational cost on path p.

The disaster relief supply chain network optimization problem can be expressed as follows. The organization seeks to determine the optimal levels of the disaster relief item processed on each supply chain link as well as the optimal amounts of the time deviations on paths, subject to the minimization of the total operational cost while satisfying the uncertain demand as closely as possible. Therefore, the optimization problem is constructed as:

$$\text{Minimize} \sum_{p \in \mathcal{P}} \hat{C}_p(x) + \sum_{k=1}^{n_R} \left(\lambda_k^- E(\Delta_k^-) + \lambda_k^+ E(\Delta_k^+)\right) + \sum_{k=1}^{n_R} \gamma_k(z), \tag{15.21}$$

subject to: constraints (15.6), (15.18), and (15.19).

Next, we present the partial derivatives of the shortages and the surpluses solely in terms of path flows, which will be used later in developing the variational inequality formulation of the problem. The respective partial derivatives of the expected values of shortage and surplus of the disaster relief item at each demand point with respect to the path flows, derived in Dong et al. (2004), Nagurney et al. (2011), and Nagurney et al. (2012a), are given by:

$$\frac{\partial E(\Delta_k^-)}{\partial x_p} = P_k\left(\sum_{q \in \mathcal{P}_k} x_q\right) - 1, \quad \forall p \in \mathcal{P}_k; \quad k = 1, \ldots, n_R, \tag{15.22a}$$

and

$$\frac{\partial E(\Delta_k^+)}{\partial x_p} = P_k\left(\sum_{q \in \mathcal{P}_k} x_q\right), \quad \forall p \in \mathcal{P}_k; \quad k = 1, \ldots, n_R. \tag{15.22b}$$

Let K denote the feasible set such that:

$$K = \left\{(x, z, \omega) \middle| x \in R_+^{n_p}, \ z \in R_+^{n_p}, \text{ and } \omega \in R_+^{n_p}\right\}, \tag{15.23}$$

where x is the vector of path flows of the relief item, z is the vector of time deviations on paths, and ω is the vector of Lagrange multipliers corresponding to the constraints in (15.19).

We now derive the variational inequality of the integrated disaster relief supply chain network problem.

Theorem 1 The optimization problem (15.21), subject to its constraints (15.6), (15.18), and (15.19), is equivalent to the variational inequality problem: determine the vector of optimal path flows, the vector of optimal path time deviations, and the vector of optimal Lagrange multipliers $(x^*, z^*, \omega^*) \in K$, such that:

$$\sum_{k=1}^{n_R}\sum_{p\in\mathcal{P}_k}\left[\frac{\partial\hat{C}_p(x^*)}{\partial x_p}+\lambda_k^+P_k\left(\sum_{q\in\mathcal{P}_k}x_q^*\right)-\lambda_k^-\left(1-P_k\left(\sum_{q\in\mathcal{P}_k}x_q^*\right)\right)\sum_{q\in\mathcal{P}}\sum_{a\in L}\omega_q^*g_a\delta_{aq}\delta_{ap}\right]$$
$$\times\left[x_p-x_p^*\right]+\sum_{k=1}^{n_R}\sum_{p\in\mathcal{P}_k}\left[\frac{\partial\gamma_k(z^*)}{\partial z_p}-\omega_p^*\right]\times\left[z_p-z_p^*\right]$$
$$+\sum_{k=1}^{n_R}\sum_{p\in\mathcal{P}_k}\left[T_{kp}+z_p^*-\sum_{q\in\mathcal{P}}\sum_{a\in L}g_ax_q^*\delta_{aq}\delta_{ap}\right]\times\left[\omega_p-\omega_p^*\right]\geq 0,\quad\forall(x,z,\omega)\in K,$$
(15.24)

where

$$\frac{\partial\hat{C}_p(x)}{\partial x_p}\equiv\sum_{a\in L}\frac{\partial\hat{c}_a(f_a)}{\partial f_a}\delta_{ap},\quad\forall p\in\mathcal{P}_k;\ k=1,\ldots,n_R. \tag{15.25}$$

Proof: Consider the optimization formulation in (15.21). The convexity of the objective function follows, under the imposed assumptions, from Nagurney et al. (2012a), and since the $\gamma_k(z)$ functions are assumed to be convex.

According to Bertsekas and Tsitsiklis (1989) (page 287), the optimization problem (15.21), along with its inequality constraints (15.19), is equivalent to the below inequality which is resulted from the Karush–Kuhn-Tucker (KKT) (cf. Karush 1939; Kuhn and Tucker 1951) conditions:

$$\sum_{k=1}^{n_R}\sum_{p\in\mathcal{P}_k}\left[\frac{\partial}{\partial x_p}(\hat{C}_p(x^*)+\lambda_k^-E(\Delta_k^-)+\lambda_k^+E(\Delta_k^+)+\gamma_k(z^*))+\sum_{q\in\mathcal{P}}\sum_{a\in L}\omega_q^*g_a\delta_{aq}\delta_{ap}\right]$$
$$\times\left[x_p-x_p^*\right]$$
$$+\sum_{k=1}^{n_R}\sum_{p\in\mathcal{P}_k}\left[\frac{\partial}{\partial z_p}(\hat{C}_p(x^*)+\lambda_k^-E(\Delta_k^-)+\lambda_k^+E(\Delta_k^+)+\gamma_k(z^*))-\omega_p^*\right]\times\left[z_p-z_p^*\right]$$
$$+\sum_{k=1}^{n_R}\sum_{p\in\mathcal{P}_k}\left[T_{kp}+z_p^*-\sum_{q\in\mathcal{P}}\sum_{a\in L}g_ax_q^*\delta_{aq}\delta_{ap}\right]\times\left[\omega_p-\omega_p^*\right]\geq 0,\quad\forall(x,z,\omega)\in K.$$
(15.26)

Substituting the partial derivatives in (15.26), and using (15.25), (15.22a), and (15.22b), one obtains the variational inequality (15.24). □

Variational inequality (15.24) can be put into standard form (Nagurney (1999)) as follows: determine $X^*\in\mathcal{K}$ such that:

$$\langle F(X^*), X - X^* \rangle \geq 0, \quad \forall X \in \mathcal{K}, \tag{15.27}$$

where $\langle \cdot, \cdot \rangle$ denotes the inner product in n-dimensional Euclidean space. If the feasible set is defined as $\mathcal{K} \equiv K$, and the column vectors $X \equiv (x, z, \omega)$ and $F(X) \equiv (F_1(X), F_2(X), F_3(X))$, where:

$$F_1(X) = \left[\frac{\partial \hat{C}_p(x^*)}{\partial x_p} + \lambda_k^+ P_k \left(\sum_{q \in \mathcal{P}_k} x_q \right) - \lambda_k^- \left(1 - P_k \left(\sum_{q \in \mathcal{P}_k} x_q \right) \right) \sum_{q \in \mathcal{P}} \sum_{a \in L} \omega_q^* g_a \delta_{aq} \delta_{ap}, \right.$$
$$\left. p \in \mathcal{P}_k;\ k = 1, \ldots, n_R \right],$$

$$F_2(X) = \left[\frac{\partial \gamma_k(z)}{\partial z_p} - \omega_p,\ p \in \mathcal{P}_k;\ k = 1, \ldots, n_R \right],$$

and

$$F_3(x) = \left[T_{kp} + z_p - \sum_{q \in \mathcal{P}} \sum_{a \in L} g_a x_q \delta_{aq} \delta_{ap}, \quad p \in \mathcal{P}_k;\ k = 1, \ldots, n_R, \right], \tag{15.28}$$

then variational inequality (15.24) can be re-expressed as standard form (15.27).

We utilize variational inequality (15.24) for our computations to obtain the optimal path flows and the optimal path time deviations. Then, we use (15.8) to calculate the optimal link flows of disaster relief items in the supply chain network.

15.2.3 An Illustrative Example and Two Variants with Sensitivity Analysis

We now present an illustrative numerical example as well as two variants along with their solutions, accompanied by some sensitivity analysis, before proceeding to the solution algorithm, which can be applied to solve large-scale disaster relief supply chain networks in practice.

15.2.3.1 Illustrative Example

Consider the simple disaster relief supply chain network topology in Fig. 15.2. The organization is assumed to possess a single procurement facility and a single storage facility, and aims to deliver the relief goods to one demand point through one arrival portal and one processing facility in the affected region. We allow two modes of transportation from the storage facility to the portal of the affected region. The links are labeled as in Fig. 15.2, i.e., a, b, c, d, e, f, and g, where links d and e represent ground and air transportation, respectively.

The total operational cost functions on the links are:

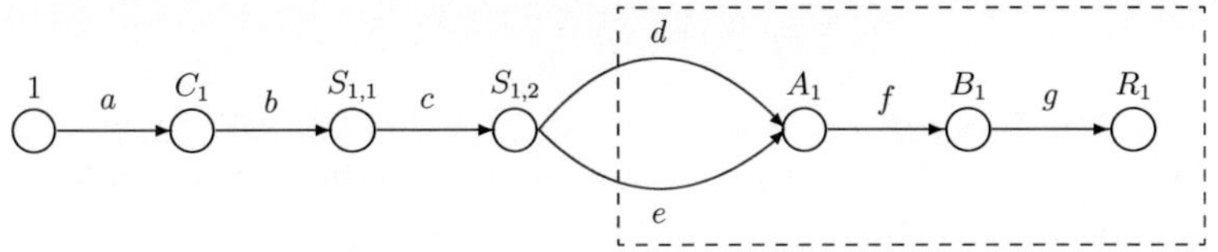

Fig. 15.2 Supply chain network topology for the illustrative numerical example

$$\hat{c}_a(f_a) = 3f_a^2 + 2f_a, \quad \hat{c}_b(f_b) = f_b^2 + 3f_b, \quad \hat{c}_c(f_c) = 2f_c^2 + f_c,$$
$$\hat{c}_d(f_d) = 4f_d^2 + 3f_d, \quad \hat{c}_e(f_e) = 7f_e^2 + 5f_e,$$
$$\hat{c}_f(f_f) = f_f^2 + 4f_f, \quad \hat{c}_g(f_g) = 3f_g^2 + 2f_g.$$

There are two paths in this network defined as: $p_1 \equiv (a, b, c, d, f, g)$ and $p_2 \equiv (a, b, c, e, f, g)$. The set of paths, $\mathcal{P}$, is identical to the set of paths connecting the origin 1 to the destination, R_1, i.e., $\mathcal{P}_{R_1}$, where

$$\mathcal{P} = \mathcal{P}_{R_1} = \{p_1, p_2\}.$$

The demand for the relief item at the demand point followed a uniform distribution on the interval [5,10]; therefore, the probability distribution function of demand at the demand point is:

$$P_{R_1}(v_{R_1}) = \frac{v_{R_1} - 5}{10 - 5} = \frac{x_{p_1} + x_{p_2} - 5}{5}.$$

The unit shortage and surplus penalties were: $\lambda_{R_1}^- = 5000$ and $\lambda_{R_1}^+ = 100$.

The organization is interested in the pre-positioning strategy; i.e., it wishes to determine the amount of the relief item that should be stored beforehand. Thus, the organization will only ship the pre-positioned supplies of relief goods and will not procure post the disaster. Consequently, the completion time on links a, b, and c (procurement, transportation, and storage) is set to zero:

$$\tau_a(f_a) = \tau_b(f_b) = \tau_c(f_c) = 0.$$

The completion time functions on the rest of the links were:

$$\tau_d(f_d) = 9f_d + 6, \quad \tau_e(f_e) = 2f_e + 2,$$
$$\tau_f(f_f) = 1.5f_f + 2, \quad \tau_g(f_g) = 5f_g + 4.$$

The target time at demand point R_1 was 72 h:

$$T_{R_1} = 72, \quad \forall p \in \mathcal{P}_{R_1}.$$

The decision-maker assigned a higher tardiness penalty to p_2 in that the expectation of on-time delivery from the path with the air transportation link was higher, so the tardiness penalty function at the demand point was:

$$\gamma_{R_1}(z) = 3.5z_{p_1}^2 + 8z_{p_2}^2.$$

Solution: For the variational inequality (15.24) to hold for all $(x, z, \omega) \in K$, at the optimal solution (x^*, z^*, ω^*), we must have that the term in each of the three left-hand-side brackets is equal to zero, assuming that the optimal value of each of the variables is positive. Thus, we have the following six equations with $x_{p_1}^*, x_{p_2}^*, z_{p_1}^*, z_{p_2}^*, \omega_{p_1}^*$, and $\omega_{p_2}^*$ as unknowns:

$$\frac{\partial \hat{C}_{p_1}(x^*)}{\partial x_{p_1}} + \lambda_{R_1}^+ P_{R_1}\left(x_{p_1}^* + x_{p_2}^*\right) - \lambda_{R_1}^+ \left(1 - P_{R_1}\left(x_{p_1}^* + x_{p_2}^*\right)\right)$$

$$+ \omega_{p_1}^* \left(g_d + g_f + g_g\right) + \omega_{p_2}^* \left(g_f + g_g\right) = 0,$$

$$\frac{\partial \hat{C}_{p_2}(x^*)}{\partial x_{p_2}} + \lambda_{R_1}^+ P_{R_1}\left(x_{p_1}^* + x_{p_2}^*\right) - \lambda_{R_1}^+ \left(1 - P_{R_1}\left(x_{p_1}^* + x_{p_2}^*\right)\right) + \omega_{p_1}^* \left(g_f + g_g\right)$$

$$+ \omega_{p_2}^* \left(g_e + g_f + g_g\right) = 0,$$

$$\frac{\partial \gamma_{R_1}(z^*)}{\partial z_{p_1}} - \omega_{p_1}^* = 0,$$

$$\frac{\partial \gamma_{R_1}(z^*)}{\partial z_{p_2}} - \omega_{p_2}^* = 0,$$

$$T_{R_1 p_1} + z_{p_1}^* - \left(x_{p_1}^* \left(g_d + g_f + g_g\right) + x_{p_2}^* \left(g_f + g_g\right)\right) = 0, \quad \text{and}$$

$$T_{R_1 p_2} + z_{p_2}^* - \left(x_{p_1}^* \left(g_f + g_g\right) + x_{p_2}^* \left(g_e + g_f + g_g\right)\right) = 0.$$

Note that, in the first two and the last two equations above, $g_a = g_b = g_c = 0$, and several δ_{ap}'s and δ_{aq}'s are zero. Also, note that:

$$T_{R_1 p_1} = T_{R_1} - h_{p_1} = T_{R_1} - \left(h_d + h_f + h_g\right) = 60$$

and

$$T_{R_1 p_2} = T_{R_1} - h_{p_2} = T_{R_1} - \left(h_e + h_f + h_g\right) = 64.$$

Next, using (15.25) to calculate the partial derivatives of the total path costs, and substituting the given parameters into the above equations, we obtain the following system of equations:

$$\begin{cases} 1048x^*_{p_1} & +1040x^*_{p_2} & & & +15.5\omega^*_{p_1} & +6.5\omega^*_{p_2} & = 10085 \\ 1040x^*_{p_1} & +1054x^*_{p_2} & & & +6.5\omega^*_{p_1} & +8.5\omega^*_{p_2} & = 10083 \\ & & 7z^*_{p_1} & & -\omega^*_{p_1} & & = 0 \\ & & & 16z^*_{p_2} & & -\omega^*_{p_2} & = 0 \\ -15.5x^*_{p_1} & -6.5x^*_{p_2} & +z^*_{p_1} & & & & = -60 \\ -6.5x^*_{p_1} & -8.5x^*_{p_2} & & +z^*_{p_2} & & & = -64. \end{cases}$$

Solution of the above system yields:

$$x^*_{p_1} = 1.04 \, \text{and} \, x^*_{p_2} = 7.50.$$

Hence, the optimal values of link flows are:

$$f^*_a = f^*_b = f^*_c = f^*_f = f^*_g = x^*_{p_1} + x^*_{p_2} = 8.54, \quad f^*_d = x^*_{p_1} = 1.04, \text{ and } f^*_e = x^*_{p_2} = 7.50.$$

As seen above, the optimal flow of the disaster relief item on link e (air transportation to the affected region) was considerably higher than that on link d (ground transportation). This is because the humanitarian organization, in this example, chooses the quicker mode of transportation, but at a higher cost. The optimal time deviations on paths p_1 and p_2 with respect to the target of 72 h are:

$$z^*_{p_1} = 4.85 \text{ and } z^*_{p_2} = 6.47.$$

Neither of the two transportation modes to the affected area would be able to satisfy the target time requirement. Interestingly, the time deviation is higher on the path that contains the air route, which is due to the majority of the load being allocated to this mode.

The value of the projected demand at point R_1 was:

$$v^*_{R_1} = x^*_{p_1} + x^*_{p_2} = 8.54,$$

which is the amount that needs to be pre-positioned at the storage facility. The projected demand was closer to the upper bound of the uniform distribution range of the demand for the relief item at point R_1. If the organization seeks to reduce the consequences of the shortage of relief items in the affected region, the decision-maker should assign a higher unit shortage penalty so as to be able to better meet the uncertain demand within the given time limit. This obviously will result in a higher operational cost and, yet, in a lower social cost.

The optimal values of the Lagrange multipliers corresponding to the time goal constraints were:

$$\omega^*_{p_1} = 33.97 \text{ and } \omega^*_{p_2} = 103.55.$$

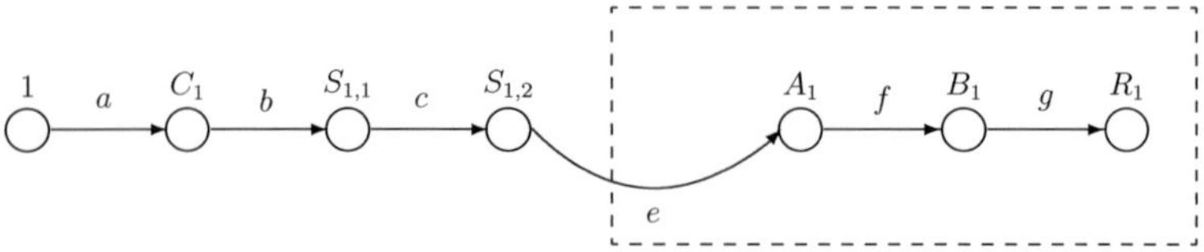

Fig. 15.3 Supply chain network topology for the illustrative numeric al example—Variant I

15.2.3.2 Illustrative Example—Variant I

We then considered the following variant of the above illustrative example. We assumed a worst case scenario in the form of unavailability of ground transport. Hence, link d would no longer be available in the supply chain (See Fig. 15.3). The data were as above with the expressions (and constraints) associated with path p_1 and link d removed.

The new solution had to satisfy the following equations, under the assumption that $x^*_{p_2} > 0$, $z^*_{p_2} > 0$, and $\omega^*_{p_2} > 0$:

$$\frac{\partial \hat{C}_{p_2}(x^*)}{\partial x_{p_2}} + \lambda^+_{R_1} P_{R_1}\left(x^*_{p_2}\right) - \lambda^-_{R_1}\left(1 - P_{R_1}\left(x^*_{p_2}\right)\right) + \omega^*_{p_2}\left(g_e + g_f + g_g\right) = 0,$$

$$\frac{\partial \gamma_{R_1}(z^*)}{\partial z_{p_2}} - \omega^*_{p_2} = 0, \text{ and}$$

$$T_{R_1 p_2} + z^*_{p_2} - \left(x^*_{p_2}\left(g_e + g_f + g_g\right)\right) = 0.$$

Similar to the original example, substitution of the partial derivatives and the numerical values into the above equations leads to the following system of equations:

$$\begin{cases} 1054x^*_{p_2} & & + 8.5\omega^*_{p_2} & = 10083 \\ & 16z^*_{p_2} & - \omega^*_{p_2} & = 0 \\ -8.5x^*_{p_2} & + z^*_{p_2} & & = -64. \end{cases}$$

Solution of the above system of equations yields:

$$x^*_{p_2} = 8.50.$$

Hence,

$$f^*_a = f^*_b = f^*_c = f^*_e = f^*_f = f^*_g = 8.50.$$

Also, the optimal value of the projected demand at demand point R_1, $v^*_{R_1} = 8.50$. Therefore, removal of the ground transportation link slightly decreased the projected demand.

The new optimal time deviation on path p_2 was:

$$z^*_{p_2} = 8.26,$$

which is higher than the value in the preceding example. The value of the tardiness penalty function, $\gamma_{R_1}(z^*)$, has increased from 438.39 in the illustrative example to 579.61 in its variant. Also, removal of the ground transportation link results in a higher total operational cost in that the sole mode of transportation to the affected region would be the costlier air transport mode.

The optimal value of Lagrange multiplier on path p_2 was:

$$\omega^*_{p_2} = 132.12,$$

which shows a 28 % increase.

15.2.3.3 Illustrative Example—Variant II

The second variant of the original example addresses the post-disaster procurement strategy—as opposed to pre-positioning of the supplies. We assumed that the organization did not store disaster items beforehand. In other words, the organization would procure relief items only once a disaster struck. Thus, the network did not include links a, b, and c—corresponding to pre-disaster procurement, shipment, and storage of relief goods—and instead incorporated link h representing post-disaster procurement, as depicted in Fig. 15.4.

The total cost functions and the time completion functions on links d, e, f, and g remained the same as in the illustrative example. As for link h, we had:

$$\hat{c}_h(f_h) = 5f_h^2 + 3f_h \text{ and } \tau_h(f_h) = 3f_h + 3.$$

In this example, the total operational cost on the post-disaster procurement link, h, was higher than that on the pre-disaster procurement link a in the example in Sect. 15.2.3.1. That is because the organization is assumed to be able to procure the relief items at a lower price under the pre-positioning strategy given the ample time that it has for preparedness. In contrast, procuring the relief goods after a disaster hits may leave the organization with fewer purchasing options since large quantities of goods have to be purchased/procured quickly (and, depending on the sources, there may even be competition for purchasing). In addition, in the network topology of this example, the post-disaster procurement activity may also include some transportation/shipment, further raising the total operational cost on link h as compared to that on link a in the previous examples.

The set of paths in this problem was $\mathcal{P} = \{p_3, p_4\}$ where $p_3 = (h, d, f, g)$ and $p_4 = (h, e, f, g)$. The demand distribution, the shortage and surplus penalties, as well as the target time were identical to the previous problems. The tardiness penalty function at the demand point was the same, except that now it was a function of the time deviations on new paths:

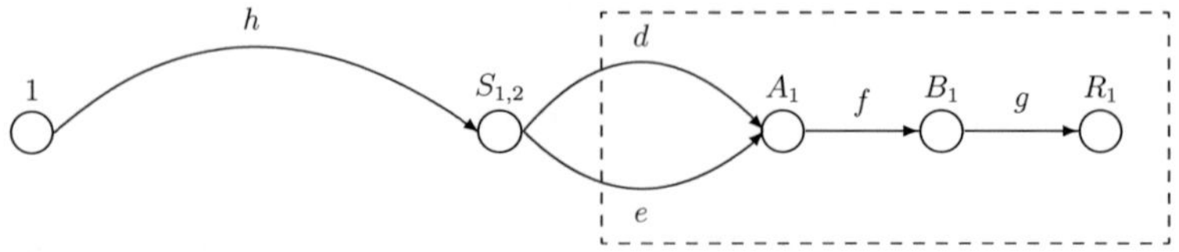

Fig. 15.4 Supply chain network topology for the illustrative numerical example—Variant II

$$\gamma_{R_1}(z) = 3.5z_{p_3}^2 + 8z_{p_4}^2.$$

Under the assumption of positive optimal solutions, the following system of equations had to be satisfied:

$$\frac{\partial \hat{C}_{p_3}(x^*)}{\partial x_{p_3}} + \lambda_{R_1}^{+} P_{R_1}\left(x_{p_3}^* + x_{p_4}^*\right) - \lambda_{R_1}^{-}\left(1 - P_{R_1}\left(x_{p_3}^* + x_{p_4}^*\right)\right)$$
$$+ \omega_{p_3}^*\left(g_h + g_d + g_f + g_g\right) + \omega_{p_4}^*\left(g_h + g_f + g_g\right) = 0,$$

$$\frac{\partial \hat{C}_{p_4}(x^*)}{\partial x_{p_4}} + \lambda_{R_1}^{+} P_{R_1}\left(x_{p_3}^* + x_{p_4}^*\right) - \lambda_{R_1}^{-}\left(1 - P_{R_1}\left(x_{p_3}^* + x_{p_4}^*\right)\right) + \omega_{p_3}^*\left(g_h + g_f + g_g\right)$$
$$+ \omega_{p_4}^*\left(g_h + g_e + g_f + g_g\right) = 0,$$

$$\frac{\partial \gamma_{R_1}(z^*)}{\partial \gamma_{p_3}} - \omega_{p_3}^* = 0,$$

$$\frac{\partial \gamma_{R_1}(z^*)}{\partial \gamma_{p_4}} - \omega_{p_4}^* = 0,$$

$$T_{R_1 p_3} + z_{p_3}^* - \left(x_{p_3}^*\left(g_h + g_d + g_f + g_g\right) + x_{p_4}^*\left(g_h + g_f + g_g\right)\right) = 0,$$

$$T_{R_1 p_4} + z_{p_4}^* - \left(x_{p_3}^*\left(g_h + g_f + g_g\right) + x_{p_4}^*\left(g_h + g_e + g_f + g_g\right)\right) = 0,$$

where

$$T_{R_1 p_3} = T_{R_1} - h_{p_3} = T_{R_1} - \left(h_h + h_d + h_f + h_g\right) = 57,$$
$$T_{R_1 p_4} = T_{R_1} - h_{p_4} = T_{R_1} - \left(h_h + h_e + h_f + h_g\right) = 61.$$

After substitution of the partial derivatives and the other numerical values into the above system, we have the following six equations:

$$\begin{cases} 1046x^*_{p_3} + 1038x^*_{p_4} + 18.5\omega^*_{p_3} + 9.5\omega^*_{p_4} = 10088 \\ 1038x^*_{p_3} + 1052x^*_{p_4} + 9.5\omega^*_{p_3} + 11.5\omega^*_{p_4} = 10086 \\ 7z^*_{p_3} - \omega^*_{p_3} = 0 \\ 16z^*_{p_4} - \omega^*_{p_4} = 0 \\ -18.5x^*_{p_3} - 9.5x^*_{p_4} + z^*_{p_3} = -57 \\ -9.5x^*_{p_3} - 11.5x^*_{p_4} + z^*_{p_4} = -61. \end{cases}$$

Solution of the above yields:

$$x^*_{p_3} = 0.33 \text{ and } x^*_{p_4} = 6.26.$$

which yields the following optimal link flows:

$$f^*_h = f^*_f = f^*_g = x^*_{p_3} + x^*_{p_4} = 6.59, \; f^*_d = x^*_{p_3} = 0.33, \text{ and } f^*_e = x^*_{p_4} = 6.26.$$

Note that, under a post-disaster procurement strategy, the optimal flow on the ground transportation link is significantly lower than that in the pre-disaster procurement strategy. This is due to the fact that the duration of the procurement activity is now being incorporated, and the organization wants to avoid excessive delays caused by assigning large quantities to the ground transportation path. In addition, the optimal flow on the air transportation link has experienced a 17 % decrease which is a consequent of increased operational costs.

The value of the projected demand is:

$$v^*_{R_1} = x^*_{p_3} + x^*_{p_4} = 6.59,$$

which is 23 % lower than that for the example in Sect. 15.2.3.1. Furthermore, the projected demand is now closer to the lower bound of the demand uniform distribution. Therefore, in the case that pre-positioning of supplies is not available, the decision-maker must be careful in assessing the shortage penalties. A sensitivity analysis on the shortage penalty value and the impact on the solution is presented in Sect. 15.2.3.4.

The optimal time deviations on paths p_3 and p_4 are:

$$z^*_{p_3} = 8.54 \text{ and } z^*_{p_4} = 14.09,$$

which are higher than their respective values in the first example. The incurred tardiness penalty value, $\gamma_{R_1}(z^*)$, is equal to 1, 844.16, i.e., a 321 % increase from that in the first example.

Finally, the optimal Lagrange multipliers on paths p_3 and p_4 are:

$$\omega^*_{p_3} = 59.77 \text{ and } \omega^*_{p_4} = 225.49.$$

15.2.3.4 Sensitivity Analysis for Variant II

In the illustrative example in Sect. 15.2.3.1 and its variants in Sects. 15.2.3.2 and 15.2.3.3, we investigated the impact of the shortage penalty on the optimal solution.

Table 15.1 Sensitivity analysis of the optimal solution to the unit shortage penalty at the demand point for Variant II

$\lambda^-_{R_1}$	$x^*_{p_3}$	$x^*_{p_4}$	$z^*_{p_3}$	$z^*_{p_4}$	$\omega^*_{p_3}$	$\omega^*_{p_4}$	Value of objective function (21)
2,500	0.50	5.56	5.09	7.66	35.66	122.58	5,081.96
5,000	0.33	6.26	8.54	14.09	59.77	225.49	8,440.02
7,500	0.20	6.79	11.18	19.02	78.25	304.39	11,021.81
10,000	0.09	7.22	13.26	22.91	92.80	366.49	13,035.31
12,500	0.01	7.56	14.94	26.05	104.57	416.72	14,655.25

Table 15.1 displays the optimal values of the path flows, $x^*_{p_3}$ and $x^*_{p_4}$, the path time deviations, $z^*_{p_3}$ and $z^*_{p_4}$, and the Lagrange multipliers, $\omega^*_{p_3}$ and $\omega^*_{p_4}$, as the unit shortage penalty, $\lambda^-_{R_1}$, is increased from 2,500 to 12,500 in Variant II.

As seen in Table 15.1, as the shortage penalty increases, the organization will be fulfilling a higher projected demand, $x^*_{p_3} + x^*_{p_4}$, by assigning higher quantities to the path that uses air transportation to the affected region. At $\lambda^-_{R_1} = 12,500$, the optimal path flow on the ground transportation path is almost zero, which means that the organization relies on the air transport mode. Handling larger volumes of goods increases the congestion on paths which, in turn, worsens the lateness of deliveries to the region.

15.3 The Algorithm

In this section, we recall the Euler method, which is induced by the general iterative scheme of Dupuis and Nagurney (1993). Its realization for the solution of disaster relief supply chain network problems governed by variational inequality (15.24) (and (15.27)) yields subproblems that can be solved explicitly and in closed form. Specifically, recall that at an iteration τ of the Euler method (see also Nagurney and Zhang 1996) one computes:

$$X^{\tau+1} = P_{\mathcal{K}}\big(X^T - a_\tau F(X^\tau)\big), \tag{15.29}$$

where $P_{\mathcal{K}}$ is the projection on the feasible set $\mathcal{K}$ and F is the function that enters the variational inequality problem: determine $X^* \in \mathcal{K}$ such that

$$\langle F(X^*), X - X^* \rangle \geq 0, \quad \forall X \in \mathcal{K}, \tag{15.30}$$

where $\langle \cdot, \cdot \rangle$ is the inner product in n-dimensional Euclidean space, $X \in R^n$, and $F(X)$ is an n-dimensional function from $\mathcal{K}$ to R^n, with $F(X)$ being continuous (see also (15.28)).

As shown in Dupuis and Nagurney (1993); see also Nagurney and Zhang (1996), for convergence of the general iterative scheme, which induces the Euler method, among other methods, the sequence $\{a_\tau\}$ must satisfy: $\sum_{\tau=0}^{\infty} a_\tau = \infty$, $a_\tau > 0$,

$a_\tau \to 0$, as $\tau \to \infty$. Specific conditions for convergence of this scheme can be found for a variety of networkbased problems, similar to those constructed here, in Nagurney and Zhang (1996) and the references therein.

15.3.1 Explicit Formulae for the Euler Method Applied to the Disaster Relief Supply Chain Network Variational Inequality (15.24)

The elegance of this procedure for the computation of solutions to the disaster relief supply chain network problem modeled in Sect. 15.2 can be seen in the following explicit formulae. Indeed (15.29) for the supply chain network problem governed by variational inequality problem (15.24) yields the following closed form expressions for the product path flows, the time deviations, and the Lagrange multipliers, respectively:

$$x_p^{\tau+1} = \max\{0, x_p^\tau + a_\tau \left(\lambda_k^- \left(1 - P_k \left(\sum_{q \in \mathcal{P}_k} x_p^\tau \right) \right) \right.$$

$$\left. - \lambda_k^+ P_k \left(\sum_{q \in \mathcal{P}_k} x_p^\tau \right) - \frac{\partial \hat{C}_p(x^\tau)}{\partial x_p} \sum_{q \in \mathcal{P}} \sum_{a \in L} \omega_q^\tau g_a \delta_{aq} \delta_{ap} \right) \},$$

$$\forall p \in \mathcal{P}_k; \; k = 1, \ldots, n_R, \tag{15.31}$$

$$z_p^{\tau+1} = \max\left\{0, z_p^\tau + a_\tau \left(\omega_p^\tau - \frac{\partial \gamma_k(z^\tau)}{\partial z_p} \right) \right\}, \quad \forall p \in \mathcal{P}_k; \quad k = 1, \ldots, n_R, \text{ and} \tag{15.32}$$

$$\omega_p^{\tau+1} = \max\left\{0, \omega_p^\tau, + a_\tau, \left(\sum_{q \in \mathcal{P}} \sum_{a \in L} g_a x_p^\tau \delta_{aq} \delta_{ap} \quad T_{kp} - z_p^\tau \right\},$$

$$\forall p \in \mathcal{P}_k; \quad k = 1, \ldots, n_R. \tag{15.33}$$

In the next section, we solve additional disaster relief supply chain network problems using the above algorithmic scheme.

15.4 A Larger Numerical Example and Variant

The scenario for the first numerical example is built on the possibility of another earthquake striking Haiti. We then construct a variant.

Figure 15.5 displays the disaster relief supply chain network topology corresponding to the case of a Haiti earthquake. Node 1 is assumed to represent the American Red Cross. We assumed that the Red Cross could utilize two of its disaster aid zones in the US, one in Maryland—representing the Northeast and the

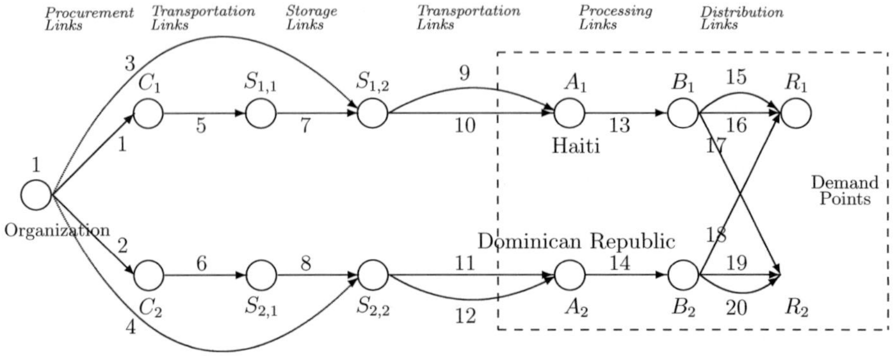

Fig. 15.5 Network topology of the larger disaster relief supply chain numerical example

East—and the other one in Florida—the closest state to the Caribbean. Each of the two zones is assumed to possess a single procurement facility, a single storage facility, and a single departure portal.

Both locations have the ability to start procuring the relief goods after an earthquake strikes when and if the need arises (links 3 and 4). These procured/purchased goods at the zones are directly sent to their respective departure portals bypassing the storage phase. Next, from these departure facilities, the collected items—whether pre-positioned beforehand or just procured—are sent via air or sea to the affected region. We assumed that the facility $S_{1,2}$—hypothetically, Maryland—would cover the arrival facility A_1 in Haiti, and the facility $S_{2,2}$ would serve the arrival port A_2 in the Dominican Republic. Links 9 and 12 represent air transportation whereas links 10 and 11 correspond to marine transportation.

After the arrived cargo is sorted and processed, relief items are distributed to the points of demand, R_1 and R_2, both located in Haiti. Links 15 and 20 correspond to the distribution of goods by helicopter, whereas links 16 through 19 represent ground distribution. Note that we allow for each of the two processing facilities located in Haiti and the Dominican Republic, i.e., B_1 and B_2, to ship to both of the two demand points in Haiti.

Now we present the available information as well as the solution to the problem. For purposes of transparency and reproducibility, we provide both the input and the output data.

Table 15.2 presents the total operational cost functions and completion time functions corresponding to the 20 links of the network. Similar to the illustrative example and its variants, the completion time functions on links 1, 2, 5, 6, 7, 8—pre-disaster procurement, transportation and storage—are set equal to zero since these activities are assumed to have been completed before the occurrence of the disaster.

There are 24 paths in the supply chain network connecting node 1 to the two demand points. The set of paths connecting the origin to demand points R_1 and R_2, respectively, are denoted by $\mathcal{P}_{R_1}$ and $\mathcal{P}_{R_2}$, and each contains 12 paths as shown in Table 15.3.

Table 15.2 Total operational costs, completion time functions, and the optimal flows on links in the numerical example

Link	$\hat{c}_a(f_a)$	$\tau_a(f_a)$	f_a^*
1	$3f_1^2+2f_1$	0	19.22
2	$2f_2^2+2.5f_2$	0	20.02
3	$5f_3^2+4f_3$	$3f_3+1$	0.00
4	$4.5f_4^2+3f_4$	$4f_4+1$	0.00
5	$f_5^2+2f_5$	0	19.22
6	$f_6^2+.5f_6$	0	20.02
7	$2.5f_7^2+3f_7$	0	19.22
8	$3.5f_8^2+2f_8$	0	20.02
9	$7f_9^2+5f_9$	$2f_9+2$	19.22
10	$4f_{10}^2+6f_{10}$	$10f_{10}+6$	0.00
11	$2.5f_{11}^2+4f_{11}$	$7.5f_{11}+5$	0.23
12	$4.5f_{12}^2+5f_{12}$	$1.5f_{12}+1.5$	19.79
13	$2f_{13}^2+4f_{13}$	$2f_{13}+2$	19.22
14	$f_{14}^2+3f_{14}$	$1.5f_{14}+1$	20.02
15	$4f_{15}^2+5f_{15}$	$3f_{15}+3$	13.95
16	$2.5f_{16}^2+2f_{16}$	$5f_{16}+4$	5.28
17	$3f_{17}^2+4f_{17}$	$6.5f_{17}+3$	0.00
18	$4f_{18}^2+4f_{18}$	$7f_{18}+5$	6.85
19	$3f_{19}^2+3f_{19}$	$4f_{19}+5$	5.68
20	$3.5f_{20}^2+5f_{20}$	$3.5f_{20}+4$	7.49

R_1 is assumed to have a higher demand for relief goods due to a larger population and its potential higher vulnerability to the disasters as compared to R_2. The demand for the relief item at R_1 and at R_2 is assumed to follow a uniform distribution on the intervals [25,45] and [10,20], respectively.

The unit shortage and surplus penalties at the demand points are:

$$\lambda_{R_1}^- = 10,000, \ \lambda_{R_1}^+ - 100,$$
$$\lambda_{R_2}^- = 7,500, \ \lambda_{R_2}^+ = 150.$$

The target times of delivery at demand points R_1 and R_2 are:

$$T_{R_1} = 72, \ T_{R_2} = 70.$$

Using (15.15), the T_{kp}'s are also shown in Table 15.3.

Recall that in selecting the penalties and the target times, the decision-maker takes into account such data as the population, accessibility, strategic location, etc. In addition, the decision-maker can prioritize certain paths by assigning higher coefficients in the tardiness penalty function. In this example—unlike the illustrative example—all paths are assumed to have equal tardiness penalty weights:

Table 15.3 Path definitions, target times, optimal path flows, optimal path time deviations, and optimal lagrange multipliers for the numerical example

	Path definition	T_{kp}	x_p^*	z_p^*	ω_p^*
$\mathcal{P}_{R_1}$: set of paths corresponding to demand point R_1	$p_1=(1, 5, 7, 9,13,15)$	65	13.95	53.66	321.99
	$p_2=(1, 5, 7, 9,13,16)$	64	5.28	39.23	235.39
	$p_3=(1, 5, 7,10,13,15)$	61	0.00	19.32	115.90
	$p_4=(1, 5, 7,10,13,16)$	60	0.00	4.83	28.99
	$p_5=(2, 6, 8,11,14,18)$	61	0.06	18.67	112.03
	$p_6=(2, 6, 8,12,14,18)$	64.5	6.79	43.12	258.75
	$p_7=(3, 9,13,15)$	62	0.00	56.66	339.99
	$p_8=(3, 9,13,16)$	61	0.00	42.23	253.39
	$p_9=(3,10,13,15)$	58	0.00	22.34	134.05
	$p_{10}=(3,10,13,16)$	57	0.00	7.84	47.03
	$p_{11}=(4,11,14,18)$	59	0.00	20.71	124.24
	$p_{12}=(4,12,14,18)$	62.5	0.00	45.24	271.46
$\mathcal{P}_{R_2}$: set of paths corresponding to demand point R_2	$p_{13}=(1, 5, 7, 9,13,17)$	63	0.00	13.87	83.25
	$p_{14}=(1, 5, 7,10,13,17)$	59	0.00	0.00	0.00
	$p_{15}=(2, 6, 8,11,14,19)$	59	0.13	0.00	0.00
	$p_{16}=(2, 6, 8,11,14, 20)$	60	0.04	0.00	0.00
	$p_{17}=(2, 6, 8,12,14,19)$	62.5	5.55	19.91	119.44
	$p_{18}=(2, 6, 8,12,14, 20)$	63.5	7.45	22.40	134.43
	$p_{19}=(3, 9,13,17)$	60	0.00	16.90	101.41
	$p_{20}=(3,10,13,17)$	56	0.00	0.00	0.00
	$p_{21}=(4,11,14,19)$	57	0.00	0.00	0.00
	$p_{22}=(4,11,14, 20)$	58	0.00	0.00	0.00
	$p_{23}=(4,12,14,19)$	60.5	0.00	21.96	131.77
	$p_{23}=(4,12,14, 20)$	61.5	0.00	24.48	146.85

$$\gamma_{R_1}(z) = 3\left(\sum_{p\in\mathcal{P}_{R_1}} z_p^2\right), \quad \gamma_{R_2}(z) = 3\left(\sum_{p\in\mathcal{P}_{R_2}} z_p^2\right).$$

The Euler method (cf. (15.31)–(15.33)) for the solution of variational inequality (15.24) was implemented in FORTRAN on a PC at the University of Massachusetts

Amherst. We set the sequence as $\{a^\tau\} = .1(1, \frac{1}{2}, \frac{1}{2}, \ldots)$, and the convergence tolerance was 10^{-6}. In other words, the absolute values of the differences between each pair of path flows, the path time deviations, and the Lagrange multipliers in two consecutive iterations were less than or equal to this tolerance. We initialized the algorithm by setting all variables equal to zero.

The computed optimal path flows, time deviations, and Lagrange multipliers are reported in Table 15.3 and the optimal link flows in Table 15.2.

As seen in Table 15.2, the optimal flows on links 3 and 4, i.e., the post-disaster procurement links, are zero. Hence, given the demand information and the cost and time functions on the supply chain network links, the organization would be better off by adopting the prepositioning strategy. In addition, links 10 and 11, corresponding to marine transportation of goods from the US to the affected region have zero or very small flows. Such an outcome is due to the importance of timely deliveries, and, thus, the organization needs to ship via air to minimize the lateness on the demand end. Similarly, among the distribution links, the ones representing shipments by helicopter (links 15 and 20) are assigned relatively higher loads whereas link 17 corresponding to one of the ground distribution links will not be utilized. Also, note that the optimal flows being almost equal on links 1 and 2 suggests an even split of pre-positioning of the load between the two US aid regions.

Table 15.3 points out that among the 24 paths in the supply chain network, fewer than onethird have considerable positive flows since the others involve links that are either costlier or more time-consuming. From the optimal values of time deviations on paths, one can observe that significant deviations from the target times have occurred on several paths in the network. This seems to be more of an issue in the paths connecting the origin to demand point R_1, i.e., the hypothetically more vulnerable location. Such an outcome may mandate additional investments on critical transportation/distribution channels to R_1 which can be done in accordance with the optimal values of respective Lagrange multipliers. The higher the value of the Lagrange multiplier on a path, the more improvement in time can be attained by enhancing that path which, in turn, leads to a more efficient disaster response system.

15.4.1 Variant

We then considered the following variant of the previous example. We assumed that the organization will now procure the items locally and, hence, the time functions associated with the direct procurement links 3 and 4 are now greatly reduced and are given in Table 15.4—the remainder of the input data remains as in the previous example. The computed optimal link flow pattern for this variant is also reported in Table 15.4.

As can be seen from Table 15.4, now both the storage links for pre-positioning (links 7 and 8) and for post-disaster procurement (links 3 and 4) have positive flows.

Table 15.4 Total operational costs, completion time functions, and the optimal flows on links in the variant

Link	$\hat{c}_a(f_a)$	$\tau_a(f_a)$	f_a^*
1	$3f_1^2+2f_1$	0	12.02
2	$2f_2^2+2.5f_2$	0	11.21
3	$5f_3^2+4f_3$	$.1f_3+1$	7.35
4	$4.5f_4^2+3f_4$	$.1f_4+1$	8.88
5	$f_5^2+2f_5$	0	12.02
6	$f_6^2+.5f_6$	0	11.21
7	$2.5f_7^2+3f_7$	0	12.02
8	$3.5f_8^2+2f_8$	0	11.21
9	$7f_9^2+5f_9$	$2f_9+2$	19.37
10	$4f_{10}^2+6f_{10}$	$10f_{10}+6$	0.00
11	$2.5f_{11}^2+4f_{11}$	$7.5f_{11}+5$	0.24
12	$4.5f_{12}^2+5f_{12}$	$1.5f_{12}+1.5$	19.86
13	$2f_{13}^2+4f_{13}$	$2f_{13}+2$	19.37
14	$f_{14}^2+3f_{14}$	$1.5f_{14}+1$	20.10
15	$4f_{15}^2+5f_{15}$	$3f_{15}+3$	14.04
16	$2.5f_{16}^2+2f_{16}$	$5f_{16}+4$	5.33
17	$3f_{17}^2+4f_{17}$	$6.5f_{17}+3$	0.00
18	$4f_{18}^2+4f_{18}$	$7f_{18}+5$	6.84
19	$3f_{19}^2+3f_{19}$	$4f_{19}+5$	5.72
20	$3.5f_{20}^2+5f_{20}$	$3.5f_{20}+4$	7.53

Hence, with the new data (and decision to procure locally) the organization should engage in both strategies. The optimal solution suggests to the organization how much of the relief good should be stored and in which location and how much should also be procured (and from where) once the disaster strikes.

Summary and Conclusions

In this paper, we developed a network optimization model for the supply chain management of a disaster relief (humanitarian) organization in charge of procurement and distribution of relief items to a geographic region prone to natural disasters such as earthquakes or hurricanes.

Our proposed framework spans such activities associated with disaster relief supply chains as the pre-disaster procurement, shipment, storage, post-disaster procurement, transportation, and processing, and the ultimate distribution of aid items to the people in need. Our system-optimization approach minimizes the total operational costs of the aforementioned activities subject to the uncertain demand being satisfied as closely as possible. We also utilized goal programming to minimize the time deviations on paths with respect to the pre-determined time targets for deliveries.

The disaster relief supply chain network model has the following features:

- it allows for the integration of two distinct policies by disaster relief organizations, one being pre-positioning the supplies beforehand, as well as the procurement of necessary items once the disaster has occurred;

- it includes penalties associated with shortages/surpluses at the demand points with respect to the uncertain demand, and
- it enables the decision-maker to prioritize the demand points based on the population, geographic location, etc., by assigning different time targets.

We used a variational inequality approach for the model formulation. The developed model was then verified through illustrative numerical examples followed by the sensitivity analysis of the optimal solution to the model parameters. We also used variational inequality theory to present a solution algorithm with nice computational features and solved two larger examples.

Our constructed framework can assist disaster relief organizations to plan their procurement and distribution strategies before the occurrence of unexpected natural catastrophes. The organization's management can also evaluate the timeliness of its service to the survivors of disasters under various scenarios of demand, time targets, and costs. Possible extensions of our framework may include the incorporation of uncertainty associated with link time functions, as well as additional empirical work, which might include the construction of alternative penalty functions and time completion functions. The modeling of cooperation as well as competition among disaster relief agencies could also be addressed. Our variational inequality approach in this paper provides the foundation for such extensions. Indeed, we recognize that having multiple disaster relief agencies—rather than a single organization—may be quite beneficial, provided that the collaboration among the agencies is rigorously planned well ahead of time. Moreover, here the focus was on planning for a single (the next) disaster—it would also be interesting to consider the possibility of multiple disasters in neighboring regions.

Acknowledgments This paper is dedicated to the memory of Professor Walter Isard, the founder of Regional Science, whose vision, research and scholarship, energy, kindness, and mentorship will never be forgotten.

The authors are grateful to the anonymous reviewer for helpful comments and suggestions as well as to the Editors for their work in putting this volume together.

References

Alexander D (1993) Natural disasters. Kluwer, Dordrecht

Alvendia-Quero R (2012) Linkages for disaster management. Business World Online, April 11. http://www.bworldonline.com/content.php?section=Opinion&title=Linkages-for-disaster-management&id=49763

Balcik B, Beamon BM (2005) Facility location in humanitarian relief. Int J Logist Res Appl 11 (2):101–121

Barbarosoglu G, Arda Y (2004) A two-stage stochastic programming framework for transportation planning in disaster response. J Oper Res Soc 55(1):43–53

Beamon B, Kotleba S (2006) Inventory management support systems for emergency humanitarian relief operations in South Sudan. Int J Logist Manag 17(2):187–212

Beckmann MJ, McGuire CB, Winsten CB (1956) Studies in the economics of transportation. Yale University Press, New Haven, CT

Bertsekas D, Tsitsiklis J (1989) Parallel and distributed computation: numerical methods. Prentice Hall, Englewood Cliffs, NJ

Blaikie P, Cannon T, Davis I, Wisner B (1994) At risk: natural hazards, people's vulnerability and disasters. Routledge, London

Borenstein S (2012) Climate change report: Miami, Mumbai must prepare for natural disasters now. Huffington Post, March 28. http://www.huffingtonpost.com/2012/03/29/climate-change-report-miami-mumbain1385173.html

Charles A, Lauras M (2011) An enterprise modelling approach for better optimisation modelling: application to the humanitarian relief chain coordination problem. OR Spectrum 33:815–841

Chicago Tribune (2013) Typhoon Haiyan: Desperate Philippine survivors turn to looting, November 13. http://www.chicagotribune.com/news/chiphilippines-typhoon-haiyan-20131113,0,4099086,full.story

Cho S, Gordon P, Moore JE II, Richardson HW, Shinozuka M, Chang SE (2001) Integrating transportation network and regional economic models to estimate the costs of a large urban earthquake. J Reg Sci 41(1):39–65

Dacy DC, Kunreuther H (1969) The economics of natural disasters: implications for federal policy. Free, New York, NY

Dafermos SC, Sparrow FT (1969) The traffic assignment problem for a general network. J Res Natl Bur Stand 73B:91–118

Denning P (2006) Hastily formed networks. Commun ACM 49(4):15–20

Dong J, Zhang D, Nagurney A (2004) A supply chain network equilibrium model with random demands. Eur J Oper Res 156:194–212

Dupuis P, Nagurney A (1993) Dynamical systems and variational inequalities. Ann Oper Res 44:9–42

Falasca M, Zobel CW (2011) A two-stage procurement model for humanitarian relief supply chains. J Humanitarian Logist Suppl Chain Manage 1(2):151–169

Fugate W (2012) The State of FEMA—Leaning forward: go big, go early, go fast, be smart. A report by the US Department of Homeland Security. www.fema.gov/pdf/about/state of fema/state of fema.pdf

Greenberg MR, Lahr M, Mantell N (2007) Understanding the economic costs and benefits of catastrophes and their aftermath: a review and suggestions for the U.S. federal government. Risk Anal 27(1):83–96

Grubesic TH, Matisziw TC, Murray AT, Snediker D (2008) Comparative approaches for assessing network vulnerability. Int Reg Sci Rev 31(1):88–112

Haghani A, Oh SC (1996) Formulation and solution of a multi-commodity, multi-modal network flow model for disaster relief operations. Transport Res A 30(3):231–250

Hale T, Moberg C (2005) Improving supply chain disaster preparedness: a decision process for secure site location. Int J Phys Distrib Logist Manag 35(3):195–207

Ham H, Kim TJ, Boyce D (2005) Assessment of economic impacts from unexpected events with an interregional commodity flow and multimodal transportation network model. Transport Res A 39:849–860

Huang M, Smilowitz K, Balcik B (2012) Models for relief routing: equity, efficiency and efficacy. Transport Res E 48:2–18

Isard W, Azis IJ, Drennan MP, Miller RE, Saltzman S, Thorbecke E (1998) Methods of interregional and regional analysis. Ashgate, Aldershot

Israelevich PR, Hewings GJD, Sonis M, Schindler GR (1997) Forecasting structural change with a regional econometric input-output model. J Reg Sci 37(4):565–590

Karush W (1939) Minima of functions of several variables with inequalities as side constraints. M. Sc. Dissertation. Department of Mathematics, University of Chicago, Chicago, IL

Kuhn HW, Tucker AW (1951) Nonlinear programming. In: Proceedings of 2nd Berkeley symposium. University of California Press, Berkeley, CA, p 481–492

Kunreuther H (1967) The peculiar economics of disaster. Pap Nonmarket Decis Making 3(1):67–83

Kunreuther H, Michel-Kerjan E (2012) Policy options for reducing losses from natural disasters: Allocating $75 billion. Revised version for Copenhagen Consensus 2012

Lin Y-H (2010) Delivery of critical items in a disaster relief operation: centralized and distributed supply strategies. Ph.D. thesis in Industrial and Systems Engineering, The University at Buffalo, State University of New York

Liu Z, Nagurney A (2013) Supply chain networks with global outsourcing and quickresponse production under demand and cost uncertainty. Ann Oper Res 208(1):251–289

MacKenzie CA, Barker K (2011) Conceptualizing the broader impacts of industry preparedness strategies with a risk-based input-output model. In: Lahr M, Hubacek K (eds) Proceedings of the 19th international input-output conference. Alexandria, VA, June 13–17

Mete HO, Zabinsky ZB (2010) Stochastic optimization of medical supply location and distribution in disaster management. Int J Prod Econ 126:76–84

Nagurney A (1999) Network economics: a variational inequality approach, 2nd and revised edn. Kluwer, Dordrecht

Nagurney A, Masoumi A (2012) Supply chain network design of a sustainable blood banking system. In: Boone T, Jayaraman V, Ganeshan R (eds) Sustainable supply chains: models, methods and public policy implications, vol 174, International series in operations research & management science. Springer, London, pp 49–70

Nagurney A, Qiang Q (2009) Fragile networks: identifying vulnerabilities and synergies in an uncertain world. Wiley, Hoboken, NJ

Nagurney A, Qiang Q (2012) Fragile networks: identifying vulnerabilities and synergies in an uncertain age. Int Trans Oper Res 19:123–160

Nagurney A, Ramanujam P (1996) Transportation network policy modeling with goal targets and generalized penalty functions. Transport Sci 30(1):3–13

Nagurney A, Yu M (2014) A supply chain network game theoretic framework for timebased competition with transportation costs and product differentiation. In: Butenko S, Floudas CA, Rassias ThM (eds) Optimization in science and engineering—in honor of the 60th birthday of Panos M. Pardalos. Springer, New York, NY

Nagurney A, Zhang D (1996) Projected dynamical systems and variational inequalities with applications. Kluwer, Boston, MA

Nagurney A, Thore S, Pan J (1996) Spatial market policy modeling with goal targets. Oper Res 44 (2):393–406

Nagurney A, Yu M, Qiang Q (2011) Supply chain network design for critical needs with outsourcing. Pap Reg Sci 90(1):123–143

Nagurney A, Masoumi A, Yu M (2012a) Supply chain network operations management of a blood banking system with cost and risk minimization. Comput Manag Sci 9(2):205–231

Nagurney A, Yu M, Qiang Q (2012b) Multiproduct humanitarian healthcare supply chains: a network modeling and computational framework. In: Proceedings of the 23rd annual POMS conference, Chicago, IL

Nagurney A, Yu M, Masoumi AH, Nagurney LS (2013) Networks against time: supply chain analytics for perishable products. Springer Business + Science Media, New York, NY

CBS News (2013) Typhoon Haiyan slams into northern Vietnam, November 10. http://www.cbsnews.com/news/typhoon-haiyan-slams-into-northern-vietnam

Okuyama Y (2003) Economics of natural disasters: a critical review. Paper presented at the 50th North American meeting, regional science association international Philadelphia, Pennsylvania

Okuyama Y (2004) Sequential interindustry model (SIM) and impact analysis: application for measuring economic impact of unscheduled events. In: Okuyama Y, Chang S (eds) Modeling spatial and economic impacts of disasters. Springer, Heidelberg

Okuyama Y, Hewings GJD, Sonis M (1999) Economic impacts of an unscheduled, disruptive event: a Miyazawa multiplier analysis. In: Hewings GJD, Sonis M, Madden M, Kimura Y (eds) Understanding and interpreting economic structure. Springer, Heidelberg

Oloruntoba R, Gray R (2006) Humanitarian aid: an agile supply chain? Suppl Chain Manag 11 (2):115–120
Ortuño MT, Tirado G, Vitoriano B (2011) A lexicographical goal programming based decision support system for logistics of humanitarian aid. TOP 19(2):464–479
Ortuño MT, Cristóbal P, Ferrer JM, Martin-Campo FJ, Muñoz S, Tirado G, Vitoriano B (2013) Decision aid models and systems for humanitarian logistics: a survey. In: Vitoriano B, Montero J, Ruan D (eds) Decision aid models for disaster management and emergencies, vol 7, Atlantis computational intelligence systems. Springer Business + Science Media, New York, NY, pp 17–44
Qiang Q, Nagurney A (2012) A bi-criteria indicator to assess supply chain network performance for critical needs under capacity and demand disruptions. Transport Res A 46(5):801–812
Reggiani A, Nijkamp P (2009) Simplicity in complex spatial systems: introduction. In: Reggiani A, Nijkamp P (eds) Complexity and spatial networks: in search of simplicity. Springer, Heidelberg
Rose A (2009) Economic resilience to disasters. Final report to Community and Regional Resilience Institute, CARRII Research Report 8, Oak Ridge National Lab, Tennessee
Rose A, Liao S-Y (2005) Modeling regional economics resilience to disasters: a computable general equilibrium analysis of water service disruptions. J Reg Sci 45(1):75–112
Rose A, Benavides J, Chang SE, Szczesniak P, Lim D (1997) The regional economic impact of an earthquake: direct and indirect effects of electricity lifeline disruptions. J Reg Sci 37(3):437–458
Rottkemper B, Fischer K, Blecken A (2012) A transshipment model for distribution and inventory relocation under uncertainty in humanitarian operations. Socioecon Plann Sci 46:98–109
Schultz C, Koenig K, Noji E (1996) A medical disaster response to reduce immediate mortality after an earthquake. N Engl J Med 334:438–444
Sheppard K (2011) 2011: The year of the natural disaster. Mother Jones, June 22. http://motherjones.com/mojo/2011/06/our-disaster-disaster
Sheu J-B (2010) Dynamic relief-demand management for emergency logistics operations under large-scale disasters. Transport Res E 46:1–17
Simpson N, Hancock P (2009) The incident commander's problem: resource allocation in the context of emergency response. Int J Serv Sci 2(2):102–124
Tomasini R, Van Wassenhove L (2009) Humanitarian logistics. Palgrave Macmillan, Basingstoke
Tzeng G-H, Cheng H-J, Huang T (2007) Multi-objective optimal planning for designing relief delivery systems. Transport Res E 43(6):673–686
UNHCR (2007) United Nations High Commissioner for Refugees UNHCR Handbook for emergencies, 3rd edn
US Department of Homeland Security (2012) National Preparedness Report. March 30. https://www.fema.gov/library/viewRecord.do?id=5914
USAID (2005) United States Agency for International Development, Office of Foreign Disaster Assistance, Field operations guide for disaster assessment and response, Version 4.0
Vitoriano B, Ortuño M, Tirado G, Montero M (2011) A multi-criteria optimization model for humanitarian aid distribution. J Global Optim 51:189–208
Vitoriano B, Montero J, Ruan D (eds) (2013) Decision aid models for disaster management and emergencies, vol 7, Atlantis computational intelligence systems. Springer Business + Science Media, New York, NY
Walton R, Mays R, Haselkorn M (2011) Defining "fast": factors affecting the experience of speed in humanitarian logistics. In: Proceedings of the 8th International ISCRAM Conference, May, Lisbon, Portugal
West CT, Lenze DG (1994) Modeling the regional impact of natural disaster and recovery: a general framework and an application to Hurricane Andres. Int Reg Sci Rev 17(2):121–150
Zhenling L (2009) Integrated supply chains of the natural disaster relief substances. International Conference on Management and Service Science (Mass'09), Wuhan, China

Regional Dynamics Under Adverse Physical and Behavioral Shocks: The Economic Consequences of a Chlorine Terrorist Attack in the Los Angeles Financial District

16

James Giesecke, William Burns, Adam Rose, Tony Barrett, and Marnie Griffith

Abstract

Emergency management decision makers must make contingency plans for a wide range of threat scenarios. In undertaking ex-ante cost/benefit evaluations of contingency plans, they must understand the economic benefits of threat deterrence and reduction. Appropriate emergency response and recovery activities ex-post can attenuate business interruption (BI) impacts. Regional economic modeling can provide quantitative input to these evaluations. In this paper, we

The authors are, respectively, Research Professor, Centre of Policy Studies, Victoria University, Melbourne, Australia (CoPS); Research Scientist, Decision Research, Eugene, Oregon, and Faculty Affiliate, Center for Risk and Economic Analysis of Terrorism Events (CREATE); Research Professor, Price School of Public Policy, and Coordinator for Economics, CREATE, University of Southern California, Los Angeles; Senior Risk Analyst, ABS Consulting, Arlington, Virginia, and Director of Research, Global Catastrophic Risk Institute; and Research Fellow, CoPS.

J. Giesecke (✉)
Centre of Policy Studies, Victoria University, 10/300 Flinders Street, Melbourne, VIC, Australia
e-mail: James.Giesecke@vu.edu.au

W. Burns
Decision Research, 1201 Oak Street Suite 200, Eugene, OR 97401, USA
e-mail: bburns@csusm.edu

A. Rose
Price School of Public Policy, RGL 230, 650 Childs Way, Los Angeles, CA 90089, USA

University of Southern California (USC), Los Angeles, CA 90089, USA
e-mail: adam.rose@usc.edu

T. Barrett
ABS Consulting, 1525 Wilson Blvd., Suite 625, Arlington VA 22209, USA
e-mail: abarrett@absconsulting.com

M. Griffith
Centre of Policy Studies, Monash University, Melbourne, VIC, Australia
e-mail: Marnie.Griffith@buseco.monash.edu.au

P. Nijkamp et al. (eds.), *Regional Science Matters*, DOI 10.1007/978-3-319-07305-7_16,

use a large-scale dynamic regional computable general equilibrium (CGE) model of the Los Angeles economy to perform an economic consequence analysis of a terrorist attack with chlorine gas. We divide the event's direct effects into resource losses (injuries, BI) and behavioral reactions stemming from fear. We provide a decomposition of aggregate economic effects in terms of these various loss components, allowing us to elucidate the relative sizes of potential loss channels. We also discuss the effect of geographic shifts of economic activity within the affected region and in neighboring regions in estimating the losses. Our analysis can assist risk managers in developing plans for pre-event mitigation and post-event resilience.

16.1 Introduction

By elucidating the benefits of threat reduction and post-event response, economic consequence analysis can be an important input to the contingency planning of emergency management decision makers. In this paper we show how dynamic regional CGE modeling can elucidate behavioral and dynamic elements relevant to such planning. We considered several modeling approaches to perform our analysis. Methodologies for analysing the regional economic consequences of specific threats have developed from the use of regional input-output models to investigate the direct and indirect effects of physical destruction, transport disruption and BI (Gordon et al. 1998) to applications of regional CGE models to also analyse the consequences of behavioral responses (Rose et al. 2009).

Input-output models are practical tools but lack behavioral content, the crux of the issue at hand. They also lack the ability to effectively model the workings of markets through factor and product price changes. Econometric models are a valuable approach, especially for forecasting but less so for simulation. They are highly dependent on time series or cross-sectional data, and less conducive to the incorporation of simulation results. Moreover, econometric models typically are based on aggregate relationships, as opposed to micro behavioral responses of consumers, firms, and providers of factor services. Agent-based modeling on the other hand is especially adept at analyzing individual decision-makers and their interactions. However, they are not yet able to scale up to the macro level and to analyze the full effects of the workings of markets. As we expand upon below, CGE models have behavioral content, are able to mimic the role of markets and prices, and can readily be recalibrated with simulation data. Moreover, they have been successfully applied to topics such as the focus of this chapter [see, e.g., Giesecke et al. 2012)].

Regional CGE models embody much detail describing economic structure (e.g. production technologies, and regional resource constraints); the behavior of economic agents (e.g. household preferences, and investor return requirements); and policy variables (e.g. the instruments of government taxing and spending). This detail has allowed these models to be applied to a wide range of issues relating to the structural, behavioral and policy drivers of regional growth and decline and the consequences of policy interventions. While regional CGE modeling has a three decade history of providing input to evidence-based policy making, research exploiting the exogenous structural, behavioral and policy detail in these models to carry shocks describing the direct effects of catastrophic events is more recent.[1] Early applications examined natural disasters. Rose and Liao (2005) examined the effects of water utility disruption following an earthquake, noting the role of pre-event mitigation and post-event resilience in influencing potential impacts. More recently, the CGE method has been turned to terrorism threats. Rose et al. (2009) examined the consequences of the 2001 World Trade Center attacks, investigating BI costs, and behavioral impacts via reduced air travel. The importance of behavioral responses within the overall economic consequences of a terrorist event was examined more generally in Giesecke et al. (2012). Investigating the consequences of a radiological dispersion device (RDD) attack in the financial district of downtown Los Angeles (LA), they compare economic costs arising from behavioral responses with the direct resource costs arising from casualties, property damage and BI.

In this paper we use a dynamic regional CGE model to perform a consequence analysis of a terrorist chlorine attack with a focus on behavioral impacts and dynamic outcomes. We perform a decomposition of the event's various loss components to explain and compare behavioral impacts with more standard resource loss effects over time. Given the potential for multiple economic loss channels, risk managers would be well advised to distinguish the many types of consequences of a terrorist attack or natural disaster. Unfortunately, most consequence analyses have not offered such decomposition analyses of the broad range of components. Furthermore, whereas previous studies have used comparative static models, our model is dynamic. A dynamic model offers a number of benefits relative to its comparative static counterpart: first, by providing a more plausible time-path for key regional stock variables (like population and capital) by allowing for gradual adjustment in regional wages, migration, and investment; and second, by facilitating a better matching of time-specific inputs to the CGE model with the projected time-path of peak and decay in behavioral responses to a particular hazard or disaster.

To provide a comparison with the scenario in Giesecke et al. (2012), we choose the same downtown LA area as the attack site: the heart of the financial district. The attack involves detonation of a chlorine storage tank, leading to the formation of a large chlorine plume. Consistent with the approach outlined in Giesecke et al., we divide the event's direct effects into two broad sets of inputs to the CGE model: (1) reduction in effective resource supply (the resource loss effect) and (2) shifts in the perceptions of economic agents (the behavioral effect). The resource loss effect describes the event's physical destructiveness, subsuming such direct impacts as deaths and injuries, BI during evacuation, clean-up and repair, and medical

[1] Recent reviews of regional CGE model applications to the analysis of public policy and regional development are provided by Partridge and Rickman (2010) and Giesecke and Madden (2013).

expenses. These shocks represent either a reduction in effective resource supply to the regional economy, directly reducing regional GDP, or in the case of medical expenses, a diversion of regional income to cover otherwise avoidable costs. The behavioral effect relates to economic adjustments stemming from fear and risk perception on the part of firms, households and government. Behavioral effects generate regional economic loss additional to the resource loss effects. Employees may require compensating higher wages to work, businesses may require compensating higher returns to invest, and consumers may switch their preferences away from goods produced in the affected region. Such responses will increase production costs in the affected region, while simultaneously reducing demand for the region's output.

Scientists, public officials and laypeople have long recognized society's vulnerability to large-scale disasters. This has been especially evident with respect to disasters involving potentially toxic substances such as radiological and chemical materials. Hazards that are poorly understood and that are potentially catastrophic carry for the public a high risk signal value (Slovic 1987). In fact, such hazards tend to inspire fear and heightened perceived risk well beyond what risk experts would expect based on their assessed likelihood of occurrence. Public reaction to radiological risks has been well documented since the accident at Three Mile Island, but these same concerns extend to chemical hazards as well. Kraus et al. (1992) studied public judgments of chemical risks and found that laypeople: (1) largely view chemicals negatively and see little benefit to them; (2) believe harmful exposure to potentially toxic chemicals is all or nothing rather than dose specific and (3) feel any contact with toxic chemicals is contaminating. These findings are important for understanding the consequences of a chlorine attack. Chlorine is a highly toxic chemical capable of killing large numbers of people. It also dissipates quickly if released in an outdoor area; hence, hours later it poses little risk. However, as noted above, toxic chemicals are thought by the public to contaminate objects and areas they touch and so people may be reluctant to resume normal activities in an area impacted by such an attack regardless of what officials and experts believe. This contrast between expert and public assessment poses a challenge for local authorities wishing to communicate about public safety following such events. Public reaction to different hazards also poses a challenge to risk managers who need to gauge the economic impact of a potential mishap.

16.2 Attack Description

16.2.1 Attack Scenario and Direct Loss Estimation

Our scenario is based on a combination of DHS' National Planning Scenario 8, "Chemical Attack—Chlorine Tank Explosion" (HSC 2004) and the chlorine tank truck attack scenarios of Barrett and Adams (2011). In our scenario, the attacker drives a tank truck of pressurized chlorine into the financial district of downtown LA (zip code 90071). Detonation of the tank releases tons of chlorine, creating a cloud that is poisonous by inhalation. The attack causes 182 fatalities, 104 serious injuries, and 1,040 minor injuries. Significant chlorine contamination covers approximately 36 city blocks, as in the RDD attack scenario of Giesecke et al. (2012). The number of deaths is selected to be similar to the 180 fatalities of the RDD scenario, and the numbers of serious and minor injuries are scaled proportionally to the number of fatalities in the HSC (2004) chlorine scenario. Chemical vapor settling on outdoor and indoor surfaces causes authorities to close the area until they can complete enough decontamination and remediation.

Official decisions on when to reopen an area after chemical contamination may not be simply dictated by existing safety standards. Decisions on clean-up options may involve tradeoffs between public safety levels and the costs and economic impacts of decontamination. With BI a cost of chlorine attack, we discussed our scenario with City of Los Angeles public officials, to understand their assessment of the range of clean-up options. Less decontamination would typically be required after a chlorine attack than after an RDD attack. Surfaces affected by chlorine potentially could be re-opened after dilution with water and neutralization with sodium bicarbonate. More extensive technical decontamination could take longer, as could law enforcement investigation. Our discussions with LA officials indicated that the decision to reopen would take into account both the desires of local businesses (who might press for rapid reopening, to minimize business impacts) and the concerns of the public (who desire not only access but also safety in the face of an unusual hazard). To reflect a medium-cost chlorine clean-up effort, in Sect. 16.4 we model a 3-day shutdown of zip code 90071 (one-tenth of the 30 day shutdown for the RDD event modeled in Giesecke et al. 2012).

16.2.2 Medical Expenses and Lost Labor Inputs Via Fatalities and Injuries

We assume that the incident generates medical expenses of $4.0 million. We develop this from assumptions for case-specific expenditures for the HSC (2004) categories of fatalities, serious injuries and minor injuries.[2] We assume that "serious injury" means hospitalized and "minor injury" means emergency room or clinic outpatient visit (as in Van Sickle et al. 2009). Our estimates of per person treatment costs for these categories are based on the influenza treatment costs for the 18–49 age group in Molinari et al. (2007), who distinguish treatment costs by risk of serious complications, with 25 % of the population assumed to be at high risk of serious complications, and 75 % assumed to be non-high-risk. We noted that the definitions and proportions given by Molinari et al. are consistent with the chlorine exposure "vulnerable population" definitions and proportions given by Withers and Lees (1985), i.e., with 25 % of the chlorine scenario population in the vulnerable/ high risk category. As such, we used the Molinari et al. complication risk

[2] Calculations are available from the authors on request.

weightings in calculating per person treatment costs for our case categories. We assumed that one in nine fatalities will first be hospitalized, based on the pattern of the Graniteville chlorine incident (Van Sickle et al. 2009). Patients who die are assumed to have incurred "high risk" hospitalization costs in line with Molinari et al.

Based on 182 fatalities and 1,144 injuries, we estimate the resource loss associated with the dead and the injured using the method outlined in Giesecke et al. (2012). This generates $9.5 million (m.) of lost labor inputs, comprising $1.9 m. from injuries and $7.6 m. from fatalities.

16.3 LA-DYN: A Dynamic CGE Model of the LA-County Economy

16.3.1 Overview of the Structure of LA-DYN

LA-DYN is a dynamic CGE model of the LA-County economy.[3] The model's theoretical structure begins with the comparative-static CGE model ORANI-LA, an LA-County implementation of the single U.S. region model ORANI-R, documented in Giesecke (2011) and used in Giesecke et al. (2012) to examine the short-run and long-run consequences of an RDD event. ORANI-LA is a single-region sub-national variant of the well-known single country models ORANI-G (Horridge 2003) and ORANI (Dixon et al. 1982). The model is implemented using IMPLAN data for LA-County (Minnesota IMPLAN Group 1997) and relevant parameter values from the large-scale CGE model of the U.S., USAGE.[4]

Production and capital formation is modeled for 72 sectors, along with commodity- and agent-specific demands for 15 "margin" commodities.[5] Consistent with Isard et al. (1998), who argue for the importance of modeling transport margins (the cost of transportations, exclusive of the goods transported) in regional CGE models, 3 of the 15 margins relate to transport services differentiated by mode.[6] Decision-making by firms and households is governed by optimizing behavior. Each representative industry is assumed to minimize costs subject to

[3] The model is solved using GEMPACK (Harrison and Pearson 1996).

[4] IMPLAN is a widely-used and accepted resource for small region U.S. input-output tables. A balanced input-output table is required as an initial solution to the LA-DYN system of equations. USAGE is a detailed, dynamic CGE model of the U.S. It has been developed at the Centre of Policy Studies in collaboration with the U.S. International Trade Commission. Prominent applications of USAGE by the U.S. International Trade Commission include USITC (2004 and 2007).

[5] The starting point for the LA-DYN model is a comparative static LA County model, implemented with IMPLAN data at the finest level of disaggregation—440 sectors. While computationally uncomplicated for comparative statics, 440 sectors is not practical for dynamic modeling. At 72 sectors, our aggregation is based on the standard IMPLAN 64 sector aggregation scheme, but with expanded detail for trade margins, consistent with a model with a business district focus.

[6] Namely, truck transport, air transport and other transport.

nested constant returns to scale production technologies and given input prices. Household commodity demands are modeled via a representative utility-maximizing household. Units of new industry-specific capital are assumed to be cost minimizing combinations of commodities sourced from the local region, the rest of the U.S. and overseas. Imperfect substitutability between local, rest-of-U.S. and foreign varieties of each commodity are modeled via CRESH functions.[7] Inter-regional and foreign export demands for local commodities are modeled via commodity- and destination-specific constant elasticity export demand schedules. The model includes the consumption of commodities by state and federal government, funded by direct and indirect taxation instruments. Commodity markets are assumed to clear and to be competitive. Purchasers' prices differ from basic prices by the value of indirect taxes and margin services.

Three dynamic processes distinguish LA-DYN from ORANI-LA: two describing stock/flow relationships between capital and investment, and between population and migration; and one describing a process of lagged adjustment in regional wages to changes in regional labor market conditions. Broadly, these mechanisms draw together the investment theory of Dixon and Rimmer (2002), and the regional labor market and migration theory of Giesecke and Madden (2013). Before describing these mechanisms, we first distinguish two types of dynamic simulation: baseline and counterfactual (Dixon and Rimmer 2002; 15). The baseline simulation is a business-as-usual forecast of the LA County economy. The counterfactual simulation is identical to the baseline simulation in all respects other than the addition of shocks describing the issues under analysis (in this case, chlorine release). The distinction between baseline and counterfactual is important for two reasons. First, the theory governing the regional employment rate relies on the distinction. Second, we present model results as percentage deviations in the values of variables in the counterfactual simulation away from their corresponding values in the baseline.[8]

16.3.2 Investment and Capital Accumulation

LA-DYN carries the assumption that investment undertaken in industry i in year t becomes operational at the beginning of year $t + 1$. That is:

---

[7] This specification allows us to use a broad range of substitution possibilities among inputs (Hanoch 1971).

[8] In this paper we are concerned with reporting the impact of a chlorine gas attack, not with the baseline forecast for the LA economy. As such, we report all results in terms of deviations in the values of variables in the attack scenario away from their baseline (no attack) forecast values. While details of the baseline are unimportant for the present application, this need not be the case for all simulations. Dixon and Rimmer (2013) note that baseline details can be important when: (1) the aim is to supply CGE forecasts to business or government; (2) the counterfactual shocks are heavily weighted towards very fast- or slow-growing sectors; (3) the focus is an evaluation of the adjustment costs of policy change. These are not relevant considerations in the present application.

$$K_{i,t+1} = K_{i,t}(1 - D_i) + I_{i,t} \tag{16.1}$$

where

$K_{i,t}$ is industry i's capital stock in year t;

D_i is industry i's depreciation rate; and,

$I_{i,t}$ is the quantity of new capital created for industry i during year t.

Investment is a function of the expected rate of return on capital, via:

$$K_{i,t+1}/K_{i,t} - 1 = F_{i,t}[EROR_{i,t}] \tag{16.2}$$

where

$EROR_{i,t}$ is the expected rate of return on investment in industry i in year t; and

$F_{i,t}$ [·] is an increasing function of the expected rate of return.

In implementing (Eq. 16.2), we assume that $F_{i,t}$ takes the inverse-logistic form described in Dixon and Rimmer (2002, pp. 190–195). This is the functional form used in the USAGE model, whose investment specification is discussed in Dixon and Rimmer (2005).

16.3.3 Regional Wage Adjustment

LA-DYN allows for limited deviations in the short-run regional wage away from its baseline forecast values. With short-run regional populations also sticky (see below), short-run labor market pressures are mainly manifested as short-run deviations in the regional employment rate. More explicitly, the path of the regional wage in the chlorine attack simulation is governed by:

$$\left(W_t^{(C)}/W_t^{(B)} - 1\right) = \left(W_{t-1}^{(C)}/W_{t-1}^{(B)} - 1\right) + \alpha\left(ER_t^{(C)}/ER_t^{(B)} - 1\right) \tag{16.3}$$

where

$W_t^{(B)}$ and $W_t^{(C)}$ are year t values for regional nominal wages in the baseline and counterfactual (chlorine) simulation respectively

$ER_t^{(B)}$ and $ER_t^{(C)}$ are regional employment rates (1—the unemployment rate) in the baseline and counterfactual simulations respectively; and

α is a positive parameter.

With (Eq. 16.3) activated in the chlorine simulation, the deviation in the regional wage grows (declines) as long as the regional employment rate remains above (below) its baseline level. Equation 16.3 represents an implementation at the regional level of the national sticky wage mechanism described in Dixon and Rimmer (2002, p. 205) and implemented in the USAGE model of the U.S. We choose a value for α consistent with that in USAGE.[9] In practice, this ensures that

[9] We set $\alpha = 0.6$.

the bulk of the regional employment rate effects of a shock in year t are eliminated by year $t+5$.

16.3.4 Interregional Migration

While (Eq. 16.3) gradually returns the regional employment rate to baseline, this is not the same as assuming that regional employment returns to baseline. We allow for endogenous movements in the size of the regional workforce via interregional migration, using a single-region variant of the migration theory described in Giesecke and Madden (2013). This models interregional migration as a function of per-capita regional income. Hereafter, we call the measure of income relevant to the migration decision "migration income", defined as:

$$Y_t^{(M)} = W_t \cdot ER_t \tag{16.4}$$

where
$Y_t^{(M)}$ is migration income
W_t is the LA-County real consumer wage; and
ER_t is the LA-County employment rate.

We define disequilibrium in the regional migration income measure away from a level consistent with a trend rate of net interregional migration via:

$$Y_t^{(M)} = Y_t^{(Diseq)} \cdot F_t^{(M)} \tag{16.5}$$

where
$Y_t^{(Diseq)}$ is a measure of disequilibrium in migration income
$F_t^{(M)}$ is a shift-variable for calibrating (Eq. 16.5).

A plausible initial parameterisation of (Eq. 16.5) is $Y_t^{(Diseq)} = 1$ and $F_t^{(M)} = Y_0^{(M)}$, where $Y_0^{(M)}$ is the initial (base period) value for $Y_t^{(M)}$. As we shall see, with such a parameterisation of (Eq. 16.5), we have assumed that the base period migration income measure is consistent with the trend level of net interregional immigration.

In (Eq. 16.5), $F_t^{(M)}$ will normally be exogenous, while $Y_t^{(M)}$ is determined by (Eq. 16.4) on the basis of regional labor market conditions (that is, outcomes for W_t and ER_t). Hence, (Eq. 16.5) determines $Y_t^{(Diseq)}$. We assume that a rise in $Y_t^{(Diseq)}$ will generate a rise in the net rate of inter-regional immigration of persons of working age (IMR_t) via:

$$IMR_t / IMR^{(Trend)} = G_t\left[Y_t^{(Diseq)}\right] \tag{16.6}$$

We use an inverse logistic function to describe G (Fig. 16.1). For modeling a region's net immigration rate, this function has three useful properties. First, it allows us to limit the minimum and maximum values for IMR_t within historically-

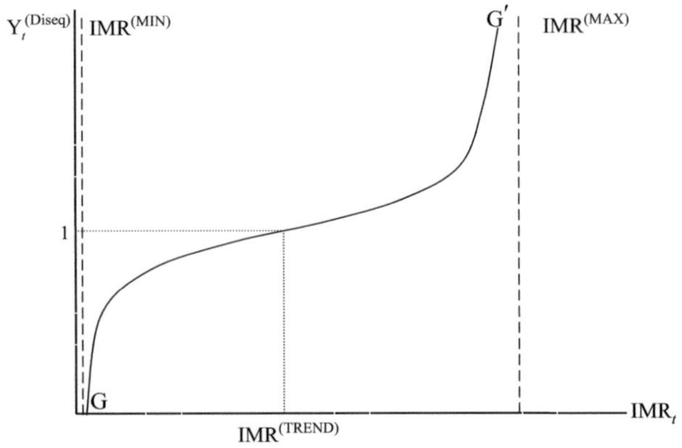

Fig. 16.1 The regional net immigration rate

observed bounds ($IMR^{(MIN)}$ and $IMR^{(MAX)}$ respectively). Second, by establishing a positive relationship between IMR_t and $Y_t^{(Diseq)}$, the function reflects the insight of Whalley and Trella (1986) that, because individuals differ in their preferences for residing within given regions, greater movements in relative regional wage rates are required to bring forth higher levels of interregional migration. Third, it allows us to model the autonomous component of net interregional immigration by establishing a trend value for IMR_t ($IMR^{(TREND)}$) consistent with a value for $Y_t^{(Diseq)}$ of 1.

To translate movements in IMR_t to movements in net interregional immigrant numbers, we multiply IMR_t by the year t regional working age population. The resulting population flow affects the start-of-year working age population in year $t+1$.

16.3.5 Economic Closure for the LA-DYN Model Under a Chlorine Attack Scenario

We outline here the main elements of the model's closure as they relate to factor markets and the expenditure side components of regional GDP.[10]

Within any given year of the simulation, we assume that the regional working age population is given. However, the regional working age population adjusts through time to changes in the migration income measure, in accordance with the migration theory described in Sect. 16.3.1.

Regional employment is the product of the working age population and the regional employment rate. In the short-run, the employment rate is endogenous, and

[10] This follows closely the closure described in Giesecke and Madden (2013, pp. 443–445), establishing an environment in which short-run capital stocks, population and the real wage are sticky (and rates of return, regional income relativities, and the employment rate are flexible), transitioning to a long-run environment in which rates of return, regional income relativities and employment rates are sticky.

the regional wage is sticky. As such, short-run labor market pressures are mainly reflected in changes in the employment rate. In the medium- to long-run, regional wage adjustment drives the employment rate towards its baseline level. As such, deviations in long-run employment are largely determined by deviations in the long-run working age population.

Within any given year, industry-specific capital stocks are given. Hence, in the short-run, demand-side pressures on capital stocks are largely reflected in changes in rates of return. Over time, movements in rates of return induce changes in investment, and eventually capital supply, that gradually return rates of return towards normal levels.

We assume that household consumption spending is a fixed proportion of household income. That is, the household savings rate is exogenous and unshocked. A potential behavioral effect that we have not modeled is a short-run rise in precautionary savings. Regional and federal government real public consumption spending is exogenous.

16.4 Simulation Design

16.4.1 Direct Resource Loss Effects

We summarize the direct resource loss effects of the scenario at:

1. 182 deaths, leading to lost labor input in the event year of \$7.6 m., measured in wage bill terms.
2. Lost labor input due to injuries in the event year valued at \$1.9 m., measured in wage bill terms.
3. BI, arising from evacuation and clean-up of 90071, valued at \$140 m. of lost output in the event year.[11]
4. Medical expenses of \$4.0 m. in the event year.

We model the direct impact of deaths as a reduction in regional population sufficient to withdraw \$7.6 m. of labor from the LA County economy. We model lost labor input from injury as a decrease in regional labor productivity, calibrated to reduce effective labor input during the event year by \$1.9 m. for a given level of employment. In translating this to industry-specific changes in labor productivity, we use Bureau of Labor Statistics (BLS) Quarterly Census of Employment and Wages data together with sectoral ratios for LA County as a whole from the LA-DYN database, to estimate sector-specific values for output and payments to

[11] Giesecke et al. (2012) estimate the value of gross output in 90071 at \$16.8 billion. We assume the event causes a three day shutdown of activity in 90071. This corresponds to \$140 m. of lost output.

labor and capital for 90071.[12] We use the 90071 sectoral wage bill estimates to translate the $1.9 m. labor productivity loss into sector-specific shocks.

We model BI as deterioration in industry-specific, all-input-using technical efficiency. As discussed in the preceding paragraph, we use BLS and LA-DYN data to estimate output by sector for zip code 90071. BI results in 3 days of lost output in 90071. We translate this to sector-specific productivity losses for LA County as a whole based on industry-specific 90071 shares in LA County output by industry.

We model the $4.0 m in direct medical expenditure as an expansion in final demand for medical services. The medical resources used to treat the victims of the chlorine attack form part of the attack's economic costs. To make this explicit, we require that the medical spending be financed via a decrease in general government consumption spending.

16.4.2 Behavioral Effects

Our estimates of the behavioral inputs to the economic model are based on results from a nationwide survey in which 624 people participated. Our approach to hazard scenario construction and survey design was consistent with well-documented methods of querying people about perceived risk and risk-related behaviors (Fischhoff et al. 1978; Slovic 1987; Burns and Slovic 2010; Giesecke et al. 2012). Respondents were drawn from a panel developed by Decision Research in 2008 based on funding from the National Science Foundation.[13] Panelists were recruited online by a number of means (e.g., Google ads) to gather a diverse assortment of Americans. We statistically reweighted our sample using U.S. Census data to better reflect the U.S. population. In this paper we focus on the sub-group of panelists who reviewed a scenario with 182 chlorine deaths to invite comparisons of the economic impacts with the RDD attack, which had a similar number of deaths. The sample size for this group was 286.

Respondents were invited to participate and were given 1 week to review our chlorine scenario and answer our online survey.[14] Prior to reading the scenario respondents were asked their perceptions regarding the likelihood and severity of such an attack during the next 12 months. They then read a 500 word news article with headline "**Chlorine Bomb Rocks Financial District of Los Angeles! 182 Dead and Hundreds Potentially Exposed to Poison Gas as Mayor Requests Downtown to Seek Shelter for Hours".** The article included a statement by the Mayor describing the details of the attack, the fact that the contaminated area had

[12] Notes 12 and 13 of Giesecke et al. (2012) estimate sector-specific values for output and payments to labor and capital for 90071.

[13] The format for describing the survey results here conforms very closely with Giesecke et al. (2012) to facilitate behavioral comparison between the chlorine and RDD attacks.

[14] Available at http://www.decisionresearch.org/pdf/ChorineAttackAug2012.pdf

been blocked off and safety officials were assessing the risk (see online appendix for details). Respondents were then asked a number of questions pertaining to their initial reaction to the attack. Following this, respondents read a follow-up report describing the findings and actions of safety officials and the mayor over the previous 3 days with headline **"Chlorine Gas Levels Throughout Los Angeles Pose Little Threat says Panel of Health Officials! Today the Mayor Received a Reassuring Report from a Team of Chemical Experts Regarding Long-term Health Risks. The Downtown to Re-Open".** The report emphasized that the mayor received reassuring news about the area's safety and that the downtown area is cleared to resume normal activities.

Initial questions involved perceived risk (very low risk to very high risk), fear (not fearful to very fearful), attention to media coverage (none to more than 8 h/day during the first week), worry (not worried to very worried) and trust in first responders (no trust to very high trust). These were followed by questions pertaining to respondents' willingness to buy products or work near the financial district. Respondents were asked to indicate how long they would delay buying products, using services or working in this area ranging from no intention to delay to indefinite delay. Specifically, we focused on professional services, food products unique to LA, electronic products and taking a vacation near the downtown area. We also asked what incentives might be needed (e.g. product discounts, higher wages) to immediately resume buying products or working near the downtown area, Table 16.1 reports survey results for respondents who evaluated the chlorine attack.

Notice that willingness to use services, purchase goods, or work in downtown LA immediately following the Mayor's "all clear" is greatest for professional services and electronic products and least for food and vacation (first percentage in each cell). Using these responses we calculated the percentage of people who remain unwilling to use services, buy products or work near the financial district at different points in time (percentage in parentheses). For example, after 6 months, at least 18 % of respondents indicated they still would not consider economic transactions in the financial district. The greatest reticence involved food products and vacations with 11 and 14 % respectively saying they would never buy food or vacation near the financial district in the future. In comparison, Giesecke et al. (2012) investigated an RDD attack on the same area with similar casualties. The authors found that after 6 months, at least 41 % of respondents indicated they still would not do business in the financial district. The greatest reservation again involved food products and vacations with 17 and 12 %, respectively, saying they would never buy food or vacation near the financial district in the future. As we describe below, we use the percentages within each column to approximate the decay in perceived risk over time for each product type or job.

We also asked respondents what percentage price reduction or wage increase they would require to consider consuming or working in downtown LA immediately following the Mayor's "all clear" (see Table 16.2). Notice that almost half of the respondents said they would not buy food near the financial district right away

Table 16.1 Survey results depicting percentage wait times before doing business in the financial district

Wait time (at least)	Professional services	Vacation	Food	Electronic products	Job
Right after "All Clear"	45 %[a] (55 %)[b]	20 % (80 %)	22 % (78 %)	45 % (55 %)	38 % (62 %)
One month	26 % (29 %)	24 % (56 %)	18 % (61 %)	20 % (36 %)	25 % (37 %)
Six months	11 % (18 %)	18 % (38 %)	19 % (42 %)	14 % (22 %)	9 % (28 %)
Twelve months	8 % (9 %)	15 % (23 %)	19 % (23 %)	8 % (13 %)	7 % (21 %)
Thirty six months	4 % (5 %)	7 % (16 %)	8 % (15 %)	7 % (6 %)	8 % (13 %)
Sixty months	2 % (4 %)	2 % (14 %)	4 % (11 %)	1 % (6 %)	1 % (11 %)
Never in future	4 %	14 %	11 %	6 %	11 %
Sample size (*N*)	286	286	287	285	287

[a]Percentage indicating they would not wait to do business in the financial district
[b]Percentage who are still waiting

Table 16.2 Levels of required incentives to consume or work in the financial district right away

Required incentive	Professional services	Vacation	Food	Electronic products	Job
0 %	35 %	18 %	18 %	35 %	25 %
2 %	4 %	3 %	3 %	3 %	3 %
4 %	10 %	2 %	4 %	4 %	5 %
8 %	0 %	8 %	5 %	7 %	7 %
15 %	10 %	10 %	7 %	9 %	11 %
25 %	8 %	15 %	8 %	13 %	15 %
50 %	17 %	17 %	8 %	9 %	8 %
100 %	NA	NA	NA	NA	5 %
No amount is enough	18 %	28 %	48 %	21 %	20 %
Mean %[a]	15	20	14	13	20

[a]Average required % incentive for those who would consider interacting with the financial district right away

for any level of incentive. Whereas less than one fifth indicated they would not accept any incentive to use professional services right away. For those willing to accept some level of incentive the average percentage required is given in the bottom row of Table 16.2 and range from 13 % for electronic products to 20 % for vacations. Incentive for jobs may have been a little higher because respondents had an opportunity to select a 100 % wage increase. In the RDD study, again almost half of the respondents indicated they would not buy food near the financial district right away for any level of incentive. By comparison, less than one-fourth indicated they would not accept any incentive to use professional services right away. For those willing to accept some level of incentive the average percentage required range from 15 % for food and electronic products to 19 % for vacations.

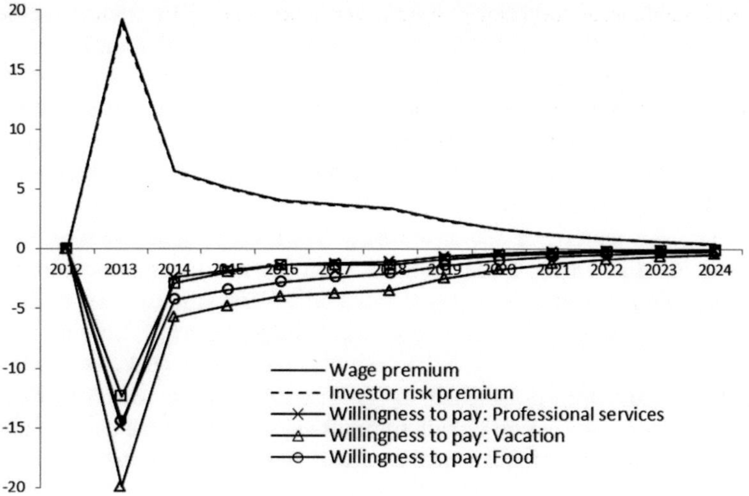

Fig. 16.2 Time paths for behavioral shocks in zip code 90071 (% deviation from baseline)

Figure 16.2 reports our translation of the Tables 16.1 and 16.2 results into shocks to behavioral variables related to activity in zip code 90071. For the event year (year 1) our behavioral shocks are based on the survey results for the mean incentives required to immediately resume business in the financial district (final row of Table 16.2). We interpret the results for incentives required to purchase professional services, vacation, food, and electronic products as elucidating vertical shifts in willingness to pay for these products, scaled to reflect 90071's share in LA County output of these commodities.[15] We interpret the job incentive as the wage premium required to maintain a given labor supply to the affected region. We implement this in LA-DYN as time-specific shocks to $F^{(M)}$, scaled to reflect 90071's share of the LA-County wage bill.

In calculating the time paths for our model shocks, we begin by noting that the Table 16.2 results for mean required incentives relate to the event year only. Our simulation runs over 12 years. To infer behavioral shocks for the remaining years, we use the decay rates for the "percentage who are still waiting" reported in Table 16.1. In calculating the decay rate between 2013 and 2014, we use Table 16.1

[15] LA-DYN models demand for LA County commodities by agents in three regions: LA County, the rest of the U.S., and the rest of the world. Demands by agents located outside LA County are modeled via constant elasticity demand schedules. Following Giesecke et al. (2012), we model declines in willingness to pay by these agents as vertical shifts in these schedules. For agents located within LA County, we follow Giesecke et al. (2012) in modeling the product incentive shifts described in Fig. 16.2 as fear-induced wedges driven between willingness to pay for LA County goods and willingness to pay for the competing product sourced from the rest of the U.S. or the rest of the world.

results for "Right after 'all clear'" and "Twelve months".[16] In calculating the decay rate between years 2014 and 2016, we use results for "Twelve months" and "Thirty six months".[17] Similarly, in calculating the decay rate between 2016 and 2018, we use results for "Thirty six months" and "Sixty months". After 2018, we assume a smooth decay rate for the behavioral shocks based on the annual average percentage change in Table 16.1 results for "percentage who are still waiting" between the "all clear" and "sixty months" marks.

The Fig. 16.2 pattern of response decay has been seen across several studies. Burns and Slovic (2007), using findings from the risk perception literature and expert elicitation, constructed a system dynamics simulation model to investigate public response to a terrorist attack. The results indicated that following news of an attack perceived risk and fear escalate rapidly, reaching a peak, and then decline quickly before levelling off slightly above pre-attack levels. Burns et al. (2011b) corroborated these simulation results by tracking public response to the attempted attack on Northwest Flight 253. Consistent with the simulation results, fear of flying and the intent to postpone air travel declined quickly before levelling off. Burns et al. (2011a) also surveyed Americans in March and April of 2011 in response to the triple disaster in Japan. They found fear of earthquakes, tsunamis and nuclear accidents declined markedly even within a month of the disaster. Most recently, Burns (2013) investigated public fear and perceptions of risk in reaction to the Boston attack. They conducted three surveys from April to July 2013 and found that public concern in July was a third of what it was in April.

The survey results do not directly elucidate the potential size of the investor risk premium, which would require an investor survey. Giesecke et al. (2012) addressed this by examining the literature on the asset price impacts of stigma related to various contamination risks. They noted an average implied risk premium in the vicinity of 20 %. This was in line with their RDD survey results for compensating wage premium. Based on this, they assumed that the two suppliers of primary factors (labor and capital) seek identical percentage movements in compensation. Like the RDD survey results in Giesecke et al., the Table 16.2 chlorine results also signal a 20 % increase in required wage compensation. Hence, we assume that firms require the same premium to invest in 90071 in the event year, and that thereafter the investor risk premium decays at the same rate as the wage premium (Fig. 16.2). We implement the increase in the risk premium in LA-DYN as increases in industry-specific required rates of return, scaled to reflect the share of each industry's capital payments explained by activity in 90071.

[16] For example, the 2013 deviation in willingness to pay for professional services is −15 %. Hence the 2014 deviation is assumed to be −2.4 % (= −15 % × 9/55).

[17] For example, the 2015 deviation in willingness to pay for professional services is assumed to be −1.8 % (= −2.4 % × (5/9)^0.5).

16.5 Simulation Results

Our explanation of the economic modeling results is based around a series of decomposition figures explaining outcomes for regional macroeconomic variables in terms of the individual contributions made by the eight sets of shocks. The decomposition figures are created by running the CGE model nine times: one full simulation in which all eight sets of shocks are implemented simultaneously, and a further eight simulations in which each of the eight sets of shocks is implemented individually. This allows us to explain total impacts in terms of the individual contributions made by each of the behavioral and resource shocks.[18] We explain the figures in a logical sequence, relying on references to economic mechanisms within the LA-DYN model to support our narrative.[19] We begin our discussion with the event year (2013). There are two main points of entry to understanding the 2013 results: negative deviations in the regional terms of trade (Fig. 16.3), and investment (Fig. 16.4). In Fig. 16.3 we see that the dominant contributor to the 2013 terms of trade deviation is the decline in willingness to pay for LA County goods. This exerts a direct effect on the regional terms of trade, depressing prices of LA County goods relative to competing imports. In Fig. 16.4 we see three shocks exert a strong negative influence on 2013 investment: investor risk premium, willingness to pay and BI.

The rise in required rates of return on LA County capital directly affects regional investment, depressing capital formation relative to baseline for any given rate of return (e.g. by about 0.2 % in the event year).

[18] The sum, for any variable, of results from the eight individual simulations is close to, but not exactly equal to, the results from the full simulation. This is because the model is non-linear, and interactions between the individual shocks that are captured by the full simulation are missed when the shocks are implemented individually. The difference between the sum of the eight individual simulations and the full simulation is reported as "Residual". The value for this is small for all variables in all years.

[19] Dixon and Rimmer (2013) describe eight ways in which CGE model results can be benchmarked or validated. Not all the methods they outline are required for every application. Rather, they advocate tailoring the validation procedure to the purpose at hand. In our discussion of results, we use the third of Dixon and Rimmer's procedures: qualitative validation via a narrative relying on economic mechanisms within the model (pp. 1297–1298). At the same time, while not reported in this paper, we have also relied on the first two of their validation methods (test simulations for which the results are known a-priori, and within-simulation cross-checks of national accounts identities). Their remaining methods (particularly vi–viii, p. 1272) are well beyond the scope of the present paper, representing independent CGE validation modeling exercises in their own right. For example, Dixon and Rimmer discuss validation of CGE results through out-of-sample forecasting. Examining the question with a 500 sector model of the U.S. economy, they find their CGE model forecasts over a seven year period are more accurate than trend extrapolation. They go on to argue that CGE forecasts can be improved further with better forecasts for macro and trade variables, and greater use of publicly available information on plausible future paths for commodity-specific and industry-specific variables relating to tastes, technologies and policy (Dixon and Rimmer 2013, pp. 1314–1324).

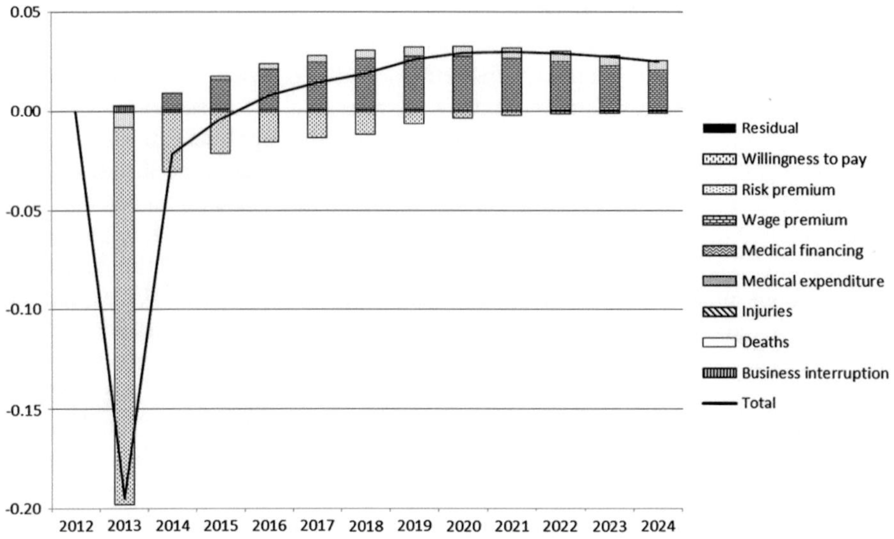

Fig. 16.3 Decomposition of the deviation in the LA County terms of trade (% point contributions to total deviation)

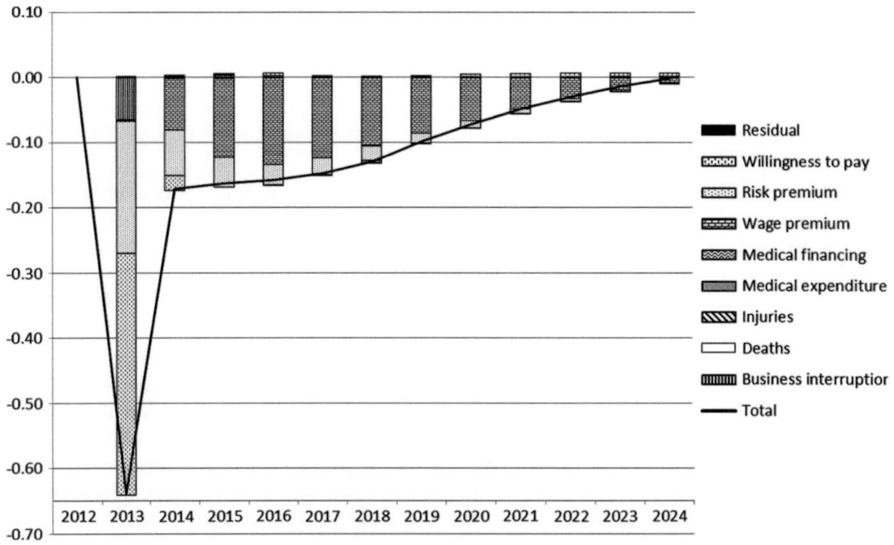

Fig. 16.4 Decomposition of the deviation in LA County real investment (% point contributions to total deviation)

The decline in willingness to pay exerts an indirect influence on regional investment, via its effect on the terms of trade (Fig. 16.3). In 2013, for a given level of the capital stock, the negative deviation in the terms of trade causes the value of the marginal product of capital to decline relative to baseline. This in turn

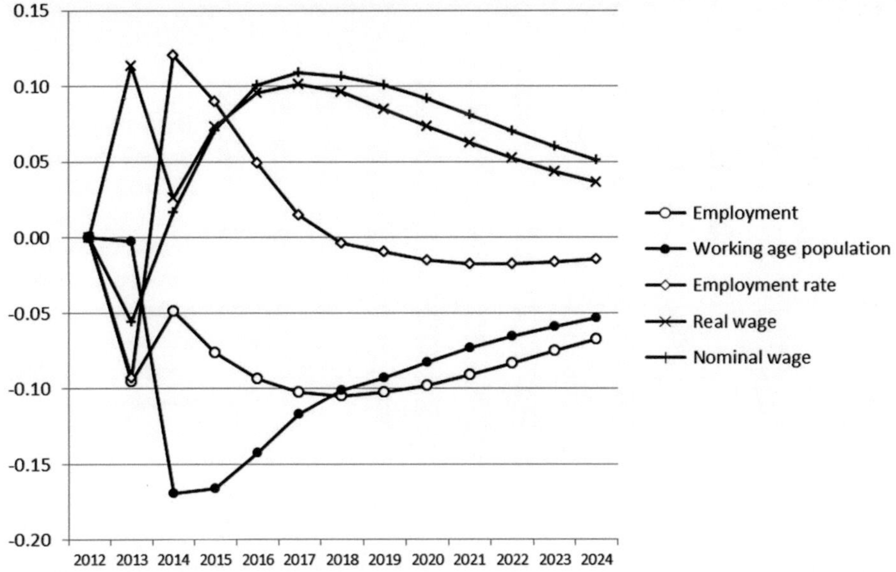

Fig. 16.5 LA County employment, working age population, employment rate and nominal and real wages (% deviation from baseline)

depresses the capital rental price, and with it, the rate of return. As is clear from Fig. 16.4, it is the decline in willingness to pay, via its effects on the terms of trade and ultimately the rate of return on capital, which has the largest impact on 2013 investment, reducing it by approximately 0.37 % relative to baseline.

BI is the third-largest contributor to the 2013 investment deviation (Fig. 16.4). As discussed in Sect. 16.4.1, BI is modeled as deterioration in input-using efficiency in industries in 90071. This causes the marginal physical product of capital to fall relative to baseline. Hence, for a given level of capital, the rate of return also falls relative to baseline. This accounts for BI's −0.06 % contribution to the 2013 investment deviation (Fig. 16.4).

Figure 16.5 reports five labor market variables. Outcomes for these variables are presented jointly to facilitate our explanation of the dynamic relationships between each. We begin with the 2013 outcome for the real wage and employment. Recall from Sect. 16.3.1 that nominal wages are modeled as sticky in the short-run. This causes the initial negative deviation in the terms of trade (Fig. 16.3) to generate a positive deviation in the real wage (Fig. 16.5). It is this initial rise in the real wage that accounts for much of the initial negative deviation in employment. This is confirmed by Fig. 16.6 which shows that the reduction in willingness to pay (via the terms of trade and real wage paths) makes the largest contribution to the 2013 employment deviation.

There is a small negative deviation in population in 2013 (Fig. 16.5). In the first year, the only influence on the size of the working age population is event-related fatalities. This is clear in our population decomposition (Fig. 16.7), in which deaths

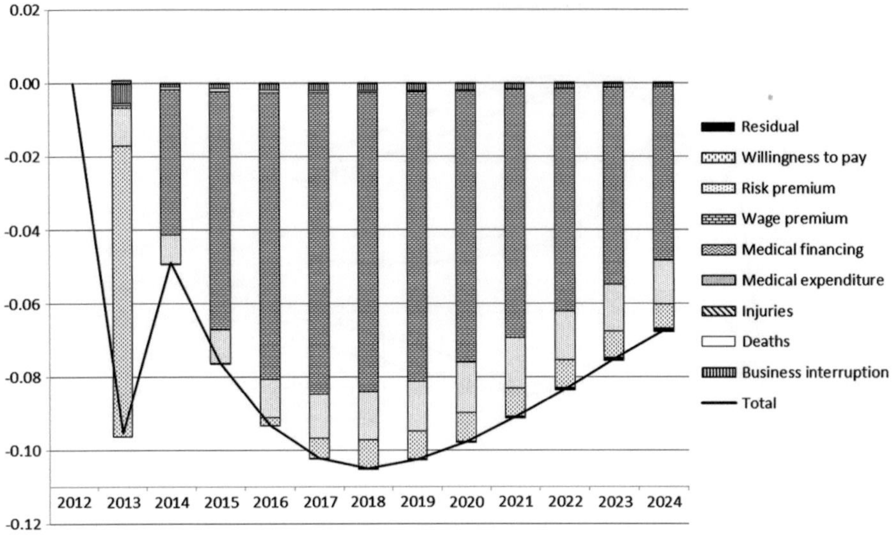

Fig. 16.6 Decomposition of the deviation in LA County employment (% point contributions to total deviation)

explain all of the 2013 deviation. Returning to Fig. 16.5, with employment sharply below baseline in 2013, and only a small negative deviation in the population, the regional employment rate must fall relative to baseline. As discussed in Sect. 16.3.1, the wage mechanism imposes sticky, but not fixed, nominal wages. With the 2013 employment rate below baseline, the sticky wage mechanism generates a negative deviation in the 2013 nominal wage (Fig. 16.5).

Figure 16.8 presents our GDP decomposition. In 2013, GDP is projected to be 0.08 % below baseline. Much of this decline is due to the negative deviation in 2013 employment, which, as discussed above, is largely due to the willingness-to-pay-induced reduction in the terms of trade. However, productivity loss, particularly via BI, also makes a substantial negative contribution to the 2013 GDP deviation (−0.03 %).

Turning to the post-event phase, we begin with Figs. 16.4 and 16.9. The negative deviation in 2013 investment (Fig. 16.4) is expressed in 2014 as a negative capital deviation (Fig. 16.9). From 2014 onwards, the magnitude of the negative investment deviation attenuates, eventually returning to baseline by the simulation's end. This reflects the gradual return of the investor risk premium to baseline (recall Fig. 16.2). However, the investment deviation lies below the capital deviation up to 2020, and, as a result, the negative capital deviation continues to grow to that year. Thereafter, as investment steadily returns to baseline, the negative capital deviation gradually attenuates.

In 2013, population falls slightly due to event-related fatalities (Fig. 16.5). In 2014, the regional population deviation reaches its lowest point (−0.17 %),

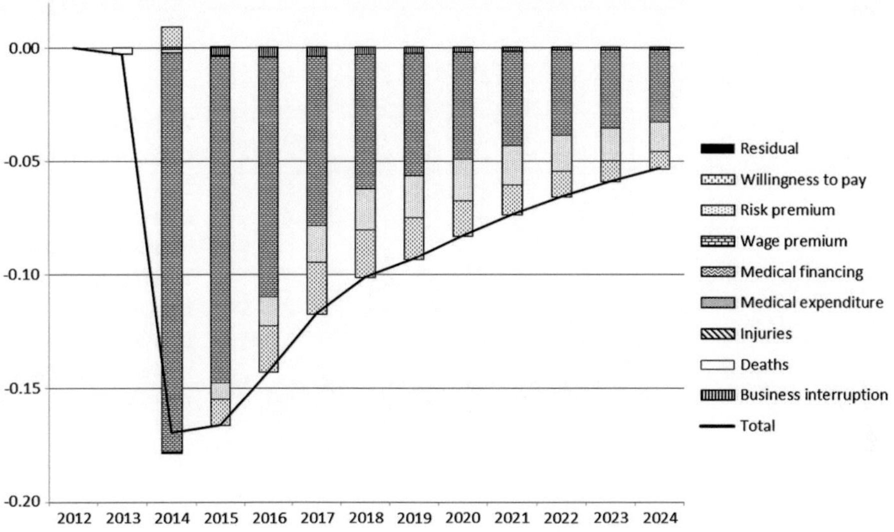

Fig. 16.7 Decomposition of the deviation in LA County working age population (% point contributions to total deviation)

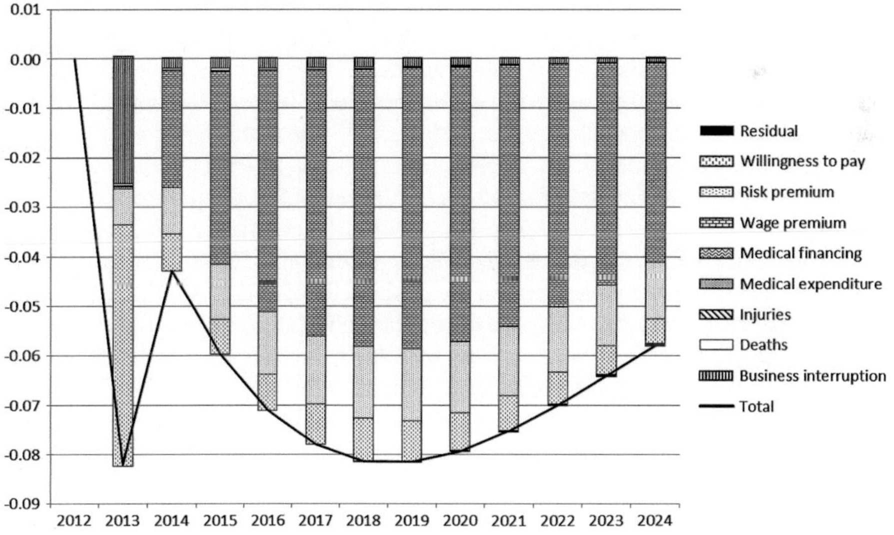

Fig. 16.8 Decomposition of the deviation in LA County real GDP (% point contributions to total deviation)

gradually returning to baseline thereafter. As shown in Fig. 16.7, for the years immediately following the event, the wage premium shock makes the largest contribution to the negative deviation in population. The wage premium operates directly on net inter-regional immigration via $F^{(M)}$ in equation (Eq. 16.5). In our

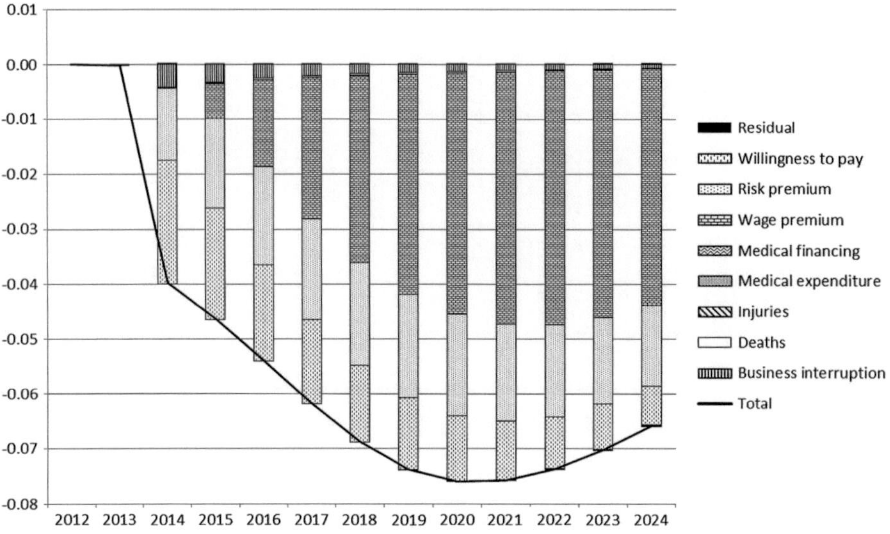

Fig. 16.9 Decomposition of the deviation in LA County capital stock (% point contributions to total deviation)

simulation, the deviation path for $F^{(M)}$ follows the path for the wage premium behavioral effect reported in Fig. 16.2.[20] With $Y^{(M)}$ determined in the regional labor market via (Eq. 16.4), the positive deviation in $F^{(M)}$ causes a negative deviation in $Y^{(Diseq)}$. Via (Eq. 16.6) this causes the net inter-regional immigration rate to fall, and with it, the regional population. After 2016, as $F^{(M)}$ gradually returns to baseline, the negative contribution of the wage premium shock to the regional population deviation gradually attenuates (Fig. 16.7). By the end of the simulation period, $F^{(M)}$ has almost returned to baseline (Fig. 16.2). Nevertheless, in Fig. 16.7, we see the wage premium continues to make a sizeable contribution to the negative population deviation. This reflects two legacies of the positive deviations in the wage premium in earlier periods. First, the working age population is a stock variable. At any given point in time, deviations in its value from baseline reflect accumulated past flows of deviations in net inter-regional migration. Second, over time, deviations in population and capital interact. By reducing the size of the regional population, and with it, regional employment, the wage premium reduces the marginal physical product of capital by lowering the labor/capital ratio. This explains the sizeable contribution made by the wage premium to the capital deviation in Fig. 16.9. However, causation also runs in the other direction. A lower capital stock, for any given level of the working age population and the employment rate, lowers the marginal physical product of labor, and with it the wage. This damping effect on the regional wage via

[20] With the size of the shock scaled to reflect the share of the wage bill in zip code 90071 in the total LA County wage bill.

the negative capital deviation feeds back into to the persistent negative deviation in the working age population in Fig. 16.7.

Returning to Fig. 16.5, the negative deviation in the working age population exerts a damping effect on regional employment in the post-event phase. In 2014, the deviation in the working age population is −0.17 %. If wages were fully flexible, the 2014 employment deviation would also be −0.17 % because the deviation in the employment rate would be zero. Recall however that we model short-run nominal wages as sticky. As such, the 2014 wage does not fully adjust to its new market clearing level, which accounts for the positive deviation in the 2014 employment rate. The model's wage mechanism requires the deviation in the nominal wage to grow so long as the deviation in the employment rate is positive. This accounts for why, in Fig. 16.5, we see the nominal wage deviation growing up to 2017, gradually driving the employment rate back towards baseline. Thus, we see the deviations in employment and the working age population tracking closely together from 2018 onwards (Fig. 16.5).

As discussed in Sect. 16.4.2, by the end of the simulation period, the deviations in willingness to pay, risk premium and wage premium have returned to close to their baseline values (Fig. 16.2). Despite this, in Figs. 16.7 and 16.9, we see that these behavioral shocks continue to make negative (although tapering) contributions to the deviations in two key stock variables: population and capital. The explanation lies in the short—to medium-run impact of these shocks on the regional capital stock, and the interaction between the capital stock deviation and the regional population and employment deviations. For example, consider the investment decomposition (Fig. 16.4). Here we see the contribution to the investment deviation made by the risk premium shock falling to zero by the end of the simulation period. This reflects the gradual return of the risk premium back to baseline, as the behavioral effect dissipates. In 2013, the risk premium shock makes a sizeable contribution to the investment deviation. In 2014, this is manifested as a negative contribution by the risk premium to the stock of capital (Fig. 16.9). In 2014, the risk premium makes a much smaller contribution to the negative investment deviation (Fig. 16.4). Nevertheless, in Fig. 16.9, the negative contribution of the risk premium to the capital deviation persists. This is so for two reasons.

First, investment and capital accumulation is modeled as a gradual process: it takes some years of steady investment to recover from the negative capital deviation generated by the 2013 risk premium shock. Second, the capital deviation and the employment deviation interact. In 2014 the risk premium shock makes a negative contribution to the real wage deviation because it lowers the 2014 capital stock relative to baseline (Fig. 16.9), lowering the marginal physical product of labor.[21] The inter-regional immigration function links year t net immigration to the year t expected wage, and links year $t+1$ population to year t net immigration. Via these mechanisms, with the risk premium making a negative contribution to the real

[21] This accounts for the temporary dip in the real wage deviation in 2014 (See Fig. 16.5). A decomposition diagram of the real wage deviation is available from the authors on request.

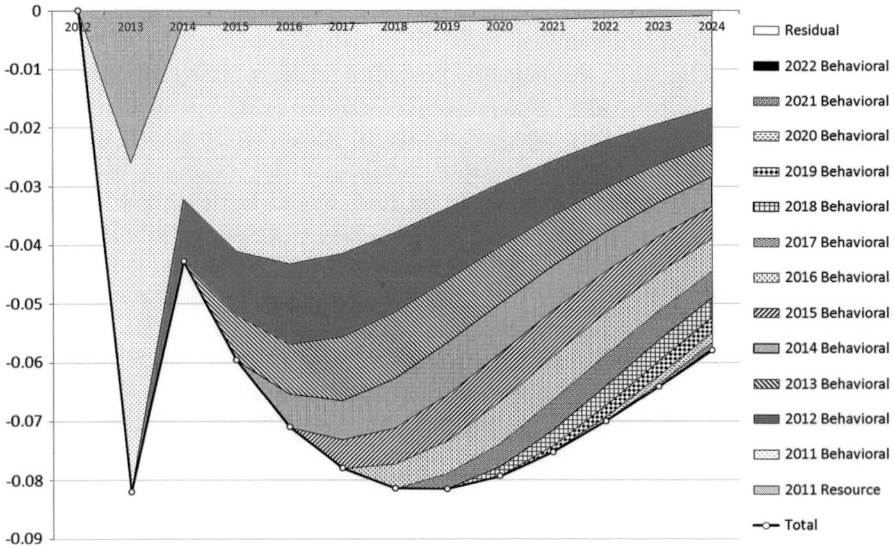

Fig. 16.10 Real GDP decomposition by year (% deviation from baseline)

wage in 2014, it also makes a negative contribution to the 2015 working age population (Fig. 16.7). For a given employment rate, this generates a negative contribution to 2015 employment (Fig. 16.6). By lowering the marginal product of capital relative to baseline in 2015, this attenuates what would otherwise be a positive investment response to the fact that the deviation from baseline in the risk premium in 2015 is significantly smaller than in 2014 (Fig. 16.2).

Examining Fig. 16.2, we see that the values of our key behavioral variables have returned close to baseline by the simulation's final year. However in Fig. 16.8, we see a persistent (but tapering) GDP deviation. The enduring GDP deviation at the end of the simulation period reflects the legacy of shocks in earlier periods, particularly those relating to behavioral variables. To make this clearer, Fig. 16.10 presents a new decomposition of the GDP deviation, distinguishing the contributions of the shocks within any given year to the GDP deviations of current and future years.[22]

For the event year, Fig. 16.10 distinguishes the effects of behavioral and resource shocks, while latter years identify only behavioral consequences (there

[22] Figure 16.10 is constructed from 14 simulations: (1) a 2013 resource loss simulation, in which all resource loss shocks are applied in 2013, and the simulation is run out to 2024; (2) 12 behavioral simulations, in which the values for shocked behavioral variables in year t are moved to their year t deviation values, returned to their baseline values in year $t+1$, and the remainder of the simulation run to 2024; and (3) the full simulation. The sum of the results of simulations (1)–(2) are very close to, but not exactly equal to, the results of simulation (3). This is due to interactions between the shocks, captured by simulation (3), but missed when the shocks are modeled as a sequence of unconnected simulations. We report the difference as "residual".

are no direct resource loss shocks after 2013). Examining the legacy of the event year shocks, it is clear that while the enduring effects of the resource loss shocks have largely dissipated by 2024, this is not the case for the behavioral shocks. In the event year, the resource loss shocks depress GDP by −0.026 % relative to baseline. By 2024 this contribution has fallen to −0.0009 % (1/28th of the 2013 resource result). The event year behavioral shocks lower GDP in 2013 by −0.056 % relative to baseline. Examining Fig. 16.10, we see a tapering contribution by the 2013 behavioral shocks to the GDP deviations of future years. Nevertheless, by 2024, the 2013 behavioral shocks are contributing −0.016 % to the GDP deviation: just over 1/4th of the 2013 behavioral impact. Examining the time paths of the decomposition results for each of the post-event years, Fig. 16.10 shows a similar pattern of gradual tapering of the enduring effects of the deviations from baseline in the year $1+n$ values for our risk premium, wage premium and willingness to pay variables. The persistence of the behavioral contributions to the GDP deviations of future years can be traced to our previous explanation of the dynamic interaction between the deviations in capital and population. That is, the interaction between these two variables, via their respective impacts on the marginal product of labor (and thus wages and inter-regional immigration) and the marginal product of capital (and thus rates of return and investment) damps the speed with which these variables return to baseline.

The advantages of a dynamic model in elucidating the long-run consequences of the behavioral responses arising from a terrorism event are demonstrated by a comparison of Fig. 16.10 with the estimated long-run real GDP consequences of an RDD attack as reported in Giesecke et al. (2012). As discussed in Sect. 16.3.1, the latter paper used a comparative static version of the dynamic model used in this paper. Because their model was not dynamic, Giesecke et al. inferred annual results between years $t+1$ and $t+10$ by assuming a linear transition of GDP impacts from short-run (event-year) results, through to long-run results for the year $t+5$, and then to an assumption of a return to baseline (i.e. zero GDP impacts) by year $t+10$.[23] However, their assumption of no enduring impacts by year $t+10$ appears, in the light of Fig. 16.10, to under-estimate the potential for long-run adverse regional GDP consequences via behavioral effects.

Table 16.3 provides a comparison of event year and long-run real regional GDP outcomes in terms of contributions by BI, other resource losses, and behavioral effects. Consistent with Giesecke et al., Table 16.3 highlights the importance of behavioral effects in both the event year and the long-run. The total real regional GDP loss in the event year is $447 m., of which $311 m. arises from behavioral effects alone. Much of the remainder is due to the direct and indirect flow-on consequences of BI ($81 m. and $50 m., respectively). These BI losses are significantly lower than those calculated for an RDD event in Giesecke et al., in large part

[23] See Fig. 3 in Giesecke et al. (2012).

Table 16.3 Summary comparison of real regional GDP outcomes

Impact	Category	
(1) Event year	Direct business interruption (BI). (Output loss, \$m.)	−\$140
(2) Event year	Direct business interruption (BI). (GDP loss, \$m.)[a]	−\$81
(3) Event year	Indirect business interruption (BI). (GDP loss, \$m.)	−\$50
(4) Event year	Other resource loss. (GDP loss, \$m.)[b]	−\$3
(5) Event year	Medical expenditure & financing. (GDP loss, \$m.)	−\$1
(6) Event year	Behavioral effects. (GDP loss, \$m.)	−\$311
(7) Event year	Total short-run. (GDP loss, \$m.)	−\$447
(8) Long-run	Average annual long-run behavioral. (GDP loss, \$m.)[c]	−\$478
(9) Long-run	Total ten-year behavioral. (GDP loss, \$m.)	−\$4,780
(10) NPV	NPV (at 5 %) of total ten-year behavioral. (GDP loss, \$m.)	−\$3,622
(11) Ratio: [(2) + (3)]/(2)	S-R total BI/S-R direct BI	1.62
(12) Ratio: [(2) + (3) + (4) + (5)]/(2)	S-R ordinary loss/S-R direct BI	1.67
(13) Ratio: (8)/(2)	L-R one-year/S-R direct BI	5.89
(14) Ratio: (8)/(6)	L-R average behavioral/S-R behavioral	1.54
(15) Ratio: (9)/(2)	Total ten-year behavioral/S-R direct BI	58.9
(16) Ratio: (9)/[(2) + (3) + (4) + (5)]	Total ten-year behavioral/ordinary loss	35.3

[a]Based on a value-added/output ratio for zip code 90071 of 0.58. With BI generating lost output of \$140 m., this is equivalent to a direct GDP loss of \$81.2 m (=\$34 m × 0.58)
[b]Via deaths and injuries
[c]Annual average of the behavioral impacts in the ten years following the event

because the chlorine clean-up period is much shorter (3 days compared with 30 days). In the post-event phase, average annual real GDP losses from the behavioral effects are \$478 m., approximately 1.5 times the event year behavioral loss (row 14). This ratio is approximately half that found in the RDD study.[24] This highlights another advantage of dynamics over comparative statics. Under the long-run comparative static closure used in Giesecke et al., regional population and capital are assumed to fully adjust to the long-run behavioral shocks, at going regional wage rates and rates of return. However, in the present dynamic modeling, both regional population and capital respond slowly, not fully adjusting to the peak behavioral effects before these shocks begin returning to baseline.

[24] Giesecke et al. (2012) report an annual long-run behavioral loss of \$2,628 m. and an event year loss of \$889 m., a 3:1 ratio.

16.6 Spatial Analysis

16.6.1 General Considerations

In this section we discuss the spatial aspects of external shocks to an economy in general and with special reference to insidious terrorist attacks. The analysis thus far has been performed in a relatively aspatial manner. We have identified the contaminated area as the site of the impacts and translated the impacts into changes in factor availability and prices for firms located there. This affects firm competitiveness and hence imports and exports of the LA economy as a whole by averaging the direct impacts across all firms in the County. But several other spatial aspects have been omitted. Some of these relate to economic resilience, or how BI losses can be muted by ordinary and adaptive responses to a disaster (Rose 2009a, b). The prime example is the response to the 2001 attacks on the World Trade Center, in which 95 % of businesses and government agencies in the affected area relocated within several weeks, primarily to mid-town Manhattan or Northern New Jersey, thus avoiding 72 % of potential BI losses (Rose et al. 2009). That study indicates how the omission of resilience can lead to an overestimate of economic losses.

Walter Isard, perhaps more than any other American social scientist, injected a spatial dimension into the economy (see, e.g., Isard 1951, 1956; Isard et al. 1969). In the presentation below, we provide an overview of a systematic framework for the spatial analysis of the impacts of a terrorist attack, primarily with the objective of improving the accuracy of the estimates of economic consequences. We also indicate how the inclusion of a broader set of spatial dimensions would affect our results.

We begin by noting two special features of a chlorine attack that distinguish it from other types of disasters. The first relates to fear and stigma effects already analyzed in prior sections. Additionally, because of the uncertainty regarding the spread of chlorine gas or other insidious weapons whose dispersion is related to weather conditions and are difficult to detect, we should also consider that the fear/stigma will not halt abruptly at the financial district boundary. It is reasonable to consider a fringe zone where these behavioral considerations may spill over and have impacts, though likely less intensive than in the core area.

A behavioral consideration not addressed thus far relates to the likelihood and pace of business relocation. The 9/11 example indicates that the response is likely to be rapid and not far from the original site. Both of these responses were conditioned somewhat upon broader aspects of resilience, in the form of demonstrating to terrorists that they cannot defeat their intended targets (Flynn 2008). There is every reason to believe that this "we will show them" attitude would prevail in LA as well.

The spatial dimension is rarely if ever addressed in full in estimating the regional economic consequences of terrorism, as well as many other perturbations. Most analyses are undertaken with single region models, and relocation within them or across regions is rarely analyzed explicitly. When inter-regional or multi-regional models are used, some consideration of relocation is made but primarily with respect to overall economic activity levels, rather than the explicit movement of individual businesses. For example, Gordon et al. (2005) use a multi-regional input-output modeling approach with inter-regional trade flows. They divide the greater LA area into sub-regions, such that impacts do have spatial delineation. However, there is no explicit attention to the relocation of firms.

16.6.2 Spatial Analysis Framework

Various adjustments in economic models have been offered in relation to the relocation of businesses. For example, Isard and Kuenne (1953) provided various ad hoc adjustments to reflect an agglomeration effect associated with the entrance of a new steel mill into a region. Adjustments referred to the likelihood that support industries would develop in the area, thereby increasing the basic multiplier effects. Our analysis deals with the opposite situation, i.e., business exit. Hence, further adjustment of general equilibrium supply-chain linkages might be in order, particularly to the extent that the model's endogenous capital supply responses understate the magnitude of these effects.

Another complication is that relocation may not be entirely out of the region for which the impact analysis was performed but may also take place within the region, as in the case of World Trade Center area firms moving to Midtown Manhattan. Also, business activity in cyberspace and tele-commuting have increased significantly in recent years, further blurring boundaries. Finally, we have the longstanding issue of the ready ability to shift economic activity among branch plants of the same company. Given all these considerations, a quarantine plus geographic averting behavior may not result in losses as great as initially predicted.

Below we discuss in detail various aspects of the potential spatial realignment of economic activity in relation to the chlorine attack scenario:

Business Relocation Alternatives First we must consider relocation out of the financial district. This would potentially include: (1) actual physical relocation; (2) a shift of activity to other branches of the firm; and/or (3) work primarily in cyberspace.

Physical relocation is likely to approach zero in a case where quarantine/decontamination lasts only a few days. Shift of activity to branch offices is a possibility, especially for firms involved in banking, finance and insurance. However, we lack data on the extent of this opportunity and the extent to which businesses will exploit it. Work in cyber space, including tele-commuting, deserves special attention here because it is becoming more prevalent, especially in the banking and finance sectors that are predominant in the area affected by the chlorine attack. This activity may not be affected in any significant way by the attack. Increasingly, businesses are backing-up their systems such that even if main computers are located in the financial district, relevant files can be accessed from elsewhere. Similarly, data are being increasingly stored on various types of "cloud."

The prevalence of these cyber options warrants adjustment of direct impact estimates when data become available.

We must also distinguish shifts of locations *within* LA County from those going elsewhere. As in the NYC shift to Mid-town, there are several advantages to relocating in close proximity to the original site. Doing so would not lead to any reduction in economic activity within LA County, all other things being equal. If we were to apply 9/11 findings, we would estimate that 95 % of the businesses would in fact relocate, with most of them within the County. However, the 9/11 physical moves averaged 6–8 weeks, and thus are not pertinent to this case. If for some reason the decontamination were to take much longer, a good deal of equipment and material could readily be moved, since chlorine gas, unlike radiation, is not a lingering contaminant.

Of course, business relocation does have its costs. If this significantly increases the cost of doing business at the new location, this would have to be factored into the analysis, and would lower the level of economic activity in LA County by affecting its competitiveness.

Economic Activity Shift Out of the Region This aspect does not pertain to the actual physical movement of firms, but rather to their activity levels in place. It relates to an increase in their cost of doing business due to increased wage demands and increased investor rates of return to compensate for increased risk. It also relates to their likely reduced profits as they have to provide customer discounts. These have been addressed in our modeling approach above and also in Giesecke et al. (2012). However, both of these analyses implicitly assume no business relocation, but simply a decrease in economic activity. The modeling can be made more accurate if an adjustment is made for those firms actually relocating elsewhere in LA County and relocating outside the County. In the case of the later, the base for the activity shifts decreases (fewer firms to which to apply what are essentially declines in competiveness or product demand that lead to reduced sectoral output).

Temporal Dimension of Spatial Shifts It is important to distinguish between business relocation at various points in time. The implications of these decisions differ significantly for the consequence estimates between the initial contamination, clean-up, risk amplification, and stigma phases. While firms may not have time to physically relocate during the short-run response (decontamination), opportunities to do so increase over time, though the incentives to do so decline as well (see the discussion above of the decay rate for fear). Thus, some relocations decisions (including branch offices and cyber space activity) affect ordinary losses, and others affect behavioral losses estimated in this study. All will typically reduce the basic loss estimates, though the extent to which they affect the ratio of behavioral to ordinary BI losses is not known a priori.

One also needs to consider the potential for a reverse movement of businesses. Abadie and Dermisi (2011) are investigating the movement of WTC area firms who left Lower Manhattan right after 9/11 but are returning. The same phenomenon could take place for the type of terrorist attack that we have analyzed, though it is less likely because of (misplaced) fear of lingering contamination. One would also need to consider the extent to which the attack site is seen as a prime target for future attacks, contributing to persistent stigma.

Conclusions

We have developed a regional CGE model to analyze the time-path of the economic impacts of a terrorist attack. The model was applied to estimating the standard resource loss and behavioral impacts of a chlorine gas attack on the LA financial district. The results indicate that behavioral effects, stemming from the social amplification of risk and stigma affects, dominate. The ratio of Total Ten-Year Behavioral to Ordinary Losses is greater than 35. Moreover, the ratio of Total Ten-Year Behavioral/to Short-Run Direct Business Interruption is nearly 60. The model was designed to capture critical economic relationships pertaining to severe shocks, including migration, capital stock damage, and investment. The extent to which the relocation of business activity could affect the results was also explored. Behavioral responses were incorporated into the model by estimating the impacts of worker wage premiums, investment premiums, and shopper discounts required to induce economic agents to return to the impacted area. The empirical basis for the behavioral analysis was a survey of respondents reacting to a simulated chlorine gas attack. Adverse behavioral impacts are large but may be less costly to mitigate than are the various types of interdiction of terrorism and hardening of targets. Improved risk communication regarding the severity of the chlorine gas threat and its remediation has the potential to significantly quell fears that would otherwise translate into sizable negative economic impacts.

References

Abadie A, Dermisi S (2011) World trade center tenant relocation patterns after September 11th, 2001. Paper presented at the annual meeting of the American Economic Association, 5 Jan 2011

Barrett AM, Adams PJ (2011) Chlorine truck attack consequences and mitigation. Risk Anal 31 (8):1243–1259

Burns WJ (2013) Public response to disasters: current approaches to exploring perceived risk, economic impacts and the mitigating role of risk communication. Presentation given to the risk-based security group at the TSA, Arlington, VA, 24 Jul 2013

Burns WJ, Slovic P (2007) The diffusion of fear: modeling community response to a terrorist strike. J Defense Model Simulat Appl Methodol Technol 4:298–317

Burns WJ, Slovic P (2010) Predicting and modeling public response to a terrorist strike. In: Slovic P (ed) The feeling of risk: new directions in the perception of risk. Earthscan Publishing, London

Burns WJ, Reilly C, Slovic P (2011a) The attack on Flight 253, the Haiti earthquake, and the Japanese disaster: a longitudinal look at emotional reactions, risk-related behaviors, and support for policy measures. The CIP Report 10(6): 22–24, 34

Burns WJ, Slovic P, Peters E (2011b) Public response to recent crises: a longitudinal look at fear, risk-related behaviors and support for different security measures. 23rd subjective probability, utility, and decision making biennial conference, August 26–28

Dixon PB, Rimmer MT (2002) Dynamic General Equilibrium Modelling for Forecasting and Policy: a Practical Guide and Documentation of MONASH. Contributions to Economic Analysis 256, North-Holland Publishing Company

Dixon PB, Rimmer MT (2005) Mini-USAGE: reducing barriers to entry in dynamic CGE modelling. The 8th annual conference on global economic analysis. Lübeck, Germany, Centre of Policy Studies, Monash University, Melbourne

Dixon PB, Rimmer MT (2013) Validation in computable general equilibrium modeling. In: Dixon PB, Jorgenson D (eds) Handbook of computable general equilibrium modeling. North-Holland

Dixon PB, Parmenter BR, Sutton J, Vincent DP (1982) ORANI: a multi-sectoral model of the Australian economy. North-Holland, Amsterdam

Fischhoff B, Slovic P, Lichtenstein S, Read S, Combs B (1978) How safe is safe enough? A psychometric study of attitudes towards technological risks and benefits. Policy Sci 9:127–152

Flynn SE (2008) America the resilient; defying terrorism and mitigating natural disasters. Foreign Aff 87(2):2

Giesecke JA (2011) Development of a large-scale single US region. CGE model using IMPLAN data: a Los Angeles County example with a productivity shock application. Spatial Econ Anal 6(3):331–350

Giesecke JA, Madden JR (2013) Regional computable general equilibrium modeling. In: Dixon PB, Jorgenson D (eds) Handbook of computable general equilibrium modeling. North-Holland

Giesecke JA, Burns W, Barret A, Bayrak E, Rose A, Slovic P, Suher M (2012) Assessment of the regional economic impacts of catastrophic events: CGE analysis of resource loss and behavioral effects of an RDD attack scenario. Risk Anal 32(4):583–600

Gordon P, Moore JE II, Richardson H, Pan Q (2005) The economic impact of a terrorist attack on the twin ports of Los Angeles-Long Beach, pp 262–268. In: Richardson HW, Gordon P, Moore JE II (eds) The economic impacts of terrorist attacks. Edward Elgar, Northampton, MA

Gordon P, Richardson H, Davis B (1998) Transport-related Impacts of the Northridge earthquake. J Transport Stat 1(2):21–36

Hanoch G (1971) CRESH production functions. Econometrica 39:695–712

Harrison W, Pearson K (1996) Computing solutions for large general equilibrium models using Gempack. Comput Econ 9:83–127

Horridge JM (2003) ORANI-G: a generic single country computable general equilibrium model. http://www.monash.edu.au/policy/ftp/oranig/oranig03.zip

HSC (2004) National planning scenarios: executive summaries, homeland security council. http://www.globalsecurity.org/security/library/report/2004/hsc-planning-scenarios-jul04_exec-sum.pdf

Isard W (1951) Interregional and regional input-output analysis: a model of a space economy. Rev Econ Stat 33:318–328

Isard W (1956) Location and space economy. MIT Press, Cambridge, MA

Isard W, Kuenne R (1953) The impact of steel upon the greater New York-Philadelphia industrial Region. Rev Econ Stat 35:298–301

Isard W, Smith TE, Isard P, Tung TH, Dacey M (1969) General theory: social, political, and economic and regional. MIT Press, Cambridge, MA

Isard W, Azis IJ, Drennan MP, Miller RE, Saltzman S Thorbecke E (1998) Methods of interregional and regional analysis. Ashgate, Aldershot

Kraus N, Malfors T, Slovic P (1992) Intuitive toxicology: expert and lay judgments of chemical risks. Risk Anal 12(2):215–232

Minnesota IMPLAN Group (1997) IMPLAN system (data and software), Stillwater, MN 55082, http://www.implan.com

Molinari NAM, Ortega-Sanchez IR, Messonnier ML, Thompson WW, Wortley PM, Weintraub E, Bridges CB (2007) The annual impact of a seasonal influenza in the US: measuring disease burden and costs. Vaccine 25(27):5086–5096

Partridge MD, Rickman DS (2010) Computable general equilibrium (CGE) modeling for regional economic development analysis. Reg Stud 44:1311–1328

Rose A (2009a) A framework for analyzing and estimating the total economic impacts of a terrorist attack and natural disaster. J Homeland Secur Emerg Manage 6(1):Article 4

Rose A (2009b) Economic resilience to disasters, community and regional resilience institute report no 8. Oak Ridge, TN

Rose A, Liao SY (2005) Modeling regional economic resilience to disasters: a computable general equilibrium analysis of water services disruptions. J Reg Sci 45(1):75–112

Rose, A., G. Oladosu, B. Lee and G. Beeler Asay. (2009). The Economic Impacts of the 2001 Terrorist Attacks on the World Trade Center: A Computable General Equilibrium Analysis, *Peace Economics, Peace Science, and Public Policy* 15(2): Article 6.

Slovic P (1987) Perception of risk. Science 236:280–285

United States International Trade Commission (2004 and 2007). The economic effects of significant US import restraints: fourth and fifth updates. Investigation No. 332–325, publications 3701 and 3906

van Sickle D, Wenck M, Belflower A, Drociuk D, Ferdinands J, Holguin F, Svendsen E, Bretous L, Jankelevich S, Gibson J, Garbe P, Moolenaar R (2009) Acute health effects after exposure to chlorine gas released after a train derailment. Am J Emerg Med 27(1):1–7

Whalley J, Trela I (1986) Regional aspects of confederation. University of Toronto Press, Toronto

Withers RMJ, Lees FP (1985) The assessment of major hazards: the lethal toxicity of chlorine; part 2. J Hazard Mater 12:283–302

Part IV

Urban Agglomerations

Framework for Spatial Models for Future Cities

17

Tschangho John Kim

Abstract

The chapter is to reflect Professor Walter Isard's scientific reasoning and contributions to formulating spatial models for urban and regional systems and to shed light on developing a framework for spatial modeling for future cities. We need alternative spatial models for cities and regions more than ever now in light of rapidly changing socio-economic and environmental challenges that we have been confronted with and more to face in the near future. Among others, I would identify the following four major recent developments that force us to consider alternative spatial models:

(a) Rapid growth of Global Urbanization
(b) Unlimited computing capability
(c) Pervasive use of instantly available Information and Communication tools
(d) Shift from Personal to Communal Ownership of Automobiles

Thus, the goal of this chapter is to shed light on developing framework for alternative spatial models for future cities and regions in the same spirit as what Isard advocated more than a half century ago (Isard, Location and space-economy, The MIT Press, Cambridge, 1956; Isard, Methods of regional analysis: an introduction to regional science, The MIT Press, Cambridge and Wiley, New York, 1960), and yet still inspiring us to undertake these changes and challenges.

T.J. Kim (✉)
Urban and Regional Systems, University of Illinois at Urbana-Champaign, Champaign, IL 61822, USA

Department of Civil, Environmental and Infrastructure Engineering, George Mason University, Fairfax, VA 22030, USA
e-mail: tjohnkim@illinois.edu

P. Nijkamp et al. (eds.), *Regional Science Matters*, DOI 10.1007/978-3-319-07305-7_17,

17.1 Introduction

This chapter is to reflect Professor Walter Isard's scientific reasoning and contributions to formulating spatial models for urban and regional systems and to shed light on formulating a framework for spatial modeling for future cities. We need alternative spatial models for cities and regions more than ever now in light of rapidly changing socio-economic and environmental challenges we have been confronted with and more to face in the near future. Among others, I would identify the following four major recent developments that force us to consider alternative spatial models:

(a) Rapid growth of Global Urbanization
(b) Unlimited computing Capability
(c) Pervasive use of instantly available Information and Communication Tools
(d) Shift from Personal to Communal Ownership of Automobiles

Thus, the goal of this chapter is to shed light on developing a framework for alternative spatial models for future cities in the same spirit as what Isard advocated more than a half century ago (Isard 1956, 1960), and yet still inspiring us to undertake these changes and challenges.

17.2 Implications of Global Urbanization on Spatial Modeling

17.2.1 Resources

According to the United Nations (2011), more than half of the world's population lived in urban areas in 2008 and the level of urbanization is expected to rise from 52 % (3.6 billion) in 2011 to 67 % (6.3 billion) in 2050.[1] The alarming rapid urbanization trend alerts to urban and regional scientists for searching for new ways to accommodate additional 2.7 billion urban populations by 2050. To accommodate the additional urban population, either existing cities need to be expanded or new cities have to be created, equivalent to 2,700 new cities of one million population.

A critical question arises. How can we provide much needed resources to accommodate the additional urban population? The resources needed to accommodate these new urban inhabitants would be enormous. For example, providing enough water for 2.7 billion urban residents, we would need about 10 of Hoover Dam capacity[2] and need to supply about 20,000 billion kWh of electricity that is equivalent to about 4,500 of Hoover Dam power generating capacity.[3]

[1] http://esa.un.org/unpd/wup/CD-ROM/Urban-Rural-Population.htm. Accessed on May 1, 2013.

[2] Assuming 50 gal are needed for gallons of water per day for drinking, cooking, bathing, and watering their yard, 2.7 billion people would consume about 50 billion gallons of water per year. See http://www.pacinst.org/reports/water_fact_sheet/for more information about water consumption. Hoover Dam's capacity is reported to be about 5 billion gallons per year. See http://en.wikipedia.org/wiki/Hoover_Dam for Dam's capacity related information.

[3] Assume 20 kWh are needed for per person per day, which is less than the US average, but a bit higher than the EU average. See URLs below for detailed information http://en.wikipedia.org/

17.2.2 Megacities

One additional impact of global urbanization trends is the rapid increase the number of megacities in which more than ten million inhabitants reside. There were only 8 cities in the world with population more than ten million; however, the number of megacities grew to 27 by 2012.[4] Of these, ten megacities are in developing countries, where overpopulated slums exhibit high rates of disease due to unsanitary conditions, malnutrition, and lack of basic health care. It took 150 years for London to see population grew from one million to eight million, while the population of Mexico City grew from 1 million to over 15 million people in only 50 years. Many countries in the developing world neither have the legal nor the financial capacity to deal with the rapid growth of megacities.

Megacities presents enormous challenges to governments and the inhabitants of the megacities in both developed and developing countries. In the US, there are many incorporated and political entities in a megacity. The Los Angeles megacity has 122 independent jurisdictions and minor civil districts (MCDs)[5] each of which has own decision-making power for maximizing its welfare within the jurisdiction by the home-rule delegated by the State of California. In exercising their power, individual jurisdiction's interests often conflict with those of the megacity as a whole.

A few critical questions arise. How can we reconcile different, often conflicting interests of the metropolis as a whole and of the individual member cities of the metropolis? Can an urban area be sustainable? Having implemented a numerous number of public policy and spent enormous amount of budget for mitigating congestion (Boyce and Kim 1987), has congestion in urban areas been mitigated? What went wrong with planning for cities and regions in the past?

17.3 Implications of Unlimited Computing Capabilities on Spatial Modeling

17.3.1 Large Scale Spatial Models

At the time when Isard proposed integrated spatial models in early 1960s (Isard 1960), computing capacity was limited for solving even a modest size of spatial models.[6] The evolution of computer and computing power opened new dimensions for spatial models and model buildings.

wiki/Electricity_consumption and http://en.wikipedia.org/wiki/List_of_countries_by_electricity_consumption

[4] http://l-lists.com/en/lists/2mzkgs.html. Accessed on May 2, 2013.

[5] According to the U.S. Census Bureau, Greater Los Angeles consists with the Los Angeles-Long Beach-Riverside Combined Statistical Area, and the entirety of San Bernardino and Riverside counties.

[6] A great number of excellent literatures on large-scale spatial models have been published that include Wilson (1974), Batty (1976), Boyce (1988), Kim (1989), Wegener (1994).

In 1946, ENIAC (Electronic Numerical Integrator and Computer) was first introduced in with a speed of one thousand times that of electro-mechanical machines, heralded in the press as a "Giant Brain".[7] The humble beginning of computing technologies evolved successively to Cloud Computing of these days via Centralized Computing, Distributed Computing, Mobile Computing, and Ubiquitous Computing. At the core of these evolutions lies the rapid development of the computing power of chips. In 1965, the Intel co-founder Gordon Moore announced "Moore's Law," saying that the computing power of chips doubles roughly *every 2 years*. In 2002, "Hwang's Law," originated by the former head of Samsung Electronics Semiconductor Business replaced the Moore's Law stating that memory density doubles *every year*. In 2008, however, Samsung Electronics, the world's biggest memory chip supplier, has dropped its use of the "Hwang's Law," for technical reasons.[8]

Owing to its evolution in computing powers, solution for large-scale spatial models that once considered not solvable because of the limit of computing powers,[9] become no longer an issue at least as far as computing powers are concerned.[10]

17.3.2 Implications of Pervasive Use of Instantly Available Information and Communication Tools

Constantly changing and ever evolving alternative information and communication technologies (ICTs) have prevailed in past decades that force us to consider in developing alternative spatial models for future cities and regions. Ubiquitous access technologies provide relatively new sets of concepts, practices, and standards. "The emergence of these technologies need to be understood within a broader context of both paradigm changes in computing technologies—from centralized to distributed, mobile, and ubiquitous computing—and the advancement of ICTs including ubiquitous geographic information (UBGI), radio-frequency identification (RFID), location-based services (LBS), global positioning systems (GPS), and sensing technologies (Kim 2013, p. 36)."

These technologies opened doors for providing tailored services to each individual (Kang et al. 2007). As an example for implications of utilization of these technologies, location-based service technologies enable drivers searching parking spaces without driving around the block. "Driving around the block looking for a

[7] http://en.wikipedia.org/wiki/ENIAC. Accessed in May 30, 2013.

[8] http://www.koreatimes.co.kr/www/news/biz/2008/09/123_30944.html. Accessed in October 30, 2010.

[9] Lee (1973, p. 169) cited one of reasons for discarding large-scale urban models is due to the limit of computing powers.

[10] Example of large scale spatial models and how it can be solved by the increasing computing power are examined in Sect. 17.5 below.

parking space is now a thing of the past in Santa Monica with the release of the Santa Monica Parking app for iPhone. Santa Monica Parking—Powered by ParkMe provides real-time parking information for the city's 28 lots, 12 structures, and 5,967 on-street metered parking spaces. The app also includes information on privately owned parking facilities in Santa Monica."[11]

Utilizing the ubiquitous access and the emergence of pervasive ICTs could conserve scarce resources and possibly provide means for existing cities to grow in a more sustainable and intelligent manner (Kang and Kim 2010). The rapid convergence of ubiquitous and cloud computing technologies, and ICTs is raising the possibility of a dramatic transformation in the way we perceive and move about the urban environment, and how we interact with each other in urban spaces (Kang and Kim 2011). I identified the following seven significant trends shifting away from the traditional urban space to the ubiquitous technology space in the information age:

1. From street-oriented to information-oriented,
2. From centralized information to distributed information,
3. From standardization to customization,
4. From efficiency-oriented to quality-oriented,
5. From regularity to flexibility,
6. From distance-oriented to time-oriented, and
7. From individual ownership of transportation mode to shared ownership.

In order to realize ubiquitous technology space in cities and regions, both necessary and sufficient conditions should be met. Ubiquity is not obtained through a property of the information itself, but by the information systems, technology and services that make it accessible to the user and other applications and appliances. Just as automobiles are ubiquitous due to highways and roads, electricity is ubiquitous today because of the network infrastructures of the utilities that support it. Thus a necessary condition for prevailing ubiquitous technology space is the availability of infrastructure ubiquitously as highways and electricity networks.

A new technology being available at anywhere and at anytime implies that services become available to help the general public utilize it without much a priori knowledge of the technology. When automobiles first became available, only trained professional drivers were hired to drive them. Ubiquitous access technologies should be made and available for the general public with no professional knowledge of them just as almost everyone with driver's license can drive automobiles these days. Thus availability of ease-of use technologies becomes a sufficient condition for a ubiquitous technology space to be realized.

[11] http://www.smgov.net/departments/transportation/parking.aspx. Accessed on May 29, 2013.

17.4 Implications of Shift from Personal to Communal Ownership of Automobiles on Spatial Modeling

Rifkin's claim in 2000 on changes in automobile ownership behavior is now being realized in many cities in the world. He wrote, "In an era where the automobile has been central to our way of life, our economy, and our sense of personal identity, its metamorphosis from something people own to something they lease is a sign of the dramatic changes taking place in the organizing of economic relationships (Rifkin 2000, p. 74)."

"Carsharing began in Switzerland and Germany in the 1990s and spread to the United States via Portland, OR, in 1998. By 2012, there were more than 806,000 members in over 26 carsharing organizations in the United States."[12] Robin Chase, co-founder of Zipcar, said, "I don't think Americans have forgotten how to share, and now as the cost of owning your own car has increased, the desire to share has gone up."[13]

Zipcar provides car sharing and car club service as an alternative to traditional car rental and car ownership. In fact, in 170 plus cities in the world Zipcar services available as of May 2013.[14] A similar service is being provided in Madison, WI, called Community Car. Community Car is a Madison-based carsharing company that rents cars by the hour to individuals and organizations. Members share access to a fleet of high gas mileage vehicles located in reserved parking spaces throughout the city.[15]

In fact, carsharing has helped reduce the number of cars in an urban area and supported a shift away from automobile dependency and towards sustainable mobility. A recent study (Martin and Shaheen 2011) found that between 9–13 vehicles have been either eliminated or avoided (not purchased) for every carsharing vehicle in the fleet.[16] The study also found that on average, drivers traded vehicles with a fuel economy of 23 mpg for the use of shared vehicles with a fuel economy of 33 mpg. While car sharing organizations offer a variety of vehicle types to members, the majority are highly efficient hybrids, sedans, and compact cars.

Among many implications of the popular use of carsharing, elimination of 9–13 vehicles is very significant impacts on land use since it means the possibility of using 9–13 parking spaces for alternative purposes such as open space, if not opt for other activities, particularly in the urban area where land is one of the scarcest resources.

---

[12] http://www.communitycar.com/faq/. Accessed on June 25, 2013.

[13] http://www.mnn.com/green-tech/transportation/stories/communal-cars. Accessed on April 25, 2013.

[14] http://www.zipcar.com. Accessed on May 1, 2013.

[15] http://www.communitycar.com/. Accessed on June 25, 2013.

[16] Martin, Elliot and Susan Shaheen, The Impact of Carsharing on Household Vehicle Ownership, Access, 2011, No. 38, accessed Internet site, http://www.uctc.net/access/38/access38_carsharing_ownership.pdf on June 23, 2013.

17.5 Improving Scientific Reasoning and Methods for Conventional Spatial Models

17.5.1 Three Dimensionality of Urban Activities

Among many, the two most scare resources in any urban areas could be "land" and "capital". One of the distinctive features that cities are different from rural areas is the intensive use of urban land. The intensive use of urban land is implicitly analyzed by earlier founding fathers of location theory using land rent functions (see e.g. Isard 1956; pp. 172–206). This fact alone manifests that the traditional two-dimensional urban activity models cannot describe nor analyze the urban issues adequately because urban activities prevail in three-dimensional spaces and thus they have to be analyzed by three-dimensional spatial activity models. The three-dimensional urban activities can be presented in a spatil urban model by introducing three-dimensional input-output models as shown in Mills (1972, 1974, 1975), Kim (1978a, b, 1979, 1983, 1997) and in Rho and Kim (1989). Figure 17.1 show how three-dimensional urban activities cab be represented in an input-out model context.

Conventional wisdom says that a compact city is more energy efficient than a city of similar size with dispersed patterns of land use and lower densities. The transportation costs for horizontal movements would certainly be higher in the latter. However, the conventional wisdom needs to be carefully analyzed for developing spatial analysis for future cities.

In any economic establishment, lower structures have been preferred over high-rise buildings as long as the cost of access to work places is not prohibitive (Gordon and Richardson 1997). And yet the reason that skyscrapers exist in a city is due to the scarcity of land. High-rise buildings require additional resources to maintain and operate the elevators, deliver water to the full height of the building, and remove large concentrations of waste. For example, the activities conducted within the Empire State Building use 181,000 gal of water, produce about 35 t of waste, and consume 0.1 million kWh each day. The annual cost for energy for the building is estimated to be $11.6 million[17] (Kim 2013, pp. 39). Willis Tower in Chicago uses a total annual energy equivalent to 187,500 barrels of crude oils.[18] Two-dimensional models of urban planning are unable to properly analyze urban activities since two-dimensional models cannot capture the three-dimensional nature of urban activities (Kim 1989; Rho and Kim 1989).

[17] http://www.esbnyc.com/tourism/tourism_facts.cfm?CFID=26581266&CFTOKEN=60379676. Accessed on January 20, 2012.

[18] http://01941e2.netsolhost.com/icon/documents/Icon%20News%20Release%20-%20Announcement.pdf. Accessed on January 20, 2012.

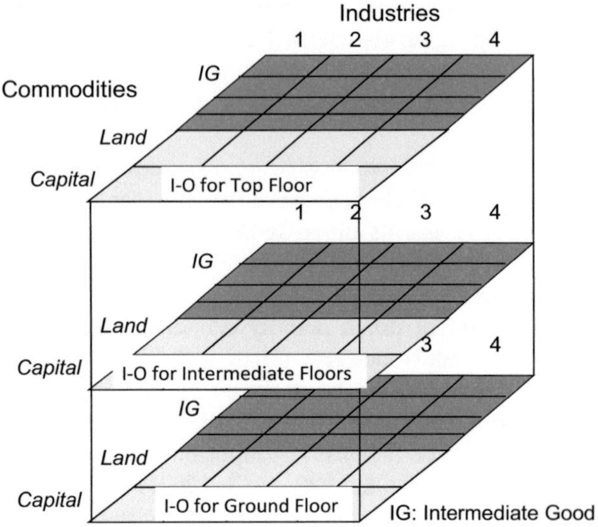

Fig. 17.1 Conceptual three-dimensional urban input-output framework

17.5.2 Multi-Jurisdictional Decision Making Processes in a Megacity

As described above, all metropolitan areas consist with multiple independent jurisdictions which make own welfare decisions posing difficult for deriving system wide decision in which all individual jurisdiction's interests combined. The issue has been identified as early as in 1956 by Isard who wrote, "...we need to know how much more about the structure of social groups, ranging from household (family) through neighborhood, community, towns, metropolis, major region. . ...." (Isard 1956, p. 684). Continuing, he further wrote, "Speaking even more broadly, if considerations of welfare (efficiency) and equity (as reflected to some extent in local community participation) can be linked together even in a very partial manner, further advances is to be anticipated." (1956, p. 687).

There are a host of matters to be coordinated for the system as a whole in a megacity. For example, transportation networks do not stop at city boundaries, air pollution does not recognize jurisdictional boundaries, and work and home locations are frequently not in the same city. The implications for building for spatial analysis models become clear: how can we identify the interests of the system and individual parts in an urban spatial model? We need a model for holistic solutions across multi-jurisdictions balancing between two overriding concerns: the welfare of the whole megacity and the interests of Individual Jurisdiction. To make effective plans for megacities, scientific reasoning needs to be exercised and methods must be developed to answer many critical questions including the following:

- How are holistic solutions planned across multi-jurisdictions?
- How can a balance between two conflicting concerns, i.e. the welfare of the megacity and the interests of the individual city, be planned for?

In answering those questions, scientific reasoning and methods developed in game theory and operations research could shed light on many issues. Many megacity issues can be formulated by using the theoretical framework of Stackelberg's leader-follower game where such organization as the metropolitan planning organization (MPO) is the leader and each city is a follower. Many such issues could also be formulated in multi-level programming problems (e.g. see Ben-Ayed et al. 1992; Suh and Kim 1992).[19]

The most effective transportation plans at the megacity level are not necessarily the sum of the most effective transportation plans for the individual cities. For example, the total minimum travel cost of a system as a whole in a megacity is not the sum of the minimum travel costs of the individual cities. Knowing that each individual city's transportation plans are usually based on findings regarding a traveler's perceived minimum cost travel patterns, an example of a multi-level programming model could be:

- Upper Level Objective Function: Minimize Total System's Travel Time
- Lower Level Objective Function: Minimize Individual City's Total Travel Time
  - Subject To: Minimize Each Traveler's Perceived Travel Time

Under this formulation, the system's optimum can be found assuming that MPOs and/or COGs function as the leader while preserving each individual city's functions as a follower, but maximizing each individual city's welfare.

17.5.3 Investment Decision Making Processes for Transport Infrastructure

Currently, making decisions about investing in transport infrastructure are based on cost-benefit analysis results by comparing the anticipated benefits and costs. The costs usually include land acquisition, construction, and maintenance cost over the life span of the infrastructure. The benefits are more difficult, but normally include savings in travel time and maintenance costs, accident reductions and environmental benefits. Since the benefit is assumed to occur to each individual traveler, the more travelers use the infrastructure, the more benefits.

Often, a number of alternative scenarios are evaluated by estimating the net benefit and the scenario that results in the highest net benefit has been chosen for implementation. If there is only one alternative scenario, the net benefits are estimated for each case with and without the scenario, providing the decision maker clues for go or no-go for the project. Frequently, transportation demand models are used for the estimating, forecasting number of passenger-car unit equivalent unit (PCU) for each link in the study area for the cost-benefit analysis.

In the past, planners and decision makers focused on the elimination of congestion as the main goal for solving transportation problems by expanding

[19] The entire Volume 34 of the Journal of Operations Research, 1992 includes articles on the development and use of multi-level programming issues.

transportation facilities, particularly constructing more and more highways (Boyce and Kim 1987). Planners frequently used transportation demand models that are based on the "four-step" approaches for searching solutions for congestion. The four-step models always result in more transportation facilities since the models require the following inputs and trying to increase accessibility between origins and destinations:

- Economic activities represented by land use.
- Number of households and/or housing units as origins for trips.
- Number of jobs and/or floor space of industries as destination for trips.

Once the number of origin and destination trips are exogenously given and then ask for minimizing travel time between the two, one of the obvious and the most frequently suggested solutions is the increase of transportation facilities to connect the two activity sites. The four-step models also produce highway biased mode choice since

- The mode choice algorithms are based on probabilities for choosing transportation mode by travelers,
- Probabilities are calculated based on the out-of-pocket trip cost, and
- The out-of-pocket cost does not represent the true social cost since gasoline price in the USA does not represent the true economic cost.

In a typical analysis model for transportation demand such as in the four-step models, modelers assume that travelers are rational and choose the alternative that gives them the highest utility, frequently measures by travel time and out-of pocket costs. Since travel time savings tend to be the largest element of benefits, the alternative that gives the largest travel time saving is probably going to give the largest benefit. In this process, automobile is usually perceived superior choice for any given journey by the traveler since the door-to-door travel time is usually shorter for automobile drivers due to a number of transfers, walking, and waiting time which are a part of total travel time for mass transit. The optimal solution is likely to be the one that attract more use of automobiles.

The fact that results from the trip generation step are used as given to the consequence steps in mode choice, trip distribution and route choice implies that transportation facilities should meet demand originating from residential areas wherever and whenever developers choose to build. There is no policy variable affecting the shape of urban form and structure in it nor there is any feed-back mechanism how the four-step process results would affect land use decisions. In fact, Boyce (2002) proved the counter-productivity of the four-step transportation planning processes.

The inelasticity of travel cost with respect to the commuting distance (Litman 2013) implies that people usually do not take the reduction in travel time as an opportunity to spend the time saved in other activities: instead they tend to travel further, sometimes as part of a long term decision about where to live. One could argue that this has led to much greater benefits to those with high incomes than those with low incomes and has contributed to the decentralization of urban areas which makes it difficult to encourage people to switch from the car to alternative modes. Thus, the conventional transportation planning model would most likely recommend increase of highway capacities to connect the origin and destination, further affecting the use of automobiles. “Why are we happy to use a methodology that encourages investment in schemes that help those with highest incomes travel further and faster, but place little or no value on helping those with low incomes or who are excluded from society to reach local opportunities to meet basic necessities in life such as shopping and social interaction?” (Mackett 2009, pp. 7). Alternative approaches need to be developed that allocate scarce urban land resources efficiently to accommodate both growing demand for transportation and land use activities. An example of such approaches has been explained above in Sect. 17.5.1, alluding need for building and solving three-dimensional urban activity model (Kim 1989).

17.6 A Framework for Spatial Models for Future Cities

A major challenge for developing spatial models for future cities is to recognize the rapid developments in socio-economic variable surrounding us and to learn how to integrate these changes into conventional spatial models. Critical questions raised above in relation to the rapid changes in the socio-economic activities include:

- How can we provide much needed resources to accommodate the additional urban population?
- How can a balance between two conflicting concerns, i.e. the welfare of the megacity and the interests of the individual city, be planned for?
- How can we utilize ITCs and pervasive computing technologies in building spatial models for future cities?
- How can we accommodate the behavioral changes of automobile ownership trends?

At the minimum, spatial models for future cities should be able to include the following criteria and variables:

- Criteria
  - Criteria should include to maximize the welfare of the metropolis as a whole and, and at the same time, that maximize the welfare of individual member jurisdictions.
  - Welfare criteria should be able to evaluate policy objectives for the benefit of the citizens in a city.

- Resource Efficiency
  - Spatial models for future cities should be the one that minimize consumption of scarce resources including land, capital, and other environmental resources including energy.[20] The models should be able to accommodate three-dimensionality of urban activities in order to minimize the consumption of the scarce urban resources.

[20] An excellent work on urban resource efficiency can be found in, http://sustainabledevelopment.un.org/content/documents/1124SustainableResourceEfficientCities.pdf. Accessed on February 24, 2014.

- Infrastructure
  - Recognizing that congestion is a result of resource allocation or misallocation, alternative spatial models should be able to allocate resources optimally among all urban activities.
  - Alternative models should be able to accommodate impacts of pervasive use of instantly available information and communication tools.
  - New trends of communal ownership of automobiles need to be reflected in the models.

This chapter stops short of formulating and presenting a spatial model for future cities accommodating all those criteria and variables suggested above. Rather it stops at identifying various variables to be considered and presented a framework for such a model. As a starting point for such building spatial models for future cities that could embrace all or a few of those critical questions described above, a spatial three-dimensional urban activity model is shown in the Appendix. While the original model was presented in the regional context,[21] it has been altered to fit in urban context. The model is able to analyze three-dimensionality of urban activities and simultaneously solves three-dimensional land uses and transportation requirements while consuming for the least resources to produce, given amount of final demand within the city.

The model in Appendix could accommodate the following challenges raised above:

- Criteria
  - The objective function is a welfare criterion, which is able to evaluate policy objectives for the benefit of the citizens in a city.
- Resource Conservation
  - The model minimizes consumption of scarce resources including land, capital, and other environmental resources including energy. The model analyzes the three-dimensionality of urban activities.
- Infrastructure
  - The model able allocates resources optimally among all urban activities.

However, the model in Appendix could not answer the questions nor accommodated critical challenges raised above as and further research on development of alternative urban spatial models is warranted.

- Criteria
  - Evaluative criteria should be able to maximize the welfare of the metropolis as a whole and, and at the same time, that maximize the welfare of individual member jurisdictions. However, The model can be modified to include criteria that enhance welfare of both metropolis and individual member jurisdictions by formulating in bi-level programming framework as shown in Kim (1990).
- Infrastructure
  - Alternative models should be able to accommodate impacts of pervasive use of instantly available information and communication tools.
  - New trends of communal ownership of automobiles need to be reflected in the models.

[21] The original model has been published in Ham et al. (2005).

Appendix: Integrated Three-Dimensional Urban Activity Model: Formulation, Optimality Conditions and Solution Algorithm[22]

Model Formulation

This model formulation may be regarded as integrating a Leontief-Strout-Wilson (Leontief and Strout 1963; Wilson 1970) interregional commodity shipment model with a transportation network model. Exports and imports of commodities in an urban area are given by the urban input-output models for the period of analysis taken as 1 year. Factor input includes labor so that commuters are included in the model as well as commodity flows among traffic zones. Then the problem is to predict the shipments by sector/commodity between each pair of zones and the transportation network flows resulting from these shipments.

Let f_a^w denote the total flow (tons) on link a of mode w over an effective distance (miles) of $d_a^w(f_a^w)$. For this prototype model, link distances are either considered to be fixed, or to be effectively longer than the actual length in the case of congested road links, representing the additional operating costs of congestion. The traditional Bureau of Public Roads (BPR) link performance function is utilized for roads to represent this effective distance, which increases as a function of the flow to capacity ratio raised to a positive power. Sensitivity analysis results are reported for powers of 0 (no congestion), 4 and 8. For mass transit modes including railways, link distances are fixed, corresponding to the assumption of unlimited capacity.

Let h_{ijr}^{mw} denote the flow (tons) of output of sector m from region i to region j by mode w on route r, and $\sum_s x^{smw}$ denote the shipment ($) of output of sector m (including housing outputs assuming one per each labor[23]) produced in s story building from zone i to zone j by mode w. To define the relationships among these variables and other quantities, consider the following definitions:

1. Link flow equals the sum of the route flows on all routes between all zones using link a of mode w;

$$\sum_m \sum_{ijr} h_{ijr}^{mw} \varphi_{ijr}^{aw} = f_a^w, \text{ for all links } a \text{ and modes } w. \tag{17.1}$$

   where φ_{ijr}^{aw} is 1, if route r from i to j uses link a of mode w, and 0, otherwise.

2. The materials balance constraint states the total shipment of commodity m into zone j from all zones i is equal to the use of that commodity for producing all commodities (intermediate demand) in s story of production facilities plus zonal final demand.

[22] The original paper title is "Implementation and estimation of a combined model of interregional, multimodal commodity shipments and transportation network flows", as described in Footnote 24.

[23] Assuming each household requires a house, m denotes labor in input sector, but housing outputs in output sector.

$$\sum_{si} x_{ij}^{sm} = \sum_{sn} a_j^{smn} \sum_k x_{jk}^n + \sum y_j^m, \text{for all stories } s, \text{sectors } m \text{ and regions } j. \tag{17.2}$$

where a_j^{smn} is the quantity of inputs from sector m required to produce one unit of output of sector n at *s-intensity* of land use (i.e. s-story building) in zone j, the quantities being measured in value terms, and y_j^m is the final demand for the output of sector m in zone j. The materials balance constraint represents the conservation of shipments between origin and destination zones. Zones are assumed to be indifferent regarding the source of the supply of commodities.

3. The sum of shipments of sector m from zone i to zone j by all modes equals the total shipment of sector m from zone i to zone j.

$$\sum_{sm} x_{ij}^{smw} = x_{ij}^m, \text{for all sectors } m, \text{and regions } i, j. \tag{17.3}$$

4. The total flow (tons) of sector m on all routes r of mode w between regions i and j equals the total shipment ($) of sector m from i to j by mode w divided by an exogenous ratio gm ($/ton) that converts commodity shipments from value of commodity shipments to weight of network flows.

$$\sum_r h_{ijr}^{mw} = \frac{x_{ij}^{mw}}{g^m}, \text{for all sectors } m, \text{all modes } w, \text{and regions } i, j \tag{17.4}$$

The ratio is derived from the 1993 Commodity Flow Survey by dividing the total value of shipments ($) by total weight of the shipments (tons) throughout the U.S.[24]

5. Finally, route flows must be nonnegative, which implies that links flows are also nonnegative:

$$h_{ijr}^{mw} \geq 0, \text{for all sectors } m, \text{and all routes } r \text{ of mode } w \text{ and zones } i, j \tag{17.5}$$

Having defined the constraints on interregional and route flows, the objective function is as follows. On the assumption that shippers collectively desire to minimize their total shipment distances and thus the dispersion of shipments across origins and destinations and modes can be represented by interzonal and modal dispersion (entropy) functions, the objective function is defined as follows:

$$\begin{aligned} \min_{h,x} (h, x) = &\sum_{aw} \int_0^{{}^w_a} d_a^w(\omega) d\omega \\ &+ \sum_m \left[\frac{1}{g^m} \sum_{ij} x_{ij}^m \ln \left(\frac{x_{ij}^m}{X_i^m} \right) \right] + \left[\sum_m \frac{1}{\alpha^m g^m} \sum_{ijws} x_{ij}^{smw} \ln \left(\frac{x_{ij}^{smw}}{x_{ij}^m} \right) \right] \end{aligned} \tag{17.6}$$

[24] The values of these ratios are shown in Table 3 of Ham et al. (2005).

The exogenous parameters and variables are: α^m, the modal distance sensitivity parameter for sector m; g^m, the ratio of monetary shipment value to weight used to convert the output of sector m from dollars to tons ($/ton); and X_i^m, the total output of sector m in zone i ($).

The total cost to be minimized is represented by the sum of the link flows times link distances by mode, the interzonal commodity shipments by mode, and the freight and passenger flows by routes and links within modes. Minimization of the objective function also results in the optimal use of land in zone i represented by x_i^{smw}, in three-dimensional variable of s. Not all of these commodity shipments and flows, however, are uniquely determined by the solution.

At the same time, there exist caveats for entropy maximization models and spatial models with gravity formulation (Nijkamp and Reggiani 1988). Nijkamp (1975) proved that "entropy results lead to an underestimation, and linear programming results in an overestimation of intraregional flows. For irrterregional flows, however, the entropy outcomes appear to overestimation: the actual flows, whereas the linear programming out- comes appear to underestimate the actual flows. Both methods appear to lead to considerable prediction errors, which can be mainly explained from the heterogeneity of the commodities in question."

Lagrangian Analysis and Optimality Conditions

The Lagrangian function for deriving the optimality conditions is constructed using objective function (6), constraints (1–5), and Lagrangian multipliers as follows:

$$\min_{(h,x,\mu,\theta,\delta)} L(h,x,\mu,\theta,\delta) = \sum_{aw} \int_0^{w_a} d_a^w(\omega)d\omega + \sum_m \left[\frac{1}{g^m}\sum_{ij} x_{ij}^m \ln\left(\frac{x_{ij}^m}{X_i^m}\right)\right]$$

$$+\left[\sum_m \frac{1}{\alpha^m g^m}\sum_{ijws} x_{ij}^{smw}\ln\left(\frac{x_{ij}^{smw}}{x_{ij}^m}\right)\right] + \sum_{mj}\mu_j^m\left[\sum_{sn} a^{smn}\left(\sum_k x_{jk}^n\right) + y_j^m - \sum_j x_{ij}^m\right]$$

$$+\sum_{mij}\theta_{ij}^m\left(x_{ij}^m - \sum_m x_{ij}^{mw}\right) + \delta_{ij}^m\left(\frac{x_{ij}^{mw}}{g^m} - \sum_r h_{ijr}^{mw}\right)$$

$$s.t. \qquad h_{ijr}^{mw} \geq 0 \tag{17.7}$$

The Lagrange multipliers are $(\mu_j{}^m), (\theta_{ij}{}^m), (\delta_{ij}{}^{mw})$ in vector form. The Karush-Kuhn-Tucker optimality conditions are obtained by taking partial derivatives with respect to the unknown variables as follows:

$$\frac{\partial L}{\partial h_{ijr}^{mw}} = \sum_a d_a(f_a^w)\phi_{ijr}^{aw} - d_{ij}^{mw} \geq 0 \;\forall m,w,r,i,j \tag{17.8}$$

$$\frac{\partial L}{\partial x_{ij}^{smw}} = \frac{1}{\alpha^m g^m}\left(\ln\frac{x_{ij}^{smw}}{x_{ij}^m} + 1.0\right) - \theta_{ij}^m + \frac{\delta_{ij}^{mw}}{g^m} \geq 0 \;\forall m,w,i,j,i \neq j \tag{17.9}$$

$$\frac{\partial L}{\partial x_{ij}^m} = \frac{1}{g^m}\left(\ln\frac{x_{ij}^{mw}}{X_i^m} + 1.0\right) + \sum_n \mu_i^{nm} a^{nm} - \mu_j^m + \theta_{ij}^m - \frac{1}{\alpha^m \gamma^m}\sum_w \frac{x_{ij}^{mw}}{x_{ij}^m} \geq 0 \forall m,i,j \tag{17.10}$$

The complementary slackness conditions are:

$$h_{ijr}^{mw}\frac{\partial L}{\partial h_{ijr}^{mw}} = h_{ijr}^{mw}\left\{\sum_a d_a(f_a^w)\phi_{ijr}^{aw} - d_{ij}^{mw}\right\} = 0 \quad \forall\; mwrij \tag{17.12}$$

$$x_{ij}^{mw}\frac{\partial L}{\partial h_{ijr}^{mw}} = h_{ijr}^{mw}\left\{\frac{1}{a^m g^m}\left(\ln\frac{x_{ij}^{mw}}{x_{ij}^m} + 1.0\right) - \theta_{ij}^m + \frac{d_{ij}^{mw}}{g^m}\right\} = 0\forall\; mwij, i \neq j \tag{17.13}$$

$$x_{ij}^m\frac{\partial L}{\partial x_{ij}^m} = x_{ij}^m\left\{\frac{1}{g^m}\left(\ln\frac{x_{ij}^m}{X_i^m} + 1.0\right) + \sum_i \delta_i^n \sum_s a^{snm} - \delta_j^m + \theta_{ij}^m - \frac{1}{a^m g^m}\sum_w \frac{x_{ij}^{mw}}{x_{ij}^m}\right\} = 0 \;\forall mij \tag{17.14}$$

Conditions representing route flows and distances may be derived from Eqs. (17.8) and (17.12) as follows:

$$\begin{aligned} &1. if\; h_{ijr}^{mw} > 0, then \sum_a d_a^w(f_a^w)\varphi_{ijr}^{aw} = \delta_{ij}^{mw};\\ &2. if\; h_{ijr}^{mw} > 0, then \sum_a d_a^w(f_a^w)\varphi_{ijr}^{aw} \geq \delta_{ij}^{mw}; \end{aligned} \tag{17.15}$$

The interpretation of these conditions is as follows: if the route flow ${h_{ijr}}^{mw}$ is positive, then $\sum_a d_a^w(f_a^w)\varphi_{ijr}^{aw}$, the cost of using route r is equal to the equilibrium shipment cost ${d_{ij}}^{mw}$ from i to j by mode w for sector m; if the route flow ${h_{ijr}}^{mw}$ is zero, then the cost of using route r is not less than the equilibrium shipment cost. Hence, Lagrange multipliers ${\delta_{ij}}^{mw}$ have an interpretation of equilibrium shipment cost. These optimality conditions are equivalent to the equilibrium conditions proposed by Wardrop and Whitehead (1952) for user-optimal route choice of drivers in road networks.

Derivation of Mode Choice and Composite Cost Functions

Assuming zonal passenger and commodity shipments by mode x_{ij}^{mw} are positive, the following model may be obtained from Eq. (17.13):

$$x_{ij}^{mw} = x_{ij}^{m}\left[\exp\left(a^m g^m \theta_{ij}^{m} - 1.0\right)\right]\exp\left(-a^m \delta_{ij}^{mw}\right)\forall mwij, i \neq j \tag{17.16}$$

By substituting Eq. (17.16) into Eq. (17.3), we obtain:

$$x_{ij}^{mw} = \sum_{w} x_{ij}^{mw} = x_{ij}^{m}\left[\exp\left(a^m g^m \theta_{ij}^{m} - 1.0\right)\right]\sum_{w}\exp\left(-a^m \delta_{ij}^{mw}\right) \tag{17.17}$$

Therefore,

$$\exp(a^m g^m - 1.0) = \frac{1}{\sum_{w}\exp\left(-a^m \delta_{ij}^{mw}\right)} \tag{17.18}$$

$$x_{ij}^{mw} = x_{ij}^{m}\left[\frac{\exp\left(-a^m \delta_{ij}^{mw}\right)}{\sum_{w}\exp\left(-a^m \delta_{ij}^{mw}\right)}\right]\forall mwij, i \neq j \tag{17.19}$$

Equation (17.19) becomes a typical transportation mode choice equation based on the probability of choosing mode w among all available modes.

Interregional Interaction Functions

Again assuming that $x_{ij}^{m} > 0$, from Eqs. (17.3) and (17.15) and we obtain:

$$x_{ij}^{m} = X_{i}^{m}\left[\exp\left(-g^m \sum_{n}\mu_i^n \sum_{s} a_i^{snm} - 1.0\right)\exp\left(g^m \mu_j^m\right)\exp\left(-\tilde{d}_{ij}^{m}\right)\right]\forall mij$$
$$defining\, \tilde{d}_{ij}^{m} = -\frac{1}{a^m}\ln\left[\sum_{w}\exp\left(-a^m \delta_{ij}^{mw}\right)\right] \tag{17.20}$$

By defining the following balancing factors,

$$\beta_i^m = \exp\left(-g^m \sum_{n}\mu_i^n \sum_{s} a_i^{snm} - 1.0\right)\ \forall m, i \tag{17.21}$$

$$\varepsilon_j^m = \exp\left(g^m \mu_j^m\right) \qquad \forall m, j \tag{17.22}$$

A simpler form for x_{ij}^m is:

$$x_{ij}^m = X_i^m \beta_i^m \varepsilon_j^m \exp\left(-\tilde{d}_{ij}^m\right) \forall mij, i \neq j \tag{17.23}$$

By substituting Eq. (17.23) into Eq. (17.19), an interregional commodity shipment function by mode can be obtained:

$$x_{ij}^{mw} = X_i^m \beta_i^m \varepsilon_j^m \exp\left(-\tilde{d}_{ij}^m\right) \left[\frac{\exp\left(-a^m \delta_{ij}^{mw}\right)}{\sum_w \exp\left(-a^m \delta_{ij}^{mw}\right)} \right] \forall mwij, i \neq j \tag{17.24}$$

The amount of interzonal flows including passengers and freight in Equation (17.24) depends on the magnitude of the Lagrange multiplier associated with zone i (β_i^m) and zone j (ε_i^m), and the probability of choosing mode for shipments of passengers and freight. A comparable result is found in the auto occupancy factors of multiclass travel forecasting models in Boyce and Bar-Gera (2003).

Solution Algorithm

A generalization of the algorithm proposed by Evans (1976) was used to solve the optimization problem. In this case each iteration of the algorithm uses Wilson's (1970) iterative balancing method to generate the subproblem interregional commodity shipments, and the all-or-nothing assignment method to find the subproblem network link flows.

The convergence of Wilson's iterative balancing method can be judged by the relative error between the observed and calculated final demands for each sector and each region, or the relative change of balancing factors.[25] Unlike the experience of Rho et al. (1989), convergence was satisfactory.

References

Batty M (1976) Urban modelling: algorithms, calibrations, predictions. Cambridge University Press, Cambridge, MA

Ben-Ayed O, Blair CE, Boyce DE, LeBlanc LJ (1992) Construction of a real-world bilevel linear programming model of the highway network design problem. Ann Oper Res 34:1–4

[25] The former convergence criterion was used successfully with 0.001 as the stopping value in Ham et al. (2005).

Boyce DE (1988) Presidential address: renaissance of large-scale models. Pap Reg Sci Assoc 65:1–10
Boyce D (2002) Is the sequential travel forecasting procedure counterproductive? ASCE J Urban Plan Develop 128:169–183
Boyce D, Bar-Gera H (2003) Validation of multiclass urban travel forecasting models combining origin-destination, mode, and route choices. J Reg Sci 43:517–540
Boyce DE, Kim TJ (1987) Role of congestion of transportation networks in urban land uses and travel choices of households and firms. Transportation 1987(14):53–62
Evans SP (1976) Derivation and analysis of some models for combining trip distribution and assignment. Transport Res 10:37–57
Gordon P, Richardson HW (1997) Are compact cities a desirable planning goal? J Am Plan Assoc 63(1):95–105
Ham H, Kim TJ, Boyce DE (2005) Implementation and estimation of a combined model of interregional, multimodal commodity shipments and transportation network flows. Transport Res B 39(1):65–79
Isard W (1956) Location and space-economy. The MIT Press, Cambridge, MA
Isard W (1960) Methods of regional analysis: an introduction to regional science. The MIT Press and Wiley, Cambridge, MA, New York
Kang S, Kim TJ (2010) Combined-cost and bi-objective approaches to multipurpose location-based services using genetic algorithms. Transportation research record. J Transport Res Board, no 2160, Transportation Research Board of the National Academies, Washington, DC, pp 40–49
Kang S, Kim TJ (2011) Adaptive dynamic programming approach to a multi-purpose location-based concierge service model. IET Intell Transport Sys 5(4):277–285
Kang S, Kim TJ, Jang S-G (2007) Location-based services: enabling technologies and a concierge service model. In: Miller H (ed) Societies and cities in the age of instant access. Springer, Berlin, pp 231–233
Kim TJ (1978a) A model of zoning for a metropolis. Environ Plan A 10(6):1035–1047
Kim TJ (1978b) Effects of subways on urban form and structure. Transport Res 12–4:231–239
Kim TJ (1979) Alternative transportation modes in a land use model: a general equilibrium approach. J Urban Econ 6–2:197–215(trans: into Japanese in Kosokutoro to Jidosha (Expressways and Automobiles), 12–23 and 4, 1980)
Kim TJ (1983) A combined land-use transportation model when zonal travel demand is endogenously determined. Transport Res B 17B–6:449–462
Kim TJ (1989) Integrated urban systems modeling: theory and applications. Kluwer Academic Publishers, Norwell, MA
Kim TJ (1990) Advanced transport and spatial systems models: applications to Korea. Springer, New York
Kim TJ (1997) A combined land use-transportation model when zonal travel demand is endogenously determined. In: Berechman J, Kohno H, Button K, Nijkamp P (eds) Transport and land use, 2nd volume of modern classics in regional science. Edgar Elgar, Lyme, NH. The article was originally published in Transport Res B 1983, 17B-6:449–462
Kim TJ (2013) Scientific reasoning and methods in urban planning. Int Reg Sci Rev 36(1):36–43
Lee D (1973) Requiem for large scale urban models. J Am Inst Plan 39:163–178
Leontief WW, Strout A (1963) Multiregional input–output analysis. In: Barna T (ed) Structural interdependence and economic development. Macmillan, London
Litman T (2013) Understanding transport demands and elasticities: how prices and other factors affect travel behavior. Victoria Transport Policy Institute
Mackett RL (2009) Why is it so difficult to reduce car use? Paper presented at the European transport conference. Leeuwenhorst Conference Centre, Noordwijkerhout, near Leiden, Netherlands, 5–7 October
Martin E, Shaheen S (2011) The impact of carsharing on household vehicle ownership. Access, No 38

Mills ES (1972) Markets and efficient resource allocation in urban areas. Swed J Urban Econ 74:100–113
Mills ES (1974) Mathematical model for urban planning. In: Brown A (ed) Urban and social economics in market and planned economies. Prager, New York
Mills ES (1975) Planning and market processes in urban models. In: Crieson RE (ed) Public and urban economics: essays in Honor of William Vickrey. Heath, Lexington
Nijkamp P (1975) Reflections on gravity and entropy models. Reg Sci Urban Econ 5:203–225
Nijkamp P, Reggiani A (1988) Entropy, spatial interaction models and discrete choice analysis: static and dynamic analogies. Eur J Oper Res 36(I):186–196
Rho JH, Kim TJ (1989) Solving a three-dimensional urban activity model of land-use intensity and transportation congestion. J Reg Sci 29(4):595–613
Rho JH, Boyce DE, Kim TJ (1989) Comparison of solution methods for Wilson's interregional commodity flow model. J Geogr Anal 21(3):259–267
Rifkin J (2000) The age of access. Ken Tarcher/Putnam, New York
Suh S, Kim TJ (1992) Solving nonlinear bilevel programming models of the equilibrium network design problem: a comparative review. Ann Oper Res 1992(34):203–218
United Nations (2011) Department of economic and social affairs, population division (2011): world population prospects: the 2010 revision, New York (Extended Dataset on CD-ROM, ST/ESA/SER.A/308, Sales No. 11.XIII.7)
Wardrop JG, Whitehead JI (1952) Some theoretical aspects of road traffic research. ICE Proceedings: Engineering Divisions 1(5):767
Wegener M (1994) Operational urban models: state of the art. J Am Plan Assoc 60(1):17–30
Wilson AG (1970) Interegional commodity flows: entropy maximizing procedures. Geogr Anal 2:255–282
Wilson AG (1974) Urban and regional models in geography and planning. Wiley, London

Rise of Megalopolis as a Mega Knowledge Region: Interactions of Innovations in Transport, Information, Production, and Organizations

18

T.R. Lakshmanan, William P. Anderson, and Yena Song

Abstract

This chapter argues that the passage of the Megalopolis from a declining industrial economy to a Mega Knowledge Region in the last three decades has been made possible by a *four-part 'Knowledge Infrastructure'*: The first two components of (a) *Transport and* (b) *Information and Communication Infrastructure (ICT)* are general purpose technologies which *physically link* and provide *access* (at low and declining costs) for economic actors in a liberalized global economy, promoting economic structural changes. The passage to the Knowledge Economy requires, in addition, two other components of the 'knowledge infrastructure', namely (c) innovations in *production and service technologies that nurture and commercialize new Industrial products and operations and the structural evolution towards a networked Knowledge-Intensive Service Economy*. The final component (d) is a set of *Institutional Innovations*—which support *physical, relational and institutional proximity* among far-flung economic agents—thereby promoting *new knowledge and value generation* in enterprises in the Megalopolis. The Boston metropolis has arrested its decline, reinventing itself as an entrepreneurial knowledge economy, focusing on knowledge-intensive production and services—Biotechnology, Scientific instruments, Software, Finance services, Producer services, Medical and Educational services—achieving this remarkable reinvention by building on its *vast stocks and variety of knowledge assets, networking opportunities, rivalry*

T.R. Lakshmanan (✉)
Boston University, 675 Commonwealth Ave., Boston, MA 02215, USA
e-mail: laksh@bu.edu

W.P. Anderson
University of Windsor, Windsor, Canada
e-mail: bander@uwindsor.ca

Y. Song
Engineering and the Environment, University of Southampton, Highfield, Southampton, UK
e-mail: Y.Song@soton.ac.uk

P. Nijkamp et al. (eds.), *Regional Science Matters*, DOI 10.1007/978-3-319-07305-7_18,

and trust-promoting contexts, and its history of institutional innovations and social cooperation. The rise of 'Tradable' and 'Non-Tradable' sectors in the Globalizing Boston metro economy is, however, widening income inequalities.

18.1 Introduction and Overview

This paper documents *the nature and scope of and the factors underlying the economic structural evolution over the last three decades towards a Mega Knowledge Region* along the dynamic settlement corridor stretching from Boston to Washington, DC in the U.S., christened by Jean Gottman (1957, 1961) as the *Megalopolis*.

The Megalopolis, stretching from the Boston Metropolitan Region in the north to the Washington Metropolitan Region in the south, comprises some of the country's oldest settled areas, locales of earliest industrialization, a key component of the 'US Manufacturing Belt', and of early urbanization and of the densest mega–urban region by mid-twentieth century. In the quarter century following 1950, the metropolitan centers of the Megalopolis experienced industrial restructuring and decline (with industrial enterprises spreading to other regions in US and abroad) with declining shares of national manufacturing and total employment. However, in the subsequent three decades to the present day, there has been a remarkable *reversal* of regional economic decline and indeed a notable economic resurgence in the metro areas of the Megalopolis. A variety of knowledge-intensive production and service sector enterprises have arrived and continue to grow in the several metro areas of the Megalopolis, which now represent large concentrations of high quality human, cultural, and organizational capital, and of some vibrant contemporary 'Knowledge Metropolises'.

Specifically the aim of the paper is *first*, to delineate the nature and scope of this recent economic dynamism in the Boston-Washington, D.C. (BOSWASH) corridor, and *second*, to identify the role of improvements in transport (physical and institutional) infrastructures in the above economic evolution over the last three to four decades. The paper outlines the many economic mechanisms through which transport investments in various transport modes support economic growth and development along these transport corridors. While the important role of such transport-induced improvements in corridor economic evolution is presented, that is only part of the story of economic evolution in the Megalopolis. The *third* aim of the paper is to present the argument that the economic evolution in the BOSWASH corridor, *in addition to transport factors,* reflects the effects of several other ongoing economic structural change processes. These include:

(a) The *rise* in the last three decades of *Globalization*—an explosive expansion of cross-country economic interactions, division of labor, complex webs of production chains, and a globally distributed production system—promoted by institutional innovations promoting freer trade and investment and made possible by physical and organizational innovations in Transport and

Information and Communication (ICT) technologies, and the emergence of corporate central organizational service functions (financial, legal, accounting, and other business and professional services) that permit business operations in multiple countries,

(b) A twofold shift to a *'Knowledge Economy'*, comprising of *first* a shift away from production that is dependent on material resources, physical capital, and low skill labor to one which increasingly exploits knowledge as the key ingredient of competitiveness and innovation, and *second, the emergence of* knowledge-intensive business services, which promote a growing shielding of manufacturing products in a 'service jacket', which exploit the economies of scale in human capital, and which promote knowledge dissemination and productivity effects in other parts of the economy and,

(c) The growing role of *spatial proximity* in offering in large metropolitan areas knowledge-rich environments. Such knowledge-intensive environments offer (1) service (urban) spaces where enterprises can function as knowledge, innovation, and expertise transfer agents [as in (b) above], and (2) where a disproportionate concentration of the corporate coordinating central functions [noted in (a) above] thrive in large metro areas in the Megalopolis.

Thus as Economic History and Geography suggest, economic evolutionary developments in the Megalopolis over recent decades are *joint consequences* of innovations in transport and Information and Communication technologies (ICT), the emergence of global economy, and the onset of dynamic new knowledge in the form of physical and institutional technologies in the manufacturing and service economies. This paper aims to offer a rich nuanced treatment of this complex subject of the multi-dimensional economic structural change and evolution in the Megalopolis.

Part II of the paper focuses on the rise and evolution of a multivalent transport system in the Megalopolis. Over recent decades, U.S. transport systems have undergone a major transformation, induced by new transport technologies and major institutional reforms pertaining to overall economic governance of transport, in the context of new information technologies. Such changes have promoted new transport logistics capacities in an increasingly global economy, where goods are sourced and markets serviced globally. In this context, the historically modally organized transport system in the Megalopolis has transformed itself into a *multi-valent and multimodal transport system*, which in turn unleashes various economic mechanisms and processes underlying the broader economic consequences of transport investments over time. Part II organizes such economic mechanisms into (1) Gains from Trade, (2) Technology Diffusion, and (3) Gains from urban agglomerations induced by transport improvements.

Part III provides a survey of the demographic, urban, and economic evolution over recent decades in the BOSWASH corridor. An analysis of urban development patterns in the I-95 Corridor over a long time (six decades) suggests different patterns of urban development of *urban decline and resurgence.* This evolution ranges from the dense and compact *industrial era spatial pattern* (pre 1957—manufacturing production and working class households concentrated in dense central cities) to the *Decentralized Spread City* (1957 to the early 1980s) in an increasingly affluent economy, to the contemporary (post 1985) *trends towards agglomeration and metropolitan clustering of 'Knowledge Economy' activities* along the metropolitan regions in the Megalopolis.

Part IV offers a brief survey of (a) the various interactive technological and institutional processes underlying the transformation of the Megalopolis and the reinvention of major metro areas in the Megalopolis in recent decades as Knowledge economies and hubs of creativity and (b) how such prepositions on structural change and economic evolution offered here hold up in an empirical review of the Boston Metropolitan Region's economic experience and its passage to a 'Knowledge economy' in the last three decades.

The literature on such economic transformation processes is highlighted here in terms of (1) current theoretical formulations of the rise of the 'Knowledge production economy' and (2) the contributions that the growing knowledge-intensive business services make in the form of positive knowledge and productivity spillovers to other industries in the broader economy. The scale and scope of the growth and evolution of such knowledge-intensive production sectors over recent decades in the Megalopolis and its component five large metro areas are presented. Next, the vibrant growth and evolution of knowledge-intensive business sectors and their functioning as knowledge, innovation, and expertise transfer agents in the Megalopolis are highlighted.

Part V of this paper implements for the Megalopolis the Spence-Hlatshway (2011) *national analysis* of economic performance in the last two decades of "Tradable" and "Non-Tradable" sectors to capture their differential economic performance in terms of how value-adding chains in an economic sector are structured by global trade requirements and thus affect employment and value-added per employee by sector. The performance of these two classes of sectors in the Megalopolis metropolitan areas, which have high levels of knowledge-intensive manufacturing and service sectors, and of global trade participation is presented. There emerge *adverse income distribution implications* of this two sector economic evolution and some inferences of the *widening income inequalities* in the Megalopolis metro areas.

Part VI concludes the paper.

18.2 Transport Networks in the Megalopolis: Their Evolution and Economic Contributions

18.2.1 Network Evolution

Over the last four to five decades, the highways, railroads, waterways, ports, and the aviation system in the Megalopolis have evolved from those serving the dense city-based transport networks of the immediate post-World War industrial era to the contemporary nationally integrated multimodal transport network enabling seamless transportation, serving a nationally integrated production system, and more recently a globally distributed increasingly knowledge-intensive production economy.

This transformation of the transport system in recent decades has been driven by (1) *three* key transport *technical* innovations—*The Interstate Highway System, Containers, and Jet aircraft—and by* (2) *Organizational Innovations—Deregulation and Privatization—which greatly enhance the economic governance of transport, physical flows and border controls* (Lakshmanan and Anderson 2002). The Interstate Highway system vastly enhanced interurban and intra urban mobility and led to steep drops in regional and national passenger and freight times and costs. Containers, first introduced in 1966 in the Megalopolis by U.S. shipping lines on the North Atlantic routes, have greatly enhanced shipping efficiency and rapidly grown and spread worldwide, facilitating intermodal transport of freight (Levinson 2006). The technical evolution of Jets and the aviation sector have led to sharp drops in travel times, air freight prices, and to growing volumes of high value-added goods being shipped by air domestically and globally. The institutional innovations in the form of deregulation and privatization of transport have enabled a broad range of transport service and process innovations and logistical improvements. The reform of rules governing transport physical flows (e.g. vehicle size/wt. rules, reinvented inspection processes) have not only enlarged transport capacities on routes and terminals and logistical potential, but have stimulated new "service opportunities",[1] which are critical to the operation of a globalized economy (Lakshmanan et al. 2009).

18.2.2 Mechanisms Linking Transport Improvements and Economic Evolution

As transport infrastructure and service improvements lower costs and increase accessibility to various market actors—input suppliers, labor, and customers—market expansion, increased integration and mutually sustaining economic growth can ensue (Lakshmanan and Chatterjee 2005). The underlying economic mechanisms, which are illustrated in Fig. 18.1, are:

1. ***Gains from Trade:*** As transport improvements expand markets for firms across regions, there is a shift from local and regional autarky to increasing specialization and trade, the consequent surge in productivity, and "Smithian" growth. Export expansion will lead to higher levels of output and efficiencies, and increasing imports put competitive pressures on local prices, removal of monopoly rents, thereby improving efficiency. Firms are able over time to draw from a broader labor pool, thus improving labor supply at lower costs. Similar effects in land and other factor markets are likely as transport improvements open up new land for

[1] For example, Precise scheduling (just in-time systems) market expansion, and spatial agglomeration of activities.

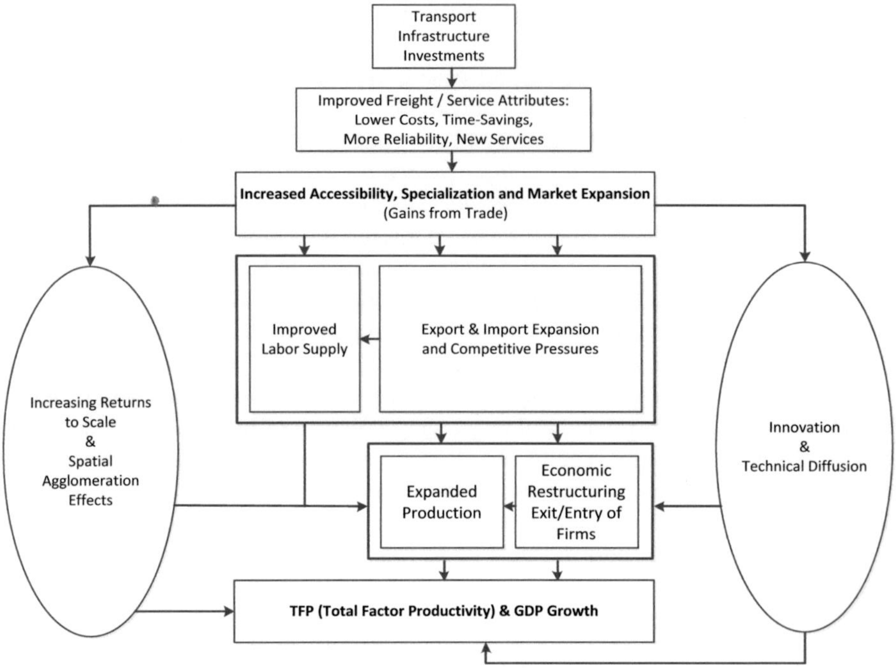

Fig. 18.1 Transport infrastructures and economy-wide benefits

economic activities, thereby promoting a highly integrated national economy and over time a globalized economy (Anderson and Lakshmanan 2007).

2. ***Technology Diffusion**:* Associated with the growth of interregional and cross-country trade there has been an upsurge in intra-industry trade and the exchange of intermediate goods. There are opportunities here for adopting new technical knowledge associated with imports. There is potential for knowledge and technology growth in the process of 'learning by doing' and 'learning by using', *and*
3. ***Gains from Agglomerations made possible by transport***: In an increasingly knowledge-intensive economy, some urban agglomerations enjoy *increasing returns* in the form of dynamic location advantages, enabling innovation and dynamic competitiveness of these cities. The three broad modeling approaches which try to capture these advantages are:
   (a) *Economic Geography and Location Economics Models:* The micro foundations of the Marshall's (1890) agglomeration economies suggest (a) *input sharing* among firms, (b) *matching* in metro areas of workers' skills with the diverse requirements of employers, and (c) *knowledge spillovers or learning* in these dense locations, with workers being the primary vehicles of these transfers—such ideas further elaborated by Hoover (1948) and Isard (1956).
   (b) *New Economic Geography Model*: Krugman (1991) applied a general equilibrium modeling framework to the geography of the economy under

conditions of increasing returns to scale and labor mobility, reinterpreting the findings of Marshall on agglomerations. In the resulting 'New Economic Geography' (NEG) model, spatial concentration and dispersion emerge (Fujita et al. 1999). The NEG model, however, accounts for only pecuniary economies, makes no mention of either human capital or technological spillovers, and has no answers for the contemporary knowledge creation process and innovation-led growth in urban agglomerations.

(c) *Innovation-led Agglomerations*: This class of models views the increasing returns associated with regional agglomerations or metros as dynamic location advantages attributable to: (1) *Physical Proximity* among economic actors, facilitating interactions and enabling access to appropriation and sharing of tacit knowledge, thus promoting innovation (Gertler 2003), (2) *Relational Proximity* of economic agents, facilitating cooperative behavior, collective learning, and socialization of innovation risk, (3) *Institutional Proximity* among the firms in the urban agglomeration in terms of shared rules, codes and norms of behavior which will promote cooperation in interactive learning processes (Camagni 2005; Capello 2011; Amin and Cohendet 1999), and (4) Lowering of *Adaptive Costs* among firms competing in an environment of rapid pace of change of knowledge (Lakshmanan and Button 2009).

18.2.3 Improvements in Metro Transit, High Speed Rail, and Knowledge-Sector Agglomerations

In the 2000–2007 period, there was a pronounced shift from earlier patterns in New York from auto commuting to public transportation, an increase in daily subway ridership, and an upsurge in bicycling (Fig. 18.2). It was also a period of considerable growth in New York in the spatially agglomerated knowledge-intensive sectors such as Finance and Business services.

Over the last decade, a high speed rail service (Acela Express) has operated between Boston and Washington, DC with a maximum speed of 241 kph and a daily passenger load of 28,000 (in 2010) along a 585 km, 14 station corridor characterized by an average number of 178,645 jobs and 219,925 residents (2008) within a 5 km range (Murakami and Cervero 2012). Murakami and Cervero's analysis of the economic stimuli provided around High Speed Rail (HSR) stations in the Tokyo-Osaka and the BOSWASH corridors (and in London a major global economy node) appears to support the symbiotic relationships (noted above) between HSR improvements and the location and growth of knowledge-intensive economic sectors around major HSR stations in metros such as New York, a top node in the global economy.

Contemporaneous with the vast improvements in the physical and institutional transport infrastructures noted here, there were major improvements in Information and communication technologies (ICT). As has been noted widely, ICT contributes greatly to productivity, profitability and growth at the firm level. Thus Transport

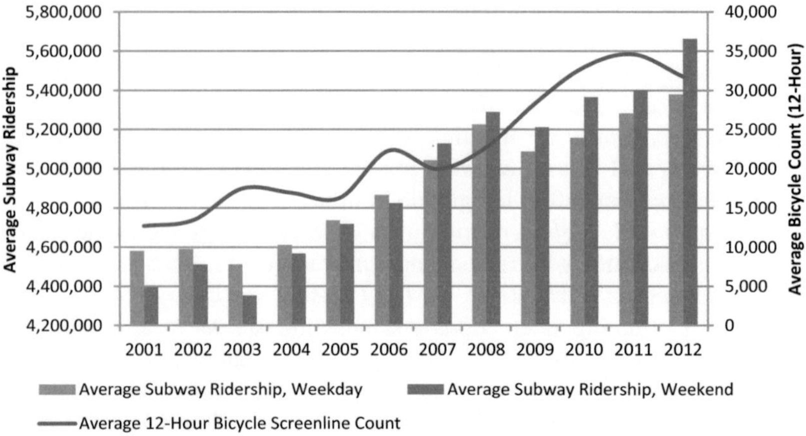

Fig. 18.2 Public transit and bicycle trends in New York Metropolitan Area (*Source*: Subway Ridership, Metropolitan Transportation Authority. Bicycle Screenline Count, Department of Transportation, New York City)

and Information innovations function as *General Purpose technologies (GPTs)*, which support and undergird a variety of *production and organizational technologies*—which jointly drive structural change and economic evolution towards the *Knowledge economy* in the Megalopolis.[2]

Finally, a major consequence of such innovations in both transport and ICT infrastructures has been the economic structural change of *Globalization* of the US and the Megalopolis economies, namely the ability to decompose and organize production into many value-adding components, many of which are fabricated in a variety of locations around the world. It appears that *the Transport and ICT systems and the emerging economy in Megalopolis both direct and adapt to change*.

18.3 Demographic and Urban/Economic Evolution of the Megalopolis

The key role of the Boston-Washington corridor or the urbanized Northeast U.S. in the national economy was anticipated by the French Geographer, Jean Gottmann (1961). He called the stretch of urban, suburban and rural areas from Southern New

[2] General Purpose Technologies (GPTs) are a few key technologies which appear to drive critical eras of technical progress and economic growth (Breshnehan and Trajtenberg 1992; David and Wright 1999). GPTs become pervasive in the economy in certain periods, as they are utilized as inputs by a large number of downstream sectors and with considerable potential for technical improvements and innovation. Thus, as GPT's improve, they spread throughout the economy, bringing about generalized productivity gains. The Steam engine and the Electric motor are commonly regarded as critical GPTs of the Industrial era.

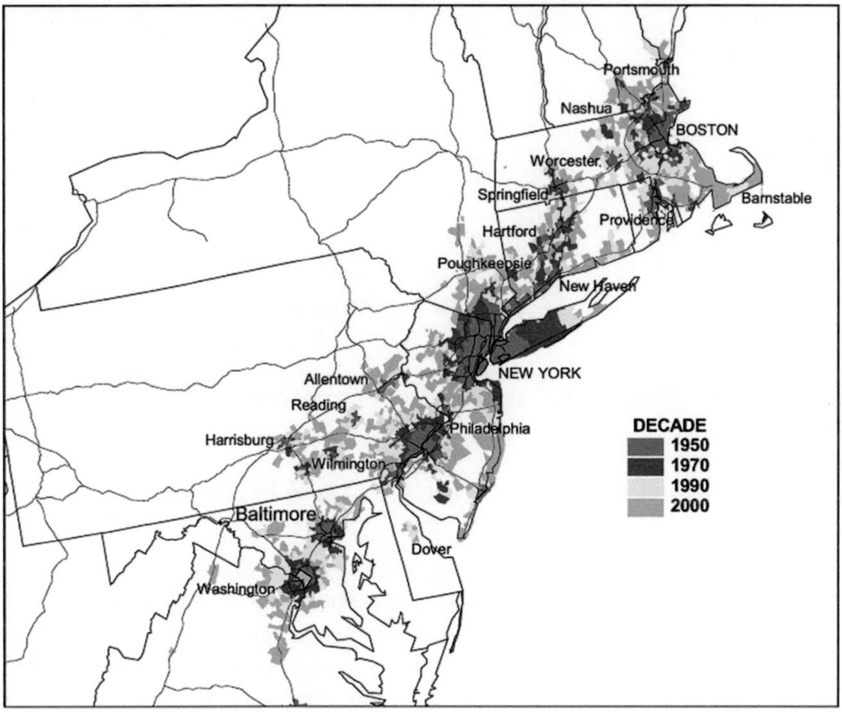

Fig. 18.3 Population and urbanized area growth in Megalopolis, 1950–2000 [*Source*: Morrill (2006)]

Hampshire to Northern Virginia and from the Appalachian foothills to the Atlantic a *Megalopolis,*[3] 'the Hinge and the Main street of the national economy'.

Figure 18.3 presents the urban evolution along the Megalopolis, differentiating the urban growth over the 1950–2000 (Morrill 2006). There is in this half century a modest growth of urban population (approximately 75 %), and a 300 % increase in the urbanized area—the dropping urban density reflecting the growth of suburbs and exurbs. In the three decades, 1972–2002, the five major metro areas in the Megalopolis had a modest growth in population. But this period actually comprised of two different sub periods. The first half (1972–1987) was a more or less stationary period for the four metro areas and one of even population loss for the

[3] Jean Gottman's work was highly influential, promoting work on large city regions. At a time when social commentators such as Patrick Geddes and Lewis Mumford weighed in against large urban agglomerations, Gottman (1961) celebrated urban size and growth, expressed his faith in the Boston-D.C. Region's robust entrepreneurialism. He believed that the dense metropolitan areas of the Megalopolis provided opportunities for experimentation, exchanges of ideas and knowledge, promoting cross-fertilization of new ideas and innovations, as this paper will show, Gottman's view of the dynamism of the Megalopolis and its role as an incubator of new economic trends was prescient.

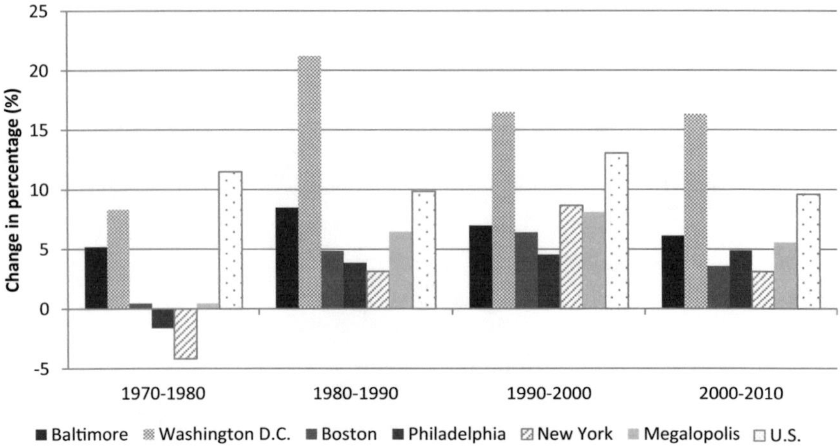

Fig. 18.4 Population change in the U.S. and five MSAs in the Megalopolis

largest metro, New York. With their economies dominated by declining manufacturing industries both metropolitan Philadelphia and Baltimore, were nearly stationary, and New York suffered population loss.

Boston, which had passed earlier through its industrial decline and had begun to restructure itself, registered minor demographic growth. In the subsequent period (1987–2002), the rise of the knowledge-intensive production and knowledge-intensive financial, business, professional, and technical services greatly enhanced economic and demographic opportunities in the Megalopolis. New York, Washington, D.C. and Boston have been particular demographic beneficiaries from these favorable economic developments.

Figure 18.4 shows that population growth has been relatively slow in the Megalopolis region, with all metro areas except Washington DC growing more slowly than the US as a whole in all decades from the 1970s to 2010. Figure 18.5 shows the progression over four decades (1970–2010) of population, employment (manufacturing, and all non-farming sectors) and total earnings (manufacturing and all non-farming sectors) in the Megalopolis in terms of national shares. The Megalopolis share of national population declined steadily in the first two decades, but this decline is at a gentler pace in the last two decades (Fig. 18.5). There is a similar pattern to the national shares of total employment, except that total employment shares in Megalopolis are always higher than that of population—showing an upward shift in per capita levels of employment in the Megalopolis as compared to the nation.

However, the manufacturing employment of this region relative to the nation drops sharply since 1990, indicating sharper regional decline of the manufacturing sector. However, the earnings shares of the Megalopolis in the manufacturing sector stay higher than the employment shares in the entire period (and more so in the last

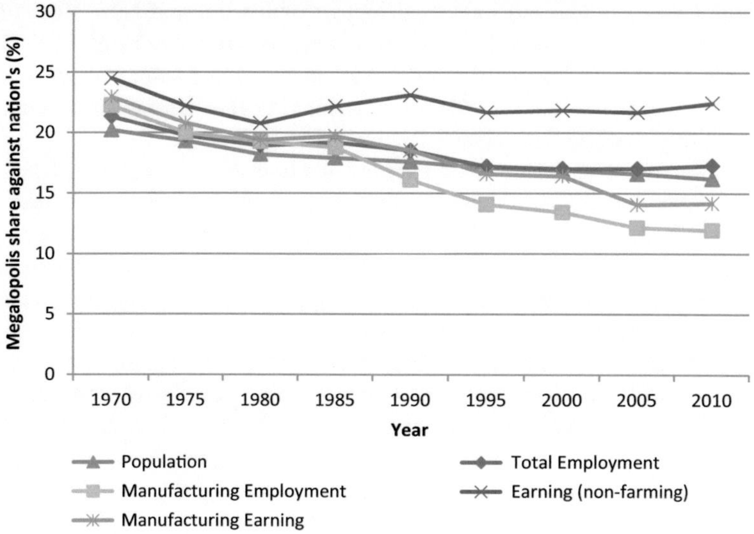

Fig. 18.5 Megalopolis: national shares of population, employment, and earnings (*Source*: Bureau of Economic Analysis)

two decades). In the last two decades, the earnings per manufacturing worker in the Megalopolis are higher, suggesting a higher proportion of high value-adding (more knowledge-intensive) sectors remaining in the region—as compared to the sector components being outsourced to the rest of the world (see Sect. 18.5).

The Megalopolis shares of national total (non-farm) earnings have dropped in the 1970–1990 period, but have since held steady, reflecting the faster growth in the knowledge-intensive service sectors. However, the Megalopolis shares of total non-farm earnings remains high—6 % higher than population shares in 2010-suggesting that earnings per employee are higher in the Megalopolis than in the nation.

18.4 The Recent Transition to Knowledge Economies in the Megalopolis

As noted above, many metropolitan areas in the Megalopolis have experienced in the recent past economic decline, as their manufacturing sectors lost their competitive edge. They had to confront the dislocations of post-industrial society and the dead ends of economic decline. Yet some of these urban regions in the Megalopolis have in the most recent quarter century reinvented themselves as hubs of creativity and development, creating jobs utilizing knowledge in technical, economic, social and cultural fields.[4]. The old industrial metros of Boston and New York are well advanced in this transition to Knowledge economies and Washington, D.C. (with large public, health, and scientific sectors) is emerging as a vibrant creative region. These and other metros are the 'Creative regions' or the 'Knowledge Economies', yielding jointly a Mega Knowledge Region of Megalopolis.

What sorts of processes have been at play and how do they interact in the birth of this Mega Knowledge Region and its continuing evolution? How did these metropolitan areas make this transition from a trajectory of decline to "creative regions"? Such questions have predictably attracted theoretical interest from a broad range of academic disciplines. To make such a transition successfully, key economic, political and social actors in the metropolitan region have to unravel the complexity of change. Such actors should be able to engage in shifting mindsets, changing behavior of various urban agents, evolving and sustaining innovation, and maintaining the dynamics of change. A tall order of evolutionary change disciplines have been harnessed to this effect: Economic Geography, Business Economics, Innovation Studies, Growth Theory, Evolutionary Economics and Urban Studies.

This paper focuses on two interactive theoretical streams of inquiry on the origin and evolution of 'Creative Regions'[5]:

1. Models of Nurture and Commercialization of Industrial Innovation and Regional Industrial Adaptation and
2. Analysis of Evolutionary Processes towards a Knowledge-Intensive Service Economy

This part of the paper can only highlight the key concepts and the evolutionary processes emphasized in this vast literature of the rise of knowledge-intensive production and services, in order to set the stage for an empirical review of the recent economic resurgence of the Megalopolis metros. Further, this empirical assessment of the models of economic structural change presented here is largely limited in this paper to the experience of the Boston metro region—supplemented by some key indicators of structural change and evolution in other major Megalopolis metropolitan areas.

---

[4] Other examples of such creative regions or knowledge economies are: Bay Area and Silicon Valley, Los Angeles, Seattle, London, Paris, Tokyo, Milan, Sydney, etc.

[5] Other theoretical formulations of *the dynamics and evolutionary response of various spatial systems–economic, social, and ecological*—have been proposed. One such is the notion of "**Resilience**". Reggiani et al. (2002) proposed the notion of 'Resilience' as a major dimension of the dynamics of spatial economic systems, especially how these systems respond to shocks, disturbances and perturbations. Martin (2011) examines the usefulness of 'resilience' as an aid to understanding the reaction of regional economies to major recessionary shocks (which are typically a shorter time scale than in our paper)—arguing that the notion of resilience can usefully be combined with that of hysteresis in order to fully capture the possible reactions of regional economies to major recessions. Rose (2007) uses the resilience notion in the analysis of natural and man-made disasters.

18.4.1 The Rise and Evolution of Knowledge-Intensive Production

Most of the literature on the emergence of knowledge-intensive production focuses largely on what may be described as innovation networks which nurture and commercialize innovation and industrial adaptation in various regions (Bathelt 2001; Scott 1998; Best 2000; Cooke et al. 2011). The key idea in this vast literature that has burgeoned over the last two decades or more is that innovation or the creation and commercialization of new knowledge in a dynamic region is based on *multidimensional interactions* among autonomous but interdependent economic agents (Camagni 2005; Capello 2011).

Such models have been already noted in Part II of this paper as "Economic Gains from Urban Agglomerations". Of the latter group, the relevant category is 'Models of Innovation-led Agglomerations'.

To recapitulate, these models attribute knowledge generation, spillovers, and accumulation in 'Creative Regions' to:

1. *Physical proximity* among economic actors, facilitating interactions and enabling access to appropriation and sharing of tacit knowledge, thus promoting innovation,
2. *Relational Proximity* of economic agents, facilitating cooperative behavior, collective learning, and socialization of innovation risk, and
3. *Institutional Proximity* among the firms in the urban agglomeration in terms of shared rules, codes and norms of regional behavior which will promote cooperation in interactive learning processes

These linkages and interactions allow firms and other economic agents to complement their core competencies with requisite knowledge and capacities from other economic actors creatively, speedily, and flexibly. Such linkages are really 'embedded in the social network'.[6] Firms (often small and medium-sized) in such regions develop flexible and interdependent relationships with suppliers and competitors and increasingly depend on intangibles, like know-how, synergies, and untraded knowledge (Von Hippel 1988, Storper 1995).

Other explanations (Glaeser 2003) for regional reinvention and structural change note the importance of human capital—suggesting that it is the broad skill base which underlies regional change and evolution. However, such discussions shed little light on the mechanisms through which varied human capital offers the region the flexibility and capacity to change significantly its economic structure and composition. Others emphasize 'Open Innovation' (Saxenian 1994; Simard and West 2006) for regional growth or the open networks, which allow the urban region to adjust frequently to new conditions and meet the periodic demand for change.

[6] Indeed, even the ('autonomous') market relations emphasized in the neoclassical economists' world are socially embedded in the sense that they depend upon assumptions, norms, and institutions shared by the actors and do not themselves derive from economic decisions (Polyanyi 1944; Granovetter 1985). The recent interest in social capital as key supporting asset of productivity has been inspired by the spatial clustering and dynamism in places such as Italy's Emilia-Romagna and California's Santa Clara Region.

Further, this paper suggests, in addition, that dynamic Knowledge Regions also exhibit, from *political and institutional perspectives*, the attributes of 'learning systems', such as entrepreneurial ability and relational skills. Such attributes of learning systems reflect the cultures of local entrepreneurial social, economic, and political agents, stimulating them to innovate *institutionally and organizationally* (Lakshmanan and Button 2009). Further, the creation of new "Governance Systems" in metro regions (e.g. inclusive urban stakeholder policy decision and implementation processes as in Boston) facilitates, among metropolitan private, public, and civil society actors, new modes of interaction, decision making, and implementation of projects and programs. Such governance changes speed up the physical adaptation of the metro region—in the form of rapid generation of new urban physical infrastructures and land uses appropriate to a transition from a mature industrial region of the late twentieth century to a globally operative contemporary Knowledge Region (Chatterjee and Lakshmanan 2009; Lakshmanan and Chatterjee 2006).

18.4.2 Evolutionary Shift to the Knowledge-Intensive Service Economy

In the U.S. where the transition to the service economy is advanced, the share of service employment has progressed from a little over 50 % in 1950 to over 75 % by year 2000. A less observed characteristic, over the past two to three decades of the US transition to a service economy, is that the share in the economy of business services has grown monotonically, as the share of the manufacturing sector has been dropping. This continuing growth of Business Services reflects *a qualitatively new stage in the structure of production and an increasingly complex division of labor between economic sectors* (Kox and Rubalcaba 2007).

The argument advanced here is that in recent decades business services (particularly, Knowledge Intensive Business Services, KIBS) have contributed heavily to the US and Megalopolis economic growth, in terms of employment, productivity and innovation. A direct growth contribution stems from the business services sector's own remarkably fast growth, while an indirect growth contribution was caused by the positive knowledge and productivity spill-overs from business services to other industries. The spill-overs come in three forms: (a) from original innovations, (b) from speeding up knowledge diffusion, and (c) from the reduction of human capital indivisibilities at the firm level. The external supply of knowledge and skill inputs of KIBS sectors exploits positive external scale economies and reduces the role of internal (firm-level) scale (dis)economies associated with these inputs.

Further, the KIBS sector includes the major corporate central organizational services, such as financial, legal, accounting, and many professional services. Such services permit corporations with headquarters in New York or Boston to coordinate their various value-adding production chains spread across the globe. Thus

Table 18.1 Knowledge intensive services, 1997 and 2007

	1997		2007	
	Number of employees (%[a])	Average wage (2007 US dollars)	Number of employees (%[a])	Average wage (2007 US dollars)
New York	381,722 (4.06)	70,141	478,409 (4.37)	82,363
Philadelphia	137,730 (4.50)	66,167	142,742 (4.11)	71,689
Boston	106,292 (3.78)	74,977	163,071 (4.60)	87,860
D.C.	241,800 (7.85)	69,414	374,383 (9.73)	84,319
Baltimore	49,021 (3.41)	57,911	80,521 (4.75)	76,012
Megalopolis	1,091,625 (4.15)	67,921	1,441,850 (4.73)	80,641
U.S.	4,092,136 (2.65)	57,436	5,696,838 (3.17)	64,675

Source: 1997 and 2007 Economic Census, US Census Bureau
[a]% of total regional/national employees in the specified sector

KIBS sectors play a crucial role in the operation of the global economy and in the facilitation of American economy's structural evolution.

This brief review of the development of knowledge-intensive business services and their role in the generation and evolution of knowledge-intensive components of the US service economy supplements the review (in the earlier section) of various models which attempt to explain the rise and evolution of knowledge-intensive production sectors.

The levels of employment and average wages in KIBS sectors in 1997 and 2007 in the nation, Megalopolis, and its five major metros are presented in Table 18.1. While the Megalopolis had 17 % of the national employment in 1997 and 2005 (Fig. 18.5), it was over represented in employment in those years in the knowledge-intensive KIBS sectors, garnering respectively 26.7 and 25.3 % of the national totals (Table 18.1.). Further, the average wage/worker is higher in the Megalopolis and component metros in 1997—with this gap widening in 2007. As noted earlier, these KIBS sectors support the arrival and evolution of Knowledge-intensive Services.

18.4.3 The Rise and Evolution of Knowledge-Intensive Production in Boston Metro Region

The initial boom in high technology industries in the Boston region was punctured in the late 1980s by the decline of the Minicomputer industry and the fall-off in military research budgets. High technology industries in the Boston Region have been, however, able to readjust and rejuvenate their product and process structures in order to sustain further innovation and growth, through a network model of complementary, vertically-disintegrated, open system firms (Bathelt 1999; Best 2000). Further, Boston's revitalization is based on its technological diversification rather than in an existing trajectory (Best 2000*), given its rich and diverse human*

		New York	Phila-delphia	Boston	D.C.	Balti-more	Megalo-polis	U.S.
Educational attainment (aged 25 or above)	Less than high school	15.2	11.3	9.4	11.1	11.1	12.7	14.1
	High school graduate	26.5	31.1	24.5	28.3	26.6	27.1	28.4
	Some college / associate	22	24.7	23	27.3	26.6	23.8	29
	Bachelor's degree or higher	36.3	32.9	43.1	33.4	35.8	36.4	28.5
occupational classes of workers	Management, business, science, arts	39.6	40.9	46.4	50.7	44.6	41.4	36
	Service	19.7	16.9	16.3	16.2	16.1	18.1	18.3
	Sales and office	24.7	25.4	22.9	20.8	23.8	24.1	24.5
	Natural resources, construction, maintenance	6.9	7.3	6.4	6.8	7.5	7.3	9.1
	Production, transportation, material moving	9	9.6	8	5.5	8	9.2	12.1

Source: 2011 American Community Survey, US Census Bureau

Fig. 18.6 Educational attainment and occupation by class of worker in Megalopolis in 2011 (%)

capital and its dynamic labor markets.[7] The Boston region has the highest proportion (43.1 %) of its workers (aged 25 or above) with a Bachelor degree or above among the Megalopolis metros—51 and 18 % above the National and Megalopolis averages (Fig. 18.6).

The biotech industry, based on genetic engineering technology, comprises small and medium-sized research firms and recently by multinational pharmaceutical firms, in an ambience of close interactions among small and large firms, university researchers, and public research centers. While Saxenian (1994) observed that 1980s IT firms in the Boston-area IT firms lacked the culture of interactivity and openness found in Silicon Valley, that culture seems to have taken hold in the biotech cluster (Breznitz and Anderson 2005).

The majority of the studies of the resurgent region of Boston often limit themselves to a discussion of the performance of mostly fabrication sectors, such as Military electronics, Microcomputer industry, Electronic components, Instruments, and the Biotechnology industry (Bathelt 2001), ignoring the larger knowledge-intensive services.

The knowledge-intensive services in the Boston Region are: Financial services, Professional services, Health Care Services, and Educational Services.

The Financial services sector is a large sector that is growing in the region, utilizing highly skilled personnel and offering high wages. New York is a globally dominant region with over 420,800 workers (13 % of the national total) with an average wage of $168,800, an average annual wage growth rate of 7.7 % in 2004

[7] Heurmann (2009) suggests that human capital externalities accrue predominantly to growing firms, which benefit from sharing, matching and learning externalities arising from a large supply of highly qualified workers in skilled labor markets.

Table 18.2 Financial services—megalopolis metropolitan areas by employment, 2004

Metropolitan area (national rank in employment)	Total employment	Share of national employment	CAGR of employment 1990–2004	Average wages	CAGR of average wages 1990–2004
New York-Northern New Jersey-Long Island **(1)**	427,296	13.0	0.44	$168,802	7.66
Boston-Cambridge-Quincy **(4)**	133,342	4.0	3.48	$114,696	7.83
Philadelphia-Camden-Wilmington **(5)**	113,112	3.4	2.21	$73,158	6.12
Hartford-West Hartford-East Hartford **(9)**	65,219	2.0	1.38	$86,851	6.52
Washington, DC **(12)**	58,098	1.8	1.99	$76,880	5.64

(Table 18.2). The Boston metro area has the fourth largest employment, third highest average wage, and the second highest wage growth rate (1990–2004) among US metros in the Financial sector in 2004. In Massachusetts, this sector accounts for 10.8 % of the Gross state Product, 5.5 % of employment, with 65 % of the workers with a post-secondary degree, and has a significant multiplier effect on Professional and Technical services. The annual growth rate of average wage in this sector over a 15 year period was in Boston (at 7.83 %), with other Megalopolis metros not far behind. The magnitude and importance of the Financial Services sector in the Boston Region are not surprising given its past history of merchant and industrial capital in earlier times and its many *financial innovations* (e.g. Merchant banking, marine insurance, 'Suffolk Bank' notes, Venture capital, Trust Bank, Mutual funds, etc.).[8] The Megalopolis metro areas of Philadelphia, Hartford, and Washington, DC draw respectively 5th, 9th, and 12th national ranks in employment levels in this sector.

The Health Services Sector is a larger sector in the Boston Region, about 2.5 times as large as the Financial Sector in employment. Table 18.3 offers an employment comparison of the Megalopolis metro areas in 2004. With many major university hospitals and research institutes, the Boston Region is not only the fifth largest national employer, but has registered a 4 % annual rate of growth over a recent 15 year period. The elite hospitals and public research institutes in this region also provide a key component of the biotechnology innovation networks.

Higher education plays a major role in the Boston Region than in most others. In 2000, Boston metro ranked first in college and university enrollment, first in degrees granted, and in per capita spending, per 100,000 people. The eight universities in the metro area, attracting research funds, partnering with firms,

[8] The "Suffolk system" was developed in Boston by the Suffolk Bank before the Civil War (by the investors who founded the textile industry). This bank issued its own notes which were redeemable at par along with those issued by any New England bank, provided those banks maintained deposits at the Suffolk Bank—thus rapidly expanding industrial capital in New England.

Table 18.3 Local health services: megalopolis metropolitan areas by employment, 2004

Metropolitan area (National rank in employment)	Total employment	Share of national employment	CAGR of employment 1990–2004	Average wages	CAGR of average wages 1990–2004
New York–Northern New Jersey-Long Island **(1)**	1,113,374	7.6	1.91	$42,631	3.64
Philadelphia-Camden-Wilmington **(4)**	373,167	2.5	1.71	$38,436	3.17
Boston-Cambridge-Quincy **(5)**	310,759	2.1	1.45	$42,180	3.95
Washington-Arlington-Alexandria **(9)**	221,485	1.5	2.29	$43,770	3.35
Baltimore-Towson **(16)**	150,055	1.0	1.86	$40,832	3.90

Source: Prof. Michael E. Porter, Cluster Mapping Project, Institute for Strategy and Competitiveness, Harvard Business School; Richard Bryden, Project Direct *Boston Region's Reinvention*

generating patents, starting companies, and training large numbers of students are major regional economic contributors. The higher education also promotes regional growth by transferring technological innovations to local firms.

The economic crisis of 2008–2009 had a negative impact in all sectors of employment, but was especially severe in financial services. Overall, Boston's recovery has been somewhat better than most. For example over the one-year period leading up to April 2013, the Boston metropolitan employment grew by 1.9 vs. 1.6 % in the US. Boston's health care, education and information services all grew at well above national levels, but the Finance services sector was a notable underperformer, (0.1 vs. 1.3 % nationwide.)[9] This is consistent with a general trend that is emerging for the recovery of financial employment to be greatest in second tier cities outside the megalopolis (Kotkin 2013). It is unclear, however, whether this represents a general spatial shift or just the fact that lower-order financial services have recovered faster.

The expanded review of the Boston's recent economic evolution in this paper suggests a robust economic recovery and export capacity across a broad range of knowledge-intensive goods and services. The five key determinants of the Boston Region's reinvention as a vibrant knowledge economy are: (1) Large and Diverse Human capital, (2) open Innovation Networks, (3) Local entrepreneurship, (4) Financial Innovations and, (5) The development of Innovative Models of Urban Governance to promote the rapid spatial reinvention of the Boston region in order to facilitate the functioning of a globally connected knowledge metropolis.

[9] Bureau of Labor Statistics press release "Boston Area Employment—April 2013" on May 30, 2013. Available at http://www.bls.gov/ro1/cesbos.pdf

18.5 Rise of "Tradable" and "Non-tradable" Sectors and Rising Income Inequalities in the Megalopolis

Globalization, made possible in recent decades by lowered international trade barriers and technical improvements in transport and information technologies, has restructured the functional and spatial organization of the economy in the Megalopolis and the US. The production and delivery of goods and services are decomposed into increasing number of value-adding components and the relevant supply chains and value-adding components of economic activities have been increasingly organized on a global basis. At the same time, innovations in Information technologies applied in service sectors have created many Knowledge-intensive Business Services (KIBS). KIBS (comprising financial, legal, accounting, information and other professional services) enable global corporations, as noted earlier, to develop and support management innovations that make possible the smooth operation of global supply chains and the integration of global corporate operations. Further, KIBS also have become tradable.

In the early stages of globalization, global corporations in the Megalopolis moved the lower wage, lower value-adding components of production to low income industrializing countries, while retaining the more knowledge-intensive components domestically. Thus the upstream knowledge-intensive activities (e.g. R&D, Product design) and downstream knowledge-intensive activities (e.g. Marketing, Brand exploitation) are retained in the Megalopolis, while tasks of fabrication of many components (except the more knowledge-intensive components) are out sourced to the newly industrializing countries. At this stage, there is a decline in low to medium wage jobs and a growth in higher wage knowledge-intensive jobs in the manufacturing and other tradable sectors in the Megalopolis. The twofold result of a loss of jobs in manufacturing and other tradable sectors and a rise in value added per job in the more knowledge-intensive value chains of production which remain in the Megalopolis in turn yields *rising income inequality* in the Megalopolis.

Over time, however, the nature of global supply chains evolves in response to the pace of economic development in the rapidly industrializing economies of the world—the recipients of the low value-adding components of industrial production outsourced from the Megalopolis. Rapidly industrializing countries (such as China) are accumulating in recent years significant levels of physical, human, and organizational capital, which permit an increasing incorporation of higher value-adding components of global chains into their manufacturing and other tradable sectors, thereby displacing such components of manufacturing and other tradable sectors in the Megalopolis. In this context of globally-linked production, the corresponding production chains disappear in the Megalopolis, which increasingly plays host largely to the high value-adding components such as R&D, design, fabrication of some knowledge-intensive components, marketing, and a few post-sales services in such sectors. While overall value added per employee rises in these sectors retained in the Megalopolis, two adverse consequences follow. Powerful market forces operate directly on the tradable sector. More medium wage jobs in these sectors disappear from the Megalopolis locations. There are indirect effects on the non-tradable economic sectors of the Megalopolis and the national economy through wage and price effects and shifting opportunities in labor markets. Income inequalities, as a consequence, widen further in the last two decades in the Megalopolis centers of economic activity.

Spence and Hlatshwayo (2011) have carried out an analysis of the evolution of tradable and non-tradable sectors in the US economy and the progress of income inequalities in the nation during 1990–2007. This part of the paper implements the Spence-Hlatswayo approach to study the evolution of tradable and non-tradable sectors and the progress of income distribution in the Megalopolis and component major metros (for a slightly longer period) during 1990–2011.[10]

The largest tradable sector in the Boston metropolitan area was manufacturing (Fig. 18.7a). In the recent 22 years (1999–2011) this sector lost about 160,000 employees (45 %) of its total. The average earnings per worker in this sector, however, climbed up from $80,000 to over $91,000 (in chained dollars) in last 11 years (Fig. 18.8a).A significant loss of employment accompanied by an upward shift of average wage in that sector in Boston over a decade replicates the pattern of increasing income inequalities predicted earlier for regions engaged in globally organized production and trade. However, two tradable sectors—Professional services and Finance and Insurance—increased in employment (about 88,000) and the high average wage per employee—over $140,000 in 2001—climbed up (in chained dollars) 23 and 13 % respectively in that decade. The overall result is a drop in tradable sector employment and increasing income inequalities in the Boston region.

Much of the growth in employment in Boston Metro in these two decades derives from the non-tradable sectors (Fig. 18.8a). The Health services sector is the largest contributor to growth, followed by Government and Retail sectors. Indeed, the greater part of the employment growth in the Boston metro region by year 2011 derives from the three non-tradable sectors of Health services, Government, and Retail. Health and the Government sectors, have average earnings per employee of about $70,000, while the earnings per employee in the third employment growth sector of Retail are low—less than $40,000 (Fig. 18.8b).

It appears that over the last two decades, employment in the Boston metro region in the tradable sectors exposed to global competition is declining significantly. This development combined with a rise in average earnings per remaining employee in those tradable sectors, leads to rising income inequalities. Further the dominant (non-tradable) employment growth sectors of Health Care, Government, and Retail evidence moderate to low earnings per employee. Clearly, evolving trends in employment levels and earnings per employee in the globalizing knowledge

[10] Spence and Hlatshwayo (2011) used a methodology developed by Bradford Jensen and Lori Kletzer. Their approach determined the tradability of an industry based on its geographic concentration—the more concentrated the industry, the higher its tradability (and vice versa). For example, take retail trade: its ubiquitous geographic presence implies that it is highly non-tradable.

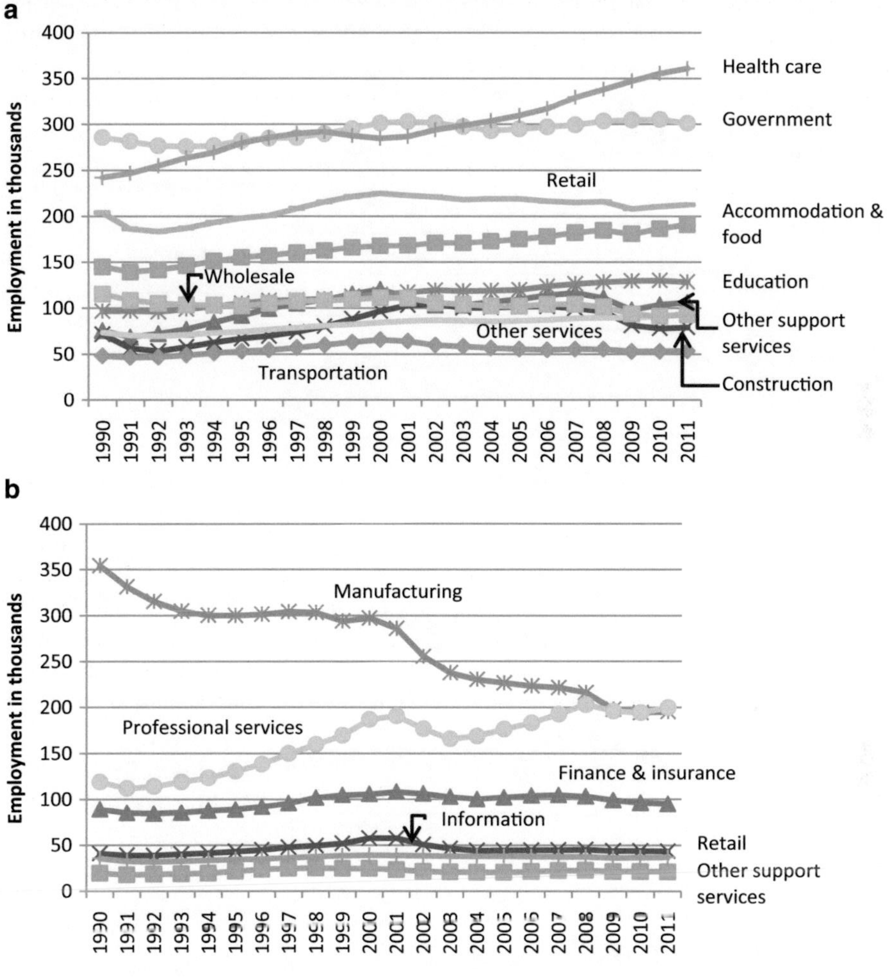

Fig. 18.7 (**a**) Non-tradable industrial employment (major sectors) in Boston Metro (**b**) Tradable industrial employment (major sectors) in Boston Metro

economy of the Boston Region over the last two decades indicate increasing income inequalities.

Similar empirical analyses conducted for New York, Washington, DC and other Megalopolis Knowledge economies (not reported here) yield comparable results of recently widening income disparities.

Figure 18.9 shows the progress of the Gini index from 1979 to 2011 in the US and in selected metropolitan areas of the Megalopolis over three decades. In 1979, income inequalities are more pronounced in the nation than in the metro areas of the Megalopolis. By 1989, when Globalization was underway, the Gini indices (income inequalities) are climbing in the nation and in the metropolitan areas of the Megalopolis. As global organization of business and a finer differentiation of

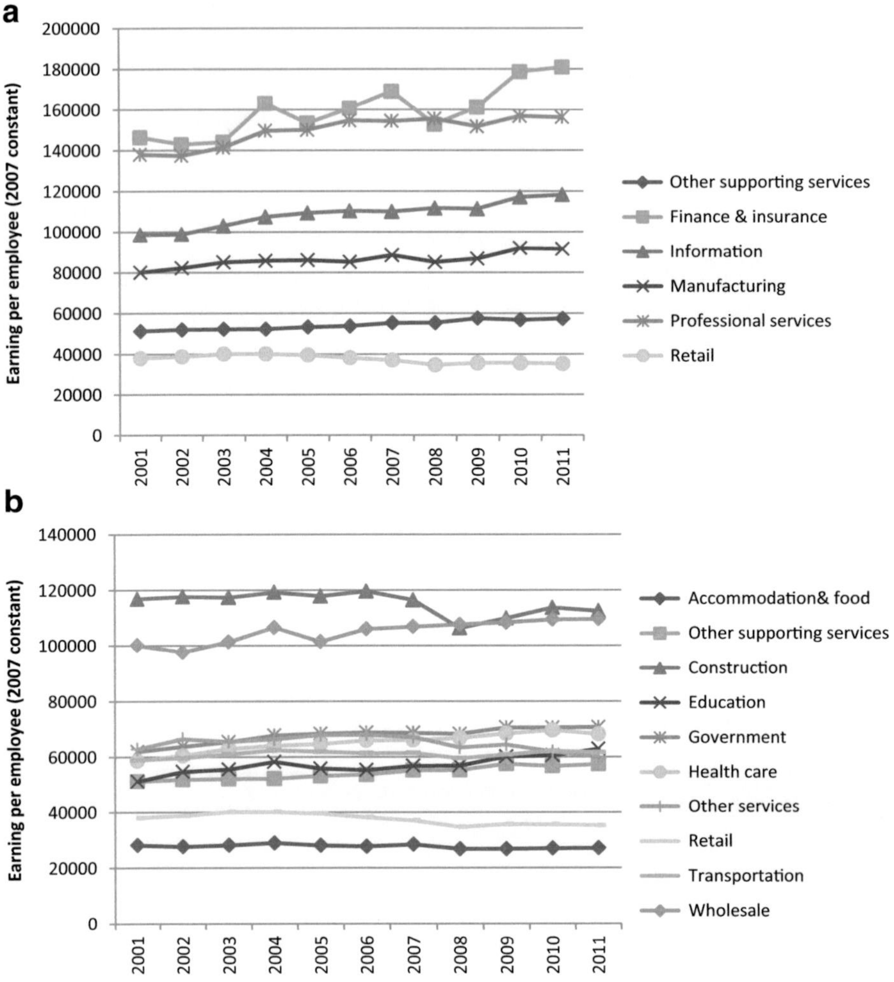

Fig. 18.8 (**a**) Tradable sectors: earnings per employee (in 2007 dollars) in Boston Metro (**b**) Non-tradable sectors: earnings per employee (in 2007 dollars) in Boston Metro

tradable and non-tradable sectors gather speed in the 1990s and in the decade of 2000s, income inequalities widen faster in the rising knowledge economies of New York, Boston and Washington than in the nation.

Concluding Comments

After experiencing robust economic growth in the quarter century following the end of World War II, the urban economies of the Megalopolis have since experienced great difficulties in reproducing the economic activities and arrangements of an industrialized society. In the last three decades, however, new economic and social relations have emerged and are being generalized in

Metropolitan Area	1979	1989	1999	2006	2011
New York	0.353	0.385	0.475	0.499	0.507
Boston	0.399	0.449	0.413	0.461	0.477
Philadelphia	0.327	0.341	0.419	0.461	0.472
Providence	0.394	0.427	0.424	0.440	0.463
Baltimore	0.391	0.412	0.405	0.437	0.452
Hartford	0.355	0.421	0.399	0.438	0.456
Washington	0.367	0.380	0.379	0.434	0.438
United States	0.404	0.431	0.458	0.470	0.475

Source: 1979 and 1989 metropolitan Gini indices from Madden, 2000, 1999 metropolitan Gini indices from Lopez, 2004 (data published online, www.diversitydata.org) and 2006 and 2011 metropolitan Gini indices and U.S.'s indices from the US Census Bureau.

Fig. 18.9 Gini Index of household income inequality in the US and for metropolitan areas of over one million in population in the Megalopolis

response to a convergence of several factors—namely a combination of new transport, logistical, information and knowledge technologies which, in the context of institutional innovations exemplified by open trade regimes and financial innovations, lead to globalization. A global organization of the production system, whose geography changes frequently in response to shifting factor prices, has emerged. *What* gets produced, *how* it is produced, and *where* it is produced change often.

In this period, *some* urban regions in the Megalopolis have reversed their economic decline and indeed, have begun to exhibit vibrant, resurgent economies. Such urban areas exhibit a major economic transition, a discontinuity, a transition from a mature industrial urban region to a newer region, being propelled by new information and knowledge technologies that are transforming their production and service activities. This evolution is towards a Knowledge Economy, towards an entrepreneurial creative urban region, where value derives from knowledge. Such urban regions offer locales and contexts for information and knowledge flows and cross-fertilization—supporting the generation and implementation of a variety of economic and social innovations that undergird the new urban creativity and entrepreneurialism.

The contemporary Boston Region offers a good example of such a metropolis which was a pioneer of the American Industrial revolution in the nineteenth century, and remained a major industrial center for over a century, only to witness the flight of its manufacturing capacity (and employment) to low cost labor locations—enduring thereby a multiple-decade urban decline. However, in the last quarter century or so, Boston has arrested this long decline and has reinvented itself as an entrepreneurial knowledge economy, focusing on knowledge-intensive production and services—Biotechnology, Scientific instruments, Software, Finance services, Producer services, Medical and

Educational services, etc. Clearly, Boston has initiated a remarkable economic renaissance, introducing technical innovations and new activities employing a high proportion of knowledge workers. But Boston also displays a 'broader' social creativity—a more social, cumulative, and collaborative account of creativity (as contrasted with the common focus in the literature on the individual spark or on the private sector changes). This collaborative creativity reflects its history, political governance traditions, and the vibrancy of its civil society institutions. Boston achieved this remarkable reinvention by building on its *vast stocks and variety of knowledge assets, networking opportunities, rivalry and trust-promoting contexts, and its history of institutional innovations and social cooperation* (Lakshmanan and Chatterjee 2006).

One must realize that a viable regeneration strategy pursued in an urban region in the Megalopolis is a very complex public good, resisting simple characterizations. However, the common focus in the larger literature on creative cities and urban evolution is on the processes underlying the generation of new knowledge and its incorporation in new goods and services, but that is only part of the story. This paper has argued that a viable urban regeneration strategy must include additionally the capacity to generate flexibility, tolerance, social capital and new governance models, which are necessary for the effective functioning and collaboration of the economic/social/political sectors in the new urban region. Finally, in resurgent Megalopolis cities such as Boston, governance involves multiple stakeholders, shared aims, interdependent resources and actions, hazy boundaries among private, public, and social sector actors. Institutional innovations and Governance models were developed in this context in Boston to facilitate the translation of new economic growth into inherited industrial urban space, and avoid market failure of spatially produced goods and services. Such capacity is created from within the urban region through collaborative innovation embracing many private, public, and social sector actors (as well as individuals and firms).

References

Amin A, Cohendet P (1999) Learning and adaptation in decentralized business networks. Environ Plan 17:87–104

Anderson WP, Lakshmanan TR (2007) Infrastructure and productivity: what are the underlying mechanisms. In: Karlsson C, Anderson WP, Johansson B, Kobayashi K (eds) The management and measurement of infrastructure, pp 147–162

Bathelt H (1999) Technological change and regional restructuring in Boston's Route 128 area. IWSG working papers 10-1999

Bathelt H (2001) Regional competence and economic recovery: divergent growth paths in Boston's high technology economy. Entrepreneur Reg Develop 13:287–314

Best MH (2000) Silicon Valley and the resurgence of Route 128: systems integration and regional innovation. In: Dunning (ed) Regions, globalization and the knowledge-based economy. Oxford University Press, Oxford

Breshnehan TF, Trajtenberg M (1992) GPTs: engines of growth. NBER working paper 4148. 43 pp

Breznitz SM, Anderson WP (2005) Boston metropolitan area biotechnology cluster. Can J Reg Sci 28(2):249–267
Camagni R (2005) Uncertainty, social capital, and community governance: the city as Milieu. In: Nijkamp P (ed) Urban dynamics and growth: advances in urban economics. Emerald Group Publishing, Bradford
Capello R (2011) Innovation and productivity: local competitiveness and the role of space. In: Philip C et al (eds) Handbook of regional innovation and growth. Elgar, Northampton, MA, pp 107–118
Chatterjee L, Lakshmanan TR (2009) The fashioning of dynamic competitive advantage of entrepreneurial cities: role of social and political entrepreneurship. In: Karlsson C, Andersson AE, Cheshire PC, Stough RR (eds) New directions in regional economic development. Springer, Berlin
Cooke P et al (2011) Handbook of regional innovation and growth. Elgar, Northampton, MA
David PA, Wright G (1999) GPTs and surges in productivity: historical reflections on the future of the ICT Revolution. Paper presented at a Symposium at Oxford University, July 2–4
Fujita M, Krugman P, Venables AJ (1999) The spatial economy. The MIT Press, Cambridge, MA
Gertler MS (2003) Tacit knowledge and the economic geography of context, or the undefinable tacitness of being (there). J Econ Geogr 3(2003):75–99
Glaeser EL (2003) Reinventing Boston, 1640–2003. NBER working paper 10166
Gottman J (1957) Megalopolis or the urbanization of the Northeastern seaboard. Econ Geogr 33 (3):189–200
Gottman J (1961) Megalopolis: the urbanized Northeastern seaboard of the United States. MIT Press, Cambridge, MA
Granovetter M (1985) Economic action and social structure: the problem of embeddedness. Am J Sociol 91:481–510
Heuremann DF (2009) Reinventing the skilled region: human capital externalities and industrial change discussion paper series No. 02/2009. Institute for labor law and industrial relations in EC(IAEEG) Trier
Hoover E (1948) The location of economic activity. McGraw-Hill, New York
Isard W (1956) Location and space-economy: a general theory relating to industrial location. Market areas, land use, trade and urban structure. MIT Press, Cambridge, MA
Kotkin J (2013) Cities that are stealing finance jobs from Wall Street. Forbes, May 31
Kox H, Rubalcaba L (2007) Business services and the changing structure of European economic growth. CPB Memorandum 183, CPB Netherlands Bureau for Economic Policy Analysis
Krugman PK (1991) Geography and trade. MIT Press, Cambridge, MA
Lakshmanan TR, Anderson WP (2002) Transport infrastructure, freight services sector and economic growth: a white paper, prepared for the U.S. Department of Transportation, January, 127 pp
Lakshmanan TR, Button KJ (2009) Institutions and regional economic development. In: Cappello R, Nijkamp P (eds) Advances in regional economics. Edward Elgar, Cheltenham
Lakshmanan TR, Chatterjee L (2005) Economic consequences of transport improvements. Access 26(Spring):28–33
Lakshmanan TR, Chatterjee LR (2006) The entrepreneurial city in the global marketplace. Int J Entrepreneur Innov Manage 6(3):155–172
Lakshmanan TR, Anderson WP, Song Y, Dan Li (2009) Broader economic consequences of transport infrastructure: the case of economic evolution in dynamic transport corridors. Center for transportation studies working paper, Boston University
Levinson M (2006) The box. Princeton University Press, Princeton
Marshall A (1890) Principles of economics. Prometheus Books, New York
Martin R (2011) Regional economic resilience, hysteresis, and recessionary shocks. Paper presented at the annual international conference of regional studies association, Newcastle, 17–20 Apr 2011
Morrill R (2006) Classic map revisited: the growth of metropolis. Prof Geogr 58(2):155–160

Murakami J, Cervero R (2012) High-speed rail and economic development: business agglomerations and policy implications. Final report UCTC-FR-2012-10, University of California Transportation Center
Polyanyi K (1944) The great transformation: political and economic origins of our time. Farrer and Rinehart, New York
Reggiani A, deGraff T, Nijkamp P (2002) An evolutionary approach to spatial economics. Netw Spatial Econ 2:211–229
Rose A (2007) Economic resilience to natural and man-made disasters: multidisciplinary origins and contextual dimensions. Environ Hazards 7(4):383–398
Saxenian A (1994) Regional advantage: culture and competition in Silicon Valley and route 128. Harvard University Press, Cambridge, MA
Scott AJ (1998) Regions and the world economy: the coming shape of global production, competition and political order. Oxford University Press, Oxford
Simard C, West J (2006) Knowledge networks and the geographic locus of innovation. In: Chesbrough H, Vanhaverbeke W, West J (eds) Open innovation: researching a new paradigm. Oxford University Press, Oxford
Spence M, Hlatshwayo S (2011) The evolving structure of the American economy and the employment challenge. Working paper, council on foreign relations, Maurice R. Greenberg, Center for Geoeconomic Studies
Storper M (1995) The resurgence of regional economies, ten years later; the region as a nexus of untraded interdependencies. Eur Urban Reg Stud 2:191–221
Von Hippel E (1988) The sources of innovation. Oxford University Press, Oxford

Gayborhoods: Economic Development and the Concentration of Same-Sex Couples in Neighborhoods Within Large American Cities

19

Janice Fanning Madden and Matt Ruther

Abstract

This paper uses census tract data from the 2000 and 2010 U.S. Censuses and the 2005–2009 American Community Survey to examine the locations of gay male and lesbian partnerships in 38 large U.S. cities. Gay men and lesbians are less segregated than African Americans and lesbians are less spatially concentrated than gay men. There is little evidence to support the common assertion that gays concentrate in more racially and ethnically diverse neighborhoods. We find evidence supporting the popular notion that concentrations of gay men lead to more rapid development of central city neighborhoods. Census tracts that start the decade with more gay men experience significantly greater growth in household incomes (and, therefore, presumably housing prices) and greater population growth over the next decade than those census tracts with fewer gay men. Census tracts with more lesbians at the start of the decade see no difference in population or income growth.

J.F. Madden (✉)
Population Studies Center, University of Pennsylvania, 3718 Locust Walk, Room 365, Philadelphia, PA 19103, USA
e-mail: madden@ssc.upenn.edu

M. Ruther
Department of Geography, University of Colorado at Boulder, 110 Guggenheim UCB 260, Boulder, CO 80309-0260, USA
e-mail: matthew.ruther@colorado.edu

P. Nijkamp et al. (eds.), *Regional Science Matters*, DOI 10.1007/978-3-319-07305-7_19,

19.1 Introduction

Many researchers have claimed that city neighborhoods where the residences of gay men and lesbians are concentrated have different characteristics and different futures than the city neighborhoods where more heterosexuals reside.[1] Most studies, however, have weak empirical bases for these claims.

A few studies have statistically analyzed the locations of gay men and lesbians. Black et al. (2002) demonstrate that partnered gay men are more likely than heterosexual couples to locate in higher amenity cities. They argue that this is the case because gays are less likely to have children, allowing more of their incomes to be spent on the amenities of a residential location than on housing square footage or other expenditures. They show that amenities, more than "gay-friendliness," attract gays to concentrate in particular cities, such as San Francisco. Black et al. explore intercity locations, but do not address the location, or concentration, of gays or lesbians across neighborhoods within cities.

Gates and Ost (2004) created an atlas showing the locations in 2000 of gay partners by state, city, zip code and urban place within metropolitan areas. They provide maps and charts showing that, while there are differences in where gay men and lesbians locate, their locations are correlated with one another and differ from those of heterosexual partners. The Gates and Ost study does not analyze the causes or effects of differences in the intra-city locations of gays, nor do they provide any summary data or statistically based comparisons.

A few earlier studies identify neighborhoods within a single city that attract concentrations of gay men or lesbians and compare the characteristics of these neighborhoods to other neighborhoods in the same city. Castells (1983) used informants, voter data on the addresses of multiple male households, addresses of gay bars and social gathering places, addresses of gay-owned businesses, and concentrations of votes for gay activist Harvey Milk to create a series of maps showing the neighborhood concentrations of gay men in San Francisco. Castells compares characteristics of these neighborhoods to the neighborhoods in the rest of the city and finds that the gay neighborhoods have fewer owner-occupied homes, fewer children, and lower rents.

Two studies pursue similar strategies in examining the concentrations of lesbians. Adler and Brenner (1992) compare the average characteristics of neighborhoods with more lesbians within one (unnamed) city to the characteristics of other neighborhoods in the same city. They use mailing lists from two lesbian organizations to identify 12 census tracts in which lesbians are concentrated. Using 1980 US Census tract data, they present comparisons similar to Castells, finding that lesbian neighborhoods have less homeownership, fewer married couples,

[1] There is an extensive literature on sexuality and space that explores the variety of reasons that gays and lesbians do or do not concentrate spatially within cities (for example, Brown 2013; Lauria and Knopp 1985; Harry 1974). We do not address those questions here. Rather, we analyze the kinds of neighborhoods that attract concentrations of gays and lesbians and the effects of those concentrations on subsequent development of the neighborhoods.

families with children, lower rents, and more women living alone in non-family households.

Ann Forsyth (1997) uses 1990 US Census tract data, as well as mailing lists and local voting data, to identify two tracts in Northampton, Massachusetts as "housing the majority of Northampton's lesbians." She uses data from the tracts with greater concentrations of lesbians to discuss a different question: do these tracts have characteristics consistent with gentrification over time. She discusses the population structure and homeownership changes within these two tracts since 1970 and finds that, as expected with gentrification, these tracts increased in educational levels and white collar workers (as did the entire state). Contradicting the hypothesis that lesbian concentrations are associated with gentrification, however, Forsyth finds that these two census tracts experienced decreases in homeownership and increases in poverty. Because Forsyth's study uses 1990 data to identify census tracts with greater concentrations of lesbian residents, the study cannot tell us whether the 1990 concentrations of lesbians in these neighborhoods *occurred* after 1970 in response to changes in these characteristics or whether these characteristics of the neighborhood changed subsequent to the neighborhoods having a greater concentration of lesbians in 1970.

Hayslett and Kane (2011) provide a more formal statistical study of intra-city locations. They investigate the correlation between the neighborhood concentrations of lesbian and gay partnered households in Columbus, Ohio in 2000 and the characteristics of the neighborhoods in the same year. They find evidence that lesbians are less spatially concentrated than gay men, although there is statistically significant positive spatial autocorrelation in their locations. They further show that while gay men are concentrated in neighborhoods with fewer family households and college graduates and with more renters, lesbian households are concentrated in neighborhoods with more family households and more foreign born residents. Both gay men and lesbians are more concentrated in neighborhoods with newer housing and with more multi-family units. Because the data are for one city and use only one year (2000 Census data) to identify gay spatial concentrations, the results can neither be generalized nor used to parse the causes and effects of concentrations of gay residents for the city studied.

Because they do not analyze the relationship of conditions within neighborhoods *prior to*, or *subsequent to*, changes in concentrations of gay men or lesbians, none of these single-city studies can attribute these characteristics to be causes, effects, or statistical correlates of other characteristics associated with the spatial concentration of lesbians or gay men within cities. Furthermore, as these studies include only a few U.S. cities, it is not clear whether the results apply more broadly. None of the studies address the concentrations both of gay men and lesbians, although Castells (1983) did argue that lesbians were less spatially concentrated.

The claim that gays are "pioneers" who move to declining central city neighborhoods and reverse the fortunes of such neighborhoods is common in discussions of the effects of gays' neighborhood concentrations (for example, Castells 1983; Lauria and Knopp 1985). After tracing the development of a gay enclave in Soho, London, Collins (2004) argues more generally that once neighborhoods reach a critical mass of gay representation, the representation accelerates. In line with the pioneer viewpoint, Collins contends that it is not amenity value, but rather cheap land values and access to night life and services, occurring in areas with extensive physical decay and vacancies, that attracts the first gay "pioneers". This initial concentration of gays subsequently attracts more gays, generating the renaissance of the neighborhood. (He also argues that in many cases the initial settlement of gays in a neighborhood is an "accident" that subsequently attracts more gays, more services for gays, and then yet more gays.) Other neighborhood characteristics that have been deemed attractive to gays include high vacancy rates (Ruting 2008) and diversity (Gates and Ost 2004).

Using 2000, 2005–2009, and 2010 US Census/American Community Survey data on census tracts, we provide the first statistical analysis and evidence on the causes and effects of the concentration of gays by gender in the neighborhoods of the largest U.S. cities. The next section discusses the data and the specification of the dependent variable, the census tract shares of the city's gay partnerships. The third section presents residential segregation indices and clustering measures for gay men and lesbians, followed by a section discussing the characteristics of census tracts that correlate with having more gay residents. The next section examines how growth in population and income (as an index of housing prices) over a decade is associated with gay representation in the census tract at the start of the decade. Conclusions are presented in the final section.

19.1.1 Data and Methods

We study 38 large central cities that are located in 35 metropolitan areas (See Tables 19.1 and 19.2 for a list of cities by metropolitan area). We examine segregation of gays by gender, how characteristics of a neighborhood in 2000 are related to shifts of gay households and of all households between census tracts within the city over the subsequent decade, and how gay presence in 2000 within census tracts is related to population and income of residents in 2009.

For our studies of segregation and spatial concentration of gay partnerships, we use the 2010 and the 2000 U.S. Census short form data, rather than the sample data provided by the Census long form or American Community Survey (ACS). The measures of segregation or concentration of a population are basically analyses of variation, and variation increases, especially for groups that are a small share of the tract population, when the sample sizes are small. Census tract data from 2000 and 2010 are based on the total census population[2] and, therefore, yield more accurate measures of segregation and concentration. Because analyses of segregation or concentration at two different points in time do not require that the spatial units (census tracts) have the same boundaries for both time periods, we compare the

[2] We use census tract data that is from the short form (SF1) answered by all households in the census. We cannot use these data to measure differences within tracts, however, because the boundaries of the census tracts changed between 2000 and 2010.

Table 19.1 Duncan indexes of dissimilarity within central city for one sex relative to two sex couples, by gender, 2000 and 2010

		Gay men			Lesbians		
Region	Central city	2000	2010	Change	2000	2010	Change
West	Denver	0.37	0.31	−0.06	0.28	0.22	−0.06
	Los Angeles	0.44	0.49	0.05	0.26	0.25	−0.01
	Phoenix	0.34	0.43	0.09	0.27	0.22	−0.05
	Portland	0.30	0.26	−0.04	0.35	0.25	−0.10
	San Diego	0.52	0.56	0.04	0.40	0.33	−0.07
	San Francisco–Oakland–San Jose	0.30	0.28	−0.02	0.24	0.21	−0.03
	Seattle	0.48	0.49	0.01	0.41	0.36	−0.05
	Mean for West	0.39	0.40	0.01	0.32	0.26	−0.06
South	Atlanta	0.50	0.41	−0.09	0.43	0.36	−0.07
	Austin	0.28	0.26	−0.02	0.30	0.21	−0.09
	Charlotte	0.34	0.37	0.03	0.30	0.22	−0.08
	Dallas-Fort Worth	0.47	0.51	0.04	0.31	0.23	−0.08
	El Paso	0.26	0.59	0.33	0.24	0.25	0.01
	Houston	0.45	0.50	0.05	0.33	0.25	−0.08
	Jacksonville	0.32	0.44	0.12	0.25	0.19	−0.06
	Memphis	0.46	0.50	0.04	0.41	0.26	−0.15
	Miami	0.44	0.44	0.00	0.29	0.23	−0.06
	Nashville	0.33	0.28	−0.05	0.31	0.21	−0.10
	New Orleans	0.52	0.47	−0.05	0.39	0.30	−0.09
	Oklahoma City	0.49	0.52	0.03	0.45	0.23	−0.22
	San Antonio	0.32	0.44	0.12	0.27	0.18	−0.09
	Mean for South	0.40	0.44	0.04	0.33	0.24	−0.09
Midwest	Chicago	0.42	0.50	0.08	0.34	0.29	−0.05
	Cleveland	0.45	0.45	0.00	0.33	0.27	−0.06
	Columbus	0.48	0.44	−0.04	0.35	0.24	−0.11
	Detroit	0.36	0.50	0.14	0.32	0.28	−0.04
	Indianapolis	0.39	0.44	0.05	0.29	0.20	−0.09
	Kansas City	0.50	0.48	−0.02	0.39	0.27	−0.12
	Milwaukee	0.41	0.43	0.02	0.34	0.27	−0.07
	Minneapolis-St. Paul	0.34	0.32	−0.02	0.30	0.29	−0.01
	St. Louis	0.37	0.35	−0.02	0.29	0.30	0.01
	Mean for Midwest	0.41	0.43	0.02	0.33	0.27	−0.06
Northeast	Baltimore	0.40	0.40	0.00	0.30	0.23	−0.07
	Boston	0.45	0.42	−0.03	0.41	0.40	−0.01
	New York	0.45	0.50	0.05	0.30	0.27	−0.03
	Philadelphia	0.47	0.50	0.03	0.33	0.25	−0.08
	Pittsburgh	0.37	0.33	−0.04	0.33	0.25	−0.08
	Washington	0.44	0.37	−0.07	0.28	0.21	−0.07
	Mean for Northeast	0.42	0.38	−0.04	0.29	0.22	−0.07

Table 19.2 Moran's I within central city for one sex relative to two sex couples, by gender, 2000 and 2010

		Gay men		Lesbians	
Region	Central city	2000	2010	2000	2010
West	Denver	0.49***	0.58***	0.32***	0.36***
	Los Angeles	0.66***	0.69***	0.13***	0.28***
	Phoenix	0.47***	0.62***	0.20***	0.22***
	Portland	0.19***	0.41***	0.58***	0.59***
	San Diego	0.62***	0.70***	0.46***	0.46***
	San Francisco–Oakland–San Jose	0.70***	0.71***	0.36***	0.41***
	Seattle	0.76***	0.78***	0.57***	0.61***
South	Atlanta	0.55***	0.56***	0.33***	0.58***
	Austin	0.34***	0.19***	0.22***	0.16***
	Charlotte	0.39***	0.37***	0.14**	0.24***
	Dallas-Fort Worth	0.46***	0.47***	0.25***	0.23***
	El Paso	0.15**	0.27***	0.05	0.10*
	Houston	0.61***	0.49***	0.09***	0.35***
	Jacksonville	0.28***	0.46***	0.14**	0.20***
	Memphis	0.51***	0.52***	0.23***	0.22***
	Miami	0.31***	0.35***	0.06	0.17***
	Nashville	0.23***	0.35***	0.11**	0.28***
	New Orleans	0.46***	0.43***	0.14***	0.15***
	Oklahoma City	0.09**	0.61***	0.24***	0.22***
	San Antonio	0.28***	0.44***	0.06	0.04
Midwest	Chicago	0.39***	0.77***	0.12***	0.38***
	Cleveland	0.22***	0.37***	0.00	0.14**
	Columbus	0.50***	0.45***	0.36***	0.22***
	Detroit	0.10***	0.04	0.01	0.06
	Indianapolis	0.43***	0.44***	0.23***	0.25***
	Kansas City	0.36***	0.56***	0.08	0.23***
	Milwaukee	0.34***	0.29***	0.07	0.26***
	Minneapolis-St. Paul	0.57***	0.47***	0.47***	0.60***
	St. Louis	0.63***	0.58***	0.20***	0.40***
Northeast	Baltimore	0.20***	0.35***	0.12**	0.23***
	Boston	0.52***	0.39***	0.55***	0.51***
	New York	0.59***	0.71***	0.17***	0.23***
	Philadelphia	0.56***	0.73***	0.45***	0.43***
	Pittsburgh	0.12***	0.20***	0.02	0.20***
	Washington	0.60***	0.65***	0.15***	0.12**

***$p < 0.001$, **$p < 0.01$, *$p < 0.05$

segregation indices and concentration statistics based on the tract data from each census year.

To study how neighborhoods (census tracts) are changing, however, it is critical to use neighborhoods that have the same boundaries over time. When boundaries shift over time, changes in household composition may be an artifact of boundary shifts and not of any change in where people live. For the analyses of neighborhood change, we use data from the 2000 Decennial Census and the 2005–2009 ACS; both use the 2000 census tract definitions. Tracts are considered part of the central city if at least 50 % of the tract population resides within the city, or if the tract is fully enclosed by the city.

In the 2000 and 2010 Censuses, as well as in the ACS, data on same-sex unmarried partners may be subject to coding errors. These errors are due to Census Bureau procedures which require recoding of Census forms that list two individuals of the same sex as spouses, and to differences in the forms and questionnaires used to collect Census data (O'Connell and Gooding 2007; Black et al. 2007; O'Connell and Felix 2011).[3] DiBennardo and Gates (2013) suggest a procedure through which the preferred (corrected) state-level estimates of same-sex partners that were released by the Census Bureau could be adjusted to the sub-state (e.g., tract) level. We rely on their procedure, as well as data and information contained in Gates and Steinberger (2009) and O'Connell and Feliz (2011), to create adjusted census tract counts of gay male and lesbian partnered households, which are used in all subsequent analyses.

We measure representation of households or other groups within a census tract by the *proportion of the city's population of the household type or other group residing in the census tract.*[4] By construction, these proportions sum to one for each group in each city in each year. Also, the mean value of this proportion for the census tracts in each city will be the same for each household group (because the proportions for each group must sum to one for the entire city, the mean proportion for each group within a city is one divided by the number of census tracts, which is the same for each group in the city.) We *do not* use the more common measure of the proportion of households in the census tract who are same-sex partner households. We use the proportion of the city population of the group residing in the census tract for the same reasons that these proportions are used for segregation indices. The city proportion measure is sensitive to relative differences in spatial concentration, *but is not sensitive to relative population sizes*. Therefore, the city

[3] For most of these time periods and locations, same-sex marriage was not legally recognized. The Census assumed in most cases of one-sex partnerships listing a spouse relationship that the spouse identification was incorrect but the gender identification was correct. The Census changed the coding of partnerships in these cases to same-sex partners. This approach to recoding was likely to have incorrectly classified some two-sex couples who incorrectly identified the gender of one partner to same-sex partnerships. Because the pool of same-sex partners accounts for less than five percent of all partnerships, any procedure that incorrectly allocates even a very small percentage of two-sex partnerships to same-sex ones leads to substantially greater bias in estimates for same-sex partners than for two-sex ones.

[4] See Madden (2014) for a study of changes in the intra-metropolitan area spatial distribution of residents by race and income using the same measurement approach.

proportion measure easily compares groups with very different sizes, allowing standardized comparisons across cities, time periods, and groups. Also, the city proportion measure "removes" or "standardizes" for swings in the population of a group due to secular economic changes in things such as revealing sexual orientation or migration within a city.

19.1.2 Are Gays Spatially Concentrated in Large U.S. Cities?

Table 19.1 reports the Duncan Index of Dissimilarity[5] for lesbian households and for gay male households, relative to all households, for large U.S. cities by region in 2000 and 2010. The Duncan index is a commonly used measure of spatial segregation that indicates whether there are neighborhoods (census tracts) within cities that include relatively more gay households than other neighborhoods. The indices show that gay men and lesbians are far less segregated than African Americans.[6] As suggested by Castells (1983), lesbians are less segregated than gay men in most cities in both years, and lesbians are becoming less segregated, as indicated by declining segregation indices between 2000 and 2010 for all but two cities. The Duncan Index of Dissimilarity measures one dimension of segregation, specifically the "evenness" of the distribution of gay households within neighborhoods in the central city. It does not, however, account for the relative spatial position, or clustering, of the neighborhoods or census tracts with similar shares of gay households. In cities with high values of the Index of Dissimilarity, gay households are concentrated within specific tracts, but these highly concentrated tracts may be distributed across the city or they may be in the same sections of the city. The global Moran's I can be used to assess whether tracts with large (or small) shares of gay households are also clustered in space.[7]

[5] The segregation index, the Duncan Index of Dissimilarity, is calculated:

$$1/2\sum i|P_i \;-\; nP_i|$$

where P_i is the proportion of the city's gay male (lesbian) households in census tract i and nP_i is the proportion of the city's non-gay households in census tract i.

The index takes on values between 0 and 1, where 0 indicates no segregation (partnerships of different sexual compositions are sorted identically across neighborhoods) and 1 indicates perfect segregation (gay partnership households and heterosexual partnership households live in completely different neighborhoods).

[6] Madden (2014) reports African American segregation indices of 0.46 for 2000 and 0.44 for 2009 for these same Western metro areas, 0.55 and 0.51 respectively for the South, 0.71 and 0.67 for the Midwest, 0.65 and 0.63 for the Northeast.

[7] Moran's I is calculated as:

$$I \;=\; \frac{N}{\sum_i\sum_j w_{ij}}\left(\frac{\sum_i\sum_j w_{ij}(X_i-\overline{X})(X_j-\overline{X})}{\sum_i (X_i-\overline{X})^2}\right) \qquad E(I) \;=\; \frac{-1}{N-1}$$

where X_i is the proportion of tract i's households that are of a given type, X_j is the proportion of tract j ($\neq$i)'s households that are of this same type, $\overline{X}$ is the mean proportion of this household type

Table 19.2 reports the Moran's I values for gay male households and lesbian households within each city. Spatial clustering of tracts with similar gay male household shares occurs for all metros in 2000 and all but one in 2010, while clustering of tracts with similar lesbian household shares does not occur in eight of these cities in 2000. The cluster measures all indicate less spatial concentration of lesbians: the Moran's I is lower for lesbians households in all but a handful of year-city cases.[8] These differences in neighborhood clustering reflect differences in the residential patterns of lesbians and gay men, as discussed below.

19.1.3 "Causes" of Gay Concentration in Neighborhoods

In order to explain these differences in residential location for heterosexual and homosexual households, we evaluate the characteristics associated with the 2009 concentration of gay male, lesbian, and all households given their 2000 levels of concentration.[9] We measure how the census tract's median household income, age of housing, population density, share of the city's single family detached housing, and vacancy rates in 2000 are related to changes in the census tract's share of the city's gay male and lesbian partnership households in 2009. We also examine the relationships of the initial demographic characteristics of the neighborhood, including the census tract's share of the city's African American, Hispanic, gay male, lesbian, and non-family households in the census tract in 2000,[10] to changes in the census tract's share of the city's gay male and lesbian partnership households.

Specifically, we model the 2009 proportion of a city's total households in a given group (gay men, lesbian, all) resident in a census tract as a function of the time invariant and of the 2000 time variant characteristics:

$$\mathrm{s}_{\mathrm{i,j,t+1}} = \alpha + \beta \mathrm{s}_{\mathrm{i,k,t}} + \gamma \mathrm{X}_{\mathrm{i,t}} + \mathrm{Z_i} + \varepsilon_{\mathrm{i}} \tag{19.1}$$

where $\mathrm{s}_{\mathrm{i,j,t+1}}$ is the city's proportion of household group *j* resident in census tract *i* in period t + 1,[11] *k* indexes the various household and other resident groups, X_{it} is a

over all tracts, w_{ij} is a matrix denoting the spatial relationship between all tracts *i* and *j*, and *N* is the total number of tracts.

[8] Moran's I varies between −1 and 1, with values increasing toward 1 indicative of higher levels of positive spatial autocorrelation and values decreasing toward −1 indicative of higher levels of negative spatial autocorrelation. A Moran's I equal to its expected value, which is approximately 0 in large samples, suggests that there is no spatial autocorrelation in the data.

[9] We use the 2005–2009 ACS data, rather than the 2010 short form Census data, because these analyses require common boundaries across the time periods, and because many of the relevant explanatory variables are not included on the Census short form.

[10] Because they have been adjusted and therefore include greater uncertainty, the effects of the 2000 locations of gay male and lesbian households are measured less precisely than those of other characteristics of census tracts.

[11] These are the census tract's population in the household or other demographic category divided by the city population in the same category.

vector of census tract i's other time variant characteristics (such as vacancy rates and median household income relative to the city median) in period t, Z_i is a vector of the census tract i's time invariant characteristics (such as its city and distance from the city center). And, α, β, γ and δ are parameters to be estimated, with ε_i as a random error term.

When analyzing spatial data, spatial dependence in the outcomes should be considered, as failure to appropriately account for a spatially dependent outcome may result in biased and/or inefficient coefficient estimates. Although Table 19.2 indicates spatial autocorrelation in the locations of gay and lesbian households, the need for spatially explicit estimation procedures is commonly assessed through analysis of the residuals from an ordinary least squares (OLS) regression.

We test the residuals of the OLS models for each household type (gay men, lesbian, and all households), nationally and within each region, using the simple Lagrange Multiplier (LM) statistics for spatial error dependence and spatial autoregressive dependence derived in Burridge (1980) and Anselin (1988), and the robust LM statistics for either type of dependence derived in Bera and Yoon (1993) and Anselin et al. (1996). The simple versions of these LM statistics test for the presence of spatial dependence in the form of a spatial autoregressive process or a spatial error process (assuming that neither is present), while the robust versions test for a spatial autoregressive process when the actual data generating process is a spatial error process, and vice versa. Based on the results described in Anselin et al. (1996), we first assess the significance of the simple LM statistics. When only one of the simple LM statistics is significant (either autoregressive or error), we proceed with estimation of that type of model. In cases where both simple LM statistics are significant, the robust LM statistics are used to determine the appropriate model.[12] When neither of the simple LM statistics is significant, a spatial model is not appropriate.

We find evidence of spatially dependent residuals in each of the three models (gay men, lesbians, all households), nationally and within each region, with the LM statistics suggesting the presence of a spatial autoregressive process in census tract shares of the city's gay and lesbian households and a spatial error process in total household shares.[13] Therefore, the estimations of gay and lesbian household shares

[12] When only one of the robust LM statistics was significant, that type of model was estimated. When both of the robust LM statistics were significant, the model with the larger test statistic was chosen.

[13] The LM statistic tests for all household shares strongly suggested the presence of a spatial error process. The LM statistic tests for gay male and lesbian household shares appeared weaker. The spatial lag model was the most consistently "preferred", although gay male and lesbian household share in the Midwest and gay male household share nationally showed no evidence of a spatial process. We estimated a spatial lag model for all regions nevertheless, to allow for easier comparisons between regions. The estimation of a spatial lag model in cases where there is no underlying spatial process should not unduly bias the results. We also tested a spatial error specification for gay and lesbian household share, and the results from these estimations were not substantively different than those from those shown. The LM statistics from each of the OLS estimations are shown in Table 19.6 in Appendix.

use a spatial autoregressive model, while the estimation of total household share uses a spatial error model.

The spatial autoregressive model used in the estimation of gay and lesbian household tract share includes a spatially lagged dependent variable as an additional covariate:

$$\mathrm{s}_{\mathrm{i,j,t+1}} = \alpha + \beta \mathrm{s}_{\mathrm{i,k,t}} + \gamma \mathrm{X}_{\mathrm{i,t}} + \mathrm{Z}_{\mathrm{i}} + \rho \mathrm{W} \mathrm{s}_{\mathrm{n}|\mathrm{i,j,t+1}} + \varepsilon_{\mathrm{i}} \tag{19.2}$$

where W is a row-standardized matrix which expresses the neighbor relationship between any two tracts *i* and *n*, and ρ measures the strength of the association between household share in tract *i* and its neighbors. Spatial autoregressive models are commonly used when the outcome of interest exhibits a diffusion or contagion process, such as households being attracted to those areas with similar households.

The model of total household tract share is estimated using a spatial error, or "nuisance", model, in which the error terms of neighboring tracts are spatially interacted. The spatial error model is expressed as:

$$\mathrm{s}_{\mathrm{i,j,t+1}} = \alpha + \beta \mathrm{s}_{\mathrm{i,k,t}} + \gamma \mathrm{X}_{\mathrm{i,t}} + \mathrm{Z}_{\mathrm{i}} + \varepsilon_{\mathrm{i}}, \quad \text{where } \varepsilon_{\mathrm{i}} = \lambda \mathrm{W} \varepsilon_{\mathrm{n}|\mathrm{i}} + \mu_{\mathrm{i}} \tag{19.3}$$

where W is again a row-standardized matrix which expresses the neighbor relationship between any two tracts *i* and *n*, and λ measures the intensity of the relationship between the errors terms in tract *i* and *n*. The spatial error model indicates that the source of the spatial dependence is non-substantive, possibly the result of the clustering of unmeasured or unobserved neighborhood features.

In all of the models estimated here, the neighbor weight matrix used is a queen contiguity matrix, in which tracts which share any common point are considered neighbors.[14] We estimate both the spatial autoregressive and the spatial error models via maximum likelihood. All analyses are carried out using the *spdep* package in R.

Table 19.3 shows the results of these estimations. The table shows regressions of the census tract's 2009 share of the central city's gay population, by gender, on the characteristics of the tract for large central cities in the nation. There are also separate regional estimates for these cities, as grouped into four regions.[15] In order to assess how the effects of various neighborhood characteristics differ for

[14] In this case, the W matrix is block diagonal, with the main diagonal blocks equal to the queen contiguity matrix within each city. The off-diagonal blocks are composed of zero matrices.

[15] Cities are assigned to regions according to Census Bureau definitions, except for Baltimore and Washington (in the South Census region, but assigned here to the Northeast); the cities in each region are displayed in Tables 19.1 and 19.2.

Table 19.3 Correlates of physical and socio-demographic characteristics on within city locations of gay couples (by gender) and all households, 2009 (Dependent Variable: Census tract proportion of city's gay male or lesbian or total households, 2009)

	National			West			South			Midwest			Northeast		
Independent variables	Gay	Lesbian	Total	Gay	Lesbian	Total	Gay	Lesbian	Total	Gay	Lesbian	Total	Gay	Lesbian	Total
Physical structure															
Distance from center (×100)	0.0007	0.0053	−0.0004	−0.0031	0.0063	−0.0021	0.0392	0.0303	0.0060	−0.0198	0.0078	−0.0135	−0.0270	−0.0058	−0.0030
	(0.11)	(0.68)	(−0.36)	(−0.40)	(0.68)	(−1.47)	(1.79)	(1.11)	(1.28)	(−0.89)	(0.29)	(−5.52)	(−2.29)	(−0.45)	(−2.64)
Distance squared (×10,000)	−0.0122	−0.0244	0.0051	0.0073	−0.0234	0.0095	−0.1768	−0.1080	0.0151	0.1417	−0.0718	0.0914	0.1510	0.0285	0.0145
	(−0.38)	(−0.63)	(0.82)	(0.24)	(−0.63)	(1.61)	(−1.43)	(−0.70)	(0.54)	(0.87)	(−0.37)	(5.07)	(2.10)	(0.36)	(2.05)
% Built before 1939 (in 2000)	0.0002	0.0010	-0.0002	-0.0006	0.0002	-0.0003	-0.0010	0.0045	0.0007	0.0023	−0.0003	−0.0004	0.0002	0.0007	−0.0001
	(0.41)	(1.69)	(−3.72)	(−0.80)	(0.22)	(−2.41)	(−0.46)	(1.73)	(2.24)	(2.27)	(−0.28)	(−4.50)	(0.31)	(1.13)	(−3.49)
Population density 2000 (×100,000)	0.0002	0.0000	−0.0001	−0.0001	−0.0003	−0.0005	0.0026	−0.0037	−0.0069	−0.0015	−0.0029	−0.0010	0.0001	0.0003	0.0000
	(0.35)	(−0.06)	(−2.25)	(−0.05)	(−0.22)	(−2.43)	(0.39)	(−0.44)	(−7.35)	(−0.60)	(−0.97)	(−4.53)	(0.33)	(0.78)	(−1.34)
% vacant (2000)	0.0029	−0.0020	0.0015	0.0075	−0.0013	0.0045	0.0059	0.0035	0.0030	0.0045	−0.0044	0.0006	0.0000	−0.0010	0.0002
	(2.13)	(−1.21)	(8.82)	(2.71)	(−0.39)	(9.51)	(1.38)	(0.67)	(5.43)	(1.57)	(−1.28)	(2.32)	(0.00)	(−0.56)	(1.52)
Prop city SFH detached 2000	0.0719	0.1761	0.0102	0.0031	0.2435	−0.0162	0.3060	0.3378	−0.0030	−0.0687	0.1502	−0.0117	0.0597	0.0921	0.0069
	(2.20)	(4.46)	(2.31)	(0.04)	(2.45)	(−1.14)	(3.08)	(2.72)	(−0.22)	(−0.63)	(1.15)	(−1.14)	(1.78)	(2.52)	(2.70)
Sociodemographic structure															
Prop city gay male HHs 2000	0.7957	0.1668	0.0123	0.7556	0.1419	−0.0011	0.9332	0.0256	0.0181	0.6250	0.4610	0.0113	0.6429	0.1581	0.0055
	(51.06)	(8.86)	(6.33)	(24.61)	(4.17)	(−0.23)	(27.68)	(0.61)	(4.03)	(17.66)	(10.87)	(3.67)	(23.91)	(5.60)	(2.95)
Prop city lesbian HHs 2000	0.0237	0.3718	−0.0054	0.0204	0.3133	−0.0080	−0.0335	0.3874	−0.0089	0.0354	0.2643	-0.0009	0.1175	0.5203	0.0009
	(1.13)	(14.68)	(−2.18)	(0.52)	(6.58)	(−1.19)	(−0.74)	(6.88)	(−1.53)	(0.75)	(4.66)	(-0.22)	(3.53)	(14.23)	(0.39)
Prop city non-family HHs 2000	0.0066	−0.0875	−0.0538	−0.1484	0.2017	−0.0739	0.0560	0.2447	−0.0684	0.0839	−0.8241	−0.0975	0.4930	−0.3824	−0.0333
	(0.11)	(−1.20)	(−6.60)	(−1.18)	(1.32)	(−3.35)	(0.42)	(1.48)	(−3.55)	(0.51)	(−4.16)	(−6.27)	(3.96)	(−2.83)	(−3.62)

Prop city total HHs 2000	0.2822	0.3706	1.0906	0.3235	0.0159	1.0788	−0.0242	-0.0638	1.1574	0.6452	1.4242	1.1649	−0.4791	0.5844	1.0378
	(2.89)	(3.14)	(82.95)	(1.51)	(0.06)	(28.80)	(−0.10)	(−0.22)	(35.11)	(2.31)	(4.24)	(43.42)	(−2.51)	(2.82)	(74.87)
Prop city black population 2000	−0.0754	−0.0312	−0.0269	−0.0185	0.1384	0.0076	−0.0906	0.0007	−0.0295	−0.1024	−0.3750	−0.0574	−0.0759	−0.2763	−0.0426
	(−4.68)	(−1.61)	(−11.18)	(−1.01)	(6.21)	(2.27)	(−2.67)	(0.02)	(−5.30)	(−2.15)	(−6.55)	(−12.00)	(−1.98)	(−6.61)	(−13.97)
Prop city Hispanic population 2000	−0.0119	−0.0345	−0.0097	−0.0533	0.0714	−0.0177	−0.0146	−0.0377	−0.0142	0.0073	−0.0795	−0.0056	−0.0272	−0.0797	−0.0038
	(−0.81)	(−1.93)	(−4.80)	(−1.53)	(1.69)	(−2.84)	(−0.42)	(−0.88)	(−2.95)	(0.26)	(−2.34)	(−2.03)	(−1.06)	(−2.84)	(−1.93)
Ratio tract med income/city med income	0.0006	0.0001	0.0002	0.0001	0.0002	0.0001	0.0007	0.0002	0.0002	0.0016	−0.0011	0.0002	−0.0001	0.0000	0.0000
	(3.34)	(0.65)	(8.41)	(0.36)	(0.58)	(1.68)	(1.39)	(0.33)	(3.30)	(3.08)	(−1.77)	(4.74)	(−0.29)	(−0.12)	(2.63)
Spatial lag (Prop gay male HHs 2009)	−0.0055	–	–	0.1539	–	–	−0.0619	–	–	−0.0421	–	–	0.0649	–	–
	(−0.39)			(5.68)			(−2.17)			(−1.41)			(2.62)		
Spatial lag (Prop lesbian HHs 2009)	–	0.0756	–	–	0.0578	–	–	0.1314	–	–	−0.0406	–	–	0.0442	–
		(5.07)			(1.82)			(4.30)			(−1.29)			(1.65)	
Lambda (spatial error coefficient)	–	–	0.4657	–	–	0.1525	–	–	0.5536	–	–	0.3099	–	–	0.3805
			(38.91)			(4.74)			(24.82)			(11.09)			(16.29)
City dummies	Yes	Yes	Yes	Yes	Yes	Yes	Yes	Yes	Yes	Yes	Yes	Yes	Yes	Yes	Yes
Pseudo R^2	0.384	0.180	0.943	0.537	0.264	0.926	0.401	0.113	0.919	0.286	0.168	0.952	0.393	0.233	0.983
N	10,450	10,450	10,450	2,290	2,290	2,290	2,444	2,444	2,444	2,524	2,524	2,524	3,192	3,192	3,192

z-score for each coefficient listed in parenthesis

gay households, we also report parallel regression analyses for the census tracts' 2009 shares of all households in the central city.[16]

The results do not support all of the hypotheses asserted in the literature on the location of gay households in cities.

Diversity Gates and Ost's contention that gays are attracted to more diverse neighborhoods is not confirmed by the patterns of shifts in gay locations between 2000 and 2009 in cities in any region of the country, at least to the extent that racial composition is used as a measure of diversity. Census tracts with greater shares of the central city's African American population in 2000 saw greater decreases in their share of the city's gay male partnered population by 2009 than in their share of all households, nationally and within each region. A similar result was found for lesbians in the Northeast and Midwest. Notably, lesbians in Western cities disproportionately shifted toward census tracts with larger African American populations.

If we measure diversity using Hispanic composition of neighborhoods, gay male partnership households are less likely than all households to shift away from census tracts with more Hispanic households. Lesbian partnership households, however, are slightly more likely to shift away from high Hispanic concentration census tracts in the Midwest and Northeast.

Vacancies and Older Housing Collins (2004) and Ruting (2008) contend that gays are attracted to neighborhoods with high vacancy rates and older housing stock. Although their discussion does not exclude lesbians explicitly, their examples apply to gay men. Table 19.3 provides some support for the hypothesis that gay men shift toward census tracts with higher vacancy rates, and of being disproportionately in census tracts with older housing in Western, Midwestern and Northeastern cities, relative to all households. While the results do not contradict the hypothesis that gay men are attracted to high vacancy neighborhoods, the differences with all households are not very big. Lesbians are also more attracted than all households to neighborhoods with older housing. Lesbian partnered households are also less likely to be attracted to high vacancy census tracts; they are significantly less likely

[16] In general, the coefficients from a spatial autoregressive model cannot be directly compared to the coefficients from a spatial error regression, as the lagged dependent variable in the autoregressive model introduces feedback and indirect impacts. While the coefficient for an explanatory variable (X) in a spatial error model is interpreted as the average impact of X on the outcome (Y), the average impact of X on Y in the spatial autoregressive model is calculated based on the coefficient, the spatial weight matrix, and the spatial autoregressive term [ρ in Eq. (19.2)] (LeSage and Pace 2009). In the results presented here, the feedback and indirect impacts in the autoregressive models are quite small, and do not make a substantive difference in the conclusions reached. As such, only the coefficients themselves are presented in the results tables.

As expected from the analysis of spatial dependence in the OLS residuals, the spatial lag terms for gay and lesbian household locations in the Midwest are not significant. However, the lag terms for the national gay male equation and the Northeast and West lesbian equations are also insignificant, suggesting that the spatial processes exhibited in these cases are not strong. The spatial error (lambda) terms included in the household share equations are significant in all cases.

In summary, the results in Table 19.3 show some interesting differences between lesbian and gay male partnered households. Because lesbian households are four times more likely to have children, and also average 20 % lower household income than those of gay men [Kyei and Madden (2012)], children and income are likely explanations for these gender differences in the central city locations of gay partnered households. The gender differences in spatial concentration among one-sex partnerships may also arise from gender differences in the use of space for political organization and activism as claimed by Lauria and Knopp (1985).

19.1.4 "Effects" of Gay Concentration on Economic Development of Neighborhoods

Planners and geographers have suggested that concentrations of gay households in a neighborhood lead to urban revitalization (Forsyth 2001; Lauria and Knopp 1985). We consider how greater concentrations of gays in a neighborhood are related to the subsequent course of economic development by examining how a census tract's population and relative income in 2009 is related to its 2000 shares of the city's gay male and lesbian households, given its population and relative income in 2000.

Collins (2004) argues that young urban professionals move into gay enclaves, leading to new construction and growth in, and change in the composition of, the neighborhood population. Florida and Mellander's (2010) study of housing prices in 331 metropolitan areas shows that gays and lesbians are associated with higher average housing prices in metropolitan areas. They argue that this is due both to the increase in amenities in the region when gays (and bohemians) migrate there and to a tolerance premium. Christafore and Leguizamon (2012) extend the Florida and Mellander study by looking at housing prices across neighborhoods in Columbus, Ohio, controlling for tolerance (using voting records on the Defense of Marriage Act in Ohio) and gay presence. They find a housing premium in "tolerant" areas and a penalty in "non-tolerant" areas.

We find support for the hypotheses that increasing representations of gay men in a census tract lead to increases in population for central cities nationally and in the Northeast, and to increases in household income nationally and in all regions. Increasing representations of lesbian households are associated with household income growth in the Northeast, but are not associated with population growth in any region. Because households residing in the neighborhood must be capable of

Table 19.4 Correlates of 2009 census tract population growth since 2000 (Dependent Variable: Census tract proportion of city's 2009 total population)

Independent variables	National	West	South	Midwest	Northeast
Distance from center (×100)	0.0034	0.0005	0.0135	−0.0062	0.0000
	(2.69)	(0.31)	(2.69)	(−2.48)	(0.00)
Distance squared (×10,000)	−0.0108	−0.0008	−0.0158	0.0488	−0.0025
	(−1.67)	(−0.11)	(−0.52)	(2.63)	(−0.33)
% built before 1939 (in 2000)	−0.0002	−0.0004	0.0009	−0.0003	−0.0002
	(−2.68)	(−2.59)	(2.57)	(−3.19)	(−3.68)
Population density 2000 (×100,000)	−0.0001	−0.0004	−0.0078	−0.0012	0.0000
	(−1.73)	(−1.76)	(−7.33)	(−4.74)	(−0.67)
% vacant (2000)	0.0014	0.0044	0.0026	0.0002	0.0005
	(7.06)	(8.06)	(4.09)	(0.60)	(3.10)
Prop city SFH detached 2000	0.0395	0.0212	0.0587	0.0020	0.0146
	(8.35)	(1.38)	(4.14)	(0.20)	(4.90)
Prop city gay male HHs 2000	0.0055	0.0047	0.0047	0.0017	0.0110
	(2.50)	(0.84)	(0.92)	(0.50)	(4.68)
Prop city lesbian HHs 2000	−0.0025	−0.0116	−0.0047	−0.0003	0.0010
	(−0.86)	(−1.48)	(−0.71)	(−0.06)	(0.34)
Prop city non-family HHs 2000	0.0142	−0.0381	0.0392	0.0049	−0.0323
	(2.70)	(−2.97)	(3.31)	(0.55)	(−5.01)
Prop city total population 2000	1.0008	1.0661	0.9948	1.0711	1.0456
	(97.61)	(34.63)	(39.20)	(57.25)	(89.57)
Prop city black population 2000	−0.0332	0.0022	−0.0292	−0.0744	−0.0727
	(−11.98)	(0.57)	(−4.49)	(−14.12)	(−19.45)
Prop city Hispanic population 2000	−0.0067	−0.0442	−0.0101	−0.0022	−0.0031
	(−2.80)	(−5.23)	(−1.74)	(−0.70)	(−1.22)
Ratio tract med income/ city med income 2000	0.0002	0.0000	0.0002	0.0002	0.0000
	(7.36)	(0.92)	(1.90)	(3.93)	(1.35)
Lambda (spatial error coefficient)	0.3954	0.1083	0.5169	0.2342	0.2184
	(30.91)	(3.29)	(22.14)	(7.98)	(8.35)
City dummies	Yes	Yes	Yes	Yes	Yes
Pseudo R^2	0.922	0.898	0.891	0.938	0.969
N	10,450	2,290	2,444	2,524	3,192

z-score for each coefficient listed in parenthesis

paying any increasing housing prices, neighborhood median household income relative to the city median must rise when housing prices increase. Tables 19.4 and 19.5 provide evidence on how the census tract's proportion of the city's gay partners are related to the rate of relative increase in the population of the census tract between 2000 and 2009 and to the rate of increase in the ratio of median census tract household income to the city median in the same period, after controlling for the same physical structure measures as included in Table 19.3.[17]

There is a positive relationship between the concentration of gay male households in 2000 and population growth for 2000–2009 nationally and in the Northeast. In contrast, the share of lesbian households has no effect on population growth, and a positive effect on income growth only in the Northeast, which is consistent with the hypothesis that their greater likelihood of having children and their lower incomes lead to intra-city locations different from those of gay men. This result may also have different implications for neighborhood economic development.

Household income growth generally decreases, however, with the census tract's share of non-family households in 2000. Non-family households include single persons (the main component), as well as adults not related by blood or marriage who share housing. We control for non-family households in these analyses because, as indicated in Table 19.3, there is a strong positive correlation between the presence of gay male households and nonfamily households in census tracts, and the explanations as to why a spatial concentration of gay households might lead to greater income or population growth seemingly apply to spatial concentrations of single people as well [see Collins (2004)]. In fact, Tables 19.4 and 19.5 indicate that the census tracts with more non-family households experience less income (and implicitly housing price) and population growth (in the Northeast and West) over the following decade. Although non-family households include gay households, gay households are a trivial share of non-family households (less than 3 % of non-family households in these cities in 2000, for example). The correlations between population or income and non-family household shares in Tables 19.4 and 19.5 reflect the relationship of population or income growth to concentrations of single person households.

19.1.5 Conclusions, Qualifications, and Next Steps

This first attempt to analyze the locations of gay male and lesbian households in large cities in the U.S. has several rather surprising results.

First, gay men and lesbians are less segregated within cities than African Americans, and gay males are more segregated than lesbians. Although gay male segregation increased slightly, on average, between 2000 and 2010 in the West, South, and Midwest, lesbian segregation during this time decreased substantially in all regions. Consistent with the finding of lower levels of lesbian segregation, lesbian households are somewhat less likely than are gay male households to be

[17] As in Table 19.3, OLS estimation is performed to test for spatial dependence in the residuals and to choose the appropriate spatial model specification. For the population growth equation a spatial error model is estimated. For the income growth equation a spatial autoregressive model is used. The LM statistics from the OLS estimations are displayed in Table 19.6 in Appendix.

Table 19.5 Correlates of 2009 census tract income growth since 2000 (Dependent Variable: Ratio of census tract median income to city median income, 2009)

Independent variables	National	West	South	Midwest	Northeast
Distance from center (×100)	−1.8514	−0.4090	−1.7434	−2.6140	−3.5589
	(−9.69)	(−1.42)	(4.00)	(−4.02)	(−6.82)
Distance squared (×10,000)	6.5559	1.7929	3.3529	9.4926	17.8020
	(6.88)	(1.55)	(1.36)	(1.99)	(5.59)
% Built before 1939 (in 2000)	0.0833	0.1275	0.1392	0.0314	0.0954
	(5.89)	(4.93)	(3.37)	(1.06)	(3.98)
Population density 2000 (×100,000)	−0.0365	−0.0859	−0.1518	−0.2165	−0.0440
	(−2.81)	(−2.17)	(−1.13)	(−2.97)	(−2.84)
% vacant (2000)	0.0046	0.1039	0.2577	0.0921	−0.1688
	(0.11)	(1.00)	(3.02)	(1.09)	(−2.29)
Prop city SFH detached 2000	4.0816	18.2020	−0.6798	6.2705	4.0859
	(4.41)	(6.33)	(−0.38)	(2.15)	(2.90)
Prop city gay male HHs 2000	2.8233	4.2528	1.2822	4.1494	3.0535
	(6.08)	(4.05)	(1.93)	(3.96)	(2.68)
Prop city lesbian HHs 2000	0.4709	−1.0847	0.1857	0.8407	2.8609
	(0.75)	(−0.74)	(0.21)	(0.60)	(1.97)
Prop city non-family HHs 2000	−3.9326	−4.7591	−4.5260	−5.5860	1.7090
	(−3.64)	(−1.96)	(−3.04)	(−2.07)	(0.53)
Prop city total population 2000	−4.3294	−15.0087	3.7082	−5.0160	−15.0497
	(−2.11)	(−2.61)	(1.16)	(−0.93)	(−2.60)
Prop city black population 2000	−3.7095	−1.8509	−2.3131	−6.7662	−5.2199
	(−7.47)	(−2.63)	(−3.28)	(−4.61)	(−3.04)
Prop city Hispanic population 2000	−2.7676	−7.1697	−3.1387	−2.5198	−0.7473
	(−5.95)	(−4.59)	(−4.34)	(−2.89)	(−0.63)
Ratio tract med income/ city med income 2000	0.8018	0.7450	0.9031	0.8166	0.7259
	(113.83)	(56.48)	(74.11)	(44.03)	(53.19)
Spatial lag of tract/ city income 2009	0.1973	0.1581	0.1754	0.2130	0.2099
	(23.68)	(9.49)	(11.93)	(10.91)	(13.10)
City dummies	Yes	Yes	Yes	Yes	Yes
Pseudo R^2	0.800	0.875	0.867	0.711	0.759
N	10,450	2,290	2,444	2,524	3,192

z-score for each coefficient listed in parenthesis

spatially concentrated. Although lesbians exhibit significant levels of spatial clustering in many cities, gay men are spatially clustered in all cities (except for Detroit in 2010).

We find little empirical support for some of the hypotheses about the characteristics of the neighborhoods that attract more gay couples advanced in case studies and more qualitatively oriented research. Relative to other city residents, gay men and lesbians do not shift toward more racially diverse neighborhoods and there is little difference in their shifts to more ethnically diverse neighborhoods. While there is some positive correlation in locations of gay men and lesbians, it is small. Nationally, gay men are somewhat more likely to reside in higher vacancy neighborhoods, but there is no evidence that is the case for lesbians. There is no evidence that centrally located neighborhoods are more attractive to gay male or lesbian partnered households than to all households. However, both lesbians and gay men are less likely to shift out of densely populated neighborhoods than other households.

Census tracts that start the decade with more gay men experience significantly greater growth in household incomes (and, therefore, presumably housing prices) over the next decade. Large central city census tracts that start the decade with more gay men experience significantly greater population growth over the decade, most notably in the Northeastern region. There is little evidence that concentrations of lesbian households are associated with future population or income growth in the neighborhood. Census tracts with relatively more lesbians at the start of the decade generally see few differences in either population or income growth from those with fewer lesbians.

There are data issues to address in future research. The biggest qualification of these results is that they rely on only one decade's data, and a potentially anomalous decade in which the U.S. experienced its most severe recession since the Great Depression. Furthermore, the measure of gay presence is complicated by the U.S. Census's erroneous recoding of unmarried partner data, so that all partnerships reporting as married and same sex are coded as unmarried same sex partnerships. Although the data used in these analyses were adjusted to account for this potential recoding error, a more precisely measured group of gay partners and a less anomalous decade may give different results.

The regression modeling approach used here may also be improved with the implementation of a more complex spatial modeling framework. The use of a spatial Durbin model, in which lagged values of the explanatory variables are included as additional covariates in the model, could lead to a better-fitting model. However, the use of such a model with census tract data is likely to involve extremely highly correlated explanatory variables, and some of the explanatory variables in the model (e.g., distance from downtown and the city dummies) cannot be appropriately included in a spatial Durbin model. This type of modeling strategy does, nevertheless, warrant further investigation.

Acknowledgement We thank Marcus Dillender, Gabrielle Fack, Gary Gates, Stephen Sheppard, and an anonymous reviewer for helpful comments on an earlier version of this paper.

Appendix

Table 19.6 Lagrange multiplier statistics from OLS regressions

	National		West		South		Midwest		Northeast	
Gay male HH share										
LM error	1.6	(0.21)	2.0	(0.16)	1.9	(0.17)	3.0	(0.08)	0.7	(0.41)
LM Lag	0.2	(0.69)	23.8	(0.00)	5.3	(0.02)	1.8	(0.18)	7.8	(0.01)
Robust LM error	3.3	(0.07)	NC		NC		1.4	(0.23)	NC	
Robust LM Lag	1.9	(0.17)	NC		NC		0.2	(0.62)	NC	
Lesbian HH share										
LM error	18.2	(0.00)	0.3	(0.61)	15.6	(0.00)	1.6	(0.21)	0.1	(0.82)
LM Lag	29.7	(0.00)	3.4	(0.06)	20.4	(0.00)	2.4	(0.12)	1.7	(0.19)
Robust LM error	14.0	(0.00)	15.1	(0.00)	4.0	(0.04)	0.4	(0.51)	19.6	(0.00)
Robust LM Lag	25.5	(0.00)	18.3	(0.00)	8.8	(0.00)	1.3	(0.26)	21.2	(0.00)
All HH share										
LM error	2281.8	(0.00)	91.2	(0.00)	591.6	(0.00)	158.9	(0.00)	285.2	(0.00)
LM Lag	629.6	(0.00)	24.3	(0.00)	213.9	(0.00)	29.2	(0.00)	0.0	(0.97)
Robust LM error	1671.3	(0.00)	66.9	(0.00)	387.5	(0.00)	130.6	(0.00)	NC	
Robust LM Lag	19.1	(0.00)	0.1	(0.81)	9.8	(0.00)	1.0	(0.32)	NC	
Population growth										
LM error	1569.3	(0.00)	54.2	(0.00)	464.1	(0.00)	83.6	(0.00)	75.6	(0.00)
LM Lag	726.8	(0.00)	44.5	(0.00)	265.6	(0.00)	8.0	(0.00)	15.2	(0.00)
Robust LM error	906.5	(0.00)	17.6	(0.00)	226.4	(0.00)	76.4	(0.00)	60.8	(0.00)
Robust LM Lag	64.0	(0.00)	8.0	(0.00)	27.9	(0.00)	0.8	(0.36)	0.0	(0.52)
Income growth										
LM error	273.6	(0.00)	40.6	(0.00)	42.9	(0.00)	82.7	(0.00)	41.9	(0.00)
LM Lag	580.1	(0.00)	93.8	(0.00)	149.0	(0.00)	123.5	(0.00)	179.8	(0.00)
Robust LM error	15.7	(0.00)	3.0	(0.08)	0.9	(0.35)	7.1	(0.01)	2.9	(0.09)
Robust LM Lag	322.3	(0.00)	56.2	(0.00)	106.9	(0.00)	47.8	(0.00)	140.8	(0.00)

All Lagrange multiplier statistics are distributed $\chi^2(1)$
p-values in parentheses
NC Not computed

References

Adler S, Brenner J (1992) Gender and space: lesbians and gay men in the city. Int J Urban Reg Res 16(1):24–34

Anselin L (1988) Lagrange multiplier test diagnostics for spatial dependence and spatial heterogeneity. Geogr Anal 20(1):1–17

Anselin L, Bera AK, Florax R, Yoon MJ (1996) Simple diagnostic tests for spatial dependence. Reg Sci Urban Econ 29:77–104

Bera AK, Yoon MJ (1993) Specification testing with locally misspecified alternatives. Econom Theor 9(4):649–658

Black D, Gates GJ, Sanders SG, Taylor L (2007) The measurement of same-sex unmarried partner couples in the 2000 U.S. census. California Center for Population Research Working Paper, PWP-CCPR-2007-023
Black D, Gates GJ, Sanders SG, Taylor L (2002) Why do gay men live in San Francisco? J Urban Econ 51:54–76
Brown M (2013) Gender and sexuality II: there goes the gayborhood? Prog Hum Geogr (first published online April 29 2013)
Burridge P (1980) On the Clifford test for spatial correlation. J Roy Stat Soc B 42:107–108
Castells M (1983) The city and the grassroots. University of California Press, Berkeley, CA
Christafore D, Leguizamon S (2012) The influence of gay and lesbian coupled households on house prices in conservative and liberal neighborhoods. J Urban Econ 71:258–267
Collins A (2004) Sexual dissidence, enterprise, and assimilation: bedfellows in urban regeneration. Urban Stud 41(9):1789–1806
DiBennardo R, Gates GJ (2013) Research note: US Census same-sex couple data: adjustments to reduce measurement error and empirical implications. Popul Res Policy Rev. doi: 10.1007/s11113-013-9289-2 (published online)
Florida R, Mellander C (2010) There goes the metro: how and why bohemians, artists, and gays affect regional housing values. J Econ Geogr 10(2):167–188
Forsyth A (1997) NoHo: upscaling main street on the metropolitan edge. Urban Geogr 18(7): 622–652
Forsyth A (2001) Nonconformist populations and planning practice: sexuality and space. J Plan Lit 15(3):339–358
Gates G, Ost J (2004) The gay and lesbian atlas. The Urban Institute Press, Washington, DC
Gates GJ, Steinberger MD (2009) Same-sex unmarried partner couples in the American community survey: the role of misreporting, miscoding, and misallocation. Paper presented to the association of public policy and management meetings, Washington, DC
Harry J (1974) Urbanization and the gay life. J Sex Res 10(3):238–247
Hayslett KL, Kane MD (2011) Out in Columbus: a geospatial analysis of the neighborhood-level distribution of gay and lesbian households. City Commun 10(2):131–156
Kyei P, Madden J (2012) Gender differences in the gay pay gap: unmeasured gender-linked characteristics, household division of labor, or greater bias against gay men? Working paper. Population Studies Center, University of Pennsylvania
Lauria M, Knopp L (1985) Toward an analysis of the role of gay communities in the urban renaissance. Urban Geogr 6(2):152–169
LeSage J, Kelley Pace R (2009) Introduction to spatial econometrics. Chapman & Hall/CRC, Boca Raton, FL
Madden J (2014) Changing racial and poverty segregation in large U.S. metropolitan areas, 1970–2009. Int Reg Sci Rev 37(3):9–35
O'Connell M, Feliz S (2011) Same-sex couple household statistics from the 2010 census. Social, economic and housing statistics division working paper number 2011–26, US Bureau of the Census
O'Connell M, Gooding G (2007) Editing unmarried couples in census bureau data. Housing and household economic statistics division working paper, US Bureau of the Census
Ruting B (2008) Economic transformations of gay urban spaces: revisiting Collins' evolutionary gay district model. Aust Geogr 39(3):259–269

An Image-Based Multi-Criteria Assessment of Sustainable Redevelopment Plans of a University Campus

20

João Romão, Peter Nijkamp, Eveline van Leeuwen, Karima Kourtit, Takao Ozasa, and Maki Ikegami

Abstract

The concept of sustainability is increasingly applied in urban facilities planning, including a University campus. A campus can essentially be seen as a living laboratory for the development of ecologically-benign practices connected to the social interactions of an academic environment. Sustainable (re)development of a University campus calls for stakeholder-based assessment tools. The present paper offers an operational framework for such an assessment, focused on the possible transformations in the landscape of the Campus and the ecological, social and functional processes of re-development. In this study, a Regime multi-criteria evaluation method complemented with a strategic choice analysis regarding the performance and sustainability of Hokkaido University campus (Sapporo, Japan) is presented. This paper serves to trace whether and how the goals and principles defined in the strategic Plans for the development of this

J. Romão • E. van Leeuwen • K. Kourtit
Department of Spatial Economics, VU University Amsterdam, De Boelelaan 1105, 1081 Amsterdam, The Netherlands
e-mail: jromao@vu.nl; jfromao@ualg.pt; e.s.van.leeuwen@vu.nl; k.kourtit@vu.nl

P. Nijkamp (✉)
Department of Spatial Economics, VU University Amsterdam, De Boelelaan 1105, 1081 Amsterdam, The Netherlands

A. Mickiewicz University, Poznan, Poland
e-mail: p.nijkamp@vu.nl

T. Ozasa
Faculty of Engineering, Division of Architectural Design, Laboratory of Urban Design, Hokkaido University, Kita13 Nishi8, Kita-ku, Sapporo 060-8628, Japan
e-mail: ozasa@eng.hokudai.ac.jp

M. Ikegami
Office for a Sustainable Campus, Hokkaido University, Kita8 Nishi5, Kita-ku, Sapporo 060-0808, Japan
e-mail: coordinator1@osc.hokudai.ac.jp

P. Nijkamp et al. (eds.), *Regional Science Matters*, DOI 10.1007/978-3-319-07305-7_20,

campus fit the needs and concerns of its users. This bottom-up approach is inter alia based on quantitative information collected in various focus group meetings with different stakeholders, followed by the use of a SWOT analysis supported by visual support tools ('images'). This approach allows us to develop an integral and quantitative assessment of the future performance of the campus, by focusing on its physical appearance, use characteristics and ecological landscape attributes.

20.1 Introduction: Scope and Aims

The current interest in the quality of places where people live and work calls for adjusted research methods to address the needs of strategic local planning (Nijkamp 2008). In this context, scenarios have become powerful tools, often in combination with complementary envisioning tools, such as Imagineering [see e.g. Ratcliffe and Krawczyk (2011)]. Such new approaches offer a systematic framework for designing and evaluating future images of cities, urban districts or places. Many—especially larger—cities offer hospitality to higher education institutes or universities. Such knowledge centres are often clearly visible local landmarks, especially if they are housed on a dedicated campus in the city. The present paper will pay attention to the way general principles for sustainable urban development can be achieved on a specific urban university campus.

Designing new urban spaces calls for vision and up-to-date knowledge. On the other hand, there is nowadays no scope anymore for blueprint planning, which strives to attain an a priori given desired endpoint. Modern spatial planning is usually interactive, with a view to flexible alternative uses of scarce spaces. In a democratic planning context, there is a need to design and judge various choice possibilities through systematic assessment methods (e.g. cost-benefit analysis, multi-criteria analysis). The standard way to handle such comparative exercises is to assign numerical figures (categorical or cardinal data) to all important attributes of such choice options or alternatives. Especially in the context of quality upscaling, it has in recent years become fashionable to use also standardized photographs or videos as a systematic tool to extract comparable judgments from stakeholders or individual choice makers. Some recent examples of this approach can be found in Neuts et al. (2012), on the evaluation of historical districts in cities, in Smit (2011), on the assessment of district visual quality of location decisions of creative entrepreneurs, or in Valencia-Sandoval et al (2010), on a participatory landscape planning process. A similar approach to sustainable campus (re)development will be adopted in the present study.

Sustainability—the ability to combine economic efficiency, social equity and environmental protection in the long run—is an important urban planning challenge: pollution, poverty, excessive consumption or depletion of resources have a far reaching impact on the local activity context. Despite the global political and institutional guidelines to frame sustainable practices or sustainable processes of development, these issues must be addressed at the real-world levels of the country, the city, the neighbourhood, the community or, in general, the places where people interact and live. Universities are no exception to this guideline: in different countries, universities are nowadays implementing dedicated strategies on sustainability, for instance, by introducing educational courses or a sustainability orientation in different disciplines (Shriberg 2002; Mcmillin and Dyball 2009), implementing processes to evaluate the sustainability of their performance (Beringer 2006; Alshuwaikhat and Abubakar 2008), developing plans for sustainable action and attaching a strong importance of the issues regarding sustainability in the processes of planning space, energy or buildings (Franklin et al. 2003; Darus et al. 2009). At an international level, various attempts to create a standard for evaluation and strategic assessment on the sustainable development of university campuses are being developed, like the Sustainability Tracking, Assessment & Rating System (AASHE 2012).

University campuses are ideal spaces where thousands of persons interact every day, performing different tasks with diverse purposes and motivations: students, researchers, teachers, managers, other professional staff and visitors use a common space for their daily activities, requiring a wide range of products and services, consuming energy, walking and relaxing in public spaces, and using common facilities. Considering the importance recently given to sustainability issues in local planning processes, the need for strategic assessment, and the implementation of academic programs in many universities, these communities are especially appropriate to be analysed as a "living lab" for the evaluation of actions and policies regarding sustainable behaviour, at individual and collective levels. This work will focus on the environmental and social dimensions of the sustainability of a concrete University Campus, analysing the perceptions of its users on the long-term implications of the physical planning of the Campus for its ecological preservation and the promotion of social interaction and cohesion.

The case study to be analysed is the campus of Hokkaido University in Sapporo, Japan. Hokkaido is a beautiful natural island and the city of Sapporo is surrounded by majestuous mountains. This situation calls for a careful planning of this university campus, which assumes a central place in the city and offers an oasis of quiet. The aim of this work is to develop and support a new orientation in campus planning, based on a stakeholder-oriented—and thus a bottom-up—approach. A preference elicitation will be organized through focus groups, while next the systematically collected information is organized through two complementary tools: multi-criteria analysis (MCA) and strategic choice analysis (SCA). This is a novel approach in the Japanese context, where stakeholder-based land use planning is rather uncommon.

20.2 Research Methodology on Sustainable Campus Development

20.2.1 General Orientation

A client-oriented approach to sustainable campus (re)development involving the users of the campus of Hokkaido University will be developed in the present study in order to identify the main success and failure factors, in particular how the physical and landscape conditions influence the performance—and hence the satisfaction level of the campus—and how the campus may look like in the future, starting with the assessment of the current campus and continuing with the identification of general strategic ideas for its future development. The basic information is based on the current Master Plan for the campus and the Action Plan for Sustainability (forecasting and backcasting). The methodology followed in this study will be applied in the future for the preparation of a new Master Plan of the Campus, including scenario analysis, choice experiments among conflictive proposals, and financial restrictions (which will not be analysed in this work).

As the campus of Hokkaido University is a rather open public space, used also by residents and tourists as a public park, this creates conditions for a close interaction between academia and the local community, but it may also prompt potential conflicts between "regular users" and occasional visitors or tourists. On the other hand, the campus is an important green area within the ecological structure of the city and its landscape (see Fig. 20.1), which reinforces the need for enhancing the importance of its planning in the context of the city plan. Examples of such approaches can be found in Kenney et al. (2005), who provide an analysis of the role of Universities in the context of a city, and in Perry and Wiewel 2005), who present different examples of interaction between University campuses and the surrounding communities.

The analysis of the sustainability of a University campus needs a systematic and comprehensive approach to the different aspects and functions daily performed by the University, systematizing different domains of analysis and considering the users of the campus as a community of different persons with different purposes sharing the same territory. A similar process of systematization has been applied in urban planning by Bourdic et al. (2012), defining three main pillars (urban form; economic and social aspects; environment) and a set of indicators for each pillar. Recent other examples of various importance categories relevant for university campus evaluation are provided by Akgün et al. (2012), proposing five systems (Physical, Social, Economic, Ecological and Institutional) in order to address "the stakeholders' multi-faceted viewpoints on future sustainable development" at a regional scale, and by Heijer (2012), suggesting five functions (Academic, Residential, Business Related, Retail and Leisure, Infrastructure) to analyse the performance of universities. Other recent examples of an application of bottom-up processes for the assessment of university campuses are provided by Abd-Razak et al. (2011), Disterheft et al. (2012), Zink et al. (2008), assuming a decentralized approach as a necessary condition for the successful implementation of sustainable action plans at community level.

The Hokkaido University Campus hosts 20,800 students and staff members (20,300 in the South and 500 in the North) in 1,776,248 m^2 and it is well known that it has two rather distinct, but mutually adjacent, areas: the South and North Campus (see map in Fig. 20.2). These areas have different characteristics, functions, users and expected developments. The two different territories of the Campus were analysed during our study, with the organization of two focus-group meetings, involving users of these two areas.

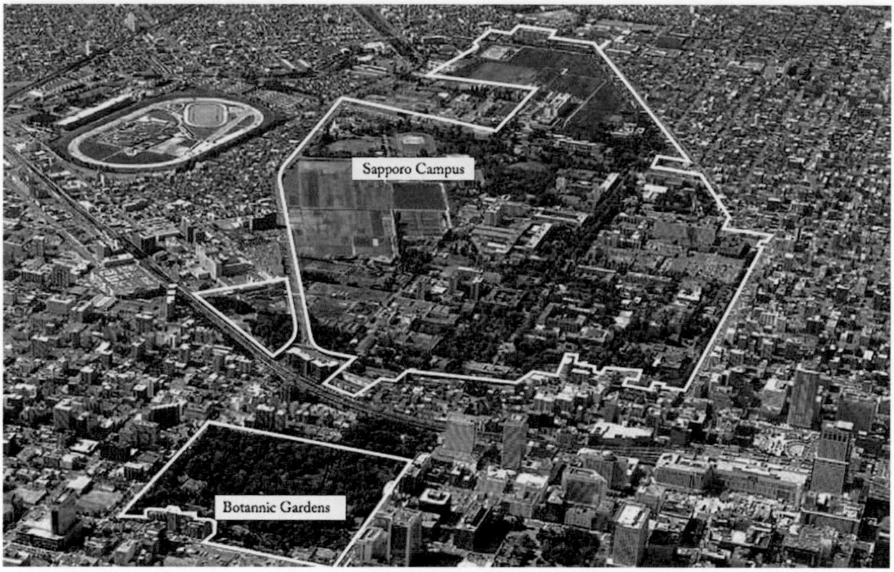

Fig. 20.1 The campus of Hokkaido University within the city of Sapporo

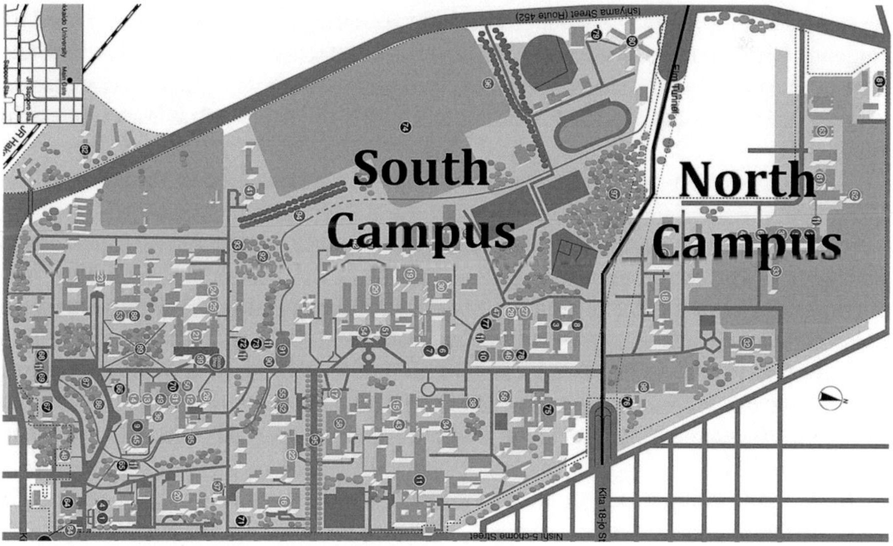

Fig. 20.2 Hokkaido University campus in Sapporo (Japan)

The North Campus (represented on the right-hand of Fig. 20.2) hosts the Hokkaido University Research & Business Park, a partnership between industry, academia and government sectors aiming at the dissemination and commercialization of research findings as the main result of cooperative alliances. Normally,

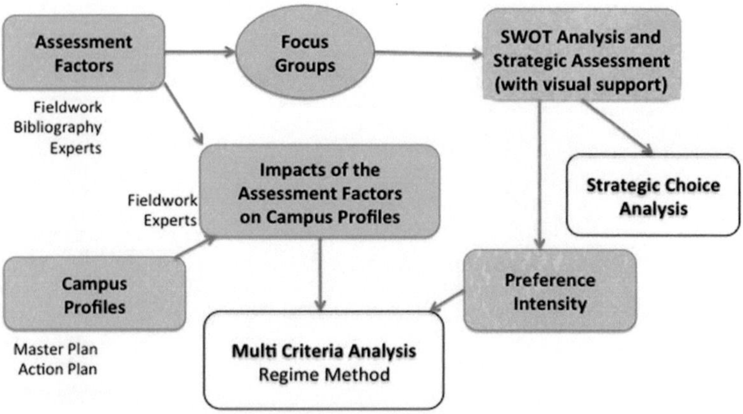

Fig. 20.3 Research methodology for campus design and planning

the users of this part of the Campus have already a relatively long presence in this place. According to the Master Plan of Hokkaido University, this area will be expanded, implying the construction of new buildings. The South Campus (the largest area, in the left-hand of the figure) is the traditional Campus area, including teaching and research facilities, libraries and other supportive academic facilities, services for students and staff, or cultural and sports equipment; it is mostly used by students, with a relatively short presence at the University; considering the land use restrictions on the expansion of the area, its development (with the implementation of new services, facilities or equipment) must be achieved without physical expansion.

The methodology adopted in the present study is represented in Fig. 20.3. In the first stage, the analytical tools are defined: the Assessment Factors—or evaluation criteria—based on fieldwork, bibliography and expert assessment (interviews with managers of different services and departments of the University, including academic, public and private institutions); and the Campus Profiles—or distinct future directions for campus design—originating from the existing plans and representing five important, mutually complementary user orientations of a Future Campus. From these two categories it was possible to generate a Matrix of Impacts relating four main Assessment Factors to the five Campus Profiles (this process is described in Sect. 20.2.2).

Next, focus group meetings were organized, in different areas of the campus and with different stakeholders (students, researchers from academic and private institutions, teachers and managers at private and public institutions; the groups were balanced in terms of gender; international students and researchers were included), generating the ingredients for a SWOT analysis supported by visual images and complemented with strategic proposals (described in Sect. 20.2.3). The information collected in these meetings is next used for the Strategic Choice Analysis (described in Sect. 20.3.2) and for the identification of a Preference

Intensity Vector which, combined with the "Matrix of Impacts", will be used for a Multi-Criteria Analysis based on the so-called Regime Method (see Sect. 20.3.1).

20.2.2 Assessment Factors and Campus Profiles

In our campus evaluation experiment, four assessment perspectives (evaluation domains) for the various activities in an integrated sustainable campus (represented in Fig. 20.2) have been distinguished and defined in order to obtain a comprehensive judgement framework for the functions to be performed by the University:

- *Science Orientation*, focused on the "traditional" domains of activity of the University, including Teaching and learning conditions, Research facilities, Libraries, Conference facilities, Special education facilities or Experiential Farms;
- *Creativity*, focused on the relation with other creative activities and with the regional economy, including academic spin-offs (incubators), Private and Collaborative R&D activities, Artistic and Creative professions, linkage to the community or Connectivity (e.g. ICT networks);
- *Environment,* focused on quality of life issues and on the use of resources, including issues related to Accessibility, Mobility, Energy consumption and production, Waste and water management, or CO2 Emissions;
- *Social Capital*, focused on the complementary services provided to the users of the campus, including the Financial support for students, Student housing, Food services, Bookstores, projects involving the local community, or Recreational, cultural and sports activities.

These four domains offer the ingredients to define interlinked assessment criteria for evaluating alternative future campus orientations or directions regarding the planning and design of the campus (re)development.

This system of evaluation is next used in order to test whether a set of five relevant campus profiles extracted from the existing Plans for the development and management of the campus are appropriate for the concerns and motivations expressed by its users. Such future profiles reflect the main principles for the campus revitalization, design, and developments and represent in a "prism" its most important appearance forms. Consequently, they are not mutually exclusive, but may complement each other. These five profiles, which correspond to a priori specified policy objectives on the Hokkaido Campus, are:

- *Experiential Campus*, covering all issues related to the possibilities of social interaction among users of the campus and the practice of other activities, like sports, culture or leisure;
- *Functional Campus*, including the management of services, facilities and equipment required for the performance of academic functions, like teaching, researching, other academic services, provision of food, or accommodation;
- *Ecological Campus*, covering all aspects related to the preservation and promotion of biodiversity, energy production and consumption, waste and water management or CO_2 emissions;

Table 20.1 Campus profiles and assessment factors

Campus profile	Policy documents: Master plan	Policy documents: Action plan	Assessment criteria: Science orientation	Environment	Social capital	Creativity
Experiential	Planning open space, frame, place making	Development of public space	2.8	2.3	4.2	2.0
Functional	Zoning	Facility design, setting and management	4.5	3.7	2.0	3.3
Ecological	Structure of natural and ecological environment	Ecosystem conservation	2.0	4.3	1.5	1.8
		Energy consumption and production				
		Waste and water management				
Accessible	Accessibility, mobility	Transport, parking, pedestrianism	3.5	3.2	3.7	3.0
Collaborative	Collaboration campus—city	Connection with regional community	2.2	1.5	3.7	5.0

- *Accessible Campus*, covering all issues related to the accessibility from the city to the University and the mobility inside the campus, including urban car traffic and public transport, bicycles, parking and public transport inside the campus, pedestrians, roads and other infrastructures for mobility;
- *Collaborative Campus*, considering the relations with the surrounding community, including enterprises, local residents, tourists or private organizations developing cooperative activities with the University.

The relations between the four assessment perspectives defined as a system for the evaluation of the campus and the five campus future profiles—projected in the Master Plan and in the Action Plan for Sustainability of the campus—are next systematized and quantified (on a rank order scale 1–5, with five the most favourable impact). The quantification of these impacts is based on fieldwork undertaken in the campus complemented by expert assessment, including eight interviews with managers of different services and departments of the University. After a consultation in various subgroups, the main impact assessment was done by seniors in the Campus development project. The average score obtained for each assessment criterion is presented in Table 20.1, while the detailed underpinning of these scores for all the sub-indicators is extensively described in Table 20.2. This matrix will be used in the Regime Multi-Criteria Analysis to be presented in Sect. 20.3.1.

The information considered in Table 20.1 can also be visualized in a so-called spider diagram (see Fig. 20.4), offering an envelope picture of various campus

Table 20.2 Impact matrix for five campus profiles

	Campus profiles				
Assessment criteria	Experiential	Functional	Ecological	Accessible	Collaborative
Science orientation	**2.8**	**4.5**	**2.0**	**3.5**	**2.2**
Teaching and learning conditions	3	5	2	4	1
Research conditions	3	5	2	4	1
Libraries	3	5	1	4	2
Conference facilities	2	5	1	4	3
Experimental Farms and production	3	2	5	1	4
Special education facilities	3	5	1	4	2
Creativity	**1.8**	**3.3**	**1.8**	**3.0**	**5.0**
Academic spin-off	1	3	2	4	5
Private R&D	1	3	2	4	5
Collaborative R&D	1	4	2	3	5
Artistic and creative professions	4	2	1	3	5
Linkage to local community	1	4	2	3	5
Connectivity (ICT networks)	3	4	2	1	5
Environment	**2.3**	**3.7**	**4.3**	**3.2**	**1.5**
Accessibility	3	4	2	5	1
Mobility	2	3	4	5	1
Biodiversity and public spaces	4	3	5	2	1
Emissions	1	4	5	3	2
Energy consumption and production	2	4	5	3	1
Water and waste management	2	4	5	1	3
Social capital	**4.2**	**2.0**	**1.5**	**3.7**	**3.7**
Financial support for students	4	2	1	3	5
Student housing	5	3	2	4	1
Various services	3	1	2	4	5
Recreation, culture and sports	5	2	1	3	4
Projects with the city	4	1	2	3	5
Bookstores	4	3	1	5	2

models represented according to the impact of the different assessment factors. Given the current information on Assessment criteria and campus profiles, a feasible campus can be designed within the boundaries as shown in this Figure.

The spider diagram in Fig. 20.4 represents the five future campus profiles according to the impacts of each factor considered in the system of evaluation. It is noteworthy that the Experiential Profile is mostly influenced by social factors, the Functional Profile is highly dependent on the scientific mission, the Ecological Profile is (not surprisingly) closely related to the environmental issues, the Collaborative Profile is strictly connected to social talent and creativity, while the than gay men, or than all households, to shift toward higher vacancy rate neighborhoods nationally.

Increasing Concentration or Acceleration Collins (2004) and Ruting (2008) contend that gays are attracted to neighborhoods with initially higher settlements of gays, prompting an "acceleration" of their representation. Table 19.3 does not provide much support for an acceleration of representation. While census tracts with more gay male, or lesbian, households in 2000 have more such households in 2009, there are no coefficients for the 2000 census tract shares of the population of gay men or lesbians greater than one, while the coefficient for the 2000 census tract shares of the population of all households is greater than one for the nation and for each region. Consistent with Table 19.1, which shows greater segregation of gay male partners than of lesbian partners and greater decreases in segregations for lesbian household over the years, the regression analyses in Table 19.3 consistently show greater dispersion (relative to 2000 locations) of lesbian partner households; the coefficient of the census tract's 2000 share of own group is less for lesbian households than for gay male households in every case, and by fairly large margins.

Closer to Downtown With fewer children and greater demand for night life [Collins (2004)], gay men and lesbians are expected to be more centrally located, in more densely populated neighborhoods, than other households. Once again, this hypothesized difference in location behavior for gay partnerships is not clearly evident in all the relevant measures. Table 19.3 shows gay male households decreasing with distance from downtown only in the Northeast region; lesbian household concentration exhibits no significant relationship with distance from downtown. In the Midwest region, gay male and lesbian household concentrations actually increase with distance from downtown, relative to all households. And relative to all households at the national level, gay male and lesbian partnership households shift toward census tracts with more single family detached homes. However, unlike all households, gay male and lesbian households do not shift away from census tracts with greater population density.

Co-Location of Gay Men and Lesbians Gates and Ost (2004) indicate that locations for gay men and lesbians, while different, are more correlated with each other than with other household types. We find inconsistent support for this hypothesis. Nationally, and in each region, gay men are no more likely to co-locate near lesbian households than near all non-family households. However, lesbians shift more toward census tracts with greater relative representation of gay male partners and away from those with more non-family households nationally and in all regions except the South. Lesbians also shift toward census tracts with greater numbers of all households, as opposed to nonfamily households or to gay male households, in the Northeast and Midwest.

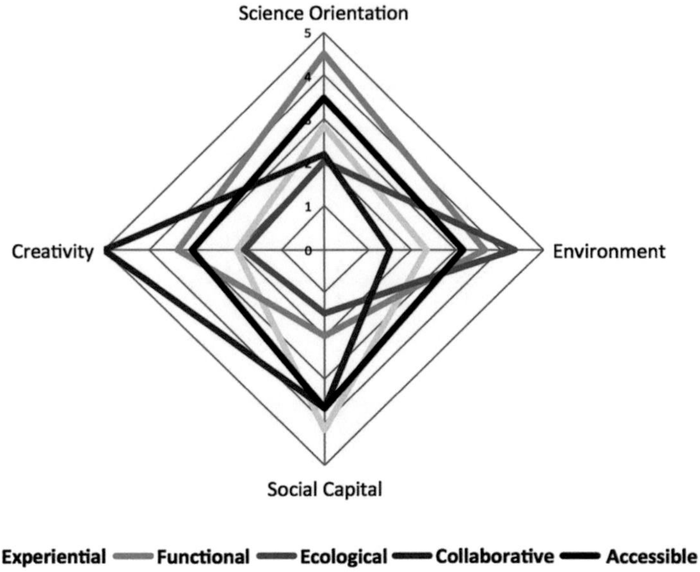

Fig. 20.4 Visual representation of five campus profiles from the perspective of four assessment factors

Accessible Profile has similar relations with all factors of evaluation. Clearly, the Functional and Accessible Profiles have higher overall impacts on the four factors under consideration, corresponding to their high importance on different aspects of the activities of the University, while the Ecological Profile has a strong relationship with the environment, but a relatively small impact on the other assessment factors. Clearly, the actual campus to be developed depends on the preference and value intensities of the actors. This will be further analysed in the next subsection.

It is important to notice that the five campus profiles are not "closed" concepts and that they are inter-related with each other. For example, issues related to the use of public space or historical buildings are important for the Experiential Profile (as they can create better conditions for the interaction among users or their fruition of the amenities of the campus), but are also important elements for the Collaborative Profile, as they can reinforce the attractiveness of the campus for the local community, visitors, tourists or even companies interested in R&D activities in cooperation with the University. On the other hand, aspects related to the accessibility to the campus, the mobility inside the campus (Accessible Profile), the implementation of new services and facilities (Functional Profile), or the openness of the campus to tourists and the local community (Collaborative Profile), have direct implications on the preservation, development or management of the green areas and public spaces inside the campus (Ecological Profile).

20.2.3 Preference Intensities

The opinions expressed by the users of Hokkaido University campus have been collected formally and informally, in various discussions involving 21 persons (students, researchers, professors and managers, balanced by gender and national origin), and ultimately been put together in two focus group meetings organized in two different areas of the campus (North and South), considering their different characteristics. The focus group meetings on both the North and South Campus followed a similar format: after a general presentation of the Plans for the Campus, an explanation of the participatory process and the assessment domains for the Campus has been provided and complemented with some essential information about the performance of the Campus and a set of characteristic pictures with positive and negative aspects. After this visualisation of a value-based metrics experiment, an open discussion among the participants took place, oriented towards a SWOT analysis structure. Photographs and visual material formed an important input for consensus building.

As part of the Living Lab experiment, the participants were asked—on the basis of various campus pictures—to write their suggestions for the SWOT analysis, identifying what they considered to be the most relevant Strengths, Weaknesses, Opportunities and Threats for the development of the campus. As a complement, they were asked to propose one or two generic strategic ideas for the future of the campus, considering the possibility to implement them in the short run and their consequences in the long run. Finally, each participant was asked to choose three typical pictures representing the most positive aspects of the campus and three pictures representing the most negative aspects. All the proposals were presented and justified by the participants.

This approach was inspired by recent applications of visual assessment for the development of different types of landscapes and territories, as proposed, among others, by Arriaza et al. (2004) in relation to rural landscapes, Smit (2011) in regard to the choice of location by creative entrepreneurs, and Johnson and Castleden (2011) addressing the design of a university campus. Examples of the pictures distributed among the participants in our case study are shown in Fig. 20.5.

After each focus group meeting, a document systematizing all the proposals (SWOT analysis, strategic assessment and visual assessment) was sent to all the participants. Each of them evaluated (through a scoring 1–5, with five being the most important) the importance of each element. After this process, all items listed in the SWOT analysis and the strategic and visual assessment were organized according to the four different assessment domains of analysis of the campus, synthetizing the preferences of the users. From this list of items it was possible to obtain Preference Intensity Vectors to be used in the subsequent Multi-Criteria Analysis (Stirling 2006; Giordano and Riedel 2008; Schetke et al 2012), for both the North and South campus.

This methodology is particularly important for the planning of sustainable development under uncertainty, as it allows for the formulation of flexible and viable strategies, combining different sustainability and performance elements.

Fig. 20.5 Examples of pictures for the visual campus assessment

A related interesting example of the importance of flexible planning for sustainability applied to the life cycle of buildings can be found in Fawcett et al. (2012). For our Strategic Choice Analysis, the suggestions from the participants in the meetings were organized according to the campus profiles, while only the most important items (according to the scores given by them) were next considered. The methodology recently developed by Kourtit and Nijkamp (2011) for a Strategic Choice Analysis based on the SWOT methodology was applied in this work.

20.3 Empirical Results on Campus Planning

20.3.1 Multi-Criteria Analysis

A Multi-Criteria Analysis (MCA) is based on a representation of the multi-faceted aspects of choice alternatives, in our case, the five campus profiles. The MCA method used is the Regime method, which presupposes a set of a priori defined alternatives and a distinct set of a priori defined evaluation criteria (assessment factors). When adding a ranking for the importance of the evaluation criteria, the relative dominance of each alternative can be assessed in the form of a success rank order—or even a rating—of the alternatives. More details can be found in Nijkamp et al. (1990), while a recent application could be found in Akgün et al. (2012).

The aim of this MCA is to identify the relative importance of each of the five future strategies (namely, the future campus profiles under consideration). Four evaluation domains (assessment criteria) were defined and decomposed into a set of operational sub-criteria. Based on literature reviews, stakeholder information and expert knowledge, a comprehensive ordinal impact matrix was next defined showing the relative contribution of each sub-criterion to the campus profiles concerned (Table 20.2). In this case the aim of the MCA is not to identify one unambiguous campus profile, but to seek for a systematically ranked set of campus profiles which in combination make up the most desirable campus image.

Next, an intensity score has been assigned to each sub-criterion, indicating how important it appears to be for the campus development (the intensity score shows how often a sub-element is mentioned in the focus-group discussions). Three distinct vectors were defined: one for the entire campus, one for the Southern part and one for the Northern part of the campus (see Table 20.3). This preference elicitation process was next used to specify weights in the Regime Analysis of an MCA. The results are shown in Fig. 20.5.

The results from Fig. 20.6 show that the functional strategy is the most important strategy for the performance of the campus, while next accessibility is an important element of redevelopment. This holds for the whole campus area and is thus a consequence of the high correlation between functionality and accessibility with all the domains of analysis of the campus (as mentioned in Sect. 20.2.1). The other three strategies are more or less equally important for the whole campus, but differ very much between North and South, according to the preferences of the users: the ecological strategy is seen as very important for the South campus, but relatively less important for the Northern part. Instead, for the North campus an experiential strategy might be more useful. For all areas holds that the collaborative strategy is the least important one; however, it should still be part of the overall plan.

To test the robustness of our findings, it is important to carry out a sensitivity analysis after the removal of one or more assessment criteria. This is a large combinatorial exercise. For the ease of presentation here, we deleted Science Orientation and repeated the analysis without the indicators related to the Science Orientation factor. Figure 20.7 shows these results, showing that, in this case, the functional strategy has become less important and the ecological strategy more important. However, the main message still holds: accessible and functional strategies are critical for the performance of the campus and, next to that, the ecological strategy is particularly relevant. Finally, the collaborative strategy becomes more important for the North and the experiential for the South parts of the campus. A differentiated balance between these five forces has to be sought in order to reach an optimal campus Image.

Table 20.3 Preference intensity vectors

	North	South	Total
Science orientation	**0**	**2**	**2**
Teaching and learning conditions	0	3	3
Research conditions	0	0	0
Libraries	0	2	2
Conference facilities	0	0	0
Experimental Farms and production	0	7	7
Special education facilities	0	0	0
Creativity	**3**	**4**	**7**
Academic spin-off (incubators)	2	0	2
Private R&D	2	0	2
Collaborative R&D	7	2	9
Artistic and creative professions	0	0	0
Linkage to local community	5	16	21
Connectivity (ICT networks)	0	5	5
Environment	**5**	**13**	**18**
Accessibility	4	4	8
Mobility	7	26	33
Biodiversity and public spaces	8	34	42
Emissions	2	3	5
Energy consumption and production	5	12	17
Water and waste management	4	0	4
Social capital	**4**	**6**	**10**
Financial support for students	2	0	2
Student housing	0	2	2
Various services	5	8	13
Recreation, culture and sports	12	8	20
Projects with the city	2	15	17
Bookstores	1	1	2

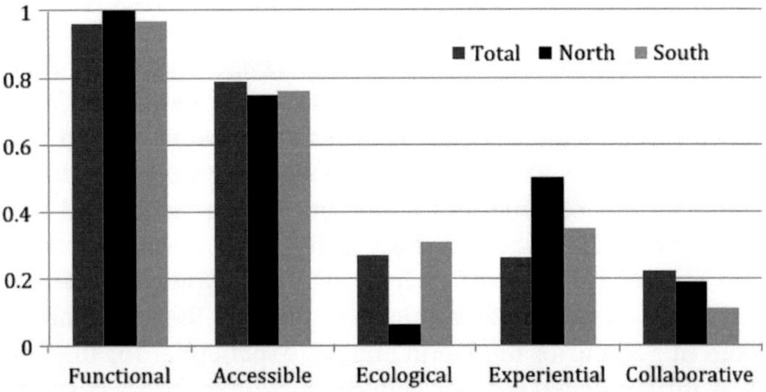

Fig. 20.6 Regime analysis: preferences for the five campus profiles

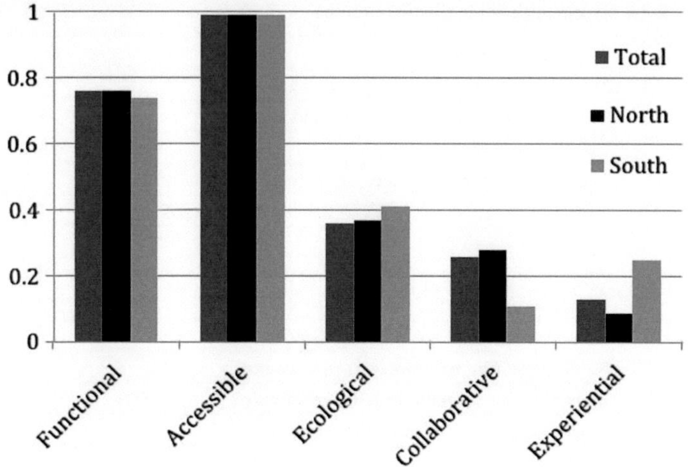

Fig. 20.7 Regime analysis with three assessment factors: creative, sustainable and social

20.3.2 Strategic Choice Analysis

20.3.2.1 Methodological Description

The next step in an analysis is the systematic comparison of various strategies. Based on the SWOT analysis and the strategic and visual assessment developed in various discussions and dedicated focus group meetings, a Strategic Choice Analysis (SCA)—as a vehicle in assessing and developing strategic policies for the development of the campus—has been developed. As this analysis is mostly oriented to the evaluation of the importance of the landscape for the performance of the campus and the coherence between the strategic ideas for its development and the needs of the users, the results are mostly focused on the physical characteristics of the campus. For each campus profile, the four most important Strengths and Weaknesses (when existing) and the two most important visual elements for positive and negative aspects of the campus (when existing) were analysed in terms of their impact on the Opportunities (SW/O) and Threats (SW/T) detected. Also, the impact of each strategic proposal on the Strengths, Weaknesses, Opportunities and Threats was evaluated (see Table 20.4).

The estimated impacts for the Accessible Profile in the South campus are presented in Table 20.4 and Fig. 20.7 (for the other profiles, similar Tables and Figures are presented in an Annex). Column “O” shows the impact of each Strength and Weakness on the Opportunities detected (considering 0 points for each non-relevant impact, 3 points for each relevant impact and 5 points for each very important impact), column “T” shows the same kind of impacts for the identified Threats and column “S” shows the impact of each Strength, Weakness, Opportunity or Threat on the strategic proposals (the value in column “S” for each strategic proposal shows its overall impact on the different items of the SWOT matrix).

Table 20.4 SCA for the accessible profile (South)

SWOT	Accessible profile	O	T	S
Strengths	Location (proximity to city centre)	10		3
	Location (accessibility to train and subway)	6		3
	Visual—mobility	15		5
Weaknesses	Lack of transportation service (especially in winter)	3		5
	Disconnection between South and North campus	3		5
	Traffic congestion	6		5
	Growth of parking areas (and decreasing open spaces)	6		3
	Visual—mobility	15		5
Opportunities	Reorganization of the parking system			3
	Implementation of a bike-borrow system			5
	Road heating systems in pedestrian ways			5
Strategies	Improvement of the transport system			**47**

20.3.2.2 Results for South Campus

The easy accessibility to the campus from the city centre of Hokkaido using the main public transport networks (Strengths) enhances the possibility to explore some Opportunities in order to integrate in the campus some new solutions being implemented in the city (bike-sharing system, new parking systems, or heating systems in pedestrian ways) and to face some Weaknesses related to internal mobility (parking, connection between North and South or congestion). The visual assessment revealed both positive (adequate trails for bicycles and pedestrians) and negative aspects (congestion). The Master Plan for the campus already defines the renovation of the infrastructures for mobility in the campus, but its adequacy to the winter conditions or the improvement in transport services should be considered in the future. These results are synthetized in Table 20.4 and Fig. 20.7.

Regarding the Ecological Profile (see Table 20.5 in the Annex), the main Strengths mentioned are related to the landscape and biodiversity of the campus, while the Weaknesses relate to the excessive consumption of energy. The strategic proposals aim to protect the natural environment and the attractiveness of the campus and this priority is followed by different measures to reduce energy consumption (and to increase energy production based on renewable sources), which can benefit from the Opportunities opened by national policies. The principles defined in the Master Plan seem to be in accordance with these proposals, trying to reorganize the mobility processes or the settlement and design of buildings.

Questions related to the lack of Wi-Fi access to internet in the campus were the most important ones regarding the Functional Profile (see Table 20.6 in the Annex), concentrating the vast majority of the Weaknesses detected and one of the strategic proposals. Nevertheless, the preservation of free empty spaces in the campus was also a strategic recommendation, considering the need for the implementation of new facilities and services. Although the provision of wi-fi services is not directly related with the Plans for the campus, the preservation of green spaces and the implementation of new services and facilities without physical expansion is already a concern expressed in the Master Plan.

Table 20.5 SCA for the ecological focus (South)

SWOT	Ecological focus	O	T	S
Strengths	Large Campus	3		6
	Green areas, trees, lawns and flowers	3		8
	Nature and landscape	3		10
	Farm and poplar avenue	5		6
	Visual—Green areas	3		10
Weaknesses	Large energy consumption	3		8
	Lack of control on energy consumption	3		9
	Lack of control on heating system	3		8
Opportunities	Increment the use of renewable energy			5
	Cultivation and consumption of vegetables			6
Strategies	Preservation of green area			**24**
	Maintaining beautiful campus			**24**
	Waste-free heating system			**13**
	Increase sustainable technology in every sector			**12**
	Utilizing renewable energy			**5**

Despite the quiet atmosphere of the campus (very close to the busy centre of the city), the existence of important elements referring to the history of the University and the availability of some places to meet and communicate, the lack of spaces for social interaction (including coffee-shops, bars, or social dinning) was mentioned as a weak point related to the Experiential Profile (see Table 20.7 in the Annex). The main strategic idea proposed was the implementation of new places for these purposes (including sports, culture or cafes) but the national laws preventing the consumption of alcoholic beverages by people below 20 years old was faced as an obstacle to the implementation of bars inside the campus. The principles defined in the Master Plan emphasize the importance of planning empty spaces in order to provide satisfactory experiences to the users of the campus and to increase social interaction.

The issues related to the existence of public spaces (mentioned as a strong point but also as a strategic aspect to be developed) or to the historical aspects of the University (simultaneously mentioned as a strong factor—when it is exposed—and as a weak factor—when there is lack of information) were considered important for the Collaborative Profile (see Table 20.8 in the Annex). The strategic guidelines proposed include the implementations of cafes or the development of cultural and entertainment activities (with general purposes or focused in the history and characteristics of the region). Combined with the opportunities opened by the existence of international exchange programs and the availability of English courses, this would contribute to increase the interaction among international students, Japanese students and the surrounding community. The opportunity to explore the proximity between academy and regional economy was also mentioned, although the relatively weak economic condition of Hokkaido can impose some limits.

20.3.2.3 Results for North Campus

Questions related to the accessibility to North campus and its geographical isolation have been considered important Weaknesses (general accessibility, bus services and connection with South campus) related to the Accessible Profile (see Table 20.9 in the Annex). In accordance, an improvement in the bus services has been proposed. This aspect can not be directly addressed by the physical plans for the campus, but must be considered as a service to be implemented.

The strategic proposals regarding the Ecological Profile (see Table 20.10 in the Annex) are mostly focused on Opportunities and Threats related to energy, an extremely important issue in Japanese society after the accident in Fukushima (2011) and the subsequent decision to close down most of the Japanese nuclear plants. A possible energy shortage would have very relevant impacts on research activities. To avoid this Threat, the available technologies to produce energy from waste or for self energy production have been mentioned as Opportunities to be explored, implying the development of adequate storage and recycling systems, combined with higher budgets for research in this area. The visual preference for green areas of the campus and the stress on the lack of forest as a weak point suggest that the Plans for the campus should consider its importance (and, in fact, the Master Plan already defines the creation of "tree-belt" around the North campus).

The Weaknesses related to the Functional Profile (see Table 20.11 in the Annex) detected in the SWOT analysis are related to the lack of different services, which should be defined in the expected expansion of the North campus, as defined in the Master Plan. Similarly to this, the Experiential Profile (see Table 20.12 in the Annex) showed an important lack of public and attractive spaces for meetings and social interaction. The visual identification of positive aspects related to the history of the University or to the quality of the public spaces in pictures of the South campus suggests that the users of North campus would like to find similar facilities.

The actual support of the University to private research projects is a Strength with relevant impact on the Opportunities detected for the Collaborative Profile (see Table 20.13 in the Annex): linkage to industry, collaboration with companies and attraction of more small companies. The implementation of more laboratories to be rented by small companies may be defined in the Plans for the expansion of the campus. The reinforcement of the elements related to the history of the University (mentioned as a Strength in the visual assessment) is seen as an important element to establish better connections, not only with companies and other organizations, but also with the surrounding communities.

Discussion and Conclusions

The results on the assessment of the Hokkaido University campus—based on a SWOT analysis, an MCA and a SCA—demonstrate that issues related to the functionality and accessibility of the campus are dominant when analysing its sustainable performance. Nevertheless, a sensitivity analysis developed without considering the "Science orientation" factor showed that the importance of the ecological questions (regarding biodiversity, landscape and energy) become more evident. In addition, increasing the opportunities for collaboration is assumed as an important aspect to be developed in the North campus (an area oriented to the interaction between academic and other institutions), while in the South campus it is possible to observe a higher importance of the aspects related to the "experiences" arising from social interaction among the users of the campus.

Clearly, the purpose of this work is not to identify one single campus profile as a priority for the development of the campus, but to identify the different viable priorities that should be given to each of them according to their impacts on the performance of the University and the problems and preferences of the users (future oriented). Therefore, in our study a more detailed observation of each different campus profile has been developed with the Strategic Choice Analysis, in order to identify what can be improved in each of these aspects of the University campus, clearly distinguishing among the North and South areas of the campus, once different activities are performed by different kind of stakeholders in these places.

The choice results from the MCA Regime Analysis call for a balanced, differentiated and well-framed future campus design, in which all elements from five campus profiles play a role, but the Functional, Ecological and Accessible criteria will be decisive for any future campus Image.

Regarding the future, an important distinction between the North and South campus must be emphasized: while the North has still land available for physical expansion allowing the implementation of new facilities or mobility infrastructures and preserving the ecological balance, the development of the South campus does not leave room for a physical expansion, in order to preserve (or increase) the quality of public spaces. This distinction is defined in the Plans for the development of the University and was expressed by the users of the campus.

Regarding the questions related to accessibility and mobility inside the campus, the disconnection between North and South and problems with mobility in winter have been mentioned several times by stakeholders from both campuses. Congestion is a major concern in the South campus, while isolation (difficult access from inside and outside the campus) is a problem in the North campus. A renovation of the circulation infrastructure is already defined in the Plans for the campus, but new regulations for mobility, an improvement in the transport services, better infra-strictures for mobility in winter and better accessibilities to North campus are required.

The facilities for the provision of "traditional" academic services (related to education and research) or to basic services (like food or convenience stores) seem adequate in the South campus, but should be improved in the North campus. On the other hand, the provision of wi-fi access to Internet all over the campus is a major demand of the users of South campus and it is not included in the Plans for the development of the campus.

Social interaction in public spaces, cafes, dining rooms, entertainment areas, leisure areas or sports facilities are not sufficient in the South campus and almost non-existent in the North campus. Although the general purposes of the Plans for the Development of the campus seem to express these concerns, concrete solutions are not implemented. On the other hand, the dynamic collaboration between academy and industry or the openness of the University to the local community and tourists require some improvements in order to maximize the potential benefits.

This process of assessment has been very motivating for the participants and has provided interesting results concerning the evaluation of the campus by its users, which can be useful for the preparation of new Plans (like the Master Plan 2016, to be developed), discussing different scenarios and possibilities for strategic development of the campus, possible conflicts among different proposals and financial restrictions. On the other hand, the range of users contributing with their opinions and proposals can also be enlarged, including other kind of "regular" (like maintenance, security or other workers in the campus) and "occasional" (local residents or tourists) users. Considering the importance of the campus in the context of the city and the necessary interconnection between them, other agents (like urban or transport planners) should be involved.

The modest costs of this planning process, the motivation of the participants, the useful and objective results and the importance to develop new processes of assessment in order to monitor the evolution of the campus and the needs of the users over time, suggest that similar initiatives may be developed at Hokkaido University in the future.

Acknowledgment This study has been supported by the Office for a Sustainable campus at Hokkaido University (Takashi Yokoyama, Tomohiro Morimoto and Yuki Matsubara) within the project "Value Metrics and Policies for a Sustainable University campus" financed by the Marie Curie Action in the Seventh Framework Programme of the European Commission. The support of the Department of Spatial Economics, VU University, Amsterdam, is gratefully acknowledged as well.

Annex: Results of Image-Based Experiments for Campus Planning

Table 20.6 SCA for the functional focus (South)

SWOT	Functional focus	O	T	S
Weaknesses	Lack of Wi-Fi internet			5
	IT infrastructure			5
	Lack of PC area 24 h open			5
	Lack of spaces to study with PC and drinks			5
Opportunities	Implement diverse educational opportunities			
Threatens	Increasing activities and decreasing open space			5
Strategies	Developing new facilities preserving free spaces			5
	Providing Wi-Fi all over the campus			16

Table 20.7 SCA for the experiential focus (South)

SWOT	Experiential focus	O	T	S
Strengths	A quiet oasis in a busy city			5
	Atmosphere around the museum			5
	Visual—history			3
	Visual—public spaces for meetings			10
Weaknesses	Lack of spaces for social interaction and to receive guests			10
	Lack of coffee shops			10
	Lack of social dining/drinking space		3	10
	Sports facilities with very restricted access			8
Threatens	National regulations prevent plans coming to fruition			3
	Governmental targets imposing not wanted policies			6
Strategies	More spaces for relaxation			25
	Winter-proof campus			9

Table 20.8 SCA for the collaborative focus (South)

SWOT	Collaborative focus	O	T	S
Strengths	Campus open for everyone	9	3	14
	Visual—history	9	0	9
	Visual—public spaces	13	0	20
Weaknesses	Lack of information about cultural heritage	9	0	9
Opportunities	Programs for international collaboration			14
	Increase English courses			8
	Interaction residents/international students			14
	Proximity between academy and business			6
Threatens	Weak economic conditions in Hokkaido			4
Strategies	Interaction international/Japanese students			4
	Cafés opened for everyone			4
	Build book café			4
	Entertainment and social activities			4

Table 20.9 SCA for the accessible focus (North)

SWOT	Accessible focus	S
Strengths	Visual—mobility	3
Weaknesses	Bus services finish very early	5
	Accessibility to North campus	3
	Connections North–South campus	3
	Visual—mobility	3
Strategies	Improvement in bus services	17

Table 20.10 SCA for the ecological focus (North)

SWOT	Ecological focus	O	T	S
Strengths	Huge land still available	0	0	0
	Visual—Green areas	0	0	6
Weaknesses	Lack of forest (makes open spaces too windy)	0	0	3
Opportunities	Self produced energy			11
	Utilization of waste to produce energy			12
Threatens	Implications of energy shortage on the research			13
Strategies	Contribute to a sustainable next generation			15
	Storage system			8
	Research budget to generate energy at HU			13
	Establishment of recycling systems			9
	Information on the cost of energy of each device			5

Table 20.11 SCA for the functional focus (North)

SWOT	Functional focus
Weaknesses	Restaurant open for short period
	Lack of book store
	Few convenience stores

Table 20.12 SCA for the experiential focus (North)

SWOT	Experiential focus	S
Strengths	Visual—history	3
	Visual—public spaces	5
Weaknesses	Lack of public spaces for communication	5
	Lack of public spaces for leisure and rest	3
	Lack of refresh and resting rooms	0
	Lack of sports facilities (like tennis court)	3
Strategies	Creation of attractive public spaces to meet	19

Table 20.13 SCA for the collaborative focus (North)

SWOT	Collaborative focus	O	S
Strengths	University support to big private projects	15	3
	Visual—history	12	5
Opportunities	Linkage to Industry		3
	Small companies can be attracted with more rental labs		6
	Research Collaboration with companies		3
	Hokkaido Collab. Centre can interact more with researchers		3
Strategies	Increasing financial support for academic activities		9
	Be a good neighbour for local people (especially kids)		0
	Services and shops related to historical buildings and landscape		8

References

AASHE (2012) STARS—sustainability tracking, assessment & rating system, version 1.2 technical manual, AASHE, Lexington

Abd-Razak M, Mustafa N, Che-Ani A, Abdullah N, Mohd-Nor M (2011) Campus sustainability: student's perception on campus physical development planning in Malaysia. Procedia Eng 20: 230–237

Akgün A, van Leeuwen E, Nijkamp P (2012) A multi-actor multi-criteria scenario analysis of regional sustainable resource policy. Ecol Econ 78(C):19–28

Alshuwaikhat H, Abubakar I (2008) An integrated approach to achieving campus sustainability: assessment of the current campus environmental management practices. J Clean Prod 16(16): 1777–1785

Arriaza M, Cañas-Ortega J, Cañas-Madueño J, Ruiz-Aviles P (2004) Assessing the visual quality of rural landscapes. Landscape Urban Plan 69:115–125

Beringer A (2006) Campus sustainability audit research in Atlantic Canada: pioneering the campus sustainability assessment framework. Int J Sustain High Educ 7(4):437–455

Bourdic L, Salat S, Nowacki C (2012) Assessing cities: a new system of cross-scale spatial indicators. Build Res Inf 40(5):592–605

Darus Z, Khalim A, Rashid A, Hashim N, Omar Z, Saruwono M, Mohamad N (2009) Development of sustainable campus: Universiti Kebangsaan Malaysia planning and strategy. WSEAS Transact Environ Develop 5(3):273–282

Disterheft A, Caeiro S, Ramos M, Azeiteiro U (2012) Environmental management systems (EMS) implementation processes and practices in European higher education institutions and top-down versus participatory approaches. J Clean Prod 31:80–90

Fawcett W, Hughes M, Krieg H, Albrecht S, Vennström A (2012) Flexible strategies for long-term sustainability under uncertainty. Build Res Inf 40(5):545–557

Franklin C, Durkin T, Schuh S (2003) The role of landscape in creating a sustainable campus, Planning for Higher Education, March–May

Giordano L, Riedel P (2008) Multi-criteria spatial decision analysis for demarcation of greenway: a case study of the city of Rio Claro, São Paulo, Brazil. Landscape Urban Plan 84(3–4): 301–311

Heijer A (2012) Managing the university campus: Exploring future models and supporting today's decisions. CELE Exchange 2012/2, OCDE

Johnson L, Castleden H (2011) Greening the campus without grass: using visual methods to understand and integrate student perspectives in campus landscape development and water sustainability planning. Area 43(3):353–361

Kenney D, Dumont R, Kenney G (2005) Mission and place: strengthening learning and community through campus design. Praeger Publishers, Westport, CT

Kourtit K, Nijkamp P (2011) Strategic choice analysis by expert panels for migration impact assessment. Int J Bus Globalisation 7(2):166–194

Mcmillin J, Dyball R (2009) Developing a whole-of-university approach to educating for sustainability: linking curriculum, research and sustainable campus operations. J Educ Sustain Develop 3(1):55–64

Neuts B, Nijkamp P, Van Leeuwen E (2012) Crowding externalities from tourist use of urban space. Tourism Econ 18(3):649–670

Nijkamp P (2008) XXQ factors for sustainable urban development: a systems economics view. Rom J Reg Sci 2(1):1–34

Nijkamp P, Rietveld P, Voogd H (1990) Multicriteria analysis for physical planning. Elsevier, Amsterdam

Perry D, Wiewel W (eds) (2005) The university as urban developer: case studies and analysis. M.E. Sharpe, New York

Ratcliffe J, Krawczyk E (2011) Imagineering city futures: the use of prospective through scenarios in urban planning. Futures 43(7):642–653

Schetke S, Haase D, Kötter T (2012) Towards sustainable settlement growth: A new multi-criteria assessment for implementing environmental targets into strategic urban planning. Environ Impact Assess Rev 32:195–210

Shriberg M (2002) Institutional assessment tools for sustainability in higher education: strengths, weaknesses, and implications for practice and theory. Int J Sustain High Edu 3(3):254–270

Smit A (2011) The influence of district visual quality on location decisions of creative entrepreneurs. J Am Plann Assoc 77(2):167–184

Stirling A (2006) Analysis, participation and power: justification and closure in participatory multi-criteria analysis. Land Use Policy 23(1):95–107

Valencia-Sandoval C, Flanders D, Kozak R (2010) Participatory landscape planning and sustainable community development: methodological observations from a case study in rural Mexico. Landscape Urban Plan 94(1):63–70

Zink K, Steimle U, Schröder D (2008) Comprehensive change management concepts development of a participatory approach. Appl Ergon 39(4):527–538

Reducing Vehicle Pollutant Emissions in Urban Areas with Alternative Parking Policies

21

Daniel Shefer, Shlomo Bekhor, and Daphna Mishory-Rosenberg

Abstract

Most metropolitan areas suffer from traffic-related air pollution. A major reason for this phenomenon is related to the increase in vehicle-kilometers traveled, which is an outcome of urban sprawl and an increase in the motorization rate. Although significant progress in vehicle technology has greatly improved air quality in developed countries, there is still a lack of policy mechanisms to mitigate air quality impacts resulting from traffic pollution.

This paper examines policy measures that can be implemented by decision-makers in order to improve urban air quality. Focusing on parking policy, this paper attempts to gain an understanding of the relationship between parking-policy enforcement in the Central Business District (CBD) and air-pollution emissions from motor vehicles. The site chosen for study is the Tel Aviv Metropolitan Area (TAMA).

The methodology presented is applied to a real-world situation, using data and models from a Mass Transit project initiated by TAMA. The traffic-related emissions are estimated from a four-stage transportation model that provides traffic volumes and travel times for each link in the network. Two policy measures are analyzed: reducing parking supply and increasing parking fees. The paper discusses the suitability of parking policies to meet environmental objectives.

D. Shefer (✉) • D. Mishory-Rosenberg
Center for Urban and Regional Studies, Technion—Israel Institute of Technology, Technion City, Haifa 32000, Israel
e-mail: shefer@technion.ac.il; galyuvaldaphna@yahoo.com

S. Bekhor
Transportation and Geo-Information Engineering, Technion—Israel Institute of Technology, Haifa 32000, Israel
e-mail: sbekhor@technion.ac.il

P. Nijkamp et al. (eds.), *Regional Science Matters*, DOI 10.1007/978-3-319-07305-7_21,

21.1 Introduction

Most metropolitan areas suffer from traffic-related air pollution. A major reason for this phenomenon is related to the increase in vehicle-kilometers traveled (VKT), which is an outcome of urban sprawl and an increase in the motorization rate. Despite positive effects brought about by technological progress in reducing emissions per unit of distance traveled (Parrish et al. 2011), the overall increase in VKT adversely affects urban air quality. Although significant progress in vehicle technology has greatly improved air quality in developed countries, there is still a lack of policy mechanisms to mitigate air quality impacts resulting from traffic pollution.

A high level of air-pollution concentrations in the Tel Aviv Metropolitan Area (TAMA), especially in the Central Business District (CBD), is the motivation for this research. Mobile sources have been determined to be the main pollutant factors in the Tel Aviv CBD (Ranmar et al. 2002; Shefer 1994).

Several factors affect air-pollution emissions from mobile sources: technology, urban form, and economic factors. Technological factors dealing with the internal mechanism of mobile sources (type of motor and the technological means employed to decrease pollutant emissions; i.e., catalytic converters, etc.) and beyond the scope of this article, which will examine the two other factors influencing air-pollution emissions from mobile sources (urban form and economic factors).

A variety of policy measures could be employed in order to decrease air-pollution emissions from motor vehicles, among them parking policy, congestion tolls, compact urban development, transit-oriented development (TOD), traffic calming, improved public transit level of service, encouraging the use of slow modes such as bicycle paths, comfortable sidewalks for pedestrians, and low-emission zones (Newman et al. 1988; Lyons et al. 1990, 2003; Katz 1994; Bernick and Cervero 1997; de Roo 1999; Burton 2000). This paper specifically focuses on parking-policy measures. The objective is to gain an understanding of the relationship between parking-policy enforcement in the CBD and air-pollution emissions from motor vehicles in the metropolitan area.

The research relates to work done on parking-policy enforcement and its effect on travel patterns, which are defined in such variables as private VKT, number of private vehicle trips, and travel time in metropolitan areas (Marsden 2006; Marsden and May 2005; Shoup 1999, 2005). The paper examines the relationship between physical and economic aspects of parking policies and its direct influence on private VKT and number of trips, as well as its indirect effect on air-pollutant emissions and travel time.

Several parking-policy measures were reviewed, focusing on measures that can have a direct impact on VKT; for example, whether a reduction in the supply of on-street and/or off-street parking and an increase in parking fees would lead to a decrease in private VKT and, hence, to a decrease in motor-vehicle air-pollutant emissions. The methodology presented is applied to a real-world situation, using data and models from the Mass Transit project initiated by TAMA. Tel Aviv, considered the business capital of Israel, attracts employees from the entire metropolitan area and beyond. According to the 2008 Israel Census, there are 400,000 working places in the CBD, which constitute 40 % of the entire TAMA jobs (Central Bureau of Statistics 2008).

Two policy measures are analyzed in this paper. The first policy measure considered is a reduction in parking supply in the CBD. Assuming fixed demand (i.e., a constant total number of motor vehicle trips), a decrease in the parking supply in the area will result in a dispersion of trips to traffic zones outside the CBD. The second policy measure studied is an increase in parking fees. Since parking fees vary across the CBD, increased fees will inadvertently affect parking search time. This assumption is related to the fact that drivers will try to look for cheaper parking places or even park far from their destination in order to save on costs (Shoup 1999). Therefore, this policy measure will be evaluated in conjunction with the decrease in parking availability.

The methodology applied in the paper is to use travel-demand elasticity values from the original transportation models to calculate the expected decrease in car trips in every traffic zone in the CBD area. Assuming differential parking fees, the traffic-assignment model produces new traffic flows and travel times. Total travel time and VKT are compared to the base case, in which parking policies are not employed.

The paper is organized as follows: The next section presents a short review of the literature on different policy measures and their effect on the environment. The methodology used to analyze the impact of alternative parking-policy measures is then presented. Subsequently we present the environmental effects of the two parking policies. Finally the last section of the paper offers a discussion on the main findings.

21.2 Literature Review

In spite of the positive effects that technological means of reducing emissions could have on the environment, urban air quality continues to deteriorate in most metropolitan areas. This phenomenon is primarily due to the continuous rise in VKT, which is an outcome of urban sprawl and the increase in the rate of car ownership (Shefer 1994; Newman et al. 1988; Lyons et al. 1990, 2003; Katz 1994; Bernick and Cervero 1997; de Roo 1999; Burton 2000).

There are numerous policies that deal with urban sprawl and the consequent rise in the number of motor vehicles and VKT. The objective of these various policies is to create conditions for travelers to the CBD to reach their destinations with a minimum use of private vehicles. Such conditions include improving access to activities such as work, shopping, etc., by means of public transit, cycling, or walking. These measures also generally impose stricter policy measures against those who drive their private vehicles. Examples of such policy measures are congestion toll and parking policies (Geurs and van Wee 2004; Levine and Garb 2002; Shoup 1999).

There is a large body of literature dealing with parking-policy measures. Several researchers stress the need for combined urban-land-use and transportation planning in order to reduce private VKT and, consequently, pollutant emissions. Several studies show that parking-policy methods have but a small influence on VKT, and therefore they are more useful when combined with urban-land-use and transportation planning policies. The main claim is that parking policies do not much affect private VKT unless the use of public transit is encouraged and the urban structure is amenable for choosing alternative modes (Marsden 2006; Marsden and May 2005).

An interesting point of view elaborated by Shoup (1999) shows the importance of parking-policy methods as a catalyst for further implementation of urban land-use-development and transportation planning policies. The author describes a vicious cycle, briefly outlined below, that enhances VKT.

This cycle starts with transportation planners who survey parking occupancy at sites that offer ample free parking but lack public transport. The parking-generation-rate manuals created by transportation engineers are used by urban planners to set the minimum number of parking spaces for all types of land uses. When the parking supply is large, most new developments will offer free parking. Transportation planners survey vehicle trips to and from sites that offer free parking, and the outcome of this procedure is a trip-generation-rate manual. The roads leading to such sites are designed and planned according to these manuals, which means that the roads and highways provide enough capacity to satisfy expected demand. Urban planners generally limit land-use density in order to prevent congestion on roads near a new project because of their limited capacity (Shoup 1999).

This vicious cycle demonstrates the huge cumulative effect that parking policy has on urban land-use and transportation planning. Shoup (1999) stresses the importance of parking policy as a means of influencing transportation land use and urban planning in a significant way, not simply as a "glue" connecting the urban land use and the transportation system (Shoup 2005). Note, as previously mentioned, that most researchers refer to parking policy as a component of the whole complex of transportation land-use-development and urban planning policies.

Since urban land-use and transportation planners seldom have the opportunity to plan from scratch in a completely new setting, joint planning of a transportation cum land-use policy is rare. Usually planning is undertaken in the framework of a new neighborhood site in an existing city. Therefore, there are usually leading land-use policies that are adopted first. Thus, in order to reduce VKT (and, hence, air-pollutant emissions), a leading policy, such as parking policy, could trigger and affect other aspects of urban form, land use, and transportation planning aimed at reducing VKT (Cervero 2002).

The problem with adopting any policy measure is that it often draws a large amount of public criticism if not backed with solutions that will alleviate, at least in part, the policy's perceived restrictions on the general public. For example, when enforcing parking-policy measures such as increased parking fees and/or decreased numbers of parking spaces, the public will demand the provision of alternative and accessible travel modes to/from the areas concerned. If there is no acceptable alternative to the use of a private vehicle, the car will continue to be the primary transport mode, despite the enforcement of the parking policy. In the public's eye, the travel mode is only a means of reaching a destination; it is not an objective in itself. That is to say, if the travel mode is comfortable, fast, efficient, and affordable, the travelers will use it, whether it is a private vehicle, public transit, or even a slow mode such as a bicycle or even one's feet (Hansen 1959; Hägerstrand 1970; Ingram 1971; Levine and Garb 2002; Geurs and van Wee 2004).

Enforcing parking-policy measures aims at reducing the number of motor vehicles on the roads, which in turn encourages the use of alternative modes. This sequence will eventually lead to a new equilibrium between the alternative transport modes and the environment. This means that the public's objections to parking-policy enforcement should not deter municipal authorities from administrating such policies. Public criticism should be stifled, though, only by creating and improving other means of reaching destinations, such as public transit or some slow mode (Kodransky and Hermann 2011).

21.3 Methodology

The main hypothesis of this paper is that physical and economic aspects of parking policy affect the demand for travel by private vehicle (and, consequently, impact on VKT and total travel time), which has a direct effect on motor-vehicle air-pollutant emissions. Thus reducing the amount of on-street and/or off-street parking and/or raising parking fee will lead to a decrease in private VKT and, hence, to a reduction in motor-vehicle air-pollutant emissions.

If this hypothesis is proved to be correct, then the use of parking policies could improve the urban environment in terms of both air quality and transportation-network performance variables. This is an important point to be mentioned, since it could be argued that adopting urban land-use and transportation policies in order to improve the urban environment, such as its air quality, may not advance both environmental and transportation objectives (Van der Schaaf 2002). Urban land-use and transportation planners might then have to make a compromise between (improved) air quality and (less) total travel time. Total travel time is usually the more important variable for transportation planners, when the objective is to achieve system efficiency based on the minimization of total travel time (Wardrop's second principle 1952).

21.3.1 Conceptual Framework

Figure 21.1 illustrates the basic interrelationship of the research components. Urban form and transportation system influence private VKT and travel time. VKT and travel time affect air-pollutant emissions from motor vehicles. The transportation

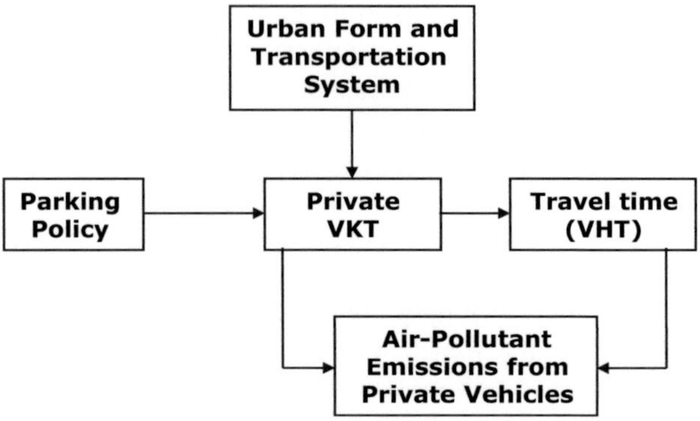

Fig. 21.1 Interrelationship of the research components

system is impacted by a set of alternative policies. As stated, parking policies will provide the focus of the present analysis.

The extent to which air-pollutant emissions can be reduced is dependent on the chosen economic and physical parking measures that are employed. These measures will directly impact private VKT and vehicle hours traveled (VHT). VKT and travel time are closely related to the rate of air-pollutant emissions.

21.3.2 Parking Policies

The first policy measure is the reduction of the parking supply in the CBD of Tel Aviv. This measure is modeled as follows. The total demand for motor vehicle trips is assumed to be constant (fixed demand). The decreased parking supply in the CBD results in a dispersion of trips to traffic zones outside the area. An application of this policy using existing transportation planning models is illustrated in the next section.

Following Marsden (2006), we assume a 20 % decrease in the total number of trips to the CBD. These trips are further assumed to be diverted to other traffic zones in the TAMA. This assumption stems from the fact that if alternative modes are not considered (i.e., the total demand for vehicle trips is fixed), the reduction of parking supply will affect mainly trip distribution. If we assume that a 20 % decrease in the total number of trips to the CBD is diverted to other traffic zones, then together with the assumption of fixed demand it will be necessary to distribute the 20 % of the trips to other traffic zones in the metropolitan area. The transportation model produces the outcome of this new situation in terms of VKT, travel times, and average speed for each link in the network.

The second policy measure dealt with in this paper is an increase in parking fees. Since parking fees vary across the CBD, an increase in parking fees will inadvertently affect parking search time (at least in the short term) because drivers will try to look for cheaper parking places or even park far from their destination in order to save on costs (Shoup 1999). Therefore, this policy measure will be evaluated in conjunction with the expected increase in parking search time (some of which may occur outside the CBD).

In contrast to the procedure used for reducing parking supply, this second measure is modeled by changing the attractiveness of private-car trips to the CBD, compared to alternative transit modes. The main assumption here is that the demand for vehicle trips is elastic; that is, the total demand for vehicle trips is not fixed. Each scenario is composed of a different increase in parking fees. The transportation model is applied to obtain new equilibrium link flows and travel times for each scenario. The pollutant emissions are calculated for each scenario, based on the VKT and travel speeds on each link of the transportation network.

21.3.3 Model Specification and Data

The transportation model used in this work was constructed by NTA—a semi-public consulting group that led the Mass Transit project for the TAMA (NTA 2001). It is a four-step model with feedback procedures for the mode-choice and trip-distribution steps; that is, travel time from the traffic-assignment model serves as input to previous steps of trip distribution and mode choice. The model converges after a number of iterations between the steps described.

Figure 21.2 presents the division of the TAMA into ten super zones, based on Israel's Central Bureau of Statistics. The transportation model calculates the VKT, VHT, and average speed for every link of the transportation network. These results are then processed by the emission model and aggregated at the super-zone level.

The trip-generation model is based on NTA's data, which were calibrated using the 1995 National Census of Population and the 1996 National Travel Habit Survey (NTHS). The Tel-Aviv metropolitan O-D matrix and network were updated to the year 2003.

The parking-policy coefficients (parking cost and parking search time) were obtained from NTA's mode-choice model, which employed a nested logit model. The point elasticity of demand in the logit model is given as follows (Ben-Akiva and Lerman 1985):

$$E = (1 - P) * \beta x \tag{21.1}$$

Where:

E = Direct point elasticity

P = Probability of traveling by private vehicle (60 % for the base case in TAMA)

β = Parking cost coefficient −0.02883 and parking search time coefficient −0.0587

x = Parking cost or parking search time variables (approximately NIS 10[1] parking per hour and 5 min parking search).

[1] NIS 4.2 = $1.

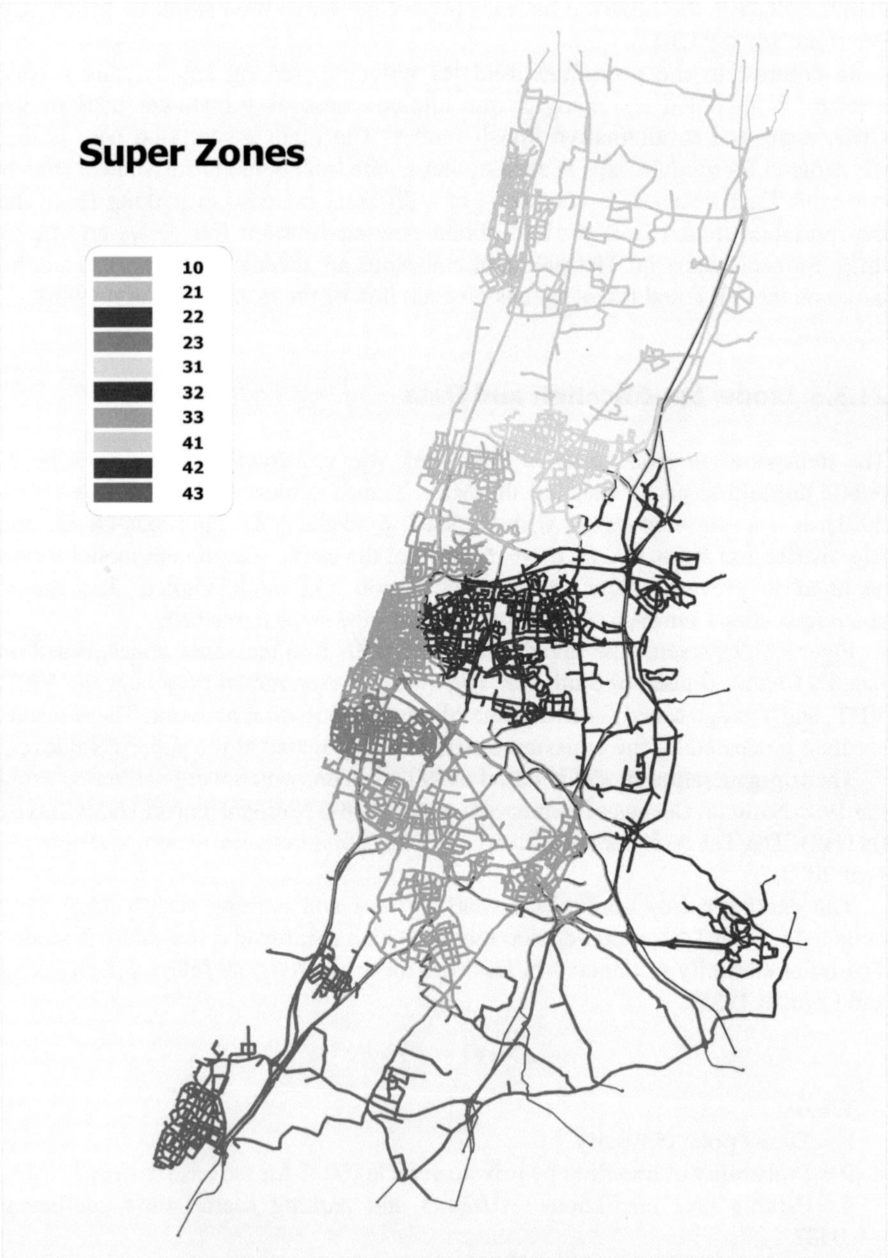

Fig. 21.2 Tel Aviv transportation network, divided into 10 super zones

Using Eq. (21.1), the point elasticity of demand for the parking fee is −0.115, and for the parking search time −0.1174. Increases of 10, 25, 50, 75, and 100 % in parking fees and parking search time were examined as was the influence of these increases on VKT, number of trips, and travel time in each and every one of the super zones and in all of them together. A comparison was made with respect to the base-case scenario.

21.3.4 Pollutant-Emission Model

The basic emission model includes the product of two variables: the emission factor of the pollutant and the level of activity that a certain type of vehicle produces. The pollutant emissions from motor vehicles depend on both variables, and the emission factor depends on vehicle speed (Parra et al. 2006).

The following emission model is based on the variables mentioned above; however, the calculation is done on the basis of emission per road link, per hour, in our case the morning peak hour. This allows us to calculate pollutant emissions at the level of traffic analysis zones or super zones, thereby creating better conditions for analyzing differences in the level of pollutant emissions among the various zones within the TAMA.

Equation (21.2) is slightly modified from Parra et al. (2006) in order to enable the calculation of average pollutant emissions per link per hour:

$$E_{i,l} = \sum_{j=1}^{n} V_{j,l} L_l F_{i,j}(S_l) \tag{21.2}$$

Where:

$E_{i,l}$—Hot exhaust emissions (expressed in g h^{-1}) of pollutant i in link l per hour (morning peak hour in our case)

$V_{j,l}$—Hourly volume of vehicle flow of category j for link l (in our case, only three categories were considered: private vehicles, buses, and trucks)

L_l—Length of link l

$F_{i,j}$—Hot exhaust-emission factor for pollutant i of vehicle category j

S_l—Average speed on link l

N—Number of categories

21.4 Results

21.4.1 Reducing Parking Supply

As presented in the methodology section, the first test dealt with the effect of reducing the parking supply in the Tel Aviv CBD. This paper assumed a parking-supply reduction of 20 %, which consequently led to a reduction of 20 % in private vehicle trips entering the CBD during the morning peak hours. The assumption is consistent with results from the 1996/1997 NTHS survey, which showed that most employees arriving at the CBD during the morning peak hour found a parking space

Table 21.1 Results of dispersion of 20 % of CBD trips to other areas

Super zone origin		Percentage change of total VKT (%)	Percentage change of total VHT (%)
Tel-Aviv city (CBD)		−10.11	−1.73
Inner ring	North	−0.89	−0.23
	Center	−2.31	−0.68
	South	−0.44	−0.61
Middle ring	North	2.32	1.26
	Center	1.84	0.33
	South	2.73	0.46
Outer ring	North	4.17	3.66
	Center	3.40	0.63
	South	4.85	0.39
Total		5.55	3.49

available; further, the majority of trips entering the area in this period are made by commuters.

As also discussed in the methodology section, we assumed that the total demand for car trips was fixed. Therefore, the 20 % of trips with a destination within the CBD that had been reduced were assumed to be diverted proportionally across the metropolitan area. This proportional dispersion would mainly increase trips to fringe areas of the CBD. The result of this process is a modified origin-destination matrix, which was assigned to the same road network as in the base case.

The link flow results of the traffic-assignment model were aggregated for each super-zone origin in the TAMA. The following variables were examined for each super zone: total vehicle kilometers traveled (VKT) and total vehicle hours traveled (VHT). Table 21.1 shows the results of this policy scenario, compared to the base case.

As expected, both VKT and VHT decrease to the greatest extent in the city of Tel-Aviv. The values of these variables decrease less intensely in the inner-ring super zones. This result is explained by the spatial distribution of the trips on the network, since trips originating in other areas have to pass through one of these areas in order to arrive at the CBD. Thus reducing parking supply in the CBD influences not only the CBD but also the fringe areas.

In contrast to the CBD and the inner ring, the middle and outer rings showed increased VHT and VKT This, too, is accounted for by the effect of trip re-distribution on the transportation network of the Tel Aviv area. Since the diverted trips were proportionally dispersed across the TAMA, most trips originating in the outer-ring zones still have to travel to fringe areas, increasing the total VKT. Therefore, an increase in the number of trips in the fringe areas has a negative impact, since in this case vehicles have to travel farther from the origin in order to reach their destinations.

The conclusion of this analysis with respect to trip distribution is that depending on the spatial distribution of trips during the morning peak hour, the dispersal may have a negative impact in terms of increasing VKT and VHT in the outskirts of the metropolitan area. However, this effect might not be so negative as far as ambient air pollution is concerned, since the emissions would spread out over a larger area. It should be noted, however, that the emissions in problematic, denser areas, such as the Tel Aviv CBD, would decrease.

21.4.2 Increasing Parking Fees

In contrast to the previous case, the demand for travel in this scenario is assumed to be elastic. Travel demand for the origin-destination (O-D) matrix is dependent on the rate of parking fees. As indicated in the methodology section, we applied the elasticity Eq. 21.1 to different parking fees and search times. In all tests, the same network as in the base case was used. The difference between each scenario is, again, due to the change in the O-D matrix. The results for parking-search time are quite similar to those for parking fees; therefore, they are not presented here.

The results obtained are quite interesting: part of the trends are linear; the CBD and inner ring super zones have less VKT, number of trips, and travel time with the parking-policy fee scenarios than with the base scenario. The outer rings have more VKT, number of trips, and travel time with the fee scenarios than with the base scenario.

Selected results are presented in Figs. 21.3, 21.4, and 21.5. The results shown are for the overall metropolitan area, the Tel Aviv CBD, and the outer ring (Netanya) super zone. Each figure displays the percentage change for each scenario compared to the base scenario with respect to three variables: VKT, VHT, and total number of trips.

Figure 21.3 shows that a small increase in parking fee will lead to an increase in the total time traveled. This result is explained by the fact that the model takes into account the combined effect of increasing parking fees: on the one hand, the total demand for travel decreases; on the other hand, parking search time increases, leading to an increase in the overall travel time.

An additional important finding is shown in Fig. 21.3. For a certain substantial increase in parking fees, total VHT decreases compared to the base scenario. In the case of the metropolitan area, this value is around 75 %; in other words, parking fees must almost double before there is a drop in total VHT. This point is important, since it represents the point at which the increase in parking fees results in a decrease of both VKT and VHT at the metropolitan level.

Figure 21.4 shows the results for the Tel-Aviv (CBD) super zone. We can observe a linear decrease in all three variables as parking cost increases. There is a steep decrease in VKT, apparently proportional to the decrease in the number of trips. In contrast, the decrease in travel time is very mild. Since the parking supply and parking fee affect mainly zones within the city itself, all three variables have values that are less than those of the base scenario, even when the parking cost is raised by only 10 %.

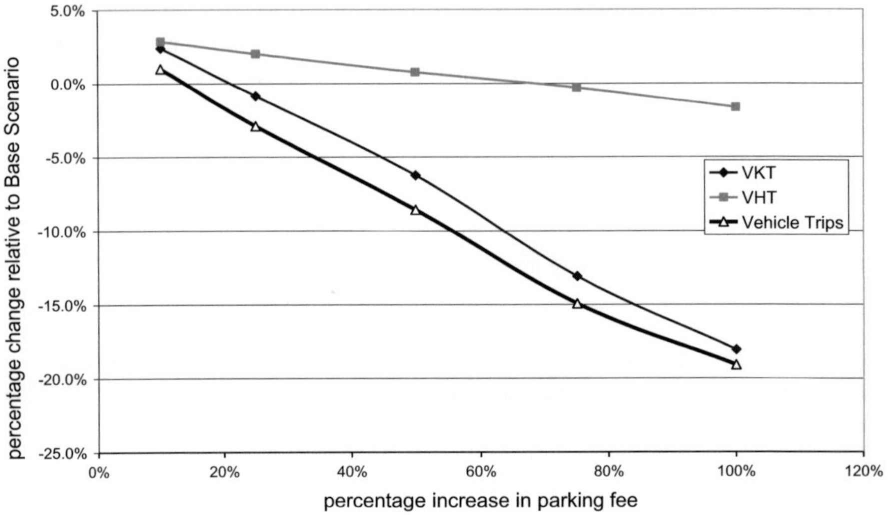

Fig. 21.3 Effect of increasing parking fees in the metropolitan area

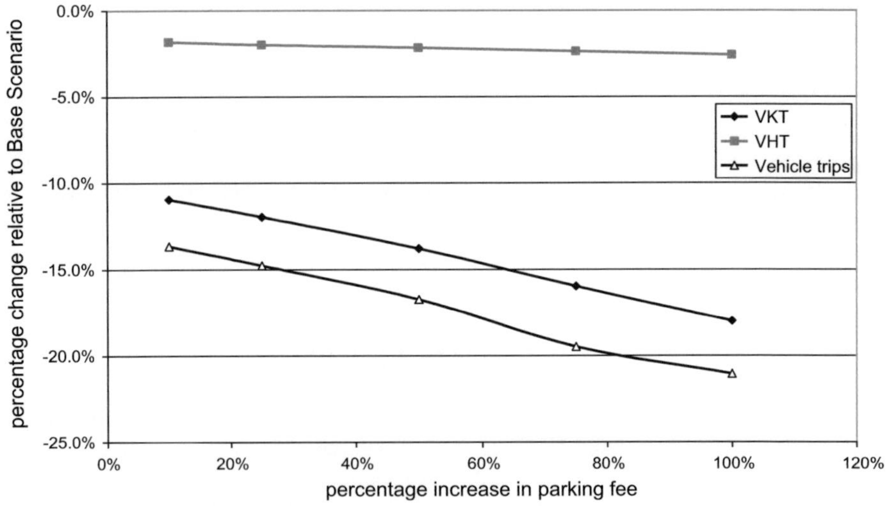

Fig. 21.4 Effect of increasing parking cost in the Tel-Aviv (CBD) super zone

Figure 21.5 shows the results for the Netanya super zone, the northern outskirts of the TAMA. A decrease may be seen in all three variables as parking cost increases; compared to the Tel-Aviv CBD super zone (Fig. 21.4), however, the decrease in VKT is less significant while the decrease in VHT is more pronounced.

It may be recalled that the effect of increasing parking fees (mainly in the Tel-Aviv CBD super zone) is modeled by applying the demand elasticity (Eq. 21.1) to every destination zone in the metropolitan area. There are many

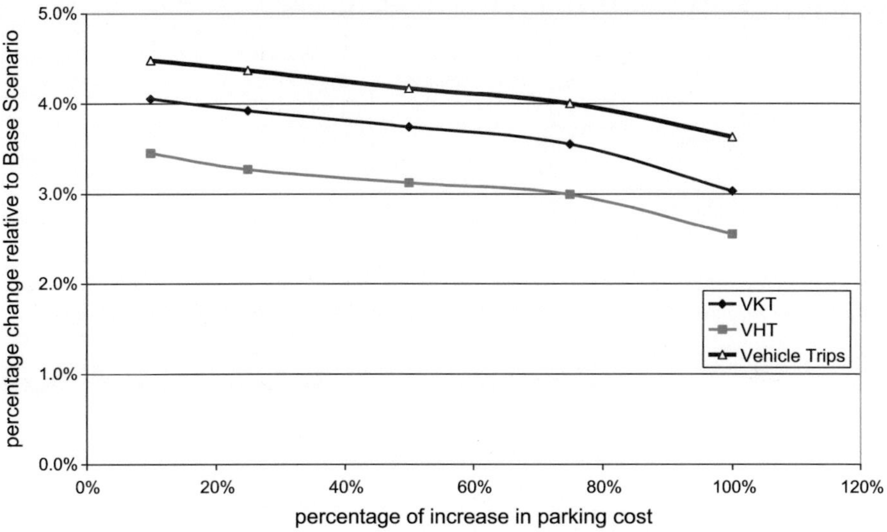

Fig. 21.5 Effect of increasing parking fees in the Netanya super zone

commuting trips from the Netanya super zone to the Tel Aviv CBD, and relatively good transit service (both bus and rail) exists between these two super zones. The outcome is a reduction in the total number of trips between them, and this could explain the reduction in total VKT and VHT for all increases in parking fees.

21.4.3 Pollutant Emission Results

This section presents results for selected types of emissions (NO_X and CO) for the base case scenario and a 75 % increase in parking fees. The results were computed by applying Eq. (21.2) to every link in the transportation network and aggregating the results for the super zone level. As was seen in the previous section (see Fig. 21.3 and the discussion following it), a 75 % increase in parking fees represents the point at which both VKT and VHT decrease at the metropolitan level. Table 21.2 presents results for the overall metropolitan area.

An examination of the mean emission values for the entire TAMA (as in Table 21.2) shows that increasing parking fees by 75 % results in only a 3 % reduction in emissions. However, this result represents the overall average. The results obtained for each super zone, as presented in Table 21.3, allow an understanding of the dynamics and the influence of urban land-use and transportation planning on vehicle pollutant emissions.

The results for the Tel-Aviv CBD super zone show a significant reduction in pollutant emissions (20 % NO_x and 28 % CO emissions). In contrast, the Netanya super zone exhibits an increase in emissions. This is due to the fact that a certain percentage of trips changed their destination from the CBD to the outskirts of the

Table 21.2 Selected emissions for the TAMA (morning peak hour)

Pollutant	Base case scenario	75 % Increase in parking fee	Percentage change (%)
NO_x emissions (kg)	1.94	1.88	−3
NO_x emissions (kg/km^2)	1.29	1.26	−3
CO emissions (kg)	3.21	3.13	−3
CO emissions (kg/km^2)	2.14	2.09	−3

Table 21.3 Selected emissions for the CBD and Netanya super zones (morning peak hour)

	CBD super zone			Netanya super zone		
Pollutant	Base case scenario	75 % parking fee increase	Change (%)	Base case scenario	75% parking fee increase	Change (%)
NO_x emissions (kg)	249.88	207.59	−20	347.18	373.39	7
NO_x emissions (kg/km^2)	4.86	4.04	−20	0.9	0.97	7
CO emissions (kg)	399.47	313.11	−28	431.44	637.24	32
CO emissions (kg/km^2)	7.78	6.09	−28	1.17	1.65	32

metropolitan area, such as the Netanya super zone (increase of 7 % NO_x and 32 % CO emissions). It should be noted that even when dispersion conditions are bad, CO concentration levels recorded in the Netanya super zone are very low in comparison to established standards. Therefore, an increase of 32 % CO emissions in this super zone will not lead to a violation of these standards; that is, the CO concentrations will still be significantly below the level permitted for this super zone.

Analyzing the level of emissions per square km emphasizes the fact that a diversion of private vehicle trips to the outskirt areas might actually have a positive effect on metropolitan air quality. The reason is that sparsely developed areas such as the Netanya super zone still have much better air quality, even after the diversion increases the total emissions. On the other hand, the diversion of private trips to the outskirts significantly improves the air quality of the Tel-Aviv CBD compared to the base case scenario.

An important point that should be emphasized is that the model assumes the same existing transit service before and after the change in parking fees. Although the model predicts that a certain percentage of trips will be diverted from the CBD to the outskirts of the TAMA (such as the Netanya super zone), most trips will continue to enter the CBD by means of either private vehicles or public transit.

Summary and Conclusions

This article stresses the need for parking-policy measures as a tool for further implementation of urban land-use and transportation-planning policies. The findings show that parking policies have significant influence on transportation and air-quality variables. Parking policies alone can reduce air-pollutant emissions in the CBD by 20–28 %. They create conditions for the distribution of emissions throughout the metropolitan area. The question of distributing emissions over a large area, such as a metropolitan area, is important in itself, since researchers usually deal with the general issue of total emission reduction, rather than in a concentrated or particular area like the CBD.

Note that the results presented in this paper suggest that a more aggressive parking policy is needed to obtain a significant reduction in pollutant emissions. For example, in the dataset used for this paper, the increase in parking fees only reduces VHT if the increase is 75 % or more.

The parking policies analyzed in this paper are assumed to change the demand for travel; the supply side (transportation network, alternative modes) was assumed to be constant. The pollutant emissions were then calculated according to link flows and travel times produced by solving the traffic assignment problem. The paper shows that the emissions may also increase in certain sections of the metropolitan area because of the redistribution of trips.

Such a solution is in accordance with the existing transit level of service in the TAMA. Public transit level of service is most efficient in high-density areas, such as the CBD and its nearby suburbs. The demand for travel in the outskirts of the metropolitan area cannot be supported by transit because of their dispersed spatial development. Therefore, such parking-policy measures as increased parking fees and/or reduced parking supply in the CBD could influence both travel destination and mode choice.

The pollutant-emission model shows that there is an advantage in implementing parking policies in the CBD, even without supportive measures such as improving public transport level of service or imposing toll roads. Decreasing overall metropolitan pollution emissions and distributing emissions so as to lower pollutants in the CBD have a positive effect on overall metropolitan air quality.

The vicious cycle in which transportation planners use parking availability for road-capacity planning should be replaced with parking-policy measures. This step could offer a basis for combining additional policy measures and examining the influence of additional transport policies, such as planning for slow modes, imposing toll roads, and improving public transport level of service, on the degree of air-pollutant emissions.

Acknowledgment This research was partially supported by the Henry Ford II Transportation Research Fund.

References

Ben-Akiva M, Lerman ST (1985) Discrete choice analysis: theory and application to travel demand. MIT Press, Cambridge, MA

Bernick M, Cervero R (1997) Transit villages in the 21st century. McGraw-Hill Co, New York

Burton E (2000) Measuring urban compactness in UK towns and cities. Environ Plan B Plan Des 29:219–250

Central Bureau of Statistics (2008) 2008 census data. http://www.cbs.gov.il/census/census/pnimi_page_e.html?id_topic=11
Cervero R (2002) Built environments and mode choice: towards a normative framework. Transp Res D 7:265–284
de Roo G (1999) Environmental conflicts in compact cities: complexity. Decision making and policy approaches. Environ Plan B Plan Des 27:151–161
Geurs KT, van Wee B (2004) Accessibility evaluation of land-use and transport strategies: review and research directions. J Transp Geogr 12:127–140
Hägerstrand T (1970) What about people in regional science? Pap Reg Sci Assoc 24:7–24
Hansen WG (1959) How accessibility shapes land use. J Am Inst Plan 25:73–76
Ingram DR (1971) The concept of accessibility: a search for an operational form. Reg Stud 5: 101–107
Katz P (1994) The new urbanism: toward architecture of community. McGraw-Hill, New York
Kodransky M, Hermann G (2011) Europe's parking U-Turn: from accommodation to regulation, ITDP (Institute for Transportation and Development Policy) New York, USA
Levine J, Garb Y (2002) Congestion pricing's conditional promise. Transp Pol 9:179–188
Lyons TJ, Kenworthy JR, Newman PWG (1990) Urban structure and air pollution. Atmos Environ 24B(1):43–48
Lyons TJ, Kenworthy JR, Moy C, dos Santos F (2003) An international urban air pollution model for the transportation sector. Transp Res D 8:159–167
Marsden G (2006) The evidence base for parking policies—a review. Transp Pol 13:447–457
Marsden G, May AD (2005) Do institutional arrangements make a difference to transport policy and implementation? Lessons from Great Britain. Environ Plann C Gov Policy 24(5):771–789
Newman PWG, Kenworthy JR, Lyons TJ (1988) Does free flowing traffic save energy and lower emissions in the cities? Search 19:267–272
NTA (2001) Mass transit Tel-Aviv metropolitan travel demand forecasting model—(MIP), Tel-Aviv
Parra R, Jimenez P, Baldasano JM (2006) Development of high spatial resolution EMICAT2000 emission model for air pollutants from the North-Eastern Iberian Peninsula (Catalonia, Spain). Environ Pollut 140:200–219
Parrish DD, Singh HB, Molina L, Madronich S (2011) Air quality progress in North American megacities: a review. Atmos Environ 45(39):7015–7025
Ranmar DO, Matveev V, Dayan U, Peleg M, Kaplan J, Gertler AW, Luria M, Kallos G, Katsafados P, Mahrer Y (2002) Impact of coastal transportation emissions on inland air pollution over Israel: utilizing numerical simulations, airborne measurements and synoptic analyses. J Geophys Res-Atmos 107(D17):ACL X-1–ACL X-14
Shefer D (1994) Congestion, air pollution and road fatalities in urban areas. Accid Anal Prev 26: 501–509
Shoup DC (1999) The trouble with minimum parking requirements. Transp Res A 33(7–8): 549–574
Shoup DC (2005) The high cost of free parking. American Planning Association, Planners Press, Chicago, IL
Van der Schaaf K (2002) Parking is maneuvering. Paper presented at the third seminar of INPRINT-EUROPE thematic network: implementing reform on transport pricing: constraints and solutions: learning from best practice, Brussels, October
Wardrop JG (1952) Some theoretical aspects of traffic research. Proc Inst Civ Eng 1(3):325–362 (Road Paper No 36)

MIX
Papier aus verantwortungsvollen Quellen
Paper from responsible sources
FSC® C105338

If you have any concerns about our products,
you can contact us on
ProductSafety@springernature.com

In case Publisher is established outside the EU,
the EU authorized representative is:
Springer Nature Customer Service Center GmbH
Europaplatz 3, 69115 Heidelberg, Germany

Printed by Libri Plureos GmbH
in Hamburg, Germany